LOCALIZATION AND DELOCALIZATION IN QUANTUM CHEMISTRY

VOLUME II

IONIZED AND EXCITED STATES

LOCALIZATION AND DELOCALIZATION IN QUANTUM CHEMISTRY

VOLUME II

Ionized and Excited States

Edited by

ODILON CHALVET and RAYMOND DAUDEL

Centre de Mécanique Ondulatoire Appliquée du C.N.R.S., 23, rue du Maroc, Paris 19ème

and

SIMON DINER and JEAN PAUL MALRIEU

Institut de Biologie Physico-Chimique, 13, rue Pierre et Marie Curie, Paris 5ème

D. REIDEL PUBLISHING COMPANY

DORDRECHT-HOLLAND / BOSTON-U.S.A.

Library of Congress Cataloging in Publication Data

Main entry under title:

Ionized and excited states.

(Localization and delocalization in quantum
chemistry; v. 2).
'Proceedings of the international seminar devoted to
localization and delocalization in quantum chemistry.'
Includes bibliographical references and indexes.
1.   Excited state chemistry—Congresses.
I.   Chalvet, Odilon.   II.   Series.
QD461.5.166   541'.28   75-42413
ISBN-13:978-94-010-1458-8   e-ISBN-13:978-94-010-1456-4
DOI: 10.1007/978-94-010-1456-4

Published by D. Reidel Publishing Company,
P.O. Box 17, Dordrecht, Holland

Sold and distributed in the U.S.A., Canada, and Mexico
by D. Reidel Publishing Company, Inc.
Lincoln Building, 160 Old Derby Street, Hingham,
Mass. 02043, U.S.A.

TABLE OF CONTENTS

PART IV / ELECTRON LOCALIZATION
AND CHEMICAL REACTIVITY

FINAL DISCUSSION

PREFACE

The second volume of the proceedings of the international seminar devoted to localization and delocalization in quantum chemistry is divided into four parts.

The first one is mainly concerned with the localizability of electrons in ionized and exited states.

The second part shows how is it possible to take advantage of the localizability of electrons to compute molecular wave-functions.

The third part of the book is an homogeneous analysis of the electronic collective excitation and of the motion of excitons in organic solids.

The last section is devoted to the study of the role of electron localizability in the chemical reactivity of molecules.

Concluding remarks are concerned with a careful analysis of the localizability concept itself in relation with a possible interpretation of the wave-mechanics.

PART I

ELECTRON LOCALIZATION IN IONIZED AND EXCITED STATES

APPLICATIONS OF PAIR DENSITY ANALYSIS

R. CONSTANCIEL and L. ESNAULT

Centre de Mécanique Ondulatoire Appliquée, Paris, France

Abstract. The method of pair density analysis is applied to various kinds of calculations. We examine the influence of the quality of the wavefunction and of the nuclear configuration; the problem of hybridization is discussed, as well as the relations between separability and excitation.

1. Introduction

In the first volume of this series we used a physical approach for describing a new method of analysing the molecular electronic wavefunctions [1] which would be able to disclose the electron group structure of any molecule. In the present paper we would like mainly to show how to deal with this method in the case of two electron groups associated with σ or π bonds, with a particular emphasis on the differences between the ground and the first excited states.

According to the general principles introduced in our first paper, we will admit that there exists a discernable two-electron group in the molecule if the wavefunction $\Psi^{(n)}$ can be written exactly as an antisymmetrized product of functions defined in the subspace $\mathcal{H}_a^{(2)} \otimes \mathcal{H}_b^{(n-2)} \subset \mathcal{H}^{(n)}$

$$\Psi^{(n)}(1, 2, \dots n) = \mathcal{A}[\Psi_A^{(2)}(1, 2)\, \Psi_B^{(n-2)}(3, \dots n)]. \tag{1}$$

The subspace $\mathcal{H}_a^{(2)} \otimes \mathcal{H}_b^{(n-2)}$ is one of the orthogonal subspaces deduced by the tensorial product from the partition $\mathcal{H}_a^{(1)} \oplus \mathcal{H}_b^{(1)}$ of the orbital space $\mathcal{H}^{(1)}$. It was proved elsewhere [2] that a necessary and sufficient condition ensuring an exact separability of the wavefunction, as in Equation (1), is that the two-particle reduced density operator $[\rho_n^{(2)}]_a$ restricted to the subspace $\mathcal{H}_a^{(2)}$ be a projection operator. In this extreme case, $[\rho_n^{(2)}]_a$ projects precisely on the two-particle function $\Psi_A^{(2)}(1, 2)$, so that we have a means of extracting from the total wavefunction the best group-function* describing the two-electron group in the field of the rest of the particles forming the molecule. Consequently, we may also easily obtain the effective energy of the considered two-electron group as was previously shown by McWeeny [3].

Barring the special methods of building the wavefunctions with the aid of chemical intuition (like the group-function method of McWeeny [3]), the quantum chemical methods derived from the widely used LCAO-MO approximation will lead to a

---

* Let us point out that this function does not reduce generally to a single determinant, but moreover can be considered as resulting from a configuration interaction limited to the subspace $\mathcal{H}_a^{(2)}$.

wavefunction defined in $\mathcal{H}^{(n)}$, and not in $\mathcal{H}_a^{(2)} \otimes \mathcal{H}_b^{(n-2)}$, so that the extreme case considered above in Equation (1) can never be realized. However, it is often possible, by choosing carefully the Hilbert space partition $\mathcal{H}_a^{(2)} \otimes \mathcal{H}_b^{(n-2)}$, to obtain through Equation (1) an highly accurate approximation of any wavefunction: in this way, we justify the existence of some quasi-discernable subsystems of electrons.

For the present purpose, we limit ourselves to the search of partitions $\mathcal{H}_a^{(2)} \otimes \mathcal{H}_b^{(n-2)}$ of the Hilbert space suggested by topological or space symmetry considerations. So, we can extract from the wavefunction analysis the existence of some subsystems having a σ or π character. These subsystems do not necessarily have a well defined local character, as they do in the Loge theory [4]; however, the topological considerations consistently lead to the choice of those subspaces $\mathcal{H}_a^{(2)}$ which are built from one electron basic functions $\varphi_{Ai}^{(1)}$ rather well localized in a given region V_A of the physical space $\mathbb{R}^3$, with a localization defect ε_A [5] defined as:

$$\int_{V_A} |\varphi_{Ai}^{(1)}(r)|^2 \, dr \geqslant 1 - \varepsilon_A, \quad \forall_i. \tag{2}$$

Thus, in some cases, a connection can be restored in a first approximation between our method of analysis and the ones working in $\mathbb{R}^3$.*

As a consequence of the non exact separability of the wavefunction, the operator $[\rho_n^{(2)}]_a$ can never be exactly identified to a projection operator on a two-particle function. Then, the diagonalization of the matrix representing this operator in the subspace $\mathcal{H}_a^{(2)}$ yields non vanishing occupation numbers corresponding to a set of two-particle functions.** It is possible to determine quantitatively to what extent the operator $[\rho_n^{(2)}]_a$ may be considered as a projector (i.e. to what extent it characterises the quasi-separability potentiality of the wavefunction, and then the quasi-discernability of the electron group) by means of the relative dispersion of the occupation numbers, or, more exactly, of the relative missing information function [6] on these occupation numbers.

As it works in functional subspaces, our procedure can be compared to the orbital 'localization' methods which use the invariance properties of a one-determinant wavefunction with respect to a unitary transformation of the molecular orbital basis [8]. From this viewpoint, it provides some substantial improvements. At first, it allows us to deal directly with two-electron group functions; and secondly, it can be applied to any wavefunction. This latter feature is of practical importance when concerned with a better description of the ground states, and also with the study of excited states. The former feature makes it possible to avoid the one-particle approximation.

---

* This holds also for an orthonormal set of one electron basic functions obtained by Löwdin's orthonormalisation from a set of atomic orbitals. Such a symmetric procedure builds functions least distorted, in a least square sense, from the initial ones localized about the lattice points [7].
** Such a situation can be compared with the density operator description of a mixed state, but not confused with it.

2. Details of the Technics

1. The first task consists in introducing a two-particle function basis defining the subspace $\mathscr{H}_a^{(2)} = \mathscr{H}_a^{(1)} \otimes \mathscr{H}_a^{(1)}$. If we limit ourselves to a minimal basis approximation, the topological considerations lead to retaining as basis functions for $\mathscr{H}_a^{(1)}$ two atomic orbitals a_1 and a_2 centered on the two nuclei between which a two-electron bond is expected to be found. After projection, we get a convenient set of six totally antisymmetrized spin-geminals of singlet or triplet type:

$$
\left.
\begin{array}{l}
a_1 a_1 \\[2mm]
\dfrac{1}{\sqrt{2}}(a_1 a_2 + a_2 a_1) \\[2mm]
a_2 a_2
\end{array}
\right\} \; \dfrac{1}{\sqrt{2}}(\alpha\beta - \beta\alpha)
$$

$$
\left.
\dfrac{1}{\sqrt{2}}(a_1 a_2 - a_2 a_1)
\right\} \;
\left\{
\begin{array}{l}
\alpha\alpha \\[2mm]
\dfrac{1}{\sqrt{2}}(\alpha\beta + \beta\alpha). \\[2mm]
\beta\beta
\end{array}
\right.
$$

$$(3)$$

In the case of π subsystems, the choice of $\mathscr{H}_a^{(1)}$ is obvious, because there is only one convenient atomic orbital per atom. For σ subsystems, however, there is some indeterminacy that can be removed *a posteriori* in a way that we will see later; in many cases, this research of the best hybrid atomic orbitals can be limited by topological considerations.

2. In a second step, we have to compute the elements of the matrix representing the two-particle density operator $\rho_n^{(2)}$ in the subspace $\mathscr{H}_a^{(2)}$. This is a problem hard to solve directly in the most general case. Some convenient formulas have been derived by K. Ruedenberg and R. D. Poshusta [9]; however, the way indicated by R. McWeeny [10], besides its simplicity of application resulting from the use of Slater's rules, seems to us the more efficient here because it allows us to derive in the same step the matrix elements of the operator $\rho_n^{(2)}$ and the electronic energy. In this way the programming effort required for the analysis procedure is greatly reduced, since an important part of the task was already achieved with a view to building the wavefunction.

The procedure can be summarized as follows: Let there be a CI wavefunction

$$\Psi = \sum_K C_K \psi_K \tag{4}$$

where

$$\psi_K = \sum_k S_{Kk} \Delta_k^K \tag{5}$$

is a convenient linear combination of determinant functions; we have:

$$\rho_n^{(2)} = \sum_{K,L} \sum_{k,l} C_K C_L S_{Kk} S_{Ll} \, \rho_n^{(2)}(KL;kl). \tag{6}$$

Then, the electronic interaction energy can be written as a linear combination of elementary contributions

$$E(KL; kl) = \langle \Delta_k^K | g | \Delta_l^L \rangle = \mathrm{Tr}_{(2)} \, g \rho_n^{(2)}(KL; kl). \tag{7}$$

By using Slater's rules, we know the expression of $\langle \Delta_k^K | g | \Delta_l^L \rangle$ in terms of atomic integrals. We get the expression of $\rho_n^{(2)}(KL; kl)$ in terms of the molecular orbital coefficients simply by identification of the last two members of Equation (7). Clearly, such a calculation can be performed by any CI programme properly adapted: we chose the programme written by G. Bessis and O. Chalvet [11] which works in any configuration basis.

3.   In the next step, we diagonalize the matrix representing $[\rho_n^{(2)}]_a$. It is necessary, for a good understanding of what follows, to detail what happens for wavefunctions which are eigenfunctions of $S_{(n)}^2$ with different values of the total quantum numbers S and M_S. For singlet states, the natural spin-geminals are eigenfunctions of $S_{(2)}^2$ and the triplet occupation numbers are triply degenerate. For triplet states with a vanishing projection of the total spin, the natural spin-geminals are also eigenfunctions of $S_2^{(2)}$, but the triplet occupation numbers are no more degenerate. At last, for triplet states with a non vanishing projection of the total spin, the natural spin-geminals are not eigenfunctions of $S_{(2)}^2$ [12].

We conclude immediately that the analysis of two triplet wavefunctions, corresponding to the same degenerate level, will lead to different results according to the fact whether $M_S=0$ or not. In effect, by our procedure, only the wavefunctions with $M_S = 0$ can be approximately separated into an antisymmetrized product of group functions which are also eigenfunctions of $S_{(2)}^2$ or $S_{(n-2)}^2$; when $M_S \neq 0$, the group functions that we get are not submitted to this constraint, so that the quasi-separability seems delusively to be better in this last case. Although this difference decreases and tends to zero for very accurately separated wavefunctions*, we decided to study only the wavefunctions with $M_S=0$ in order to preserve the continuity of the quasi-separability index with respect to the group-spin values.

4.   Finally, we use the N normalized diagonal elements n_i to determine a measure of the quasi-separability by means of the function

$$D_a = \frac{N}{N-1} \left(\sum_{i=1}^{N} n_i (1 - n_i) \right) \tag{8}$$

or, better, by means of the missing information function

$$I_a = -(\log N)^{-1} \left(\sum_{i=1}^{N} n_i \log n_i \right). \tag{9}$$

---

* Because in an extreme case, as in Equation (1), the group functions $\Psi_A^{(2)}$ and $\Psi_B^{(n-2)}$ are just eigenfunctions of $S_{(2)}^2$ and $S_{(n-2)}^2$ whatever M_s may be.

These functions* can be minimized in order to get the best subspace $\mathcal{H}_a^{(2)}$ able to describe a two-electron group: this is a means of characterizing well adapted hybrid orbitals as we have pointed out above. Care must be taken, however, that, in general, the search for the various optimized subspaces – for σ subsystems in particular – cannot be effected for each group independently, for an hybrid orbital cannot be best determined without a modification of the other hybrids centered on the same atom.

3. Influence of the Quality of the Wavefunction

The π system of Butadiene provides a good example of a system for which approximate and exact wavefunctions can be analysed and compared.

The first problem is concerned with the influence of the configuration interaction on the separability property: does the correlation destroy the separability observed for a closed-shell determinant? The results of Table I correspond to the ground-state wavefunctions obtained (i) by one-determinant SCF approximation, and (ii) by consideration of the complete configuration basis. They show that the missing information function characterizing the subspace built from the π atomic orbitals 1 and 2 is independent of the wavefunction used and sufficiently lower than that obtained for the couples 1-3, 1-4 and 2-3 to justify the usual representation of the π system:

The second problem is to know what happens for the first electronically π excited states with the same nuclear configuration as the ground state one. We see, by looking at the results of Table I, that separability does not hold any more, even in a rough approximation, whatever the wavefunction we are using may be. Such a result can be accounted for through an excitonic representation of the excited state:

$$\Psi^*(1234) = a\varphi_A^*(12)\ \varphi_B(34) + b\varphi_A(12)\ \varphi_B^*(34) \tag{10}$$

derived from the ground state partition $\mathcal{H}_a^{(1)} = \{1, 2\}$ and $\mathcal{H}_b^{(1)} = \{3, 4\}$. Here the energetic degeneracy implies that $a = b$ so that no separability can be obtained. In all cases the computed values of the usual bond orders p_{rs} (off-diagonal elements of the one-particle reduced density matrix) have been given in order to allow the comparison with the corresponding values of the missing information function. The accordance is generally good for the ground state, a result which was theoretically founded previously [13], but some divergences can be noted in the study of the excited states.

---

* Both functions D_a and I_a would vanish in the limiting case of exact separability. In all other cases, I_a provides a better index, because it takes into account all the momenta of the distribution of the n_i's.

TABLE II(a)

Trans butadiene

Bond	Ground state		1st singlet excited state			1st triplet excited state		
	SCF	CCI	SCF	Limited CI (mono excitations)	CCI	SCF	CCI ground state geometry	CCI modified geometry
r-s	$I_{rs}(p_{rs})$	$I_{rs}(p_{rs})$	$I_{rs}(p_{rs})$	$I_{rs}(p_{rs})$	$I_{rs}(p_{rs})$	$I_{rs}(p_{rs})$	$I_{rs}(p_{rs})$	$I_{rs}(p_{rs})$
1–2	0.136 (0.977)	0.138 (0.942)	0.794 (0.499)	0.775 (0.488)	0.764 (0.492)	0.851 (0.434)	0.793 (0.453)	0.857 (0.273)
2–3	0.982 (0.213)	0.974 (0.215)	0.785 (0.536)	0.770 (0.594)	0.821 (0.516)	0.653 (0.748)	0.766 (0.555)	0.358 (0.877)
1–3	1.000 (0.000)	0.994 (0.000)	0.975 (0.000)	0.982 (0.000)	0.976 (0.000)	0.898 (0.000)	0.908 (0.000)	0.896 (0.000)
1–4	0.982 (−0.213)	0.976 (−0.196)	0.794 (0.465)	0.799 (0.381)	0.828 (0.459)	0.866 (0.252)	0.920 (0.199)	0.405 (0.077)

TABLE I(b)

CIS butadiene

Bond	Ground state		1st singlet excited state			1st triplet excited state		
	SCF	CCI	SCF	Limited CI (mono excitations)	CCI	SCF	CCI ground state geometry	CCI modified geometry
r-s	$I_{rs}(p_{rs})$	$I_{rs}(p_{rs})$	$I_{rs}(p_{rs})$	$I_{rs}(p_{rs})$	$I_{rs}(p_{rs})$	$I_{rs}(p_{rs})$	$I_{rs}(p_{rs})$	$I_{rs}(p_{rs})$
1–2	0.148 (0.974)	0.145 (0.941)	0.797 (0.498)	0.765 (0.487)	0.753 (0.494)	0.826 (0.438)	0.670 (0.454)	0.857 (0.278)
2–3	0.981 (0.224)	0.973 (0.220)	0.718 (0.549)	0.773 (0.592)	0.812 (0.531)	0.657 (0.740)	0.737 (0.557)	0.359 (0.876)
1–3	1.000 (0.000)	0.994 (0.000)	0.975 (0.000)	0.984 (0.000)	0.979 (0.000)	0.898 (0.000)	0.880 (0.000)	0.896 (0.000)
1–4	0.971 (−0.224)	0.975 (−0.205)	0.794 (0.451)	0.804 (0.368)	0.825 (0.434)	0.710 (0.259)	0.775 (0.206)	0.405 (0.078)

4. Influence of the Nuclear Configuration

The analysis of the wavefunction gives different results according to which nuclear configuration we are considering.

Thus, we can approach a relatively good separable π wave-function for the triplet state of Butadiene by assuming a central bond length shorter than the border ones, a situation which is the inverse of that of the ground state. Although the missing information relative to the central bond is not quite negligible, it clearly appears that the results of the analysis for the modified nuclear configuration are rather closely related to the classical picture of π excited states usually involved in explaining photochemical reactivity. However, in contrast with the representation given by the valence bond method, we never find any well defined 1–4 bond, the existence of which would be interpreted as a feature favourable to a ring closure. The formation of this bond, leading to cyclobutene, would be emphasized only by further analysis including all valence electrons and taking into account the displacements of the hydrogen nuclei. The π analysis simply suggests that, after excitation, the Butadiene molecule relaxes to a nuclear configuration from which cyclisation can occur through a least modification of the π system:

The influence of the nuclear configuration on the results of pair-density analysis of saturated compounds has been studied for closed shell determinant wavefunctions of Methane and Ammonium ion in the CNDO/S approximation. We found that the missing information function characterizing a two-electron bond has a minimum for an internuclear distance, O–H or N–H, near the experimental one.* This suggests that, for ground states, a good separability of the electronic wave function is generally obtained in the neighbourhood of the wells of the potential energy surfaces. As a consequence, we may anticipate (i) that difficulties will be encountered in representing the transition states of two reactants with the aid of various traditional concepts of subsystems, in an analytical study of the ground state chemical reactivity, and (ii) that the separability will be generally poorer in excited states, where the potential wells are less deep, or, if it exists, it will be obtained for a nuclear configuration different from that of the ground state as we saw in the case of Butadiene.

---

* The vanishing value of I_{CH} for Methane is consistent with the result of Mc Weeny and Del Re which showed that delocalization is energetically not significant in this case, so that the 'perfect pairing' model is closely valid [14].

5. The Problem of Hybridization

This has been handled within the same approximation in the case of Ammonia. For each nuclear configuration defined by the d_{NH} distance and the $\theta_{\widehat{HNH}}$ angle, we determined the optimized value of the φ angle between two of the nitrogen atomic hybrids used to represent the NH electron pairs, for which I_{NH} is minimal. The Figure 1 shows the optimal values of I_{NH} and the corresponding values of φ for nuclear configuration near the experimental one. We see that for the most probable configurations we have $\varphi < \theta$, a result already obtained through the analysis of ab-initio studies [15]. As a consequence, we may suggest that the disappointing results obtained sometimes by the SCF-group function method come from a non optimized choice of the atomic

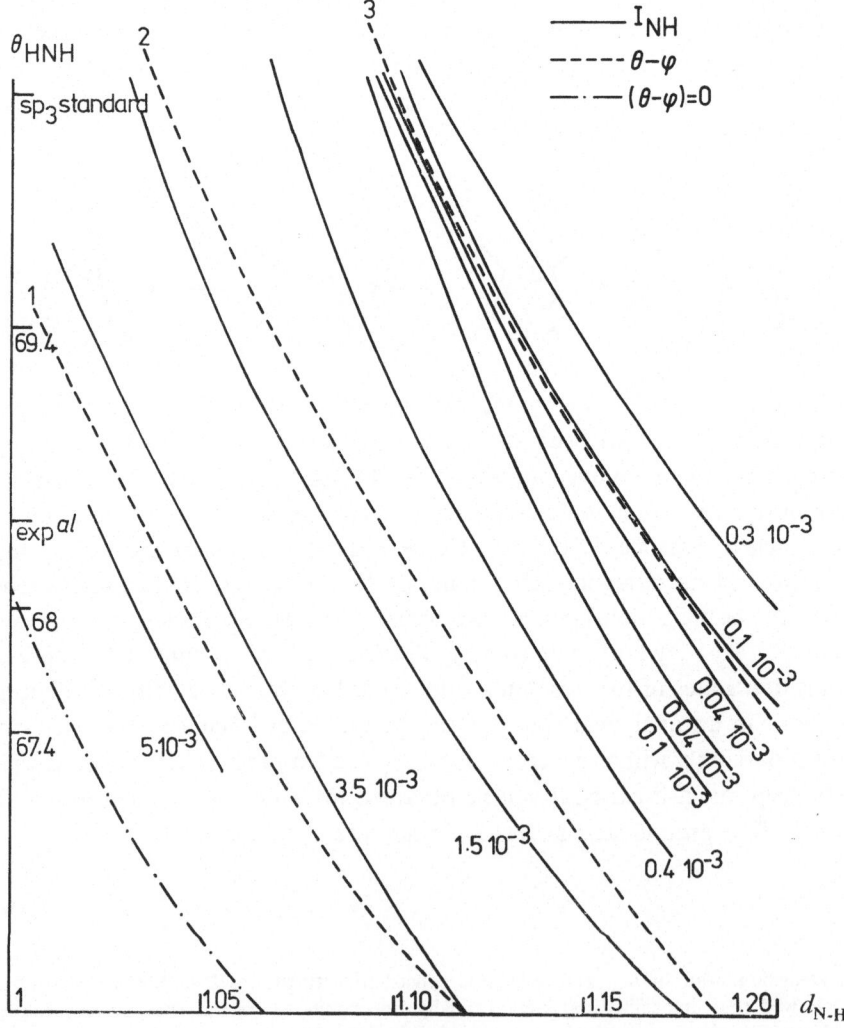

Fig. 1.   Ammonia.

hybrid orbital basis for each group: this is particurlarly the case when the molecule has some lone pairs. These conclusions are supported by the examination of the results [16] of SCGF calculations on Methane and Water molecules reported in Table II. Moreover, we see that, if for Methane the energies in the SCGF method are

TABLE II

Comparison between the SCFMO and SCGF methods for various types of approximations in the computation of integrals, from ref. [16]

	SCFMO	SCGF	
	−83.98	−83.93	CNDO
H_2O	−84.74	−84.69	NDDO
	−84.84	−84.88	All integrals
	−52.38	−52.41	CNDO
CH_4	−52.71	−52.76	NDDO
	−53.45	−53.48	All integrals

always better than those in the SCFMO procedure, that is not true for Water; thus, the quality of the SCGF approach is shown to depend on the kind of approximation made in the computation of the integrals. Clearly, this implies that the choice of optimized hybrid orbitals is dependent on the approximation involved in constructing the wavefunction.

6. Separability and Excitation

We saw that, in general, the separability is destroyed under excitation, leading sometimes to another kind of separation in groups for the electronic system. Now, we will report two cases where the partition of the atomic orbitals is conserved.

Let us first examine the results of Table III: the study of Ethylene represented by a CNDO/S wavefunction including all the monoexcited configurations shows that the effect of excitation is to modify the structure of the π subsystem which remains separated in the low-lying triplet state. That is not true for the first excited singlet state.*

We see in Table IV some results relative to formaldehyde. For the lowest $n \to \pi^*$ states, the two-electron π subsystem is no more defined, whereas the σ bonds remain defined and are separable from the rest of the molecule to a high degree of accuracy. That suggests a strong modification of the subsystems of electrons associated to the π

---

* Let us keep in mind that such a description results from the analysis of a given type of wavefunction. The separability depends essentially on the mixture of $\sigma \to \sigma^*$ and $\pi \to \pi^*$ configurations; for larger systems, such as Benzene or Phenol, we would have a better separation for singlets too.

system and to the lone pairs. Furthermore, the calculation of the number of particles in each subsystem shows that the π one now contains three particles; then, we deduce that the excited wavefunction is an event-type one [2] which differs from that of the ground state by removing an electron from the lone pair groups and adding it to the π group. It seems to us interesting that such a description of $n \to \pi^*$ transitions can be obtained without referring explicitly to the orbital concept, or, in other words, without being constrained in the framework of any particular kind of wavefunction. Naturally, for a wave function more elaborate than the CNDO/S one, or simply by introducing more than the monoexcited configurations in this function, the results of the analysis just presented could be altered in the sense that the π electron group

TABLE III

Ethylene

	Ground state	1st excited singlet	1st excited triplet
I_{CH}	0.044	0.061	0.044
$I_{\sigma C-C}$	0.000	0.205	0.000
$I_{\pi C-C}$	0.000	0.179	0.000

TABLE IV

Formaldehyde

Geometry:　C—H = 1.12 Å
　　　　　　C—O = 1.21 Å
　　　　　　$\widehat{HCH}$ = 118°
　　　　　　$\widehat{nCn}$ = 120°

	Ground state	First excited singlet state ($n \to \pi^*$)	4th excited singlet state ($\pi \to \pi^*$)
I_{C-H}	0.114	0.076	0.181
$I_{\sigma C-O}$	0.008	0.008	0.187
$I_{\pi C-O}$	0.000	0.871	0.168
Number of pairs in the π group	1	3	1

will not always contain exactly an integer number of particles. Nevertheless, it is anticipated that the above description will remain valid in a good first approximation. As we pointed out above, however, if we know that the excited wavefunction is of event type, i.e. is defined in $\mathcal{H}_\pi^{(3)} \otimes \mathcal{H}_\sigma^{(n-3)}$, we are not sure that it is separable [17]. A definite answer could be obtained by the study of the operator $\rho_n^{(3)}$ restricted to the subspace $\mathcal{H}_\pi^{(3)} = \mathcal{H}_\pi^{(1)} \otimes \mathcal{H}_\pi^{(1)} \otimes \mathcal{H}_\pi^{(1)}$.

7. Conclusion

We gave a rapid review of some problems connected with the application of the pair-density analysis to CNDO/S wavefunctions. The results presented seem to us encouraging, as our main purpose is to extract from a quantummechanical calculation a representation of the molecular system which agrees with the usual representations traditionally used by the chemists. We would like to stress again the fact that the worth of the representation obtained is determined by the quality of the wavefunction. However, as the building of the wavefunction and its analysis are two entirely distinct steps, the efficiency of the latter one (which is our main preoccupation at the present time) has nothing to do with the former. In this respect, it was natural to first analyse CNDO/S wavefunctions in consideration of their simplicity. Nevertheless, the structural indications extracted from this type of wavefunction, widely used in the field of applied quantum chemistry, seem to us sensible from an intuitive chemical viewpoint, so that it is hoped that they will be confirmed by ab-initio calculations. Such problems are now under study.

References

1. Constanciel, R.: *Localization and Delocalization in Quantum Chemistry*, Vol 1, p. 43 (1975).
2. Constanciel, R.: *Phys. Rev.* **A11**, 395 (1975).
3. McWeeny, R.: *Rev. Mod. Phys.* **32**, 335 (1960).
4. Bader, R. F. W.: *Localization and Delocalization in Quantum Chemistry*, Vol 1, p. 15 (1975).
5. Daudel, R.: *Les Fondements de la Chimie Théorique*, Gauthiers-Villars, Paris (1956) .
6. Shannon, C. E.: *Bell System Tech. J.* **30**, 50 (1951).
7. Carlson, B. C. and Keller, J. M.: *Phys. Rev.* **105**, 102 (1957).
8. Millié, Ph., Lévy, B., and Berthier, G.: *Localization and Delocalization in Quantum Chemistry*, Vol. I, p. 59 (1975).
9. Ruedenberg, K. and Poshusta, R. D.: *Adv. Quant. Chem.* **6**, 267 (1972).
10. McWeeny, R.: *Proc. Roy. Soc. London* **A253**, 242 (1959).
11. Bessis, G. and Chalvet, O.: Unpublished work.
12. Bingel, W. A. and Kutzelnigg, W.: *Adv. Quant. Chem.* **5**, 201 (1970).
13. Constanciel, R.: *Chem. Phys. Letters* **16**, 432 (1972).
14. McWeeny, R. and DelRe, G.: *Theor. Chim. Acta* **10**, 13 (1968).
15. Petke, J. D. and Whitten, J. L.: *J. Chem. Phys.* **51**, 3166 (1969).
16. Cook, D. B., Hollis, P. C., and McWeeny, R.: *Mol. Phys.* **13**, 553 (1967).
See also: M. Sanchez, These 3° cycle, Paris (1975).

QUESTIONS TO CONSTANCIEL

Malrieu: If $\psi = \hat{A}[\phi_A^{(p)}\phi_B^{(n-p)}]$, is there a fluctuation of the number of particles in each group?

Constanciel: No.

Malrieu: What about the converse?

Constanciel: It does not hold; when there is no fluctuation, the most general form for ψ is the following

$$\psi(1, 2, \ldots, n) = \hat{A}\Big[\sum_i \sum_j c_{ij}^p \phi_{A_i}^{(p)}(1, 2, \ldots p)\, \phi_{B_j}^{(n-p)}(p+1, \ldots, n)\Big]$$

$$\in \mathcal{H}_A^{(p)} \otimes \mathcal{H}_B^{(n-p)}$$

Levy: If ψ is a single determinant, there exist an obvious separability.

Constanciel: Yes, but there is no *a priori* partition for the AO's.

Durand: Concerning the localization of π-electrons in butadiene: the basis is not adapted for the apparition of this localization.

Berthier: How does one define a bond index in an CI procedure?

Constanciel: The bond indexes may be defined as the non diagonal elements of the matrix of $\rho^{(1)}$ when written on the AO's.

Berthier: This definition is not unique.

Constanciel: Of course, but it is the most usual one.

Durand: The diagonalization of $P_A \rho^{(1)} P_A$ (P_A is the projector on atom A) is the right procedure leading to the hybrids.

Constanciel: Here we use the *same* criterion for the definition of the 'best' hybrids and for the measure of group separability.

Daudel: About Levy's comment: Let us take a Slater determinant:

$$\{\phi_A\} = A \quad \{\phi_2, \phi_3, \ldots, \phi_n\} = B \quad \psi = \hat{A}[\phi_A(1)\,\phi_B(2, 3, \ldots, n)]$$

and the separability occurs.

Claverie: I agree; for me, there is a topological intention behind Constanciel's procedure: one looks for ϕ's which are *spatially localized*.

Constanciel: This is not always true: take for instance the case of the σ-π separability.

Durand: It would be interesting to analyze the correlation of the very small values of the missing information function and the use of methods where there is no overlap integrals.

Dannenberg: What are the relations, if any, between missing information and potential surfaces?

Constanciel: These relations are not obvious, but it is certain that modification of electronic configuration leads to variation of missing information.

LOCALIZATION AND LOCAL PHENOMENA
IN MOLECULAR EXCITED
AND IONIZED STATIONARY STATES

JAQUELINE LANGLET

Laboratoire de Biochimie Théorique, Institut de Biologie Physico-chimique,
75005 Paris, France

and

JEAN-PAUL MALRIEU

Laboratoire de Physique Quantique, Université Paul Sabatier,
31077 Toulouse, France

Abstract. The localization of the ground state SCF MO's gives bond and lone pair MO's without changing the H—F determinant. The delocalized symmetry adapted MO's allow the construction of symmetry-adapted determinants and remove the energetic degeneracies between them, while the degenerate, unsymmetrical locally excited determinants only may be considered as a basis for a CI (excitonic) treatment. One may express the delocalized MO's as linear combinations of localized MO's, and analyse a $n\pi^*$ or $\pi\pi^*$ delocalized Virtual Orbital Approximation description in terms of Local and Charge Transfer Excitations. It is shown that the π^* delocalized MO's are completely irrelevant for the description of an $n\pi^*$ excited state in a polyenic aldehyde, since the 'particle' should remain in the region of the 'hole'; the CI process is equivalent in this case to the relocalization of the π^* MO in the CO bond. In linear polyenes the delocalized VO Approximation overestimates the role of the long range charge transfer components of the wave function. The comparison with the excitonic treatment shows that after the CI of singly excited determinants, the $\pi\pi^*$ excited state is essentially composed of intrabond ethylenic excitations and charge transfers between adjacent bonds. The $\sigma\pi$ coupling and mixture toward the Rydberg levels had been thought to have a vanishing effect (as N^{-1}) when the dimension N of the polyene increases; the short range character of the 'hole-particle' pair implies on the contrary that these effects, so important in ethylene, remain important whatever the dimension of the conjugated system. It is shown that in butadiene and hexatriene unsymmetrically distorted geometries may be prefered in the $\pi\pi^*$ excited states; this Born-Oppenheimer geometry instability suggests one represent the adiabatic excited state as a linear combination of locally excited, locally distorted components. Most of these results are obtained through excitonic treatments using fully localized MO's. It is shown however on the case of the transition energies of linear polyenes that the tails of the localized SCF-MO's (i.e. the ground state delocalization) may have a qualitative influence upon spectroscopic features.

1. Introduction

The present review does not cover the time dependent problem of the possible localization and migration of an excitation in a molecular system, which will be treated in a further chapter of this book. The excited states studied hereafter are 'stationary' states represented through time independant Schrödinger equation solutions, using a spin-less hamiltonian, as usually done in quantum chemistry. The main scope of this work is to analyze the qualitative differences between the ground state and the excited and ionized states from the point of view of the localization. Two questions are discussed hereafter:

O. Chalvet et al. (eds.), Localization and Delocalization in Quantum Chemistry, Vol. II, 15–47. All Rights Rseerved
Copyright © 1976 by D. Reidel Publishing Company, Dordrecht-Holland

(i) Is the localization of the particles less pronounced in the excited states, referring for instance to the mean fluctuation of the number of particles per loge as a measure of the physical localizability of the system [1]?

(ii) Is it possible to speak of a localization of the excitation (resp. ionization)?

In the whole development the ionization may be considered as a special case of the excitation, in which an electron is excited towards a monoelectronic wave-function removed to infinity. The specificity of the ionization process will not be explicited at each step.

2. The Basic Difference Between Ground and Excited States

A. SYMMETRY CONSIDERATIONS

Referring to the molecular orbital description of a $2N$ electron molecule, the qualitative specificity of the ground state appears immediately. If the orthonormal MO's φ_i are eigenfunctions of a mono-electronic hamiltonian for the molecular problem (such as the Hückel hamiltonian, or the Hartree-Fock hamiltonian),

$$F\varphi_i = \varepsilon_i \varphi_i, \tag{1}$$

these 'canonical' MO's are 'symmetry adapted', ie eigenfunctions of the symmetry operations R of the problem, if F has the proper symmetry of the molecule,

$$R\varphi_i = \lambda_i \varphi_i. \tag{2}$$

With these monoelectronic wave-functions it is possible to build a single determinantal approximation of the $2N$ electron ground state wave-function,

$$\Phi_0 = |\varphi_1 \bar{\varphi}_1 \ldots \varphi_i \bar{\varphi}_i \ldots \varphi_N \bar{\varphi}_N|, \tag{3}$$

Φ_0 belonging to an eigenspace of the symmetry operators. But it is well-known now that the set $[\varphi_i]$ required to build Φ_0 as a antisymmetrized product of monoelectronic wave-functions is not unique. Any unitary transform $\mathscr{U}$ applied to the set $[\varphi_i]$ gives a new set $[\phi_r]$ of orthonormal MO's

$$[\phi_r] = \mathscr{U}[\varphi_i] \tag{4}$$

such that

$$\Phi_0' = |\phi_1 \bar{\phi}_1 \ldots \phi_r \bar{\phi}_r \ldots \phi_N \bar{\phi}_N| = \pm \Phi_0. \tag{5}$$

The single determinant built from the set $[\phi_r]$ is equal to $\pm \Phi_0$; the ground state single determinant Φ_0 is invariant under any unitary transform of the canonical MO's. But the new MO's are no longer eigenfunctions of F and R

$$F\phi_r = \sum_s \varepsilon_{rs} \phi_s \tag{6}$$

$$R\phi_r = \sum_s \lambda_{rs} \phi_s. \tag{7}$$

The 'equivalent' MO's are no longer eigenfunctions of the symmetry operators of the problem. Among these equivalent MO's, some are strongly localized in definite regions of space. They may be obtained through the maximization of a localization criterion [3–6] and may be identified as core, lone pairs and bond monoelectronic descriptions. For the ground state problem the canonical symmetrical MO's are equivalent to the localized MO's.

The spectral decomposition of the monoelectronic hamiltonian F does not give only the monoelectronic wave-functions below the Fermi level, it also gives a set of virtual MO's φ_j^* which again are symmetry-adapted.

$$F\varphi_{j*} = \varepsilon_{j*}\varphi_{j*} \tag{8}$$

$$R\varphi_{j*} = \lambda_{j*}\varphi_{j*}, \tag{9}$$

and these virtual MO's will be used to represent the excited states and negative ions. The single-determinant

$$\Phi(_i^{j*}) = a_{j*}^+ a_i \phi_0 = |\varphi_1\bar{\varphi}_1 \ldots \varphi_{j*}\bar{\varphi}_i \ldots \varphi_N\bar{\varphi}_N|, \tag{10}$$

where a_i is an annihilator of φ_i and a_j^+ a creator of φ_j^*, is also symmetry adapted.

But one might perform unitary transformations over the set of virtual MO's $[\varphi_j^*]$ as done previously for the ground state occupied MO's, applying for instance the same criterion of localization

$$[\phi_t^*] = \mathcal{U}'[\varphi_{j*}] \tag{11}$$

which gives antibonding MO's on the valence bonds and local 'Rydberg' states if the atomic orbitals basis set is not minimal. The ϕ_t^* are no longer eigenvectors of the symmetry operators. Then if we try to build a single determinant from the sets of occupied and virtual localized MO's $[[\phi_r] + [\phi_t^*]]$, the determinants

$$\Phi\begin{pmatrix} t^* \\ r \end{pmatrix} = a_{t*}^+ ar\, \phi_0' = |\phi_1\phi_1 \ldots \phi_{t*}\bar{\phi}_r \ldots \phi_N\bar{\phi}_N| \tag{12}$$

are no longer symmetry adapted

$$R\Phi\begin{pmatrix} t^* \\ r \end{pmatrix} \neq \lambda\Phi\begin{pmatrix} t^* \\ r \end{pmatrix},$$

and therefore cannot be considered as a satisfactory approximation of an excited state wave function, which must be an eigenfunction of R. The first main superiority of the canonical MO's appears there; it is possible to build single determinantal descriptions of the excited states which at least possess the right symmetry properties, while the localized MO's do not.

B. Energetic Considerations

Another superiority of the canonical MO's is energetic. Referring to the monoelectronic hamiltonian F, one sees that the mono-electronic energies ε_i of the set $[\varphi_i]$

are spread over the largest spectrum, all inessential degeneracies are removed. On the contrary the localized MO's energies

$$F_{rr} = \langle \phi_r | F | \phi_r \rangle \tag{13}$$

are strongly degenerate, since the equivalent sites of the molecules have equivalent MO's and equivalent energies. Therefore the spectrum of the localized MO's energies is a set of degenerate levels corresponding to the various types of sites; the diagonalization of the monoelectronic hamiltonian removes the degeneracies, as illustrated in Figure 1. Therefore if one uses a zeroth order hamiltonian H_0, sum of monoelectronic operators

$$H_0 = \sum_i F(i) = \sum_i \varepsilon_i a_i^+ a_i \tag{14a}$$

$$H_0' = \sum_r F_{rr} a_r^+ a_r \tag{14b}$$

the single determinants built with delocalized MO's have different energies (except for spin and essential spatial degeneracies), while the use of localized MO's leads to

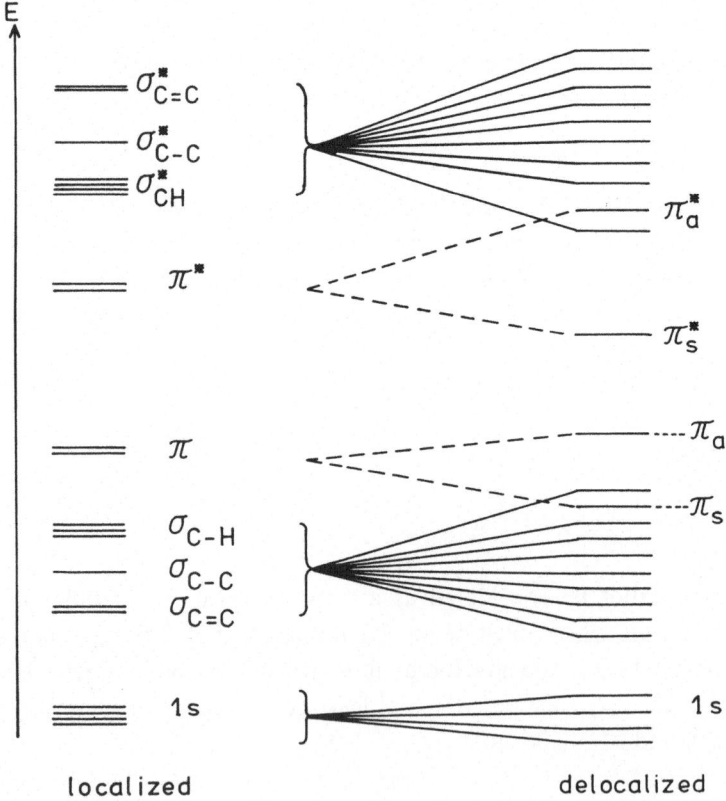

Fig. 1.   Monoelectronic energy levels distribution for localized (left) and delocalized (right) pictures (case of the butadiene molecule, minimal basis set description).

numerous degenerate energies for the equivalent single determinants. The spectrum of the single determinant energies is rather well resolved in the delocalized picture. This statement remains valid when the single determinantal energies are taken as mean values of the exact hamiltonian. Therefore the delocalized picture may give an approximate approach of the spectrum through the consideration of single determinants, while the localized picture cannot.

The delocalized MO's do not present any intrinsic advantage for the ground state description. The reason of their success and of their nearly monopolistic status in text books lies in their ability to give simple single determinantal descriptions of the ions and excited states [7]; the ionization appears as the departure of an 'electron' from a delocalized MO, the excitation as a promotion to an empty orbital. Still now, one may find experimentalists and 'theoreticians' who more or less believe in the physical meaning of this langage, and that they 'see' monoelectronic levels or probabilities. In terms of localized MO's, such a simple description of ionization or excitation is impossible.

C. LOCALIZING THE EXCITED STATE MO's

An alternative and symmetric approach to show this qualitative difference consists in localizing the MO's of the excited state determinant instead of using the ground state MO to build the excited state. One must distinguish two types of MO's in the R.H.F. excited state; the doubly occupied orbitals (closed shells) and singly occupied orbitals (open-shells). Two types of transformations may be considered. The first

TABLE I

Localized π MO's for the ground state and the first triplet state of naphtalene

Atoms	Ground state		Excited state		
			Closed shells		Open shell
	[1–2] bond	[9–10] bond	[2–3] bond	three atomic [4–5–10] central bond	[4–5] bond
1	0.677	0.157	0.299	0.103	0
2	0.677	−0.126	0.632	0.139	−0.372
3	0.159	−0.126	0.632	−0.106	0
4	−0.129	0.157	0.299	−0.462	0.601
5	−0.006	0.157	−0.072	−0.462	0.601
6	0.036	−0.126	0.030	−0.106	0
7	−0.020	−0.126	0.030	0.139	0.372
8	−0.019	0.156	−0.072	0.103	0
9	0.161	0.647	—0.069	—0.185	0
10	—0.121	0.647	—0.069	−0.675	0
τ	$\tau_2 = 91.4$	$\tau_2 = 83.6$	$\tau_2 = 79.8$	$\tau_3 = 87.1$	$\tau_2 = 72.2$

τ_i is the sum of the i highest squared coefficients.

one mixes the α (resp. β) spin orbitals, giving an UHF looking like determinant. The second solution consists in an unitary transform among the open shells, if they have the same spin, as in an $S_z = 1$ triplet state. The Ruedenberg criterion was applied in this scheme to the lowest triplet state of naphtalene in the Pariser-Parr approximation [7]. As seen in Table I, for the ground state one obtains the symmetric Kekule structure with a rather high degree of localization, despite the aromatic character of the molecule, with four external well localized ($\tau_2 = 91.4\%$) π bonds, plus a central

ground state   excited state

poorly localized ($\tau_2 = 83.6\%$) double bond. In the excited state the closed shells give two $\beta\beta'$ external double bonds with a worse degree of localization ($\tau_2 = 79.8\%$) and two *three center* bonds in the central part ($\tau_3 = 87.1\%$); the open shell are non adjacent two center 'bonds' on the geminal α summits ($\tau_2 = 72.2\%$). Here τ_i is the sum of the squares of the i highest coefficients. In the ground state the different MO's are localized on different parts of the molecule, allowing the definition of loges [8], while in the excited state a doubly occupied and a singly occupied MO spread on the *same region* of space (atoms 4–5–10). The good localizability of the ground state MO's disappears in the excited state.

One should not exaggerate however the signification of this result and the intrinsic superiority of the delocalized MO's. For several reasons:

(i) The single determinantal description which is rather satisfactory for the ground state is in general very poor for the excited states, even when a variational principle has been applied to the excited state. This is due to the fact that the Brillouin's theorem only allows the first order mixture of the ground state determinant with high energy doubly excited determinants, while the (singly excited determinants always interact with other singly (and doubly and triply) excited determinants which lie near in energy [7]. In any case the configuration interaction is necessary.

(ii) If one performs a correct CI the result is independent of the basis of MO's; the exact solution may be reached and expressed using delocalized or localized MO's, and even AO's.

If the localized and delocalized MO's are related by an unitary transform the equivalence of the CI's does not concern only the full CI but any 'excitation subspace' or union of 'excitation-subspaces'. This means that the subspace of all the p-times excited determinants *is invariant* in the unitary transform of the MO's; the localized p-excited determinants are linear combination of p-excited delocalized determinants (and vice-versa). This statement is evident since the occupied (resp. virtual) MO's are linear combinations of occupied (resp. virtual) MO's only. This means that the

CI problem (i.e. hamiltonian) restricted to, say, the singly (or singly + doubly) excited determinants, is the same in both basis.

The localized bases may be used then to get a better picture of the electronic partition and of the effect of the excitation in terms of local events. It will also give an interesting insight upon the CI effect.

3. Development of the Delocalized MO's in the Basis of Localized MO's for a Simplified Model Problem [9]

Let us consider a regular chain of equivalent 2 electron bonds. This might be a chain of H_2 molecules or the π bonds of a linear polyene. Let us call ϕ_r, ϕ_{r*} the bonding and antibonding (SCF) MO's localized on the unit r. One may build the nondiagonal Fock operator in this basis, and its diagonalization will give the delocalized MO's φ_i. Neglecting end effects and assuming that the Fock operator elements are negligible between non adjacent bonds (since they involve exponentially decreasing overlap distributions),

$$\langle\phi_r|F|\phi_r\rangle = E \; \forall_r$$

$$\langle\phi_r|F|\phi_{r+1}\rangle = \langle\phi_r|F|\phi_{r-1}\rangle = F \; \forall_r \qquad (15)$$

$$\langle\phi_r|F|\phi_{r+s}\rangle \ll F \; \forall_{r,\,|s|>1}$$

one obtains for the Fock operator a Hückel-like matrix representation, and the delocalized MO's φ_i are given by the well-known solutions

$$\varphi_i = \sum_{r=1}^{N} C_{ir}\phi_r; \qquad C_{ir} = \sqrt{\frac{2}{N+1}} \sin\frac{i\pi r}{N+1};$$

$$\varepsilon_i = E + 2F \cos\frac{i\pi}{N+1}. \qquad (16)$$

The corresponding energies increase when i varies from 1 to N, F being negative. The Fock operator for the virtual orbitals would be the same

$$\langle\phi_{s*}|F|\phi_{s*}\rangle = E^* \; \forall s$$

$$\langle\phi_{s*}|F|\phi_{s+1}^*\rangle = F^* \; \forall s \qquad (17)$$

$$\langle\phi_{s*}|F|\phi_{s+r}^*\rangle = 0 \; \forall s, \quad \text{if} \; > 1,$$

and the delocalized MO's may be written:

$$\varphi_{j*} = \sum_{s=1}^{N} C_{j*s}\phi_{s*}; \qquad C_{j*s} = \sqrt{\frac{2}{N+1}} \sin\frac{j\pi s}{N+1};$$

$$\varepsilon_j = E^* + 2F^* \cos\frac{j\pi}{N+1}. \qquad (18)$$

The corresponding energies now decrease when j varies from 1 to n because $FF^* < 0$ as may be easily understood by considering the respective phase behaviour of the adjacent occupied and virtual MO's

$$\frac{+ \ldots + \quad + \ldots +}{\phi_r \qquad \phi_{r+1}}$$

$$\frac{+ \ldots - \quad + \ldots -}{\phi_{r*} \qquad \phi_{r+1}^*} .$$

The Fock operator was supposed to be zero between occupied and virtual MO's (i.e. was supposed self-consistant), allowing independant diagonalizations of the two diagonal blocks for the occupied and virtual subsets. If the localized MO's ϕ_r are fully localized, as in the PCILO method or most of the excitonic treatments, the Fock operator would not split into two blocks, but the diagonalization of the occupied and virtual blocks would lead to delocalized MO's very close to the SCF ones, since the SCF localized and fully localized MO's are very close [6, 10–15].

4. Possible Overestimation of the Delocalization of the Excitation by the Canonical Representation: the Case of nπ* Transitions [16]

In this section we demonstrate that the use of the canonical MO's may lead to an unrealistic representation of the excited state structure, due to an overestimation of the extent of the excitation process. The example refers to an inhomogeneous system, the polyenic aldehydes $O = CH—(CH=CH)—H$. In the canonical description, the $\bar{p}$ locally antisymmetric σ lone pair is well localized on the oxygen atom, with small tails on the $C_\alpha H$ and $C_\alpha C_\beta$ bonds. It represents the highest occupied MO and its energy does not significantly depend on the dimension of the conjugated chain. On the contrary the π^* MO's energies vary significantly. If one neglects the difference between the $C=0$ and $C=C$ π bonds, the π^* MO's coefficients and energies are given by Equation (18), and the lowest empty level decreases from E^* (for H_2CO) to $E^* - 2F^*$ when N tends to infinity. The amplitude of the variation is of the order of magnitude of 4 eV. In the virtual orbital approximation, the delocalized picture leads to a single determinant for, say, the lowest triplet $\pi\pi^*$ excited state:

$$^3\Phi\left(\frac{\pi_N^*}{\bar{n}}\right) = a_{\pi_{N*}}^+ a_{\bar{n}} \, \Phi_0. \tag{19}$$

The corresponding transition energy from the ground state is:

$$^3\Delta E\left(\frac{\pi_N^*}{\bar{n}}\right) = \varepsilon_{(\pi_N^*)}^* - \varepsilon_n - J_{n\pi_N^*}$$

$$= E^* - 2F^* \cos\frac{\pi}{N+1} - \varepsilon_n - J_{n\pi_N^*}. \tag{20}$$

To determine the N-dependance of the transition energy, it is necessary to evaluate the variation of the coulombic interaction between the localized hole distribution on the lone pair (nn) and the $(\pi_N^* \pi_N^*)$ delocalized charge distribution. Neglecting the overlap distributions, one may develop this distribution into its local components,

$$\pi_N^{*2} = \sum_{r=1}^{N} C_{N*r}^2 \phi_{r*}^2 = \frac{2}{N+1} \sum_{r=1}^{N} \sin^2 \frac{\pi r}{N+1} \phi_r^{*2}. \tag{21}$$

The electrostatic interaction may be developed,

$$J_{nn_N^*} = \frac{2}{N+1} \sum_{r=1}^{N} \sin^2 \frac{\pi r}{N+1} (nn, \phi_r^* \phi_r^*), \tag{22}$$

and $(nn, \phi_r^* \phi_r^*)$ is approximately equal to $1/R_{nr}$ where R_{nr} is the distance between the bond r and the lone pair n, which is proportional to r.

$$J_{nn_N^*} = \frac{2\alpha}{N+1} \sum_{r=1}^{N} r^{-1} \sin^2 \frac{\pi r}{N+1}. \tag{23}$$

A rapid estimate of the summation shows that it grows more slowly than $\log N$, and therefore $J_{nn_N^*}$ decreases and tends towards zero when the dimension increases. This is not surprising since the center of gravity of the π_N^* orbital is in the center of the molecule and goes further and further from the lone pair when the dimension of the conjugated system increases. For formaldehyde J_{nn^*} is very large (~ 12 eV) the two distributions being almost monocentric; therefore the total variation of J_{nn^*} is larger than the variation of $\varepsilon_{\pi_N^*}^*$ and the transition energy calculated in the virtual orbital approximation using a delocalized π^* MO *increases* when the dimension of the conjugated system increases.

This is clearly an artefact, and simply shows the inadequacy of the delocalized π^* MO to reproduce the distribution of the supplementary π electron in the $n\pi^*$ state. Actually in the localized picture the determinant $^3\Phi\left(\dfrac{\phi_1^*}{\bar{n}}\right)$ corresponding to the local $n \to \pi_{CO}^*$ excitation has the same energy in the series and its interaction with the other $n \to \pi_{C=C}^*$ excited state of the same symmetry necessarily diminishes its energy. The $^3\Phi\left(\dfrac{\phi_2^*}{\bar{n}}\right)$ determinant, corresponding to an excitation towards the adjacent C=C bond, has a higher energy than $^3\Phi\left(\dfrac{\phi_1^*}{\bar{n}}\right)$ since

$$^3\Delta E\left(\Phi\left(\frac{\phi_2^*}{\bar{n}}\right)\right) - {}^3\Delta E\left(\Phi\left(\frac{\phi_1^*}{\bar{n}}\right)\right) = J_{n\phi_1} - J_{n\phi_2} \geqslant 0. \tag{24}$$

The interaction between these two determinants is non negligible,

$$\langle \Phi(\tfrac{\phi_2^*}{\bar{n}}) |H| \Phi(\tfrac{\phi_1^*}{\bar{n}}) \rangle = \langle \phi_1^* |F - J_n| \phi_2^* \rangle, \tag{25}$$

and the CI in a basis of localized determinants leads to a wave function centered on $\Phi\left(\genfrac{}{}{0pt}{}{\phi_1^*}{\bar{n}}\right)$ ($n \to \pi_{CO}^*$ transition) with a small tail on $\Phi\left(\genfrac{}{}{0pt}{}{\phi_2^*}{\bar{n}}\right)$,

$$\psi = \Phi(\genfrac{}{}{0pt}{}{\phi_1^*}{\bar{n}}) - [\langle \phi_1^*|F - J_n|\phi_2^* \rangle / (J_{n\phi_1^*} - J_{n\phi_2^*})]\, \Phi(\genfrac{}{}{0pt}{}{\phi_2^*}{\bar{n}}). \tag{26}$$

This description is almost equivalent (to the second order) to a single determinantal description ${}^3\Phi\left(\genfrac{}{}{0pt}{}{\phi_1'^*}{\bar{n}}\right)$ where the excitation occurs towards an almost localized MO $\phi_1'^*$

$$\phi_1'^* = \phi_1^* - [\langle \phi_1^*|F - J_n|\phi_2^* \rangle / (J_n \phi_1^* - J_n \phi_2^*)]\, \phi_2^* \tag{27}$$

centered on the C=O bond with a small tail on the adjacent C=C bond. In other words, *for the description of the nπ* excited state, the best π* MO is not the canonical MO, which leads to an absurd delocalization, but an almost localized MO restricted to the CO region* and its immediate surroundings, whatever the dimension of the molecule.

This result is confirmed by extensive numerical calculations [16] and is easy to understand: the localization of the 'hole' prevents a delocalization of the 'particle', while in a ππ* excitation the delocalized π* MO might be relevant. This example suggests two conclusions:

(i) The delocalized canonical MO's have no intrinsic general quality for the representation of the excited states, which may be more appropriately described through localized descriptions.

(ii) The exact solution being unique, the CI performed in the delocalized picture must relocalize the excitation. The virtual orbital approximation might be expressed in terms of locally excited determinants

$$^3\Phi\left(\genfrac{}{}{0pt}{}{\pi_N^*}{\bar{n}}\right) = \sum_{r=1}^{N} \sqrt{\frac{2}{N+1}}\, \sin\frac{N\pi r}{N+1}\, \Phi\left(\genfrac{}{}{0pt}{}{\phi_r^*}{\bar{n}}\right) \tag{28}$$

as a rather democratic mixture of all possible $n \to \pi^*$ excitations towards the various π bonds. After the CI process, the weight of the $\Phi\left(\genfrac{}{}{0pt}{}{\phi_1^*}{n}\right)$ determinant decreases from 100% in formaldehyde to 97.8% in acroleine, and remains constant for $N \geqslant 3$.

5. Local Analysis of a Delocalized ππ* Excitation, Its Short Range Character

The preceeding conclusions have to be checked on a more typical problem, where the hole is no longer compelled to be localized by a physical singularity, and we turn back to the standard test-case of the homogeneous linear polyenes.

A. Analysis of the Delocalized Virtual Orbital Approximation [9]

Referring to the linear chain of Section 3, one may consider the unoptimized single determinantal representation of the excited state (frequently called the virtual orbital approximation) obtained by the substitution of a virtual MO φ_{j*} to an occupied one φ_i. In the 2nd quantization formalism, this may be written as:

$$\Phi\binom{j^*}{i} = a_{j*}^+ a_i \Phi_0. \tag{29}$$

Expressing the canonical creation and annihilation operators in terms of local creation and annihilation operators (through Equations (16) and (18)) one gets an equivalent expression of the excitation procedure

$$\Phi\binom{j^*}{i} = \frac{2}{N+1} \sum_r \sum_s \sin\frac{i\pi r}{N+1} \sin\frac{j\pi s}{N+1} a_{\phi_s^*}^+ a_{\phi_r} \Phi_0, \tag{30}$$

where Φ_0 may expressed now as the antisymmetrized product of local MO's,

$$\Phi\binom{j^*}{i} = \frac{2}{N+1} \sum_r \sum_s \sin\frac{i\pi r}{N+1} \sin\frac{j\pi s}{N+1} \Phi\binom{s^*}{r}. \tag{31}$$

The single excitation from φ_i to φ_{j*} appears as the linear combination of local excitations from the bonds r toward the antibonding MO of the bonds s. These excitations may be described in general as electron transfers from one region to another. Some of the local excitations, for which $s = r$ do not imply such an electron transfer since they simply represent a $(\pi\pi^*)$ excitation in the ethylenic system of a (double) bond. These excitations ($r = s$) will be called local excitations (LEs) while the former ones ($r \neq s$) will be called charge transfer excitations (CTEs). The CTEs will be characterized by the amplitude of the electron jump through the quantity $p = |r - s|$ and identified as ^{p}CTE.

Let us analyze now the relative weight of local and charge transfer excitations in the (lowest) excitations. The weight of LEs is given by:

$$d_{LE}\binom{j^*}{i} = \left(\frac{2}{N+1}\right)^2 \sum_r \left(\sin\frac{j\pi r}{N+1} \sin\frac{i\pi r}{N+1}\right)^2. \tag{32}$$

Since $\sin^2 \alpha \leqslant 1$,

$$d_{LE}\binom{j^*}{i} \leqslant \frac{4}{N+1}. \tag{33}$$

A more accurate calculation gives $d_{LE}\binom{j^*}{i} = \frac{1}{N+1}[1 + f(i, j, N)/(N+1)]$, where f is bounded by a constant whatever i, j, N.

Therefore the relative weight of LEs in the virtual orbital description of excited states decreases as $1/N$ when the dimension N of the conjugated system increases.*

On the contrary the weight of the CTE increases towards 1. The excitations are treated almost democratically, whatever the distance between the departure and arrival bonds, i.e. the amplitude of the electron jump. It may be seen [9] that a jump over p intermediate bonds becomes as probable as the local excitations when N is much larger than p.

B. THE EXCITONIC MATRIX

The excitonic methods for a stationary state problem are nothing but the Configuration Interaction in a basis of localized MO's. Some excitonic treatments correspond to drastic truncations of this CI matrix but in principle the excitonic model may be derived to any degree of accuracy. Let us study first the structure of the CI matrix restricted to the singly excited determinants. One may notice that at the single determinantal level, the LEs are less energetic than the CTEs as may be seen from the zeroth-order transition energy of a single determinant,

$$\Delta E \begin{pmatrix} s^* \\ r \end{pmatrix} = \langle \phi_s^* | F | \phi_s^* \rangle - \langle \phi_r | F | \phi_r \rangle - J_{\phi_r \phi_s^*} + K_{\phi_r \phi_s^*} \pm K_{\phi_r \phi_s^*}. \tag{34}$$

In this expression the difference between the diagonal elements of the Fock operator are independent of the distance between r and s, while the charge-charge interaction $J_{\phi_r \phi_s^*}$ decreases as R_{rs}^{-1}; $K_{\phi_r \phi_s^*}$ decreases exponentially with R_{rs} but remains always much smaller than $J_{\phi_r \phi_s^*}$ and the localized transition energies may be ordered in the following way:

$$\left(\Delta E_0 = \Delta E \begin{pmatrix} r^* \\ r \end{pmatrix} \right) < \left(\Delta E_1 = \Delta E \begin{pmatrix} (r \pm 1)^* \\ r \end{pmatrix} \right) < \left(\Delta E_2 = \Delta E \begin{pmatrix} (r \pm 2)^* \\ r \end{pmatrix} \right)$$

$$< \ldots < \left(\Delta E_p = \Delta E \begin{pmatrix} (r \pm p)^* \\ r \end{pmatrix} \right) < \ldots < (\langle \phi_s^* | F | \phi_s^* \rangle - \langle \phi_r | F | \phi_r \rangle). \tag{35}$$

One may build therefore the excitonic matrix in a basis of localized excited determinants in which the LEs occur first followed by CTE excitations between adjacent bonds, and so on.

In the block of the LE, for singlet states, the extradiagonal matrix elements are essentially transition dipole-transition dipole-interactions

$$\left\langle {}^1\Phi \begin{pmatrix} r^* \\ r \end{pmatrix} | H | {}^1\Phi \begin{pmatrix} s^* \\ s \end{pmatrix} \right\rangle = 2(\phi_r \phi_r^*, \phi_s \phi_s^*) - (\phi_r \phi_s^*, \phi_s \phi_r^*). \tag{36}$$

---

* These results are not modified by considering the proper S^2 eigenfunctions for the excited state

$$\psi \begin{pmatrix} j^* \\ i \end{pmatrix}_{1,3} = 1/\sqrt{2} \left(\Phi \begin{pmatrix} j^* \\ i \end{pmatrix} \pm \Phi \begin{pmatrix} j^* \\ i \end{pmatrix} \right)$$

instead of the single determinant $\Phi \begin{pmatrix} j^* \\ i \end{pmatrix}$ of Equation (29).

The second term decreases exponentially with the distance R_{rs} between the bonds r and s, and the first one decreases as $1/R_{rs}^3$. For the triplet state the dipolar interaction disappears. In both cases one may consider that the only important off-diagonal elements occur between excitations on adjacent bonds

$$(\phi_r \phi_r^*, \phi_{r+1} \phi_{r+1}^*) = (\phi_r \phi_r^*, \phi_{r-1} \phi_{r-1}^*) = b. \tag{37}$$

The LE determinants are coupled with the CTE ones, through

$$\left\langle \Phi \begin{pmatrix} r^* \\ r \end{pmatrix} |H| \Phi \begin{pmatrix} t^* \\ s \end{pmatrix} \right\rangle = \langle \phi_r \phi_t^* | \phi_r^* \phi_s \rangle = \langle rt^* | rs \rangle. \tag{38}$$

If r, s and t are all different this matrix element may be approximated through Mulliken's approximation,

$$\left\langle \Phi \begin{pmatrix} r^* \\ r \end{pmatrix} |H| \Phi \begin{pmatrix} t^* \\ s \end{pmatrix} \right\rangle = \tfrac{1}{2} \langle \phi_s | \phi_t^* \rangle \left[\langle rs | r^* s \rangle + \langle rt^* | r^* t^* \rangle \right]; \tag{39}$$

$\langle ps | pt^* \rangle$ decreases exponentially as $\exp(-R_{st})$, and the two integrals decrease as R_{rs}^{-2} and R_{rt}^{-2} respectively. The matrix element will be negligible except when $t = r$

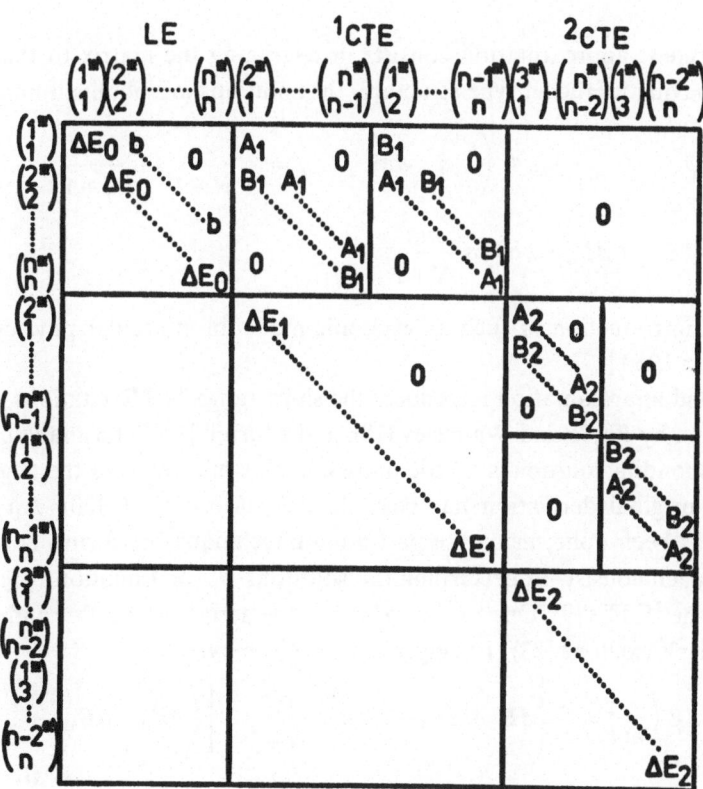

Fig. 2.   Structure of the excitonic matrix.

or $s^* = r^*$. In such a case

$$\left\langle \Phi\left(\begin{matrix} r^* \\ r \end{matrix}\right) \middle| H \middle| \Phi\left(\begin{matrix} s^* \\ r \end{matrix}\right) \right\rangle = \langle \phi_r^* | F - J_r | \phi_s^* \rangle,$$

$$\left\langle \Phi\left(\begin{matrix} r^* \\ r \end{matrix}\right) \middle| H \middle| \Phi\left(\begin{matrix} r^* \\ s \end{matrix}\right) \right\rangle = \langle \phi_r | F + J_{r^*} | \phi_s \rangle. \tag{40}$$

If s and r are adjacent bonds ($s = r+1$), these matrix elements are rather important

$$\left\langle \Phi\left(\begin{matrix} (r+1)^* \\ r \end{matrix}\right) \middle| H \middle| \Phi\left(\begin{matrix} (r+2)^* \\ r \end{matrix}\right) \right\rangle = \langle \phi_{r+1}^* | F - J_r | \phi_{r+2}^* \rangle = A_2$$

$$\left\langle \Phi\left(\begin{matrix} (r+1)^* \\ r \end{matrix}\right) \middle| H \middle| \Phi\left(\begin{matrix} (r+1)^* \\ r+2 \end{matrix}\right) \right\rangle = - \langle \phi_r | F + J_{(r+1)}^* | \phi_{r+2} \rangle = B_2. \tag{42}$$

In the CNDO hypothesis $A_1 = A_2 = \ldots A_p = B_1 = B_2 = \ldots B_p = -\beta$. The structure of the excitonic CI matrix is shown in Figure 2.

C. THE EXCITONIC WAVE-FUNCTION

This CI matrix restricted to singly excited determinants may be diagonalized. Analytical approximations of the solution may be obtained through supplementary simplifications.

(1) The crudest approximation consists in restricting the matrix to the LE block. This block having a Hückel type structure, the solutions are obtained immediately

$$\psi_m^0 = \sum_{r=1}^{N} C_{mr} \Phi\left(\begin{matrix} r^* \\ r \end{matrix}\right) \quad \text{with} \quad C_{mr} = \sqrt{\frac{2}{N+1}} \sin \frac{m\pi r}{N+1}$$

$$\Delta E_m = \Delta E_0 + 2b \cos \frac{m\pi}{N+1} \tag{43}$$

The original introduction of such an excitonic model in molecular physics was done by Simpson in 1950 [17].

(2) A second approximation introduces the short range ^{1}CTExcitations, as numerically done first by Pople and Walmsley [18], and Murrell [19]. The analytic derivation of the corresponding solution is a little more complex and requires the neglect of end effects. The original derivation has been done by Merz et al. [20]. An equivalent derivation has been done, using the partitioning technique, by Szabo et al. [21] and a third approach consists in perturbing the solutions ψ_m^0 of Equation (43) under the influence of the ^{1}CTEs [9]. One may also look at the derivation by Tric and Parodi [22] using Wannier's excitons [23]. The eigenvalue is given by:

$$\Delta E\left(\begin{matrix} 1 \\ m \end{matrix}\right) = \frac{1}{2}\left\{ \Delta E_0 + \Delta E_1 + 2\beta \cos \frac{m\pi}{N+1} + \left[\left(\Delta E_0 - \Delta E_1 + \right.\right.\right.$$

$$\left.\left.\left. + 2\beta \cos \frac{m\pi}{N+1}\right)^2 + 8\alpha^2 \cos \frac{m\pi}{2(N+1)}\right]^{1/2} \right\}. \tag{44}$$

The wave function is written

$$\psi_m = \sum_{r=1}^{N} C_m \binom{r^*}{r} \Phi \binom{r^*}{r} + \sum_{r=1}^{N-1} C_m \binom{(r\pm1)^*}{r} \Phi \binom{(r\pm1)^*}{r} \tag{45}$$

where, for the lowest excited state

$$C_m \binom{r^*}{r} = C_{rr^*} = \frac{A(\pm1)}{\sqrt{N}} \sin\left(\frac{2r+1}{2N}\pi\right) \tag{46}$$

$$C_m \binom{(r\pm1)^*}{r} = C_{r(r+1)^*} = \frac{B(\pm1)}{\sqrt{N}} \sin\frac{r\pi}{N} \tag{47}$$

The coefficients A and B are the components of the eigenvectors of the matrix

$$\begin{bmatrix} \Delta E_0 + 2b \cos\dfrac{\pi}{N} & -\sqrt{2}\beta \cos\dfrac{\pi}{2N} \\[2em] -\sqrt{2}\beta \cos\dfrac{\pi}{2N} & \Delta E_1 \end{bmatrix}$$

for the eigenvalue $\Delta E \binom{1}{1}$ of Equation (44), and satisfy the normalization condition $A^2 + B^2 = 1$. Introducing the variable φ,

$$A = \cos\varphi \qquad B = \sin\varphi,$$

one obtains

$$\tan\varphi = \frac{\sqrt{2}\beta - \cos\pi/N}{\Delta E\binom{1}{1} - \Delta E_1}. \tag{48}$$

Since $\Delta E\binom{1}{1}$ tends towards a constant (as will be discussed later on), $\tan\varphi$ tends towards constant, and so do the terms A and B. Turning back to Equation (46) and summing over all coefficients, one will find that the total weight of the LExcitations tends towards a constant. Numerical estimations of the parameters suggest that this weight would lie around 50%. Therefore in that model the weight of the Local Excitations decreases in favor of the CTEs, but it does not vanish when N increases, in contradiction with the result suggested by the analysis of the Virtual Orbital Approximation. Of course this discussion is not very convincing since this analytical derivation does not introduce the long range Charge Transfer Excitations which become more and more numerous when the dimension increases. However one may demonstrate that in the excitonic treatment the weight of the ^pCTEs decreases exponentially with the amplitude p of the Charge Transfer [9]. Morever the analysis of the coefficients numerically resulting from the diagonalization of a full CI excitonic matrix restricted to the singly excited determinants has been performed by us [24] up to seven double bonds and shows that the weight of the neutral structures (Local Excitations) tends toward a constant. With the CNDO hamiltonian the weight of LEs in the lowest singlet excited state wave-function is 100% for $N = 1$, 58.1 for $N = 2$, 51.4 for $N = 3$, 48.4 for $N = 4$, 47.3 for $N = 5$, 46.7 for $N = 6$ and 46.4 for $N = 7$, suggesting a 46% asymptotic value.

The analysis of the virtual orbital approximation in the delocalized scheme suggests a representation of the excited state as a 'democratic' mixture of CTExcitations where the long distance electron jumps become highly probable while the Local Excitations inside the bond have a vanishing weight. On the contrary the excitonic treatment describes the excited state as mainly built from local excitations in the bond (the corresponding probability decreases but tends towards a constant when the number of equivalent bonds increases) and from ^1CTE between adjacent bonds. The electron jump probabilities decrease exponentially with the amplitude of the jump.

TABLE II

Summary of the comparison between delocalized and localized pictures of the excitation

	Delocalized picture	Localized picture
MO's	Delocalized $\varphi_i \varphi_{j*}$	Localized $\phi_r \phi_{s*}$; $\{\phi\} = U\{\varphi\}$, U unitary
Ground state determinant Φ_0	identical	Φ_0
Excited determinants	$\Phi\begin{pmatrix} j^* \\ i \end{pmatrix}$	Local Excitations $\Phi\begin{pmatrix} r^* \\ r \end{pmatrix}$ Charge Transfer Excitations $\Phi\begin{pmatrix} (r\pm p)^* \\ r \end{pmatrix}$
Representation of the excited state Single determinantal approx.	Virtual orbital approximation $\Phi\begin{pmatrix} j^* \\ i \end{pmatrix}$ may be expressed in terms of $\Phi\begin{pmatrix} s^* \\ r \end{pmatrix}$ The weight of LE's $\to 0$ of PCTE's $\to 1$ } when $N \to \infty$ The fluctuation of the dipole moment in the excited state behaves like N^2	no single determinantal approximation
CI between the singly excited determinants	Usual CIS approximation. This treatment is identical to the excitonic one. The conclusions obtained for the excitonic treatment hold for the CIS approximation which therefore relocalizes the excitation with respect to the VO approximation. } $=$ {	Excitonic method The weight of LE's $\to$ constant C ($\sim$50%) CTE's $\to$ 1—C ($\sim$50%) when $N \to \infty$ The fluctuation of the dipole moment in the excited state $\to$ constant.

D. Fluctuation of the Electronic Repartition, Overestimation of the Electronic Delocalization

The consideration of charges per atom or bond only deals with mean values. A neutral system as the π system of linear polyenes gives a mean partition of two electrons per bond, or two π electrons per double bond loge. But this situation may occur from the neutral event $\boxed{\quad \cdot \ \mid \ \cdot \quad}$ or, as well, from the superposition of the 'ionic' events $\boxed{\quad \cdot \quad \mid \ \cdot \cdot \quad} \cdot \boxed{\quad \cdot \cdot \ \mid \quad \cdot \quad}$ with equal weights. The first situation with always two electrons per loge suggests a strong localization while the second one suggests a strong delocalization. The concept of fluctuation of the number of particles per volume was therefore introduced as a measure of the electronic delocalization around the most probable partition [1]. For the excited state both the virtual orbital approximation and the CI of singly excited determinants lead to a *mean* partition of two electrons per loge but the Configuration Interaction between the singly excited configurations qualitatively changes the picture of the excited state. In the orbital description, the mean number of electrons per bond is correct, but the fluctuations of electronic positions are much too large. The CI tends to bring back the electrons two by two in each loge. One may say that the orbital description of the excited state is too disordered, and that the CI brings some order in the excited state, diminishes the entropy of the description.

The implication of the overestimation of the electron delocalization through ionic structures may be illustrated by the study of the fluctuation of the dipole moment. Again in these symmetric and neutral molecules the permanent or 'mean' dipole moment is zero but the 'ionic' CTE introduces some transient dipoles, and therefore a non-zero fluctuation of the dipole moment. To simplify the derivation, instead of the wave-function (31), one may use a 'fully democratic' delocalized description of the excited state where each configuration $\phi\begin{pmatrix} s^* \\ r \end{pmatrix}$ would appear with a probability $1/N^2$,

$$\psi = 1/N \sum_r \sum_s \Phi\begin{pmatrix} s^* \\ r \end{pmatrix}, \tag{49}$$

neglecting the trigonometric phase factor.

For a linear system, the configuration $\Phi\begin{pmatrix} s^* \\ r \end{pmatrix}$ introduces a dipole moment equal to

$$\mathcal{M}_{rs} = e\mathbf{R}_{rs} = el(r-s), \tag{50}$$

where l is the standard distance between two bonds $(r, r+1)$. We shall neglect here the transition dipole moment between different CTE excitations, even when they differ by two adjacent MO's,

$$\left\langle \Phi\begin{pmatrix} s^* \\ r \end{pmatrix} |\Sigma \mathbf{R}| \Phi\begin{pmatrix} (s+1)^* \\ r \end{pmatrix} \right\rangle = \langle s^*|R| (s+1)^* \rangle \simeq 0, \tag{51}$$

as would be done in the CNDO hypotheses for fully localized MO's. All the configurations where $r-s=q$ contribute to the same value $\mathcal{M}_q = elq$, which appears with the probability

$$p_q = \frac{N - |q|}{N^2}.$$

One knows therefore the full histogram of the dipole moment of the excited state, i.e. the spectral decomposition of the dipole moment in the excited state.

The fluctuation of the dipole moment in the excited state may be calculated immediately

$$\mathcal{F} = 2 \sum_{q=1}^{N-1} \mathcal{M}_q^2 p_q = 2\Sigma(el)^2 q^2 \frac{N-q}{N^2} = 2\left(\frac{el}{N}\right)^2 \sum_{q=1}^{N-1} (Nq^2 - q^3). \quad (52)$$

Since

$$\sum_{q=1}^{N-1} q^2 \simeq \frac{N^3}{3} \quad \text{and} \quad \sum_{q=1}^{N-1} q^3 \simeq \frac{N^4}{4},$$

$$\mathcal{F} = 2(el)^2 N^2/6. \quad (53)$$

The fluctuation of the dipole moment increases as N when the dimension of the system increases. One may see easily that this feature is kept when one uses Equation (31) for the wave-function, instead of the 'fully democratic' single determinantal representation of Equation (49).

On the contrary the excitonic model, i.e. the full CI of singly excited determinants, will give a completely different histogram. Since the wave-function appears as mainly built from the LEs and ^{1}CTEs between adjacent bonds, in a ratio which tends towards a constant, the histogram tends towards a constant shape and the fluctuation of the dipole moment also increases towards a constant when N increases.

This result remains valid when one takes into account the long distance ^{p}CTEs with their exponentially decreasing weight e^{ap}. Their contribution to the fluctuation of the dipole moment is indeed proportional to

$$2 \sum_{p=1}^{N-1} p^2 e^{ap}.$$

This sum may be bracketed through the integration

$$\int_1^N p^2 e^{ap} \, dp < \sum_{p=1}^{N-1} p^2 e^{ap} < \int_0^{N-1} p^2 e^{ap} \, dp. \quad (54)$$

$\int_0^N p^2 e^{an} \, dp = e^{aN}(N^2/a - 2N/a^2 + 2/a^3)$ is bouded whatever N and therefore the fluctuation of the dipole moment is bouded, a being negative.

E. CONCLUDING REMARKS

One may say in some sense that the classical description of an excitation as the electron promotion from a delocalized level to another delocalized level above the Fermi level suggests an excessively delocalized picture of the phenomenon. The excitonic treatment leads to a picture of the excited state wave-function as a linear combination of coupled local excitations and short range excitations between adjacent bonds.

This result concerns not only the excitonic treatment but the usual Configuration Interaction between the singly excited configurations; since the canonical MO's are obtained from the localized ones by two unitary transformations in the spaces of occupied and virtual MO's, the eigenvectors of the CI matrix of singly excited determinants are the same, regardless of the localized or delocalized character of the MO's. The excitonic treatment gives therefore some insight upon the physical effect to the CI of singly excited states, which, in the delocalized scheme, seems rather intricate and difficult to analyse, since all the matrix elements lose their local character. Once again the localized picture demonstrates its higher handiness and its interpretative power.

Of course the usual statement that in an homogeneous system the excitation is delocalized is valid. But the implication of this statement is not straightforward. According to the statistical interpretation of quantum mechanics (as understood for instance by Ballentine [25]), the hamiltonian and the wave-function deal with a *set* of similarly prepared systems, not with a unique system, and one cannot know what happens under the excitation in a given molecule. The delocalization linked to the symmetry properties of the hamiltonian does not prove anything about the local or global character of the excitation in each molecular system. One may imagine however some cases where one can reject the idea of the excitation as a process *simultaneously* disturbing the whole electronic system of the concerned molecule. Let us consider a chain of N He atoms at large enough distances to allow the neglect of overlap. One may consider the wave function of the $2N$ electrons system, $\psi_0 = |1\bar{1}...N\bar{N}|$ corresponding to a regular partition of the electrons by pairs in the $1s$ loges.

One may consider an excited state corresponding to the $1s \rightarrow 2p_z$ excitation for an isolated molecule, the corresponding wave function is $\begin{pmatrix} 2p_z \\ 1s \end{pmatrix}$ and corresponds to the introduction of an electron in a sort of cylindric $2p_z$ loge $\boxed{\odot}$. For the chain of N He atoms, one may consider the corresponding excited states. As one neglects the overlap between the He atoms wave functions, no charge transfer may occur and a satisfactory description of the excited state is given by the CI of LE's,

$$\psi_m = \sum_i C_{mi} \, \Phi_i \begin{pmatrix} 2p_z \\ 1s \end{pmatrix},$$

C_{mi} being given by Equation (43).

From the point of view of the partition of the electrons in the loges, each component of this wave function corresponds to a different event. For instance as long as this wave function is correct, one may say that the event

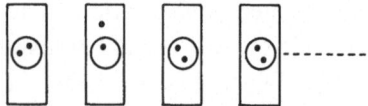

corresponds to the component $i = 2$ of the wave-function, in which the atoms $i \neq 2$ are not excited. The event

in which there is a simultaneous promotion of electrons in loges 2 and 4 does not appear. One may say therefore that in such a model there is no instantaneous delocalization of the excitation and that for an instantaneous individual observation the excitation involves a definite atom of the chain. The delocalization of the wavefunction is statistical and may perhaps give some insight about the overall partition of the excitation in each molecule if one assumes an ergodic hypothesis, which is not compulsory. In such an approach the delocalization of the excitation in a stationary state (if one may consider an excited state as stationary) would be a time phenomenon to be studied through time-evolution operators as suggested by Claverie and Diner [26]. But as long as this work has not been done, one cannot say that the delocalized form of the excited wave-function (Equations (43), (45–48)) proves that the excitation concerns the whole electronic system of each molecule.

6. Implications of the Short Range Character of the $\pi\pi^*$ Excitations

By short range character of the excitation we mean the fact that in the excitonic description the 'hole' and 'particle' are in the same chemical bond or in adjacent ones. This fact explains in a very simple manner some features of the $\pi\pi^*$ spectroscopy.

A. THE $\sigma\pi$ COUPLING IN $\pi\pi^*$ EXCITATIONS

The symmetry considerations only play a physical role at the level of the final wave-function and the distinction between π and σ electrons is only a convenient intermediate. Therefore when one says that the lowest $^1B_{1u}$ excited state of benzene results from the ground state through a $\pi\pi^*$ excitation, this is simply an approximation; when applied to the ground state $^1A_{1g}$ determinant the $\pi\pi^*$ and $\sigma\sigma^*$ excitations of correct symmetry lead to determinants of the same symmetry which interact, and therefore any $\pi\pi^*$ excited determinant interacts with $\sigma\sigma^*$ excited determinants. This may be called the $\sigma\pi$ coupling in $\pi\pi^*$ excitations, since as a first approximation the $\pi\pi^*$ determinants have lower energies than the $\sigma\sigma^*$ ones, and may be taken as zeroth order approximations. However in the lowest $\pi\pi^*$ singlet excited state of ethylene the coefficient of the component of the $\sigma\sigma^*$ singly excited determinant is large ($\simeq 0.3$) leading to a very important energy decrease (1 to 2 eV) [27–33]. This coupling is linked to the interaction

$$\left\langle {}^1\Phi\left(\begin{matrix}\pi^*\\\pi\end{matrix}\right)\middle| H \middle| {}^1\Phi\left(\begin{matrix}\sigma^*\\\sigma\end{matrix}\right)\right\rangle = 2\langle\pi\pi^*, \sigma\sigma^*\rangle - \langle\pi\sigma^*, \sigma\pi^*\rangle$$

representing the dipole-dipole interaction between the $\pi\pi^*$ and $\sigma\sigma^*$ transition dipole moments. The corresponding energetic effect may be evaluated

$$\delta E_{\sigma\pi} = \left\langle {}^1\Phi\left(\begin{matrix}\pi^*\\\pi\end{matrix}\right)\middle| H \middle| {}^1\Phi\left(\begin{matrix}\sigma^*\\\sigma\end{matrix}\right)\right\rangle^2 \middle/ \left(\Delta E\left(\begin{matrix}\pi^*\\\pi\end{matrix}\right) - \Delta E\left(\begin{matrix}\sigma^*\\\sigma\end{matrix}\right)\right). \tag{55}$$

For a small system the interaction between the $\pi\pi^*$ excitation and the surrounding $\sigma\sigma^*$ dipoles is important. If one considers a unique π bond and localizes the σ systems into bond MO's, one sees that the energetic effect of the interaction between the σ_i bond and the π system decreases rapidly (as $R_{\sigma_i\pi}^{-6}$) when the distance between the σ and π bonds increases; the effect remains almost constant if the *adjacent* bonds are not modified. One also sees that the $\left(\begin{matrix}\sigma_s^*\\\sigma_r\end{matrix}\right)$ excitations are not important since they introduce negligible overlap distributions in the matrix element.

What happens when the dimension of the π system increases?
To analyze this question one may work with
  σ and π delocalized MO's,
  σ localized and π delocalized MO's,
  σ and π localized MO's.
The first solution has never been explored and cannot be handled analytically. The second one has been studied analytically [34] and numerically [35] by Denis and Malrieu. They started with the π delocalized Virtual Orbital Approximation $\Phi\left(\begin{matrix}j^*\\i\end{matrix}\right)$ and perturbed it under its coupling with the localized $\Phi\left(\begin{matrix}\sigma_r^*\\\sigma_r\end{matrix}\right)$ excitations. The second order effect is given by

$$\delta E_{\sigma_r\pi} = \left\langle \Phi\left(\begin{matrix}j^*\\i\end{matrix}\right)\middle| H \middle| \Phi\left(\begin{matrix}\sigma_r^*\\\sigma_r\end{matrix}\right)\right\rangle^2 \middle/ \left(\Delta E\left(\begin{matrix}j^*\\i\end{matrix}\right) - \Delta E\left(\begin{matrix}\sigma_r^*\\\sigma_r\end{matrix}\right)\right)$$

$$\simeq 4\langle ij^*, \sigma_r\sigma_r^*\rangle^2 \middle/ \left(\Delta E\left(\begin{matrix}\pi^*\\\pi\end{matrix}\right) - \Delta E\left(\begin{matrix}\sigma^*\\\sigma\end{matrix}\right)\right), \tag{56}$$

neglecting the $\sigma\pi$ exchange effect, and simplifying the denominator. The matrix element may be decomposed using Equation (30), i.e. decomposing the ij^* distribution into local distributions. If one had only one σ bond per π bond one might write

$$\langle ij^*, \sigma_r \sigma_r^* \rangle \simeq C_{ir} C_{j^*r} \langle \pi_r \pi_r^*, \sigma_r \sigma_r^* \rangle$$

$$= \frac{2}{N+1} \sin \frac{i\pi r}{N+1} \sin \frac{j\pi r}{N+1} \langle \pi_r \pi_r^*, \sigma_r \sigma_r^* \rangle. \tag{5'}$$

In a more realistic case, one should introduce the various short range interactions between a π bond, the σ bond of the same double bond and the adjacent single CC and CH bonds, but it would not bring any new qualitative element. Therefore the final second order effect of the $\sigma\sigma^*$ single excitations upon the $\pi\pi^*$ transition energy is:

$$\delta_{\sigma\pi}(\Delta E_{\pi\pi^*}) = \frac{4}{(N+1)^2} \frac{\langle \pi\pi^*, \sigma\sigma^* \rangle^2}{\Delta E\left(\dfrac{\pi^*}{\pi}\right) - \Delta E\left(\dfrac{\sigma^*}{\sigma}\right)} \sum_r \sin^2 \frac{i\pi r}{N+1} \sin^2 \frac{j\pi r}{N+1}. \tag{58}$$

The summation is easy and one may demonstrate [34] that it behaves as N^{-1}

$$\delta_{\sigma\pi}(\Delta E_{\pi\pi^*}) \simeq \frac{4}{N+1} \frac{\langle \pi\pi^*/\sigma\sigma^* \rangle^2}{\Delta E\left(\dfrac{\pi^*}{\pi}\right) - \Delta E\left(\dfrac{\sigma^*}{\sigma}\right)}. \tag{59}$$

The effect of the coupling with $\left(\dfrac{\sigma^*}{\sigma}\right)$ single excitations decreases as N^{-1} when N increases.

This fact is easy to understand since in the delocalized transition distribution ij^*, the local overlap distributions $\pi_t \pi_{\sigma^*}(t \neq \sigma)$ are negligible. The only important local components of the ij^* distribution are the Local Excitations $\pi_r \pi_{r^*}$. But in virtual orbital approximation, here taken as zeroth order approximation, the weight of the LEs vanishes as N^{-1}, and so vanishes the $\sigma\pi$ coupling.

Following the third way (σ and π localized MO's), one may start from an excitonic wave-function (Equation (45)); each $\left(\dfrac{\sigma_r^*}{\sigma_r}\right)$ excitation is coupled with the $\left(\dfrac{\pi_r^*}{\pi_r}\right)$ component of the wave-function

$$\delta E_{\sigma_r \pi} = \left\langle \psi_m |H| \Phi\left(\dfrac{\sigma_r^*}{\sigma_r}\right) \right\rangle^2 \Big/ \left(\Delta E\left(\dfrac{\pi^*}{\pi}\right) - \Delta E\left(\dfrac{\sigma^*}{\sigma}\right) \right)$$

$$= C_m^2\left(\dfrac{r^*}{r}\right) \frac{4\langle \pi\pi^*, \sigma\sigma^* \rangle^2}{\Delta E\left(\dfrac{\pi^*}{\pi}\right) - \Delta E\left(\dfrac{\sigma^*}{\sigma}\right)}. \tag{60}$$

The final effect,

$$\delta_{\sigma\pi}(\Delta E_{\pi\pi^*}) = \frac{4\langle \pi\pi^*, \sigma\sigma^* \rangle^2}{\Delta E\left(\dfrac{\pi^*}{\pi}\right) - \Delta E\left(\dfrac{\sigma^*}{\sigma}\right)} \sum_{r=1}^{N} C_m\left(\dfrac{r^*}{r}\right)^2, \tag{61}$$

is equal to the effect on the $(N = 1)$ element of the series, multiplied by the weight of the LEs. As we have seen, this weight does not vanish, but rapidly decreases toward a constant value ($\sim 1/2$). Therefore, *due to an excessive delocalization of the excitation, the Virtual Orbital Approximation delocalized description suggests that the coupling between the $\pi\pi^*$ and $\sigma\sigma^*$ excitation decreases toward zero as N^{-1} when the dimension of the conjugated system increases, while the consideration of the CI wave function, relocalizing the excitation, keeps a constant non negligible coupling energy between the $\pi\pi^*$ and $\sigma\sigma^*$ excitations. The CI process introduces a representation of the excitation as linear combination of local (and short range) excitations each of them being coupled with the local $\sigma\sigma^*$ excitations in the same region of space.*

This conclusion has been verified numerically [35]; within the CNDO hypotheses the effect of the $\sigma\pi$ coupling upon the lowest singlet excited state of the linear polyenes decreases slowly from 1.05 eV for $N = 2$ to 0.95 eV for $N = 7$. This value is already obtained for $N = 5$ and represents the asymptotic value for large N.

Going deeper in this detailed analysis [36] one may notice that since the $\pi\pi^*$ LExcitations are preferentially coupled with the $\sigma\sigma^*$ excitations, their weight increases in the variational $(\sigma + \pi)$ wave function. This fact may be understood as follows. If the π wave function resulting from the π excitonic matrix diagonalization may be written

$$\psi_m = \alpha \, \text{LE} + \beta \, {}^1\text{CTE},$$

the $\sigma - \pi$ interaction lowers the transition energy by an amount which increases when α increases. Therefore the best value of α in the $(\sigma + \pi)$ treatment is higher in absolute value than in the purely π treatment. As a consequence the $\sigma\pi$ coupling diminishes strongly the charge fluctuation per bond in the excited state [36] (for the butadiene for instance, in the lowest singlet state the fluctuation in a π bond loge is 0.65 if one makes a purely π calculation, 0.57 in a $(\sigma + \pi)$ calculation). *The $\sigma - \pi$ coupling favors the localization of the excitation.*

One may wonder about the disappearance of the $\sigma\pi$ coupling in our first approach, where the VO Approximation was perturbed to the second order by the $\begin{pmatrix} \sigma^* \\ \sigma \end{pmatrix}$ singly excited determinants. Of course the final result of the CI being independent of the basis set, one must find somewhere the $\sigma\pi$ coupling when using delocalized MO's. A non vanishing $\sigma\pi$ coupling appears through higher excitations and/or higher orders of perturbations. The role of the $\begin{pmatrix} \sigma^* \pi^* \\ \sigma\pi \end{pmatrix}$ double excitations has been analytically and numerically demonstrated [34, 35] showing that they lead to a second order effect increasing toward a constant when the dimension N increases. The higher orders effect, even between singly excited determinants, would be illustrated through such third order diagrams, in which the $\begin{pmatrix} \sigma_r^* \\ \sigma_r \end{pmatrix}$ excited determinant interacts with both the zeroth order VO approximation $\begin{pmatrix} j^* \\ i \end{pmatrix}$ and another delocalized determinant $\begin{pmatrix} l^* \\ k \end{pmatrix}$ which also interacts with $\begin{pmatrix} j^* \\ i \end{pmatrix}$. The delocalization effect implicitely occurs through the π CI interaction, (ij^*, kl^*) occurring on the first line of the diagram.

B. The Reorganization of the π^* Excited Level by Mixing with Rydberg Levels; Diffuseness of $\pi\pi^*$ Excited States

A similar development may be made concerning the mixture of valence $\pi\pi^*$ excited configurations with the $\pi \to R^*$ excited configurations of the same symmetry, where R^* is a diffuse Rydberg type level. For ethylene this phenomenon is very important, the $\pi\pi^*$ excited state is highly Rydberg in character; the variational single determinantal calculations lead to a very diffuse π^* orbital [37]. The phenomenon is less pronounced when correlation effects are introduced [30, 33] since when one introduces the 'π^* electron' on a diffuse orbital its correlation with the other π and σ electrons is less important and a ζ optimization with correlated wave functions leads to a more contracted electronic distribution.

Any way, one may wonder about the generality of the result obtained for the ethylene molecule. A long time ago Silverstone and Joy [38] had studied this π reorganisation through the problem of ζ exponent optimization for the excited state and they had reached the result that the change between the best ground state and best excited state exponents decreases toward zero when the dimension of the conjugated system increases. An alternative demonstration was performed by Denis and Malrieu [39] who perturbed the $\pi\pi^*$ delocalized VO Approximation determinant under the influence of $\pi \to R^*$ determinants where the R^* levels were *atomic* localized $3p_z$ AO's. They reached the following results:

(1) Each $2p_z$ AO in the π^* MO of the $\pi\pi^*$ excited state is mixed with the corresponding $3p_z$ AO with a coefficient proportional to N^{-1}. The final energetic effect of the rearrangement of the π^* MO with the $3p_z$ AO's is proportional to N^{-1}. The effect upon the π half occupied MO is less important, with analogous N dependance.

(2) The reorganization of the π doubly occupied MO's is negligible in the singlet. The corresponding energetic effect on the triplet state is proportional to N^{-1}.

(3) The interaction of a given $\pi\pi^*$ VO Approximation determinant with *all* $\pi \to R^*$ singly excited determinants leads to an energetic correction proportional to N^{-1} for both singlet and triplet states. This is essentially a correlation effect.

(4) The essential role of the R^* levels goes through doubly and triply excited determinants. These correlation effects diminish the transition energy by an approximately constant value in the series of linear polyenes (evaluated to be 0.4–0.8 eV).

The decrease of the diffuseness of the π^* MO's in the single determinantal description is confirmed by the recent extended basis set calculations on conjugated molecules. For instance in the benzene molecule as calculated by Hay and Shavitt [40] the lowest triplet and singlet states are completely valence in character, the difference between purely valence and purely Rydberg excited state is very striking.

One may try to analyze this problem using the excitonic description. The local $\pi_r \to R_r^*$ ethylenic Rydberg excitations are strongly coupled with the $\pi_r \to \pi_r^*$ valence excitations in the same bond

$$\left\langle \Phi \begin{pmatrix} R_r^* \\ r \end{pmatrix} |H| \, \Phi \begin{pmatrix} r^* \\ r \end{pmatrix} \right\rangle = \langle R_r^* | F - J_r | r^* \rangle = a,$$

where F is the ground state Fock operator and J_1 the coulombic operator corresponding the local hole ϕ_r in the π shell. It is well known [12–13] that the matrix elements of the Fock operator in a basis of localized MO's are transferable, and so is the local coulombic operator J_r. Therefore the matrix element is practically the same as for the ethylene molecule whatever the position r of the bond in the chain and the length of the chain. The second order effect of the interaction between the Rydberg local excited state and the valence excitonic wave function is

$$\delta E_{R(r)} = \frac{\left\langle \Phi \begin{pmatrix} R_r^* \\ r \end{pmatrix} |H| \, \psi_m \right\rangle^2}{E_m - E \begin{pmatrix} R_r^* \\ r \end{pmatrix}} \tag{62}$$

and according to Equation (45)

$$\left\langle \Phi \begin{pmatrix} R_r^* \\ r \end{pmatrix} |H| \, \psi_m \right\rangle = \sum_{s=1}^{N} C_m \begin{pmatrix} s^* \\ s \end{pmatrix} \left\langle \Phi \begin{pmatrix} R_r^* \\ r \end{pmatrix} |H| \, \Phi \begin{pmatrix} s^* \\ s \end{pmatrix} \right\rangle + \tag{63}$$

$$+ \sum_{s=1}^{N-1} C_m \begin{pmatrix} (s+1)^* \\ s \end{pmatrix} \left\langle \Phi \begin{pmatrix} R_r^* \\ r \end{pmatrix} |H| \, \Phi \begin{pmatrix} (s+1)^* \\ s \end{pmatrix} \right\rangle.$$

The first type of matrix elements represent interactions between the (rather weak) $r \to R_r^*$ transition dipole toward the Rydberg level of bond r and the valence transition dipole on bond s. They rapidly decrease with the distance between the two bonds and one may keep only the local interaction $r = s$. The second type of elements imply overlap distributions which may be neglected compared to a. Thus

$$\left\langle \Phi \begin{pmatrix} R_r^* \\ r \end{pmatrix} |H| \, \psi_m \right\rangle \simeq a C_m \begin{pmatrix} r^* \\ r \end{pmatrix}, \tag{64}$$

and the Rydberg effect on bond r is

$$\delta E_{R(r)} = \frac{a^2 C_m \left(\dfrac{r^*}{r}\right)^2}{E_m - E\left(\dfrac{R_r^*}{r}\right)}. \tag{65}$$

The final effect of all Rydberg levels on all bonds is therefore equal to

$$\delta E_R = \sum_{r=1}^{N} \delta E_{R(r)} = \frac{a^2}{E_m - E\left(\dfrac{R_r^*}{V_r}\right)} \sum_{r=1}^{N} C_m \left(\dfrac{r^*}{r}\right)^2. \tag{66}$$

As the last summation, representing the weight of the Local Excitations, does not vanish but tends towards a constant ($\simeq \frac{1}{2}$), the final contribution tends toward a constant: *The effect of mixture toward the Rydberg states, even àt the level of singly excited CI, is not negligible, it should slightly diminish when going from ethylene to larger systems but should remain important. However this effect is not obtainable through the optimization of a delocalized single determinantal description of the excited state since in that case the local ethylenic excitations have a decreasing weight. The CI of singly excited configurations (the excitonic treatment for instance) suggests a representation of the excited state as a linear combination of short range and Local Excitations each of them being rather diffuse. In other words the excitonic approach introduces the representation of the excited state as a linear (delocalized) combination of local excitations accompanied by their local couplings (σ—π coupling for instance) and their local reorganizations (toward the Rydberg levels for instance).*

A lot of similar results may be obtained concerning the polarization of valence shell under a core ionization [41], or the polarization of the valence shell under a valence shell ionization, etc.

7. The Local Deformations in 'Delocalized' Excited States; a Case of Geometry Instability

The preceding discussions concerned the vertical excitations, i.e. the excitation process in which the nuclei keep the same mean position as in the ground state. However the optimal geometry for the excited state is in general different from the equilibrium ground state geometry, and one may study the nuclear rearrangements of the excited state. These studies in general accept the Born-Oppenheimer approximation in its strongest form, i.e. with a total (nuclear + electronic) wave function,

$$\psi = X(Q)\, \phi(Q, q), \tag{67}$$

simple product of a nuclear wave function by an electronic wave function depending on the nuclear coordinates Q. Then the minimization of the energy with respect to the geometry generally neglects the vibrational energy and minimizes the (electronic + nuclear repulsion) energy.

For polyatomic molecules the problem is made complex by the large number of independent coordinates and some simplifications are made; for instance the CH bond lengths are kept constant when the excitation is a $\pi\pi^*$ excitation, supposed to concern primarily the carbon skeleton. Another simplification may be inferred from symmetry considerations, assuming that the excited state geometry keeps all or some symmetry properties of the ground state. As an example one may refer to the case of the butadiene molecule in its lowest $\pi\pi^*$ excited states. One may suppose that the molecule keeps all the equilibrium ground state symmetries and only change the bond lengths and valence bond angles in a planar structure. But it is well known that ethylene undergoes a rotation around the C=C bond in its lowest $\pi\pi^*$ excited state with a minimum energy for a perpendicular form, and one may try to analyze the possibility of such rotations around the C=C bonds in butadiene. Two possibilities appear here:

(i) keep a center of symmetry (C_2) in the middle of the central bond by applying simultaneous rotations around the two terminal C=C bonds;

(ii) destroy this center of symmetry and perform independant rotations around the two terminal C=C bonds.

The former (restrictive) solution was studied by Shih, Buenker and Peyerimhoff through ab initio (SCF+CI) calculations, showing that the energy of the $^1Bu(^1B)$ state is lowered by 0.2 eV upon a 30° rotation angle [42]. The symmetry constraint had been removed in extended Hückel calculations [43] who show that butadiene and hexatriene, in the first singlet excited state, twist respectively around one terminal double bond and the central double bond. Baird and West [44] with a variant of their π semi empirical NDDO method have shown a preference of triplet states to twist about a C=C bond in butadiene, hexatriene and octotetraene.

The problem has been studied (45) using the excitonic method in the CNDO approximations over the atomic integrals, using fully localized MO's, according to the PCILO scheme (46), and the results are:

(i) For the butadiene molecule, a rotation of 50° (resp. 60°) minimizes the energy of the lowest singlet (resp. triplet) by 1.35 eV (resp. 1.08 eV). For a symmetric C_2 rotation, the minimum is obtained for a rotation of 15° (resp. 0°) leading to a energy gain of 0.08 eV (resp. 0 eV). A simultaneous optimization of bond lengths does not alter significantly these results.

(ii) For the hexatriene molecule, in agreement with the results of Hoffmann [43] and Baird [44], these calculations show that the twisting occurs about the central bond (45° and 0.9 eV for the singlet, 60° and 0.8 eV for the triplet); the rotations around the terminal bonds lead to a destabilization of the singlet state and a small stabilization of the triplet (0.4 eV).

In the distorted excited states the excitation is almost fully concentrated on the distorted bond. One may understand that two mechanisms are competing:

(i) The π delocalization effect requires a delocalization of the excitation and a planarity of the system.

(ii) The ethylenic distortion requires a concentration of the excitation and destroys the π delocalization.

This is the reason why the distortion is less pronounced in hexatriene than in butadiene, because the π delocalization effect is stronger. The problem of the geometry distortion is somewhat analogous to the problem of the σ–π coupling; the energy gain under the geometric distortion and the σ–π coupling favor the concentration of the excitation. At the Born–Oppenheimer approximation level the distorted solutions destroying the symmetry of the total hamiltonian are certainly inadequate. This is a case of geometry Born–Oppenheimer instability, analogous to the numerous Hartree-Fock instabilities. The Hartree–Fock equations are linked to a restriction on the form of the wave-function, and removing the spin symmetry or spatial symmetry of the charge distribution one may obtain lower single determinantal energies than the solution which keeps the correct final symmetry constraints. Here, in the butadiene problem, the left-distorted solution $X_1(Q_1)\,\phi(Q_1, q)$ is not the final solution of the problem, even if ϕ is the exact solution of the electronic hamiltonian, since the right-distorted solution $X_2(Q_2)\,\phi(Q_2, q)$ is degenerate. The symmetry properties of the total hamiltonian require for the final solution a form:

$$\psi = [\sum_i C_i\, X_i(Q_i)\, \phi(Q_i, q)] \tag{68}$$

i.e. a linear combination of locally excited and locally distorted states. For butadiene the interaction between the two components of Equation (68) still lowers the energy through a kind of delocalization effect. As in the case of the Hartree–Fock instability with respect to the symmetry of charge densities, the symmetric linear combination of the degenerate stable solutions goes besides the initial approximation, removes the Hartree–Fock (here Born–Opperheimer) constraints.

For the problem of the localization of the excitation, the result is very important. It suggests that for the adiabatic excited state at least, the proper representation of the excited state is not the product of an electronic wave-function where the excitation is spread over the whole skeleton, by a nuclear wave function corresponding to a small distortion everywhere. The proper representation would be a linear symmetric combination of local $\pi\pi^*$ electronic excitations inducing strong local $\sigma\pi$ coupling *and* strong local geometrical reorganizations, a linear combination of strongly localized excitations accompanied by strong local deformations of the nuclear geometry.

Again the question of the time evolution in the 'stationary' excited state would have to be studied with appropriate techniques. One may notice that the geometry instability phenomenon perhaps explains a qualitative discrepancy between theoretical predictions and experimental features. In the Born–Oppenheimer approximation the energy stabilization through nuclear rearrangements may be studied analytically in the series of linear polyenes, if one relates the modification of bond lengths to the modification of bond indexes. In such a model one gets the result that the energy gain under the geometry relaxation of the excited state tends toward zero as N^{-1} when the dimension of the conjugated system increases [47]. Some simplifications about the vibrations of the carbon skeleton establish a direct connection between the vibrational structure of the $\pi\pi^*$ absorption bond and the energy gain under the geometry reorganization [48]. This model leads to the conclusion that the ratio of the intensities of the 0–1 over 0–0 transitions (from the lowest vibrational level of the ground state to the first excited and lowest vibrational levels of the electronic excited state) should tend to zero as N^{-1} when the dimension of the system increases; the spectrum should be composed of a single 0–0 peak for large enough systems. This is in absolute contradiction with experiment, since the shape of the vibrationnal spectrum tends to be constant when N increases, $I_{0-1}/I_{0-0} \to C$. If one no longer accepts the Born-Oppenheimer approximation and refers to a linear combination of locally distorted local excitations the deformation energy no longer tends toward zero, but toward a constant and the intensity of the vibrational bands 0–1 no longer tends to zero. The form of the wave function given in Equation (68) would explain this apparent paradox.

8. Limits of a Crude Excitonic Model; the Role of the Ground State Delocalization on the Results of the Singly Excited CI [21]

Most of the results established here, especially the analytic derivations, were based upon an excitonic treatment in which the localized MO's and their matrix elements were supposed to be transferable from one place of the molecule to the other. This means for instance that the end effects were neglected, the MO's were supposed to be fully localized bond MO's which are different from the SCF localized MO's. The difference appears through the following three statements.

(i) The CI of singly excited states should give the same results in two basis sets of MO's if the subspace of occupied MO's and the subspace of virtual MO's are invariant under the transformation of the basis sets. Therefore the singly excited CI (CIS) gives the same result for SCF-delocalized and SCF-localized MO's.

(ii) The numerical results of the πCIS approximation using SCF delocalized MO's for the lowest $S_0 - S_1$ transition energy in the series of linear polyenes have been plotted as a function of the number N of conjugated double-bonds, and appears to verify the relation

$$\Delta E = A + B/N$$

whatever the parameters used for the atomic integrals and the bond alternation introduced in the model. One may notice that this is in agreement with a perturbative calculation in the Hückel method for bond alternation [49] but the main point is that the experiment also suggests this relationship [50].

(iii) The analytic expression given in Equation (44) for an excitonic model reduced to transferable matrix elements from fully localized MO's with LE and ^{1}CTE gives an asymptotic development for large values of N

$$\Delta E = A' + B'/N^2$$

which suggests a much more rapid stabilization of the transition energy in the series of linear polyenes. Numerical calculations involving all the ^PCTE's (i.e. full πCIS with fully localized MO's) are unambiguous, showing an N^{-2} dependance of the excitonic results in contradiction with both experiment and SCF-CIS results.

This discrepancy shows the qualitative role of the tails of SCF-MO's. Actually the only difference between the SCF localized MO's and the fully localized MO's lies in the small tails that the former have on the adjacent bonds, which do not appear in the second ones. The difference between the two πCIS treatments may only be attributed to these ground state tails, i.e. to the ground state delocalization. The N^{-1} component of the transition energy may be certainly obtained in a CI process from fully localized MO's, but through the introduction of higher excitations. The CIS treatment from SCF (localized) MO's introduces some contributions from all types of excitations in the fully localized excitonic treatment. This difference shows that the tails of the SCF-MO's, i.e. the ground state electronic delocalization, which seem often negligible, certainly play a qualitative role in the transition energies of conjugated systems.

9. Conclusion

The representation of excited and ionized states in the form of a single determinant has led to attributing a physical reality to the delocalized symmetry MO's. And it is actually impossible, in general, to use localized MO's for a single determinantal representation of the excited state. But we have shown in the case of the $n\pi^*$ excited state of conjugated aldehydes that the canonical π^* delocalized MO's may be too delocalized if the 'hole' is localized, and the role of the CI consists in an elimination of the longe range charge transfer components to bring back the 'particle' near the 'hole'. Any way the single determinantal description is a crude approximation. We have decomposed the delocalized Virtual Orbital Approximation for a $\pi\pi^*$ excited state in terms of locally excited determinants built from localized bonding and anti-bonding MO's and shown that this description is a democratic mixture of all kinds of local and charge transfer excitations, the long range charge transfer in which the hole and particle are separated by p bonds being as probable as the local or short range charge transfer excitations. A comparison with the excitonic treatment (CI in a basis of localized MO's) has been performed both analytically and numerically and shows that the CI of singly excited determinants here also eliminates the long range charge transfers in favor of the local excitations and CTExcitations between adjacent bonds. This means that the 'hole-particle pair' is never very extended; the excitation has at each moment a short range character even if it is statistically spread over the whole skeleton, and moves along the chain in the stationary state. From the point of view of the fluctuation of the number of electrons per volume, one may say that the CI process relocalizes the excitation, and that the Virtual Orbital Approximation is qualitatively erroneous.

The short range character of the excitation has numerous effects among which one may quote the fact that the $\sigma - \pi$ coupling between $\pi\pi^*$ and $\sigma\sigma^*$ single excitations is not negligible for large systems as previously thought. Another consequence concerns the mixture of valence and Rydberg levels in $\pi\pi^*$ excited states; it was believed that this mixture, very important in ethylene, vanished when the dimension of the conjugated system increases. Describing the excitation with a non vanishing linear combination of ethylenic excitations and short range charge transfers, one keeps an important mixture with the valence → Rydberg excitations for large systems.

The excitation appears as a linear combination of local events accompanied by local $\sigma\pi$ coupling and local reorganization towards Rydberg levels. A study of the geometry of the excited state allowing a loss of symmetry shows a Born-Oppenheimer instability with respect to the geometry, the energy minima of the butadiene and hexatriene $\pi\pi^*$ lowest excited states correspond to important twists about the $C\!=\!C$ bonds. This result corresponds to a rather localized adiabatic excitation accompanied by a local deformation. A correct representation must go beyond the Born-Oppenheimer approximation and represents the wave-function for the excited state minimum energy as a linear combination of locally distorted local excitations, a description which is strikingly different from the usual delocalized picture of the excited state.

The present contribution is somewhat paradoxal since it introduces the concept of localization or locality in the proper castle of delocalization, the domain of electronic excitation and ionization. It is clear that in general for homogeneous regular systems the excitation, from the basic principles, must be delocalized. But the use of the delocalized MO's in a too simple representation overestimates the degree of delocalization of the electronic events. The use of localized MO's allows a deep insight of the local contents of the excitation, an interpretation of the CI process (which usually appears so mysterious) as a relocalization of the 'hole-particle pairs'. It destroys a number of paradoxes about the vanishing $\sigma\pi$ coupling or Rydberg mixing in $\pi\pi^*$ excited states. It leads to a completely different description of the adiabatic excitation requiring an abandon of the Born-Oppenheimer approximation. In other words the localized approach proves to be a powerful tool for the understanding of delocalized phenomena. This success is certainly linked to the basically local character of the interactions appearing in the hamiltonian.

References

1. Cf. the contributions of R. F. Bader (p. 15), R. Daudel (p. 3) and J. P. Malrieu (p. 9) in Volume 1 of the present book.
2. Lennard Jones, J. E.: *Proc. Roy. Soc. (London)* **A198**, 114 (1949); Hall, G.G. and Lennard Jones, J. E.: *ibid.* **A202**, 155 (1950); **A205**, 367 (1951).
3. Boys, S. F.: in *Quantum Theory of Atoms, Molecules and Solid State* (ed. by P. O. Löwdin), New York, Academic Press (1966).
4. Edmiston, E. and Ruedenberg, K.: *Rev Med. Phys.* **34**, 457 (1963).
5. Magnasco, V. and Perico, A.: *J. Chem. Phys.* **47**, 971 (1967).
6. Diner, S., Malrieu, J. P., Jordan, F., and Claverie, P.: *Theoret. Chim. Acta (Berl.)* **18**, 86 (1970).
7. Malrieu, J. P.: *J. Photochem. Photobiol.* **7**, 531 (1968).

8. Daudel, R.: in *Les fondements de la Chimie Théorique*, Paris, Gauthier-Villars (1956).
9. Denis, A. Langlet, J., and Malrieu, J. P.: *Theoret. Chim. Acta* **29**, 117 (1973).
10. Masson, A., Levy, B., and Malrieu, J. P.: *Theoret. Chim. Acta (Berl.)* **18**, 193 (1970).
11. Newton, M. D., Switkes, E., and Lipscomb, W. N.: *J. Chem. Phys.* **53**, 2645 (1970).
12. Leroy, G. and Peeters, D.: Contribution to Vol. 1 of the present book; *Theoret. Chim. Acta (Berl.)* **36**, 11 (1974).
13. Berthier, G., Levy, B., and Millie, Ph.: Contribution to Vol. 1 of the present book.
14. Daudey, J. P.: *Chem. Phys. Letters* **24**, 574 (1974).
15. Diner, S., Malrieu, J. P., Claverie, P., and Jordan, F.: *Chem. Phys. Letters* **2**, 319 (1968).
16. Langlet, J.: *Theoret. Chim. Acta (Berl.)*, in press.
17. Simpson, W. T.: *J. Amer. Chem. Soc.* **73**, 5363 (1851); *ibid.* **77**, 6164 (1955).
18. Pople, J. A. and Walmsley, J. M.: *Trans. Far. Soc.* **58**, 441 (1962).
19. Murell, J. N.: *J. Chem. Phys.* **37**, 1162 (1962).
20. Merz, J. M., Straub, P. A., and Heilbronner, E.: *Chimia* **19**, 302 (1964).
21. Szabo, A., Langlet, J., and Malrieu, J. P.: to be published.
22. Tric, C. and Parodi, G.: *Mal. Phys.* **13**, 1 (1967).
23. Wannier, G. H.: *Phys. Rev.* **52**, 191 (1937).
24. Langlet, J. and Malrieu, J. P.: *Theoret. Chim. Acta (Berl.)* **30**, 59 (1973).
25. Ballantine, L. E.: *Rev. Mod. Phys.* **42**, 358 (1970).
26. Claverie, P. and Diner, S.: Contribution to the present volume.
27. Dunning, T. H. and Mc Koy, V.: *J. Chem. Phys.* **47**, 1735 (1967).
28. Bash, H. and Mc Koy, V.: *J. Chem. Phys.* **53**, 1628 (1970).
29. Whitten, J. L.: *J. Chem. Phys.* **56**, 5458 (1972).
30. Levy, B. and Ridard, J.: *Chem. Phys. Letters* **15**, 40 (1972).
31. Ryan, J. A. and Whitten, J. C.: *Chem. Phys. Letters* **15**, 119 (1972).
32. Tanaka, K.: *Int. J. Quant. Chem.* **6**, 1978 (1972).
33. Iwata, S. and Freed, K.: *J. Chem. Phys.* **61**, 1500 (1974).
34. Denis, A. and Malrieu, J. P.: *Theoret. Chim. Acta* **12**, 66 (1968).
35. Denis, A. and Malrieu, J. P.: *J. Chem. Phys.* **52**, 4769 (1970).
36. Langlet, L.: *Theoret. Chim. Acta (Berl.)* **27**, 223 (1972).
37. Dunning, T. H., Hunt, W. J., and Goddard III, W.A.: *Chem. Phys. Letters* **4**, 147 (1969); cf. also Buenker, R. J. Peyerimhoff, S. D., and Kammer, W. E.: *J. Chem. Phys.* **55**, 814 (1971); Buenker, R. J., Peyerimhoff, S. D., and Hsu, H. L.: *Chem. Phys. Letters* **11**, 65 (1971).
38. Silverstone, H. J. and Joy, H. W.: *J. Chem. Phys.* **67**, 1384 (1967); Silverstone, H. J., Joy, H. W., and Orloff, M. K.: *J. Amer. Chem. Soc.* **88**, 132 (1966); Joy, H. W. and Silverstone, H. J.: *Mol. Phys.* **13**, 149 (1967).
39. Denis, A. and Malrieu, J. P.: *J. Chem. Phys.* **52**, 6076 (1970).
40. Hay, P. J. and Shavitt, I.: *Chem. Phys. Letters* **22**, 33 (1973); *J. Chem. Phys.* **60**, 2865 (1974).
41. Denis, A., Langlet, J., and Malrieu, J. P.: *Theoret. Chim. Acta* **38**, 49 (1975).
42. Shih, S., Buenker, R. J., and Peyerimhoff, S. D.: *Chem. Phys. Letters* **16**, 244 (1972).
43. Hoffmann, R.: *Tetrahedron* **22**, 521 (1966).
44. Baird, N. L. and West, R. M.: *J. Amer. Chem. Soc.* **93**, 4427 (1971).
45. Langlet, J. and Malrieu, J. P.: *Theoret. Chim. Acta (Berl.)* **33**, 307 (1974).
46. Diner, S., Malrieu, J. P., and Claverie, P.: *Theoret. Chim. Acta (Berl.)* **13**, 1 (1969); Diner, S., Malrieu, J. P., F. Jordan, and Gilbert, M.: *Theoret. Chim. Acta (Berl.)* **15**, 100 (1969).
47. Diner, S. and Malrieu, J. P.: *Theoret. Chim. Acta (Berl.)* **7**, 15 (1967).
48. Mc Coy, E. F. and Ross, I. G.: *Austral. J. Chem.* **15**, 373 (1962).
49. Kutzelnigg, W.: *Theoret. Chim. Acta (Berl.)* **4**, 417 (1966).
50. Jaffe, H. H. and Orchin, M.: in *Theory and Applications of Ultraviolet Spectroscopy*, p. 228, New York, J. Wiley (1962) see also table 1 of ref. 20.

QUESTIONS TO MALRIEU

Daudel: In BeH, we found a good loge for the ground state and a rather poor one for the first excited state.

Malrieu: I agree, there exists a *qualitative* difference between ground and excited states. In the latter, the fluctuations are important because of the ionic terms.

Aslangul: I do not see the relation between vanishing of long range interaction and the relocalization of the excitation; it is enough to introduce a *next-nearest* neighbour interaction to obtain a quantum model where sites (atoms or molecules) loose completely their individuality giving rise to electronic collective excitations.

Anyway, the preparation of a localized non stationary excited state is not always trivial, (and it is the necessary condition for subsequent motion of excitation from site to site). On the contrary, the well known selection rule $\Delta k = 0$ for an infinite perfect crystal shows that the only state accessible by optical perturbation is the most delocalized one; furthermore, since it is a *stationary state*, the usual interpretation of Quantum Mechanics says that it does not exhibit any motion of any kind. Now, if we take a long polymer, this selection rule is not so severe, but still effectively and numerically stands for $n \gtrsim 20$. So, as long as you take a polymer of equivalent sites, you cannot discuss the problem of motion, without specifying how you construct the initial state. These points will be discussed further in a forthcoming meeting devoted to localization of excitation.

Malrieu: I have only discussed stationary states, the change of the wave-function from the ground state is therefore delocalized. If one refers to the statistical stochastic interpretation of Quantum Mechanics (as exposed by Diner and Claverie in the present volume) the wave function refers to a set of molecules, the measure may fluctuate around the mean value, and the question of the time evolution in the stationary state is an open question.

Aslangul: In my comment, I referred to the non stochastic Quantum Mechanics, where, as a definition all the average values are constant in time (mean values, as well as dispersion or higher moments).

IONIZATION AND LOCALIZATION

R. DAUDEL

Sorbonne et Centre de Mécanique Ondulatoire Appliquée du C.N.R.S., Paris, France

Professor Hohlneicher has evaluated the limits of Koopman's theorem by studying this problem from the energetic view point. The analysis of the hole which is produced during ionization provides another approach.

The simplest and also the best procedure to obtain information on that hole is to draw the electronic density map in the molecule and the corresponding map in the molecule. The difference between the two maps gives a very precise description of the hole resulting from the ionization. As the calculation of such densities is possible for any kind of wave function there is no limitation to this approach. The comparison between the hole obtained by following that procedure and the density corresponding to the best orbital ‚occupied' by the extracted electron gives a direct measurement of the validity of Koopman's theorem from the density view point. The observation of the actual hole clearly shows the localization of ionization. Figure 1* compares the electronic density in the molecule LiH and the electronic density in the positive ion LiH$^+$ (groundstate). The spheres surrounding the Li nucleus correspond to the

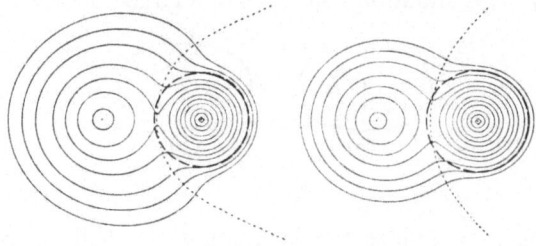

Fig. 1

core loge: the leading event corresponding to 2 electrons in the core and 2 electrons in the bond loge for LiH and only 1 electron in the same loge for LiH$^+$. It is seen that the positive hole has the shape of a crescent corresponding to a contraction of the LiH bond loge. The radius of the core loge slightly increases during ionization (1.42 AU → 1.55 AU).

Figure 2* shows that the situation in BeH is completely different. For the neutral molecule three loges are obtained: the spherical core loge ($R = 0.95$ AU), the BeH bond loge, the lone electron bond loge. (The frontier between the two latter loges is the Bader zero flux surface). The leading event corresponds to two electrons in the core, two electrons in the bond, one electron in the third loge. For the ion BeH$^+$ only two loges remain: the core ($R = 1$ AU) and the BeH bond loge. The leading event

---

* Daudel, R., Bader, R., Stephens, M., and Borrett, D.: *J. Can. Chemistry* **52**, 1310 (1974).

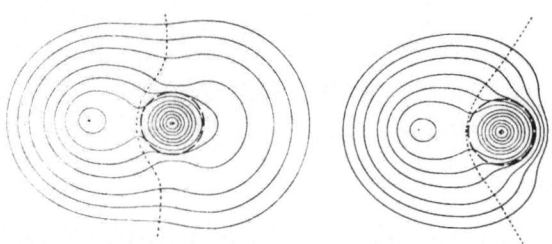

corresponds to two electrons in each loge. The ionization leads to the disappearance of the lone electron loge. The density in the BeH loge is practically unchanged. The radius of the core is slightly increased.

It must be pointed out that the relation between the behaviour of the hole and the localizability of the electron in the neutral molecule is not a simple one. Therefore the relation between the photoelectron spectrum of a molecule and the localizability of the electrons of that molecule is not obvious.

Let us consider for example* a molecule in which there are two good identical loges, each containing two electrons during the leading event.

Imagine an ionized state for which the leading event corresponds now to three electrons for the two loges. The two events denoted (a) and (b) on Figure 3 will have the same probability. This situation will leads to a large fluctuation of the number of

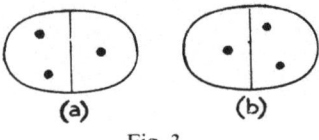

Fig. 3

electrons in each loge. To reduce the fluctuation we shall be led to suppress the frontier between the two loges. A delocalized three electron loge will appear. In conclusion it turns out that the ionization can produce delocalization. *The fact that the positive hole is delocalized is not the proof that the electrons were delocalized in the neutral molecule.*

As a conclusion we suggest one represent the effect of an ionization or an excitation of a molecule by the difference density function $\delta(M)$ which is simply the difference between the density $\rho_f(M)$ in the ionized or excited state and the density $\rho_i(M)$ in the initial state

$$\delta(M) = \rho_f(M) - \rho_i(M).$$

This function is a generalization of the usual difference density function introduced to represent the effect of bonding.**

* Aslangul, C., Constanciel, R., Daudel, R., Esnault, L., and Ludena, E.: *Int. J. Quantum Chemistry* **8**, 499 (1974).
** Daudel, R.: *Compt. Rend. Acad. Sci. Paris* **235**, 886 (1952); Roux, M. and Daudel, R.: *Compt. Rend. Acad. Sci. Paris* **240**, 30 (1955).

PART II

ELECTRON LOCALIZATION AS A STARTING
POINT TO CALCULATE WAVE FUNCTIONS

THE REALITY OF SOME BOND PROPERTIES
ADDITIVE SYSTEMS:
FARADAY EFFECT AND BOND MAGNETIC ROTATIONS

FERNAND G. GALLAIS

Laboratoire de Chimie de Coordination du CNRS, B.P. 4142, 31030 Toulouse, Cedex, France

This paper which mainly reports experimental data might appear to be out of place in a theoretical discussion on electronic localization in molecules. Still it seems justified by the fact that these data are treated as bond specific moduli, as it is clearly impossible to think of bond moduli in the absence of localized bonds or at least in the absence of privileged space zones in which the electronic charge density should have a relatively high value and to a certain extent a specific one.

In the reverse, the experimental evidence that such moduli do exist strongly supports these methods which describe molecular structures in terms of localized charges such as the *lodge* theory.

The Faraday effect is specially useful in that sense, for it is an excellent additive property which leads to very precise measurements as magnetooptic rotations* may be known up to a few hundredths of a degree [1]. But other properties may be used – and have been used – in the same way such as certain thermodynamics ones (mainly molecular formation or combustion energies), diamagnetic susceptibilities, refraction and even vectorial properties such as permanent dipole moments, etc. [2].

Difficulties which have been encountered in the use of additive properties and even failures to which such attempts have sometimes led may nearly always be ascribed to the lack of a correct definition of the bond modulus and consequently to a misunderstanding of the way through which such moduli may be deduced from experimental data which always apply to molecules.

This is particularly true when the property which is being used is mass dependant. In that case striking errors have sometimes been made in dividing the mass of an atom between the bonds in which this atom is engaged; clearly the mass fraction associated with a given bond must be a constant independant of the nature of the molecule in which this bond is present while in any molecule the mass of each atom must be taken entirely into account.

These requirements don't ask for special consideration in the field of organic chemistry where the carbon atom by making practically always use of its four valence electron keeps a constant coordination number. But the situation is entirely different

---

* The true magnetooptic invariant is not the 'rotation' ρ but the 'rotativity' $[\Omega] = (9n/(n^2 + 2)^2) [\rho]$ (*n* being the refraction index). Still, as long as no change in physical state intervenes there is no objection to the use of rotations which in the liquid state or in solution remain perfectly additive in a range of temperature covering a few degrees.

O. Chalvet et al. (eds.), Localization and Delocalization in Quantum Chemistry, Vol. II, 53–67. All Rights Reserved
Copyright © 1976 by D. Reidel Publishing Company, Dordrecht-Holland

in the field of inorganic chemistry where coordination numbers are as a rule essentially variable.

In that case, it is necessary:

to define the bond moduli in such a way that they remain always independent of the value of the coordination exerted by the atoms involved;

to take into account the activity of free atomic electrons, i.e. free valency electrons which have not yet been used for the building of inter atomic links.

The simplest way to attain these results is to divide equally the mass of an atom E between its n_E valency electrons so that a same fraction $1/n_E$ of the atomic mass be always associated with each of these valency electrons.

Under such conditions, the modulus (α) of magnetic rotation (or just as well of refraction or of magnetic susceptibility) of a single bond (σ or π) between two atoms A and B will always be represented by the symbol $(\alpha)\,(A/n_A - B/n_B)$, while if an atom X is engaged in a bond to which it contributes by more than one electron it will be associated with a heavier mass fraction; as an example, if two electrons are engaged in this bond which might be a double-bond, or a donor-acceptor bond in which X plays the part of the donor atom, the mass fraction X associated with this bond will be $2X/n_X$, etc.

By suitable addition of such bond moduli, it is always possible to calculate the molecular rotation of a definite compound inasmuch as the moduli $(\alpha)\,(e)$ of all valency electrons still remaining free are also taken into account.

Sulphur compounds in which sulphur coordination number may vary between 2 and 6 afford excellent examples for the application of such rules. In an hexacoordinated compound such as SF_6, there is of course no difficulty in dividing equally the molecular rotation into six contributions to define the rotation of one bond $(S/6 - F/1)$, and this modulus may be used directly in any molecule where sulphur has also engaged its six M electrons. This is the case for example in fluorosulphonic acid, of which the molecular rotation may be formulated as follows

$$[\alpha]_M(O_2S(OH)F) = [\alpha]\left(\frac{S}{6} - OH\right) + 2[\alpha]\left(\frac{S}{3} - O\right).$$

On the other hand, the bond $(S/6 - Cl/1)$ which plays in chlorosulphonic acid exactly the same part as the bond $(S-F)$ in FSO_3H does not appear alone in the molecule of sulphur dichloride. In that case, the existence of two lone electronic pairs must be taken into account so that the molecular rotation of SCl_2 must be formulated

$$[\alpha]_M(SCl_2) = 2[\alpha]\left(\frac{S}{6} - Cl\right) + 4[\alpha]\,(e)_S$$

and if it is found convenient for the sake of simplification to divide the molecular rotation of SCl_2 in two parts

$$[\alpha]_M(SCl_2) = 2[\alpha]\left(\frac{S}{2} - Cl\right),$$

it must always be remembered that now (α) $(S/2-Cl)$ is not the true magnetic rotation of the $(S-Cl)$ bond and that

$$[\alpha]\left(\frac{S}{2}-Cl\right) = [\alpha]\left(\frac{S}{6}-Cl\right) + 2[\alpha]\,(e)_S.$$

It is true that the magnetic rotation of free valency electrons remains unknown in many cases so that this convention leads to introducing unknown constants in formulas. Still this practice is absolutely necessary in order to avoid such errors as the one which would consist in making indifferent use of the rotation bonds $S/6-Cl/1$ and $S/2-Cl/1$. Moreover, it is already known that the activity of free valency electrons may be extremely high and, as an order of magnitude, the figures of 250 μr for phosphorus, 50 μr for sulphur and 20 μr for chlorine may be cited, so that their omission in formulas would lead to quite erroneous estimations of bond true rotations [3].

Bond moduli having been clearly defined, it remains to test the validity of this concept. Obviously, organic compounds offer the best material for this trial, not only on account of the constant coordination number of carbon, but even more because throughout these compounds the environment of a definite bond is not subject to undergoing important modifications when passing from one molecule to another one.

To start with, it seems advisable to determine, as a basis for the entire set of bond magnetic rotations those of the C—C and C—H bonds for which the alkanes furnish many valuable experimental data. It is then found that from C_4H_{10} to $C_{16}H_{34}$ the molecular magnetic rotations of these carbon hydrides is a linear fonction of the number n of their carbon atoms,

$$[\alpha]_M = a(n-2) + b$$

so that the slope a and the zero-ordinate b of the line so-defined,

$$a = 2[\alpha]\left(\frac{C}{4}-H\right) + [\alpha]\left(\frac{C}{4}-\frac{C}{4}\right)$$

$$b = 6[\alpha]\left(\frac{C}{4}-H\right) + [\alpha]\left(\frac{C}{4}-\frac{C}{4}\right)$$

lead to a direct estimation of the rotation of the C—C and C—H bonds.

It is essential to underline that this procedure does not call for the use of any arbitrary constant and that the only hypothesis which is being made is that the C—C and C—H bonds are exactly the same whether they belong to a —CH_2— or to a —CH_3 group. This of course, cannot be absolutely true but it appears that this approximation is of a high order, for the calculated values for linear alkane molecular rotations are perfectly in line with the ones which are measured as it is shown in Table I.

TABLE I

n. Alkanes

No.	Name	Formula	ρ_M measured	calculated	Diff.%
1	(Ethane)	C_2H_6		182,0	–
2	Propane	C_3H_8	257 [a]	255,0	0.8
3	Butane	C_4H_{10}	330 [a]	328,1	0.6
4	Pentane	C_5H_{12}	401	401,0	0
5	Hexane	C_6H_{14}	474	474,0	0
6	Heptane	C_7H_{16}	547	547,0	0
7	Octane	C_8H_{18}	620	620,0	0
8	Nonane	C_9H_{20}	694	693,0	0.15
9	Decane	$C_{10}H_{22}$	766	766,0	0
10	Undecane	$C_{11}H_{24}$	838* [b]	839,0	0.12
11	Dodecane	$C_{12}H_{26}$	912* [b]	912,0	0
12	Tetradecane	$C_{14}H_{30}$	1 058* [b]	1 058,0	0
13	Hexadecane	$C_{16}H_{34}$	1 200* [b]	1 204,0	0.33

[a] Measured by de Mallemann, Suhner and Grange at $-42°$ for Propane and at $-8.7°$ for butane, and corrected for $15°$.

[b] Measured in the Laboratory (J.-F. Labarre).

(In this table as in any other part of this paper, the magnetic rotation values are related to measurements made at room temperature ($\simeq 20\,°C$) on liquid substances or on solutions and for the yellow radiation of mercury, $\lambda = 5780$ Å; they are always expressed in microradians μr.)

Still it may soon be verified that the presence of the CH_3 group in a chain slightly affects the molecular rotation of the compound; for, as soon as branching occurs on a linear alkane, the molecular rotation is found to be somewhat higher than would be expected. In such a case, the estimation of this rotation must include a 'branching increment' approximately equal to $\delta_1 = +8\ \mu r$ while if the branching occurs in position 2, i.e. if the second CH_3 group comes close to the first one, the increment is twice as high, i.e. $\delta_2 = +16\ \mu r$.

So there just seems to exist a 'methyl effect' which is a rather general one inasmuch as the magnetic rotation of any functional group is always higher when this group is in position 2 in a hydrocarbon chain. The alcohols offer a good example in this respect as may be seen on Table II.

When the rotation of the (C/4 – OH) bond, specific of the alcohol function, is being deduced from the molecular rotation of the alcohols by subtracting the suitable contribution of the C—C and C—H bonds it appears to be quite constant and equal as an average to 42 μr for secondary as well as for primary alcohols. But there is an exception when the hydroxyl group is linked to a carbon in position 2. Then again the rotation is higher by some 7 μr ($\rho = 49\ \mu r$).

This by the way seems to indicate that the main difference in alcohols does not lie between primary and secondary alcohols but between alcohols in position 2 and all

TABLE II

Alcohols

No.	Formula	$[\rho]_M$	$[\rho]\left(\dfrac{C}{4}-OH\right)$
Primary alcohols			
1	(CH_3-OH)	(117)	–
2	CH_3-CH_2OH	198	43.3
3	$C_2H_5-CH_2OH$	270	42.3
4	$C_3H_7-CH_2OH$	341	40.3
5	$C_4H_9-CH_2OH$	417	43.3
6	$C_5H_{11}-CH_2OH$	487	40.3
7	$C_6H_{13}-CH_2OH$	562	42.3
8	$C_7H_{15}-CH_2OH$	636	43.3
9	$C_8H_{17}-CH_2OH$	709	43.3
			Average 42.3
Secondary alcohols in position 2			
10	$CH_3-CHOH-CH_3$	(279)	51.3
11	$C_2H_5-CHOH-CH_3$	351	50.3
12	$C_3H_7-CHOH-CH_3$	424	50.3
13	$C_4H_9-CHOH-CH_3$	493	46.3
14	$C_5H_{11}-CHOH-CH_3$	568	48.3
15	$C_6H_{13}-CHOH-CH_3$	643	50.3
16	$C_7H_{15}-CHOH-CH_3$	715	49.3
17	$C_8H_{17}-CHOH-CH_3$	788 [a]	49.3
18	$C_9H_{19}-CHOH-CH_3$	861 [a]	49.3
			Average 49.4
Secondary alcohols in a position other than 2			
19	$C_3H_7-CHOH-C_2H_5$	489	42.3
20	$C_4H_9-CHOH-C_2H_5$	562 [a]	42.3
21	$C_5H_{11}-CHOH-C_2H_5$	636 [a]	43.3
22	$C_6H_{13}-CHOH-C_2H_5$	709 [a]	43.3
23	$C_7H_{15}-CHOH-C_2H_5$	781 [a]	42.3
24	$C_8H_{17}-CHOH-C_2H_5$	855 [a]	43.3
			Average 42.8
25	$C_3H_7-CHOH-C_3H_7$	562 [a]	42.3
26	$C_4H_9-CHOH-C_3H_7$	634 [a]	41.3
27	$C_5H_{11}-CHOH-C_3H_7$	708 [a]	42.3
28	$C_6H_{13}-CHOH-C_3H_7$	782 [a]	43.3
			Average 42.3

[a] Values measured in the laboratory (J.-F. Labarre).

others. Anyhow, it is easy enough by passing from one functional series to another one to determine the magnetic rotation of all main σ bonds which carbon may form, as well as the magnetic rotation of a few others such as N—H, O—H, S—H which occur very frequently in organic chemistry.

TABLE III

linear Alkenes

No.	Formula	[E]	[A]	[E]–[A]
1	C_3H_7—CH=CH$_2$	463 [a]	401	62
2	C_4H_9—CH=CH$_2$	536 [a]	474	62
3	C_5H_{11}—CH=CH$_2$	609 [a]	547	62
4	C_6H_{13}—CH=CH$_2$	682	620	62
9	$C_{12}H_{25}$—CH=CH$_2$	1 120	1 058	62
			Average	62.0
10	C_2H_5—CH=CH—CH$_3$	452	401	51
11	C_3H_7—CH=CH—CH$_3$	525 [a]	474	51
16	C_8H_{17}—CH=CH—CH$_3$	890 [a]	838	52
			Average	50.7
17	C_9H_{19}—CH=CH—C$_2$H$_5$	973 [a]	912	61
18	C_5H_{11}—CH=CH—C$_3$H$_7$	827 [a]	766	61
28	C_5H_{11}—CH=CH—C$_4$H$_9$	901 [a]	838	63
			Average	61.6

[a] Measured in the laboratory (J.-F. Labarre).

TABLE IV

Linear Alkynes

No.	Formula	[Y]	[A]	[Y]–[A]
1	C_3H_7—C≡CH	447 [a]	401	46
2	C_4H_9—C≡CH	519 [a]	474	45
6	C_8H_{17}—C≡CH	812 [a]	766	46
			Average	45.7
7	C_3H_7—C≡C—CH$_3$	501 [a]	474	27
8	C_4H_9—C≡C—CH$_3$	576 [a]	547	29
9	C_5H_{11}—C≡C—CH$_3$	649 [a]	620	29
			Average	28.3
10	C_2H_5—C≡C—C$_2$H$_5$	519 [a]	474	45
13	C_4H_9—C≡C—C$_4$H$_9$	812 [a]	766	46
			Average	45.5

[a] Measured in the laboratory (J.-F. Labarre).

It then remains to deal with π bonds. It will be seen on Tables III and IV that for an identical number of carbon atoms the chain of an alkene or an alkyne is more active than the one of an alkane. In the last columns of these two tables effectively appears the positive difference between the rotation of the alkene E or of the alkyne Y and the rotation of the isologue alkane A.

It may be verified at the same time that this increment is sufficiently constant except when the unsaturated bond is in position 2–3. So that here again a methyl effect can be observed with that remark that it is in that case of an opposite sense to what it was in saturated compounds.

It is then possible to calculate a rotation modulus for isolated π-bonds themselves inasmuch as the rotation of the C—C or the C—H σ-bonds may be considered to remain constant whichever the hybridization state of the carbon atom they involve. This of course cannot be entirely true so that the values obtained for the rotation of the π bonds are only rough approximations. Such as they are they suffice to show that the $C.C_\pi$ bonds have a magnetic rotation which is from 4 to 6 times higher than the $C.C_\sigma$ one. This means that the magnetooptical activity of the π electrons is extremely high, a fact which is largely confirmed by further studies.

Anyhow, these π bond rotations appear to be just as exactly additive as the σ ones as long as they are perfectly localized, whether they belong to an open chain or to a ring. This, on the contrary, ceases to be true when these bonds are so close one to the other as to interact in the way which chemists call conjugation and which in fact corresponds to some extent of delocalization. Finally, the magnetic rotation of non conjugated organic compounds may be estimated by use of bond magnetic rotations such as those which appear in Table V with a very good approximation, i.e. with a difference which is usually less than 1%, with regard to the figure which is determined experimentally.

So that, obviously, in the field of organic chemistry it is entirely justified to define and to make use of bond magnetic rotation moduli. It must besides be noted that the same conclusion applies to bond diamagnetic susceptibilities such as those which appear on the same table and to bond refractions* except that they are not so precisely known as the magnetic rotations so that they lead to a slightly less satisfactory agreement between estimated and measured molecular susceptibilities.

The magnetic data are nevertheless fairly consistent with theoretical values estimated by Baudet et al. [4].

Professor Daudel will explain in the next paper under which conditions a function associated with the loges may lead in such a way to an additive system of specific values.

A last remark must be made regarding organic compounds in the sense that the additivity law of which the value has just been established is strictly limited to these numerous compounds which are colourless even in the near UV region, i.e. to these

---

* For bond refractions, the values given by Vogel et al., remain untouched (Cf. J. Chem. Soc., 1952, p. 514).

TABLE V

Magnetic susceptibilities $[\chi]_L$ and magnetic rotations $[\rho]_L$ of main bonds issued from a fourth group element (C, Si) and of N—H, O—H and S—H bonds (emu, cgs and μr)

Bond	$10^6[\chi]_L$	$[\rho]_L$	Bond	$10^6[\chi]_L$	$[\rho]_L$
$\left(\frac{C}{4}-\frac{C}{4}\right)_\sigma$	−2.90	18.5_0	$\left(\frac{C}{4}-\frac{N}{3}\right)$: (NR$_3$)	−3.4	47.0
$\left(\frac{C}{4}-\frac{C}{4}\right)_\pi$: (H$_2$C=CH$_2$)		117	$\left(\frac{3C}{4}\equiv\frac{N}{1}\right)_{\sigma+\pi}$: (nitriles)	−11.8	65.5
$\left(\frac{C}{4}-\frac{C}{4}\right)_\pi$: (HC≡CH)		77.5	$\left(\frac{C}{1}\overset{\leftarrow}{=}\frac{2N}{3}\right)_{\sigma+\pi}$: (carbylamines)	–	119
or					
$\left(\frac{C}{2}=\frac{C}{2}\right)_{\sigma+\pi}$	−3.6	135	$\left(\frac{C}{4}-\frac{Si}{4}\right)$	–	74
$\left(\frac{3C}{4}\equiv\frac{3C}{4}\right)_{\sigma+\pi}$	−10.3	173	$\left(\frac{C}{4}-\frac{Ge}{4}\right)$	−7.7	101
$\left(\frac{C}{4}-H\right)$	−4.25	27.2_5	$\left(\frac{C}{4}-\frac{Sn}{4}\right)$	−11.6	174
$\left(\frac{C}{4}-F\right)$	–	20	$\left(\frac{C}{4}-\frac{Pb}{4}\right)$	−12.8	278
$\left(\frac{C}{4}-Cl\right)$	−20.0	131	$\left(\frac{C}{4}-\frac{Be}{2}\right)$	–	190
$\left(\frac{C}{4}-Br\right)$	−29.3	267	$\left(\frac{C}{4}-\frac{Hg}{2}\right)$	−22.0	260
$\left(\frac{C}{4}-I\right)$	−46.1	566	$\left(\frac{Si}{4}-\frac{O}{2}\right)$: (Si(OR)$_4$)	–	28.9
$\left(\frac{C}{4}-\frac{O}{2}\right)$	−4.1	14.2	$\left(\frac{Si}{4}-\frac{O}{1}\right)$: (SiCl$_4$)	–	171
$\left(\frac{C}{2}=\frac{O}{1}\right)_{\sigma+\pi}$: (acids)	−7.0	30.6	$\left(\frac{Si}{4}-\frac{Si}{4}\right)$: (Si$_2R_6$)	–	325
$\left(\frac{C}{2}=\frac{O}{1}\right)_{\sigma+\pi}$: (aldehydes)	−3.5	35.5	$\left(\frac{N}{3}-H\right)$: (R$_2$NH)	−5.7	20
$\left(\frac{C}{2}=\frac{O}{1}\right)_{\sigma+\pi}$: (ketones)	−3.3	41.7	$\left(\frac{O}{2}-H\right)$: (R.OH; R.CO$_2$H)	−5.6	28.0
$\left(\frac{C}{4}-\frac{S}{2}\right)$	−10.2	126	$\left(\frac{S}{2}-H\right)$: (RSH)	−12.6	158
$\left(\frac{C}{2}=\frac{S}{1}\right)_{\sigma+\pi}$: (CS$_2$)	−21.1	385	Specific increment for the branching of a CH$_3$	−1.2	+8

compounds which have a normal magnetic rotatory dispersion law (one term only in the SERBER equation) and which are being analysed far enough from their first absorption band.

Turning now to the field of inorganic compounds it is found that an additivity law still holds in many cases as may be seen from the following examples taken from chemistry of groups III, V and VI elements. In Table VI appear moduli of most boron bonds such as may be found in boric and alkylboric esters, thioboric esters, alkylboron and their halogenated derivatives [5] together with aluminium, gallium and indium carbon bonds [5 bis]. In that case bond rotations are not just as constant as they were in the case of organic compounds and it cannot be said that they are known with a better approximation than 2 microradians when the magnetic data are known only at 0.2×10^{-6} and the refractions at 0.2 ml [6].

TABLE VI

Bond refractions $[R]_1$, magnetic susceptibilities $[\chi]_1$ and magnetic rotations $[\rho]_1$ for group III elements

Bond	$[R]_1$	$10^6[\chi]_1$	$[\rho]_1$
$\left(\dfrac{B}{3} - \dfrac{C}{4}\right)$	1.93[a]	—2.37	45
$\left(\dfrac{B}{3} - \dfrac{O}{2}\right)$	1.61	—4.85	18
$\left(\dfrac{B}{3} - \dfrac{S}{2}\right)$	5.59	—10.95	140
$\left(\dfrac{B}{3} - \dfrac{Cl}{1}\right)$	6.88	—20.0	145
$\left(\dfrac{B}{3} - \dfrac{Br}{1}\right)$	9.80	—28.5	300
$\left(\dfrac{B}{3} - H\right)$	2.70	—3.80	
$\left(\dfrac{B}{3} - \dfrac{B}{3}\right)$		—3.65	51
$\left(\dfrac{Al}{3} - \dfrac{C}{4}\right)$ [5 bis]			138
$\left(\dfrac{Ga}{3} - \dfrac{C}{4}\right)$ [5 bis]	3.6	—5.9	165
$\left(\dfrac{In}{3} - \dfrac{C}{4}\right)$ [5 bis]	5.2[b]	—9.3	307

[a] 1.88 (Gillis).
[b] 5.9 (Gillis) [6].

In Table VII group V elements have been studied in the case of trico-ordinated atoms, for example in $P(XYZ)$ molecules where X, Y, Z may be R, OR, SR, PR$_2$ [7]. With the proposed bond moduli it is possible to calculate the molecular rotations with an error of usually less than 1.5% and the refractions with an error <2% [6].

TABLE VII

Bond refractions, magnetic susceptibilties and magnetic rotations
for group V elements

Bond (1) (2)	$[R]_1$	$10^6[\chi]_1$	$[\rho]_1$
$\left(\dfrac{P}{3} - \dfrac{C}{4}\right)$	3.73	—5.1	159
$\left(\dfrac{P}{3} - \dfrac{O}{2}\right)$	3.07	—6.1	110
$\left(\dfrac{P}{3} - \dfrac{S}{2}\right)$	7.55	—13.0	364
$\left(\dfrac{P}{3} - \dfrac{Cl}{1}\right)$	8.73	—21.2	262
$\left(\dfrac{P}{3} - \dfrac{Br}{1}\right)$			550
$\left(\dfrac{P}{3} - \dfrac{P}{3}\right)$ (R$_2$P—PR$_2$)	5.83	–	349
$\left(\dfrac{As}{3} - \dfrac{C}{4}\right)$	4.5	—9.7	190
$\left(\dfrac{As}{3} - \dfrac{O}{2}\right)$	4.0	—8.6	150

(1) $[\alpha]\left(\dfrac{P}{3} - X\right) = [\alpha]\left(\dfrac{P}{5} - X\right) + \dfrac{2}{3}[\alpha]\,(e)_P$

(2) $[\alpha]\left(\dfrac{P}{3} - \dfrac{P}{3}\right) = [\alpha]\left(\dfrac{P}{5} - \dfrac{P}{5}\right) + \dfrac{4}{3}[\alpha]\,(e)_P$

And to end by group VI, (Table VIII) numerous sulphur compounds have been investigated, namely H$_2$S, thiols and alkyl sulphites, sulphur and sulphenyle chlorides, sulphoxylates, disulfanes in which sulphur remains dicoordinated while for selenium only selenites and selenols have been investigated. In that case, the additivity law holds with an approximation of 2%, sometimes even only 3% [8], [6].

It is now clear that, while an additivity law does hold in the field of inorganic chemistry it is not as well verified in that sector as it is for organic compounds. In fact, there are some cases in which an additivity law cannot be applied. As an example,

TABLE VIII

Bond refractions, magnetic susceptibilities and magnetic rotations for
group VI elements

Bond (1) (2)	$[R]_1$	$10^6 [\chi]_1$	$[\rho]_1$
$\left(\dfrac{S}{2} - \dfrac{C}{4}\right)$	4.61	—10.2	126
$\left(\dfrac{S}{2} - \dfrac{O}{2}\right)$	4.62	—6.60	105
$\left(\dfrac{S}{2} - \dfrac{Cl}{1}\right)$	10.31	—24.7	286
$\left(\dfrac{S}{2} - H\right)$			158
$\left(\dfrac{S}{2} - \dfrac{S}{2}\right)$ (RS—RS)		—14.9	319
$\left(\dfrac{Se}{2} - \dfrac{C}{4}\right)$	6.07 [a]	—14.6	200
$\left(\dfrac{Se}{2} - H\right)$	6.21 [b]	—19.9	250

[a] 6.0 (Gillis);   [b] 6.5 (Gillis) [6].

(1) $[\alpha] \left(\dfrac{S}{2} - X\right) = [\alpha] \left(\dfrac{S}{6} - X\right) + 2[\alpha] (e)_s$

(2) $[\alpha] \left(\dfrac{S}{2} - \dfrac{S}{2}\right) = [\alpha] \left(\dfrac{S}{6} - \dfrac{S}{6}\right) + 4[\alpha] (e)_s$

TABLE IX

Variations of some bond magnetic rotations

Bond		
$\dfrac{S}{2} - \dfrac{S}{2}$	319 µr in RS—SR	349 µr in S_8
$\dfrac{P}{3} - \dfrac{P}{3}$	349 µr in R_2P—PR_2	395 µr in P_4
$\dfrac{S}{2} - H$	158 µr in RSH	163 µr in H_2S
$\dfrac{O}{2} - H$	28.0 µr in ROH	35.8 µr in H_2O
$\dfrac{C}{4} - Cl$	131 µr in $R.CH_2Cl$	127 µr in $R.CHCl_2$
	125 µr in $R.Cl_3$	119 µr in $C.Cl_4$

the B—S bond may be cited the rotation of which is equal to 140 μr in thioboric esters but goes up to 197 μr in the esters R_3BSR' of dialkylthioboric acids; the B—N bond rotation is also liable to vary widely perhaps from 1 to 50 so that it has not been taken into account in the previous tables [9]. On the other hand, the X.H bonds with $X = B$, Si or P, must be supposed to undergo rather important variations if one wants to be able to consider as constants the other bond rotations in the chemistry of boron, silicon and phosphorus. It must also be stressed that the rotation of a given bond may vary slightly according to its organic or inorganic environment; such examples appear in Table IX, dealing in particular with oxygen, sulphur and phosphorus [10].

Finally, for purely organic compounds themselves, some exceptions to the additivity law do appear of which nothing has been said up to now. This is mainly the case for those compounds in which halogen atoms accumulate on the same carbon atom: in that case, there is a progressive depreciation of the rotation of the C.X bond, the C—Cl bond for example falling down from 131 μr in RCH_2Cl to 119 μr in CCl_4 [11].

Now it remains to see how such observations may be reconciled with those, much more numerous, which tend to show that bond rotations are practically constant and consequently may be used as additive values. Truly speaking, it must be said that the 'normal cases' are certainly the most surprising ones, for to suppose that a bond rotation is strictly constant amounts to admitting that the bond itself is liable to remain strictly identical to itself in any molecular structure in which it takes part, i.e. to admitting that its electronic content is not modified in the least when its environment undergoes thorough changes. This obviously cannot be absolutely true, so that one might anticipate that the rotation of this bond as well as most of its properties should, to some extent, vary under such conditions. In that respect, it has already been said that a change in the hybridization of atomic orbitals must lead to a change in bond rotation as it is the case when passing from sp^3 to sp^2 or to sp^1 carbon. We have shown in that respect that effectively a bond rotation goes down when the proportion of p character in the corresponding atomic orbital is going up [12]. But even more important as a factor of bond rotation seems to be the partial ionic character I_1 of this bond. In fact, the rotation $(\rho)_1$ is a linear function of the partial ionic character

$$(\rho)_1 = a_1 + b_1 I_1$$

and quite generally is lowered by an increase in I_1, i.e. by an increase in the intensity of the electric field in which the bond charge happens to be placed [13].

Now, it is a general rule that the effective electronegativity of a central atom does change with the nature and the number of its ligands, so that the partial ionic character of all the bonds in which this atom is taking part should change at the same time. For this reason no bond rotation should be really constant and an additivity law should nearly never be observed. If this is not the case generally, as experiment clearly shows, this is due to a phenomenon of spontaneous intra-molecular compensation. In most molecules indeed built around a central atom the changes which the I_1 – and with them the $(\rho)_1$ – undergo are of such sense that their effects cancel one another at least as a first approximation. It has been shown that this is the case when the central atom electronegativity is lower or higher than those of all the ligands between which mutual substitution intervenes, while when this condition is not satisfied this mechanism of mutual cancellation cannot take place. Boron and phosphorus bond magnetic rotations appear as additive because in the electronegativity scale boron and phosphorus are placed lower than most of the atoms or groups to which they are usually covalently linked. For carbon on the reverse with its electronegativity equal to 2.5 it may be predicted that the additivity law must fail when this atom is linked at the same time to hydrogen, the electronegativity of which (2.1) is lower than its own and to an element such as chlorine of which the electronegativity (3.0) is higher. This is why the polyhalogenated alkanes don't have a 'normal' behaviour when the halogen atoms accumulate on the same carbon atom. As an example, in the series which goes from CH_4 to CCl_4, the rotation of the C—H bond increases regularly when passing from CH_4 to $CHCl_3$, while the rotation of the C—Cl bond decreases regularly in passing from $CHCl_3$ to CH_3Cl. As a result, in an intermediate compound such as CH_2Cl_2, the C—H bond rotation as taken in CH_4 is too low and the same is true for the rotation of the bond C—Cl as taken in CCl_4, so that the estimated value of the molecular rotation is of course lower than the experimental value (293 μr against 309 μr). In the corresponding boron series which goes from BR_3 to BCl_3 the situation is different in the sense that while the rotation of the B—C bond goes up in passing from BCl_3 to BCl_2R, the rotation of the B—Cl bond comes down in passing from BCl_3 to $BClR_2$, so that for any intermediate compound the B—C bond as estimated in BR_3 is taken into account with a value which is too low while the rotation of the B—Cl bond comes in with a value which is too high. In such a case there is mutual compensation to such an extent that for BCl_2R the estimated rotation, 1132 μr, is nearly the same as the experimental one, 1130 μr, this being also true for the molecular rotation of BCl_2R (calc. 780 μr, meas. 783 μr).

As a general conclusion, it may now be said that the experimental evidence in favour of numerous approximately constant and additive bond rotations strongly supports the theories such as the 'loge theory' which describe molecular structures in terms of localized charge clouds expanding in these regions of space where chemists have always admitted the existence of definite interatomic links.

Still as bond moduli are not strictly constant, experimental evidence also confirms that any bond electronic content must vary slightly when this bond is defined in different environments.

References

1. Gallais, F. and Voigt, D.: *Bull. Soc. Chim. Fr.*, p.70 (1960).
2. Pascal, P.: *Ann. Chim. Phys.* **19**, p. 70 (1910);
   Lemoult, P.: *Rev. Gen. Sci. Pures Appl.* **19**, p. 137 (1908);
   Von Steiger, A. L.: *Ber.* **54**, p. 1381 (1921);
   Pascal, P., Pacault, A., and Hoarau, J.: *C.R.* **233**, p. 1078 (1951).

3. Gallais, F. and Labarre, J.-F.: *C.R.* **263**, p. 1202 (1966);
   Gallais, F., Labarre, J.-F., and de Loth, Ph.: *J. Chim. Phys.* **64**, p. 247 (1967).
4. Guy, J. and Tillieu, J.: *C.R.* **241**, 1955, p. 382; *J. Chem. Phys.* **24**, p. 1117 (1956);
   Baudet, J., Tillieu, J., and Guy, J.: *C.R.* **244**, p. 1756 and 2920 (1957).
5. Laurent, J.-P.: *C.R.* **252**, p. 3785 (1961);
   Laurent, J.-P. and Pasdeloup, M.: *C.R.* **256**, p. 133 (1963);
   Laurent, J.-P.: *C.R.* **258**, p. 1481 (1964);
   Gallais, F., Laurent, J.-P., and Cros, G.: *C.R.* **259**, p. 4262 (1964);
   Cros, G. and Laurent, J.-P.: *C.R.* **265** (C), p. 1065 (1967);
   Laurent, J.-P. and Bonnet, J.-P.: *Bull. Soc. Chim. Fr.*, p. 1447 (1966);
   Laurent, J.-P. and Pasdeloup, M.: *C.R.* **261**, p. 4725 (1965);
   Laurent, J.-P. and Berthalon, J.-P.: *Bull. Soc. Chim. Fr.*, p. 766 (1965).
5. *bis*. de Loth, Ph.: *Thèse Doctorat Sciences Physiques*, Toulouse, (1967);
   Haran, R. and Laurent, J.-P.: *Bull. Soc. Chim. Fr.*, p. 3454 (1966).
6. Cf., Mikhailov, B. M., Shchegoleva, T. A., and Bubnov, Yu. N.: *Izvest. Akad. S.S.S.R. Nauk*, Otdel. Khim. Nauk, p. 413 (1962).
   Gillis, R. G.: *Defence Standards Laboratories (Australia)*, Technical Note No. 51, July 1959;
   Sayre, R.: *J. Amer. Chem. Soc.* **80**, p. 5438 (1958);
7. Pascal, P., Voigt, D., Labarre, M.-C., and Fournes, L.: *C.R.* **262** (C), p. 1481 (1966);
   Voigt, D. and Labarre, M.-C.: *Bull. Soc. Chim. Fr.*, p. 1583 (1967);
   Labarre, M.-C., Voigt, D., and Gallais, F.: *Bull. Soc. Chim. Fr.*, p. 3328 (1967).
8. Gallais, F. and Voigt, D.: *Bull. Soc. Chim. Fr.*, p. 1935 (1963);
   Mila, J.-P. and Labarre, J.-F.: *C.R.* **263** (C), p. 1481 (1966);
   Mila, J.-P. and Labarre, J.-F.: *C.R.* **264** (C), p. 1157 (1967).
9. Laurent, J.-P.: *C.R.* **256**, p. 3283 (1963).
10. Gallais, F., Labarre, J.-F., and de Loth, Ph.: *J. Chim. Phys.* **63**, p. 1175 (1966); **64**, p. 247 (1967);
    Voigt, D., Turpin, R., and Labarre, M.-C.: *Bull. Soc. Chim. Fr.*, p. 3561 (1968).
11. Gallais, F. and Voigt, D.: *Bull. Soc. Chim. Fr.*, p. 70 (1960).
12. Gallais, F. and Labarre, J.-F.: *C.R.* **263** (C), p. 1202 (1960).
13. Gallais, F., Labarre, J.-F., Voigt, D., and de Loth, Ph.: *J. Chim. Phys.* **63**, p. 1175 (1966).

DISCUSSION

The factor $(n^2 + 2)^2 / 9n$ which establishes the bridge between the rotation and the rotativity amounts to neglecting the internal field. If this field is taken into account the factor becomes very near 1 for isotropic molecules.

There is a striking analogy of behaviour between the Faraday effect and diamagnetic susceptibility.

When there is no chemical change produced by a solvent the Faraday effect is approximately the same for a free molecule and for the same molecule in that solvent. The effect of hydrogen bonding is usually small.

In the case of fluorine compounds the additivity law is less satisfactory.

It would be possible to introduce moduli assuming that a double bond is made of two equivalent $\pi + \sigma$ single bonds.

THE LOCALISABILITY OF ELECTRONS
AS A STARTING POINT TO BUILD
ELECTRONIC WAVE FUNCTIONS

R. DAUDEL

Sorbonne et Centre de Mécanique Ondulatoire Appliquée du C.N.R.S., Paris, France

1. Loge functions or group functions

The possibility of using the loge theory in order to calculate a new kind of electronic wave function has been pointed out at the very beginning of the theory [1]. Let us consider a partition of the space into good loges (Figure 1). As we shall only need the topology of the partition and not the precise knowledge of the frontiers of the loges, a crude wave function, some experimental information or the knowledge of a partition for a simpler molecule of the same family can be sufficient to generate a schema like that of Figure 1.

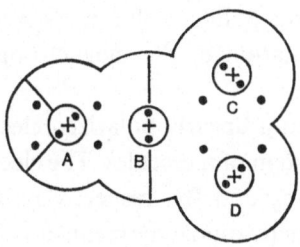

Fig. 1

Let us assume that during the leading event there are two electrons in each loge except in the larger one where four electrons are found. It becomes natural to calculate approximate electronic wave functions starting from a generating function as:

$$\gamma = K_A(1, 2) \, K_B(3,4) \, K_C(5, 6) \, K_D(7, 8)$$

$$L_A(9, 10) \, B_{AB}(11, 12) \, B_{BCD}(13, 14, 15, 16)$$

The various functions K_A, K_B ... B_{BCD} which appear in the trial function γ are called *loge functions*. They are assumed to be *roughly* localized in each loge.

To transform that generating function into the wave function we have to take account of the spin and of the Pauli principle. If σ denotes a convenient spin function and a an antisymmetrizer the wave function can be written:

$$\Psi = a\gamma\sigma.$$

Such is the general formalism. To go further a practical method of selecting the loge

O. Chalvet et al. (eds.), Localization and Delocalization in Quantum Chemistry, Vol. II, 69–74. All Rights Reserved

functions is needed. Such a powerful procedure has been proposed by McWeeny [2] under the name of *group function method*. A basis set of atomic orbitals (which can be hybrid orbitals) is chosen to represent each loge. Each loge function is built as a configuration interaction on the corresponding basis set. The calculation of the various coefficients is made by solving a kind of self consistent field equation between the various local C.I. functions starting from the classical variational equation:

$$\delta\langle\Psi|H-E|\Psi\rangle = 0.$$

According to this procedure a program has been written with the contribution, of McWeeny, Valdemoro, Dacre and Sanchez. It is based on gaussian functions and the section BASINT of IBMØL to compute molecular integrals.

The centers of gaussian functions can be put anywhere, therefore the method of floating gaussian functions can be used. In the program just described strong orthogonality conditions between the loge functions are introduced. More severe conditions have been introduced by Ludeña and Amzel [3]: they used completely localized loge functions. That is to say that each loge function vanishes outside the volume of the corresponding loge.

The loge function formalism has been used to analyze the additive properties of the Faraday effect. [4] The starting point of this work is a time dependent variation-perturbation method of calculation of the magneto-optical rotation constant established by Smet et Tillieu. [5]

A magnetostatic field $\mathbf{B}'$ and a linearly polarized electromagnetic wave of angular frequency ω interact with a system of molecules. The electric and magnetic fields of the optical wave will be denoted $\boldsymbol{\varepsilon}_0$ and $\mathbf{B}_0$, respectively. If $\mathbf{P}_M$ and $\mathbf{M}_M$ represent the macroscopic electric and magnetic dipole moments it is found that the rotation of the polarization plane of the incident wave is proportional to the intensity of the magnetostatic field $\mathbf{B}'$ through a constant V characteristic of the molecule:

$$V = \frac{2\pi\mathcal{N}\omega^2}{c}\left[\eta\frac{(n^2+2)^2}{9n} + \xi n\right]$$

This is the theoretical expression of the Verdet constant in which $\mathcal{N}$ is the number of molecules per unit volume, n the mean refraction index, η et ξ the coefficients of $\boldsymbol{\varepsilon}_0 \wedge \mathbf{B}'$ in $P_M/\mathcal{N}$ and of $\dot{\mathbf{B}}_0 \wedge \mathbf{B}'$ in $M_M/\mathcal{N}$.

It is easy to show that:

$$\xi \leqslant \eta.$$

Therefore ξ can be neglected and the calculation of V is reduced to the calculation of η.

If now we assume that the loge functions are well localized in the loges the wave function of an isolated molecule in the absence of fields can be taken as a non-anti-symmetrized product of orthogonal loge functions L_K.

In such a case it is possible to demonstrate (Daudel *et al.*, *loc. cit.*) that η is a simple sum of the various η_K associated with the various loges:

$$\eta = \sum_k \eta_k.$$

Therefore, if it is possible to associate with a molecule a good partition into loges the Verdet constant is a sum of moduli associated to the loges.

If the modulus associated to a given kind a loge remains approximately constant in a given family of molecules additivity rules will be obtained.

2. General Formalism

The loge function formalism has been recently extended [6]. It has been shown that any kind of wave function (*and therefore the exact one*) can be expressed in terms of correlated loge functions if all electronic events are taken into account.

To each electronic event λ a certain event function Ψ_λ is associated. If during the event λ there are p_l electrons in the loge l the function Ψ_λ is identical to the studied wave function Ψ when p_l points are inside of the loge l and vanishes for other organizations of the points. Therefore:

$$\Psi \equiv \sum_\lambda \Psi_\lambda .$$

Furthermore each event function is expressed in terms of correlated loge functions. If we consider for example an event corresponding to one electron in the loge A and two electrons in the loge B the generating function for the corresponding event will be written as:

$$\gamma_\lambda = L_A(1) \, L_B(2, 3) \, f(1, 2, 3)$$

where L_A and L_B are completely localized loge functions and where the function f allows electronic correlation between the two loges. The event function becomes:

$$\Psi_\lambda = \mathfrak{a}\gamma_\lambda\sigma .$$

If now the wave function Ψ is unknown and has to be calculated the various loge functions L and the various correlation functions f can be considered as variational functions and calculated by solving the classical equations:

$$\delta\langle\Psi|H - E|\Psi\rangle = 0 .$$

It can be seen (and this is shown in the paper referred to) that many classical trial functions based on the localizability of the electrons amounts to introducing some simplifications in the general formalism just described. For example the powerful P.C.I.L.O. method is related to a trial function in which correlation functions are not introduced.

3. Application to the analysis of molecular additive properties

The expectation value of any bielectronic operator $\hat{\Omega}$ associated with any wave function Ψ can be expressed as a sum of loge contributions, and loge pair contributions [7]. For a given event a partition of the operator $\hat{\Omega}$ is introduced. If for this event there are n_l electrons in the loge l all monoelectronic operators associated with n_l electronic points in the loge, all bielectronic operators between them and all operators representing the interaction between a nucleus of the loge and one of the n_l electronic points are associated with the loge l for that event λ. The sum of these operators can be denoted $\hat{\Omega}_{l,\lambda}$.

All bielectronic operators between one of the n_l electronic points in a loge l and one of the $n_{l'}$ electronic points in a loge l', all operators between a nucleus of the loge l and an electronic point of the loge l' (and vice versa) are associated with the pair of loges l and l'. The corresponding sum will be denoted $\hat{\Omega}_{l,l',\lambda}$.

It is easily seen that:

$$\langle \Psi_\lambda | \hat{\Omega} | \Psi_\lambda \rangle = \sum_l \langle \Psi_\lambda | \hat{\Omega}_{l,\lambda} | \Psi_\lambda \rangle + \sum_{l<l'} \langle \Psi_\lambda | \hat{\Omega}_{l,l',\lambda} | \Psi_\lambda \rangle.$$

Therefore:

$$\langle \Psi | \hat{\Omega} | \Psi \rangle = \sum_\lambda \langle \Psi_\lambda | \hat{\Omega} | \Psi_\lambda \rangle = \sum_l \sum_\lambda \langle \Psi_\lambda | \hat{\Omega}_{l,\lambda} | \Psi_\lambda \rangle$$
$$+ \sum_{l<l'} \sum_\lambda \langle \Psi_\lambda | \hat{\Omega}_{l,l'\lambda} | \Psi_\lambda \rangle.$$

Putting:

$$\bar{\Omega}_l = \sum_\lambda \langle \Psi_\lambda | \hat{\Omega}_{l,\lambda} | \Psi_\lambda \rangle$$

$$\bar{\Omega}_{l,l'} = \sum_\lambda \langle \Psi_\lambda | \hat{\Omega}_{l,l',\lambda} | \Psi_\lambda \rangle$$

it can be finally written:

$$\langle \Psi | \hat{\Omega} | \Psi \rangle = \sum_l \bar{\Omega}_l + \sum_{l<l'} \bar{\Omega}_{l,l'}.$$

As stated above any expectation value is rigorously divided into loge contributions $\bar{\Omega}_l$ and loge pair contributions $\bar{\Omega}_{l,l'}$.

Such a formula is a good starting point to analyze the molecular additive properties related with $\sum_l \bar{\Omega}_l$ and the deviations to additivity related with $\sum_{l<l'} \bar{\Omega}_{l,l'}$. If there are no two-particle operators the formula reduces to:

$$\langle \Psi | \hat{\Omega} | \Psi \rangle = \sum_l \bar{\Omega}_l.$$

An additivity law will be observed if the moduli $\bar{\Omega}_l$ remain constant in a given family of molecules.

References

1. Daudel, R.: *Les fondements de la chimie théorique*, Gauthier Villars (1956), (English version, *The Fundamentals of Theoretical Chemistry*, Pergamon, 1968).
2. Mc Weeny, R. C.: *Proc. Roy. Soc. (London)* **A253**, 242 (1959); **A259**, 554 (1961); *Rev. Mod. Phys.* **32**, 335 (1960); Klessinger, M. and Mc Weeny, R. C.: *J. Chem. Phys.* **42**, 3343 (1965).

3. Ludeña E. V. and Amzel, V.: *J. Chem. Phys.* **52**, 5923 (1970).
4. Daudel, R. Gallais, F., and Smet, P.: Ed., **1**, 873 (1967).
5. Smet, P. and Tillieu, J.: *Compt. Rend. Acad. Sci.* **257**, 3123, 3319 (1963).
6. Aslangul, C., Constanciel, R., Daudel, R., Esnault, L., and Ludeña, L.: *Int. J. Quantum Chemistry* **8**, 499 (1974).
7. Aslangul, C., *et al.* (*loc. cit.*).

DISCUSSION

Parks and Parr have made, before McWeeny, calculations analogous to group function calculations. They were however limited to two-electron loges and based on V.B. functions.

Whitten has also used an analogous procedure.

ON THE CALCULATION OF WAVE FUNCTIONS
IN THE THEORY OF LOGES

EDUARDO V. LUDEÑA

*Centro de Petroleo y Quimica, Instituto Venezolano de Investigaciones Cientificas,
Caracas, Venezuela*

1. Introduction

It is convenient and customary in many branches of physics to divide the physical space into subvolumes or cells for the sake of an easier handling of many-body problems. These cells, which in general have arbitrary shapes and sizes, are no more than useful mental constructs superimposed on the physical system.

In the theory of electronic structure of atoms and molecules a more interesting and deeper meaning can be given to the separation of the space into cells by resorting to the notion of loge [1]. This notion provides the means for determining, for a given wavefunction of an *n*-particle system, the volumes for which there is an optimum probability of finding a group of particles. The separation of the total space into these optimum volumes or loges has then physical significance as it is related to the square of the wavefunction in a well determined manner. In addition, the concept of localization of electron groups has been corroborated by experimental evidence obtained from Compton profile studies of molecules [2].

The notion of loge is potentially a very useful concept in quantum chemistry. By giving a prescription for an optimum partition of the physical space, it establishes the link between 'chemical' descriptions of molecules in terms of bonds, inner cores and lone pairs, and the less obvious molecular wavefunctions. But more important, the notion of loge can be used to develop variational schemes for the calculation of molecular wavefunctions.

This article discusses this type of approach to wavefunction determination. In Section 2, we deal with the completely localized loge model (CLLM) [3] where a variational procedure restricted to continuous functions is advanced. In Section 3, we discuss a variational method which is suitable for discontinuous loge functions.

2. Variational Principles for Continuous Functions in the CLLM

A. LOCALIZED AND DELOCALIZED LOGE FUNCTIONS

Let us suppose that $\mathbb{R}^3$ has been divided into K non-overlapping loges $V_1, \ldots V_k$, so that the probability of finding n_1 electrons in V_1, n_2 in V_2, $\ldots n_K$ in V_K is optimized,

O. Chalvet et al. (eds.), Localization and Delocalization in Quantum Chemistry, Vol. II, 75–90. All Rights Reserved

where for the n-electron system

$$n = \sum_{i=1}^{K} n_i. \tag{1}$$

Consider the complete set $\{\varphi\}$ of single particle functions in Hilbert space. Let us assume that from this set it is possible to construct (see Section 2c) the localized subset $\{\lambda\}$ and the delocalized subset $\{\gamma\}$. The first subset can be defined more explicitly as

$$\{\lambda\} = \bigcup_{I=1}^{K} \{\lambda_{I,i}\} \tag{2}$$

where the functions $\lambda_{I,i}$ are continuous, have continuous first derivatives and satisfy the following conditions:

$$\lambda_{I,i}(k) = 0 \quad \forall k \notin V_I \quad \text{and} \quad \in S(V_I)$$

$$\nabla\lambda_{I,i}(k) = 0 \quad \forall k \in S(V_I) \tag{3}$$

where $S(V_I)$ is the closed boundary of loge V_I. These conditions are necessary to establish the continuity of the wavefunction at the boundary.

It is clear that the set $\{\lambda\}$ does not span the complete Hilbert space and that it is a subset of $\{\varphi\}$. The delocalized subset $\{\gamma\}$ is defined as the complement of $\{\lambda\}$ and its functions need vanish only at infinity. In addition they must be orthogonal to the loge localized functions.

Having specified the general characteristics of the functions λ we can now introduce a projection operator $\mathscr{P}(1 \dots n)$ which is defined as follows:

$$\mathscr{P}(1 \dots n) = \sum_{S_1 \dots S_k} |\Psi_{S_1}(1 \dots n)\rangle \langle \Psi_{S_1 \dots S_k}(1 \dots n)| \tag{4}$$

where

$$\Psi_{S_1 \dots S_k}(1 \dots n) = \frac{\mathscr{A}}{\sqrt{n!}} \left\{ \prod_{I=1}^{K} \Lambda_{S_I}(1_I \dots n_I) \right\} \tag{5}$$

and

$$\Lambda_{S_I}(1_I \dots n_I) = \lambda_{I,i}(1_I) \lambda_{I,j}(2_I) \dots \lambda_{I,z}(n_I) \times \text{spin part}. \tag{6}$$

The subscript S_I spans all index combinations which arise from $\{\lambda_{I,i}\}$ and the particle coordinates $1_I \dots n_I$ are ascribed to the loge V_I.

It follows from this definition that $\mathscr{P}$ has the property of projecting from any wave function (either approximate or exact) that part which is localized in the loges $\{V_1, \dots V_k\}$. Hence, we can write

$$P(1 \dots n) \Psi(1 \dots n) = \Psi_L(1 \dots n) \tag{7}$$

where the subscript L stands as an abbreviation for loge localized projection. It is easy to see that

$$P^2 = P \quad \text{and} \quad P^+ = P. \tag{8}$$

An operator Q can be defined as

$$Q = I - P \tag{9}$$

where I is the identity. It also follows that Q is a projection operator and that it is orthogonal to P. The loge delocalized projection of a wavefunction Ψ is

$$Q(1 \ldots n) \, \Psi(1 \ldots n) = \Psi_D(1 \ldots n). \tag{10}$$

B. VARIATIONAL PRINCIPLES FOR THE LOCALIZED AND THE DELOCALIZED LOGE FUNCTIONS

Let us consider the following variational functional:

$$\mathsf{J}(\Psi^t) = \langle \Psi^t | H - E | \Psi^t \rangle. \tag{11}$$

An extremum of this functional is obtained for

$$\delta \mathsf{J}(\Psi^t) = 0. \tag{12}$$

When the variation of the trial wavefunction is carried out in the complete Hilbert space, Equation (12) implies, of course, the Schrödinger equation:

$$H\Psi = E\Psi. \tag{13}$$

When the variation is restricted to a class of functions which form a subspace of Hilbert space, then the extremum is obtained for a function Ψ^v which makes

$$\mathsf{J}(\Psi^v) = 0 \tag{14}$$

or equivalently

$$E^v = \frac{\langle \Psi^v | H | \Psi^v \rangle}{\langle \Psi^v | \Psi^v \rangle}. \tag{15}$$

As is well known,

$$E^v \geqslant E \tag{16}$$

where the equality sign holds only when $\Psi^v = \Psi$.

In order to apply this principle, the functions which make up the subspace must be continuous and have continuous first derivatives. [4]

It is clear that by successively enlarging the subspaces where the variation is carried out, E^v will converge toward the exact solution.

In the present case, Hilbert space is divided into two subspaces: the localized subspace generated by the functions $\{\lambda\}$ and the delocalized subspace generated by the functions $\{\gamma\}$.

The function Ψ_L is obtained by carrying out the variation of the functional

$$\mathsf{J}(\Psi_L^t) = \langle \Psi_L^t | H - E_L | \Psi_L^t \rangle \tag{17}$$

where Ψ^t is a trial wavefunction which spans the localized subspace. The function Ψ_L must necessarily include the loge localized (or intra-loge) correlation effects. As these effects are generally small, the main part of this wavefunction is made up of the uncorrelated contribution. Taking this contribution as the independent particle model wavefunction in a loge localized basis, it is clear that Pseudo-Hartree-Fock equations [5] can be obtained for the calculation of the best single particle loge functions. In the following section we discuss the variational determination of these Pseudo-Hartree-Fock equations as well as the derivation of pair equations for the evaluation of intraloge correlation effects.

If we call Ψ_L^v the particular normalized variational function which satisfies the extremum condition of Equation (17), we can then evaluate the delocalized wavefunction by varying the functional

$$\mathbb{J}(\Psi_L^v + \Psi_D^t) = \langle \Psi_L^v + \Psi_D^t | H - E | \Psi_L^v + \Psi_D^t \rangle \tag{18}$$

where Ψ_L^v is assumed to be fixed and Ψ_D^t is taken as the trial function which spans the delocalized subspace. It is necessary in this variation to observe the constraint

$$\langle \Psi_L^v | \Psi_D^t \rangle = 0. \tag{19}$$

Obviously, the convergence of this procedure strongly depends on how well the physical system is represented by the localized wavefunction. If one can consider the function Ψ_D to be simply a small correction to the localized wavefunction, then the approximation whereby we maintain Ψ_L^v fixed would have little effect on the overall calculation. If this is not the case, then the total wavefunction would have to be calculated by performing a simultaneous variation of Ψ_L and Ψ_D. Clearly then, the method advanced here relies heavily upon the assumption that the physical system can be described to a very good approximation by means of groups of electrons which are confined to loges.

C. THE PSEUDO-HARTREE-FOCK EQUATIONS

Let us consider the function Ψ_L. It is made up of two parts:

$$\Psi_L(1 \ldots n) = \Psi_L^0(1 \ldots n) + \Psi_L^C(1 \ldots n). \tag{20}$$

Ψ_L^0 is defined as the uncorrelated part of the loge localized wavefunction and Ψ_L^C as the correlated part. For simplicity, we shall consider a physical system where all loges have an occupation number of two ($n_I = 2, \forall I = 1, \ldots K$).

The results obtained here can be easily generalized to systems with arbitrary occupation numbers. The function Ψ_L^0 is the single Slater determinant

$$\Psi_L^0(1 \ldots n) = \frac{\det}{\sqrt{n!}} \left\{ \prod_{I=1}^{K} \left[\lambda_{I,1}^{\alpha}(1_I) \, \lambda_{I,1}^{\beta}(2_I) \right] \right\}. \tag{21}$$

Because of loge orthogonality, the functions $\lambda_{I,1}$ satisfy

$$\langle \lambda_{I,1} | \lambda_{J,1} \rangle = \delta_{IJ}. \tag{22}$$

If we take as the variational functional

$$\mathbb{J}[(\Psi_L^0)^t] = \langle (\Psi_L^0)^t | H - E_L^0 | (\Psi_L^0)^t \rangle \tag{23}$$

and search for its extremum by varying the loge localized functions in the localized subspace of Hilbert space [6], we obtain pseudo-Hartree-Fock equations

$$h_L^I(1) \, \lambda_{I,i}(1) = \mathscr{E}_L^{I,i} \lambda_{I,i}(1) \tag{24}$$

where

$$h_L^I(1) = I(1) + 2 V_L^I(1) + J_{I,1}^L(1) \tag{25}$$

$$I(1) = -\tfrac{1}{2}\nabla_1^2 + \sum_P \frac{Z_P}{r_{1P}} \tag{26a}$$

$$V_L^I(1) = \sum_{J(\neq I)=1}^{K} J_{J,1}^L(1) \tag{26b}$$

and

$$J_{J,j}^L(1) = \iiint_{V_J} d\Omega(2) \, \lambda_{J,j}(2) \frac{1}{r_{12}} \lambda_{J,j}(2). \tag{26c}$$

Because the orthogonality condition between orbitals is automatically fulfilled as long as the variation is carried out in the localized subspace, there is no need to include this condition via Lagrange multipliers in (23).

The general framework of the SCF procedure is directly applicable to these pseudo Hartree-Fock equations. Furthermore, the integrals appearing in these equations can be easily calculated as they represent coulombic interactions between non-overlapping charge distributions.

The choice of basis functions is important with regard to the numerical solution of this problem [7]. The functions $\{\lambda\}$ for a spherical loge (inner core) can be constructed as follows

$$\lambda_{I,i}(1_I) = R_{n_i,l_i}(r_1^I) \, Y_{l_i,m_i}(\theta_1^I, \phi_1^I) \left(1 - \frac{r_1^I}{R_I}\right)^{k_i}; \quad r_1^I \leqslant R_I$$

$$= 0; \quad r_1^I > R_I \tag{27}$$

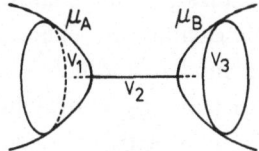

Fig. 1.

where R_I is the loge radius. The parameter k_i must be equal or greater than two in order that the conditions (3) be satisfied. Similarly, for a bond loge between atoms A and B, which is limited by the surfaces μ_A, μ_B and ξ (see Figure 1), where these surfaces are defined in terms of prolate spheroidal coordinates

$$\mu = \frac{r_A - r_B}{R_{AB}}; \qquad \xi = \frac{r_A + r_B}{R_{AB}}; \qquad \phi = \phi_A = \phi_B \tag{28}$$

the loge functions can be defined as follows:

$$\lambda_{I,i}(1_I) = F_i(\mu_1^I, \xi_1^I) \, e^{im_i\phi_1^I} \left(1 - \frac{\mu_1^I}{\mu_A}\right)^{k_i} \times$$

$$\times \left(1 - \frac{\mu_1^I}{\mu_B}\right)^{k_i} \left(1 - \frac{\xi_1^I}{\xi}\right)^{k_i}; \quad \text{for} \quad \mu_A \leqslant \mu_1^I \leqslant \mu_B \quad \text{and} \quad \xi_1^I \leqslant \xi$$

$$= 0 \quad \text{for} \quad \mu_1^I \leqslant \mu_A; \quad \mu_1^I \geqslant \mu_B; \quad \xi_1^I > \xi \tag{29}$$

D. Pair Equations and Intra-Loge Correlation

In terms of the set $\{\lambda\}$ of loge localized functions, the most general expansion for Ψ_L^C is

$$\Psi_L^C(1 \ldots n) = \sum_i \Psi_L^i \tag{30}$$

where the Ψ_L^i is the sum of ith-excited determinants which can be obtained from Ψ_L^0 by substituting the 'occupied' orbitals by the 'excited' or virtual orbitals.

Because the set $\{\lambda\}$ is the union of the non-interacting subsets $\{\lambda_{I,i}\}$, the expansion (30) does not contain all possible excitations. This follows from the fact that the virtual orbitals assigned to a group of electrons cannot appear in the expansion of another group (cf. Equation (3)). In this sense, the condition of complete loge localization is similar to that of strong orthogonality; in fact, it is a more drastic condition. From the above considerations we may conclude that Ψ_L^C describes explicitly only the intraloge correlation, although some of the interloge correlation may be implicitly included due to the form of the loge functions (8).

For the particular case of a system made up solely of doubly occupied loges, the function Ψ_L can be expanded in terms of Ψ_L^0 and all double, quadruple, etc., excitations, where because of loge localization, all quadruple and higher excitations can be written as products of double excitations. Then, as this is exactly the trial wave function for the independent pair model [9], which is given in this case in a loge localized representation, it is easy to show by a straightforward variational procedure that the pair function for a loge V_I must satisfy the equation [3b]

$$\mathcal{H}_L^I(1, 2) \, \Psi_L^I(1, 2) = \mathcal{E}_L^I \, \Psi_L^I(1, 2) \tag{31}$$

where the effective Hamiltonian is of the form

$$\mathcal{H}_L^I(1, 2) = I(1) + I(2) + \frac{1}{r_{12}} + V_L^I(1) + V_L^I(2) \tag{32}$$

the terms $V_L^I(1)$ and $V_L^I(2)$ denote the effect of the $n-2$ remaining electrons (which are described by 'occupied' loge localized functions) on the pair in loge V_I. The variational problem for pair functions can be reduced to standard form by expanding the pair functions as follows:

$$\Psi_L^I(1, 2) = \sum_p C_p \Lambda_{I,p}(1, 2) \tag{33}$$

where in general

$$\Lambda_{I,p}(1, 2) = \frac{\mathcal{A}}{\sqrt{2}} \{\lambda_{I,p}(1) \, \lambda_{I,p}(2)\} \times \text{spin part}. \tag{34}$$

The unknown coefficients are determined by the eigenvalue problem

$$[\mathbb{H} - \mathscr{E}_L^I \mathbb{S}] \, \mathbb{C} = 0 \tag{35}$$

where

$$\mathbb{H}_{p,q} = \langle \Lambda_{I,p} | \mathcal{H}_L^I | \Lambda_{I,q} \rangle \tag{36}$$

and

$$\mathbb{S}_{p,q} = \langle \Lambda_{I,p} | \Lambda_{I,p} \rangle. \tag{37}$$

The same considerations regarding integral calculations in the pseudo Hartree Fock equations are also applicable in this case.

Concerning the use of pair equations for the calculation of the intraloge correlation effects, there remains the problem of how good the independent pair model is in the present case. In general, the sum of pair correlation energies obtained by solving pair equations is only approximately close to the true correlation energy [10]. This comes from the fact that in an orbital representation it is not possible to decouple exactly the correlation part into a sum of independent pair equations.

The coupling terms which are most important in an orbital representation become zero in the present approach, because of the use of a loge localized basis. There remain, however, some higher order coupling terms whose effect for most practical purposes is estimated to be negligible [3b]. Hence it is possible to write the localized correlation energy as a sum of loge pair correlation energies

$$E_L^C \cong \sum_{I=1}^K \mathscr{E}_L^I. \tag{38}$$

The validity of this approximation is further supported by the fact that when the function Ψ_L is subjected to a perturbation theory treatment, both the second order and the third order energies can be written as a sum of pair energies [11]. In the usual manner, let

$$\Psi_L = \sum_{i=0} \mu^i \Phi_L^i \tag{39}$$

$$E_L = \sum_{i=0} \mu^i E_L^i \tag{40}$$

and

$$H = H_0 + \mu H_1 \tag{41}$$

where H_0 is a sum of suitable one-electron operators and H_1 is a sum of two electron operators:

$$H_0 = \sum_{i=1}^{n} h(i); \qquad H_1 = \sum_{i=1}^{n-1} \sum_{j=i+1}^{n} g(i,j). \tag{42}$$

From the perturbation theory equations [12] it follows that

$$E_L^3 = \langle \Phi_L^1 | H_1 - E_L^1 | \Phi_L^1 \rangle - 2E_L^2 \langle \Phi_L^0 | \Phi_L^1 \rangle. \tag{43}$$

In the loge localized representation a general expansion for the first order wave function is

$$\Phi_L^1 = \sum_{I=1}^{K} \frac{\mathscr{A}}{\sqrt{n!}} \left\{ \Psi_L^I(1_I, 2_I) \prod_{J(\neq I)=1}^{K} [\lambda_{J,i}^\alpha(1_J) \, \lambda_{J,i}^\beta(2_J)] \right\}. \tag{44}$$

The terms which couple the pair functions in (43) arise from exchange interactions. Taking into account Equation (3) it is easy to see that the third order correction to the energy can be written (for doubly occupied loges) as

$$E_L^3 = \sum_{I=1}^{K} \mathscr{E}_I^3 = \sum_{I=1}^{K} \langle \Psi_L^I(1, 2) | g(1, 2) | \Psi_L^I(1, 2) \rangle. \tag{45}$$

Obviously, in the derivation of this expression we have taken the virtual loge functions to be orthogonal to the occupied ones.

E. VARIATIONAL PERTURBATION TREATMENT FOR THE DELOCALIZED WAVEFUNCTION

The variational functional given by Equation (18) can be rewritten as

$$\mathbb{J}(\Psi_L^v + \Psi_D^t) = -E_D + 2\langle \Psi_L^v | H - E | \Psi_D^t \rangle + \langle \Psi_D^t | H - E | \Psi_D^t \rangle \tag{46}$$

where

$$E = E_L^v + E_D. \tag{47}$$

A variational perturbation expansion [3c, 12] of this functional can be obtained by letting

$$E_D = \sum_{m=0} \lambda^m E_D^{(m)} \tag{48}$$

$$\Psi_D^t = \sum_{m=0} \lambda^m \Psi_D^{(m)t} \tag{49}$$

and

$$H = H_0 + \lambda H_1. \tag{50}$$

Let us notice that the extremum of Equation (18), or equivalently, Equation (46) corresponds to

$$\delta \mathbb{J}(\Psi_L^v + \Psi_D^t) = 0. \tag{51}$$

But at this extremum point [12],

$$J(\Psi_L^v + \Psi_D^v) = 0 \tag{52}$$

where Ψ_D^v is the particular delocalized function which satisfies (51).

Introducing the expansions (48), (49) and (50), we see that Equation (46) can be rewritten as

$$J(\Psi_L^v + \Psi_D^t) = \sum_{m=0} \lambda^m J^{(m)} (\Psi_L^v + \Psi_D^{(0)t} + \dots + \Psi_D^{(m)t}). \tag{53}$$

Because of the linear independence of the power series in λ for each order in the expansion, the extremum corresponds to

$$\delta J^{(m)} = 0 \tag{54}$$

which leads to the condition

$$J^{(m)} (\Psi_L^v + \Psi_D^{(0)v} + \dots + \Psi_D^{(m)\,v}) = 0. \tag{55}$$

It is convenient to define

$$H_0 = \sum_{I=1}^{K} \sum_{i=1}^{2} h_L^I(i_I) \tag{56}$$

and

$$H_1 = \sum_{I=1}^{K} \{1/(|\mathbf{r}_1^I - \mathbf{r}_2^I|) - J_{I,1}^L(1_I) - J_{I,1}^L(2_I)\} +$$

$$+ \sum_{I=1}^{K} \sum_{J(\neq I)=1}^{K} \sum_{i=1}^{2} \sum_{j=1}^{2} \{1/(|\mathbf{r}_i^I - \mathbf{r}_j^J|) - J_{J,1}^L(i_I) - J_{J,1}^L(i_J)\}. \tag{57}$$

The perturbation potential H_1 is a sum of two particle operators which resemble the 'fluctuation' potential of Sinanoğlu's many-electron theory [9]. These operators represent the electrostatic interactions between pairs of electrons minus the potentials produced at the points r_i^I and r_j^J by the loge localized functions $\lambda_{J,1}$ and $\lambda_{I,1}$, respectively. H_1 is therefore a sum of the residual electrostatic interactions and its effect should be of importance only in the terms which contribute to the correlation energy of the delocalized wave function. Since the total correlation energy is a very small fraction of the exact total energy, we can expect that in the variational perturbation expansion, the terms which contain H_1 should contribute very little. Hence, we expect the dominant part of the delocalized wave-function to be the zeroth order contribution, namely $\Psi_D^{(0)}$. Let us consider $J^{(0)}$:

$$J^{(0)}(\Psi_L^v + \Psi_D^{(0)t}) = -E_D^{(0)} + 2\langle \Psi_L^v | H_0 - E_L^v - E_D^{(0)} | \Psi_D^{(0)t} \rangle +$$

$$+ \langle \Psi_D^{(0)t} | H_0 - E_L^v - E_D^0 | \Psi_D^{(0)t} \rangle \tag{58}$$

at the extremum, we obtain

$$E_D^{(0)\,v} = \frac{2\langle \Psi_L^v | H_0 - E_L^v | \Psi_D^{(0)\,v} \rangle + \langle \Psi_D^{(0)\,v} | H_0 - E_L^v | \Psi_D^{(0)\,v} \rangle}{1 + \langle \Psi_D^{(0)\,v} | \Psi_D^{(0)\,v} \rangle} \tag{59}$$

where we have taken into account the orthogonality condition (19). In order to analyze Equations (58) and (59) more closely it is necessary to write $\Psi_D^{(0)\prime}$ in an explicit
way. This function must span the delocalized Hilbert space generated by the orthonormal functions $\{\gamma\}$. Substituting in the single Slater determinant Ψ_L^0 of Equation (21)
the localized functions $\{\lambda\}$ by the delocalized functions $\{\gamma\}$, we can form a set of
singly-, doubly-, etc. excited determinants which constitute a basis for the expansion
of $\Psi_D^{(0)}$. Because the functions $\{\lambda\}$ and $\{\gamma\}$ are mutually orthogonal and the operator
Ho is a sum of single particle operators, the only matrix elements in Equation (58)
and (59) which do not vanish are of the form

$$\langle \gamma_i(1)|h_I^L(1)|\gamma_i(1)\rangle = \langle \gamma_i(1)|I(1)|\gamma_i(1)\rangle$$
$$+ \langle \gamma_i(1)|V_L^I(1)|\gamma_i(1)\rangle + \langle \gamma_i(1)|J_{I,1}^L(1)|\gamma_i(1)\rangle . \tag{60}$$

Hence, in the evaluation of the zeroth-order correction, we find only modified
Coulomb integrals which are easier to evaluate than the customary many-center
integrals of molecular orbital theory [7].

Let us notice, however, that in the higher orders of the variational perturbation treatment, the difficulties of the many-electron problem are unavoidably present. Nevertheless, as the perturbation Hamiltonian is a sum of fluctuation potentials, it follows
that these higher terms should contribute mostly to the correlation energy of the
delocalized wavefunction, which in our scheme should be rather negligible.

F. Preliminary Calculations for Simple Systems

The main aspect of the variational procedure advanced here for the solution of a many
particle problem can be studied in a very simple case: one electron in a coulomb field.
This calculation has been carried out for the ground state of the hydrogen atom and
for some one electron ions of the first row elements [7]. In this problem, as there are no
two-particle operators, all matrix elements can be easily computed. In addition,
these systems can be treated exactly and we have therefore exact solutions with which
to compare our results. In Table I we list some of these results. The basis functions
for the localized part were of the form

$$\lambda_{I,i}(1_I) = (r_1^I)^{l_i}\left(1 - \frac{r_1^I}{R_I}\right)^{k_i} \quad \text{for} \quad r_1^I \leqslant R_I$$

$$= 0 \quad \text{for} \quad r_1^I > R_I . \tag{61}$$

The delocalized functions $\{\gamma\}$ were spherical Slater type orbitals which were made
orthogonal to the set $\{\lambda\}$. In the first part of the calculation Ψ_L^v and E_L^v were computed
by means of Equation (17). In the second part, Ψ_D^v was computed holding E_L^v fixed as
in Equation (18). In this case, $H = H_0$ and $\Psi_D = \Psi_D^{(0)}$. The sum of E_L plus E_D yielded
exactly E, the true solution of this one particle Schrödinger equation.

TABLE I

Loge localized and delocalized energies for the hydrogen atom ground state

R	E_L(CLLM)	E_D(CLLM)	E_T(CLLM)	E_L [a]
0.5	0.151 820 13D 02	—0.156 820 13D 02	—0.5D 00	
1.0	0.246 751 92D 01	—0.296 751 92D 01	—0.5D 00	
1.5	0.472 076 67D 00	—0.972 076 67D 00	—0.5D 00	
2.0	—0.108 793 09D 00	—0.391 206 91D 00	—0.5D 00	
2.5	—0.326 641 99D 00	—0.173 358 01D 00	—0.5D 00	
3.0	—0.419 548 94D 00	—0.804 510 56D–01	—0.5D 00	—0.447 5
3.5	—0.461 956 52D 00	—0.380 434 80D–01	—0.5D 00	
4.0	—0.481 963 73D 00	—0.180 362 66D–01	—0.5D 00	—0.485 2
4.5	—0.491 509 66D 00	—0.849 033 52D–02	—0.5D 00	
5.0	—0.496 053 07D 00	—0.394 692 58D–02	—0.5D 00	—0.496 55
5.5	—0.498 192 75D 00	—0.180 724 83D–02	—0.5D 00	
6.0	—0.499 185 68D 00	—0.814 315 39D–03	—0.5D 00	—0.499 27
6.5	—0.499 639 04D 00	—0.360 957 49D–03	—0.5D 00	
7.0	—0.499 842 63D 00	—0.157 369 62D–03	—0.5D 00	—0.499 86
7.5	—0.499 932 38D 00	—0.676 240 25D–04	—0.5D 00	
8.0	—0.499 970 75D 00	—0.292 542 02D–04	—0.5D 00	—0.499 97

[a] Compressed hydrogen atom: S. R. De Groot and C. A. Ten Seldam, *Physica 12*, 669 (1946).

3. Variational Principles for Discontinuous Wavefunctions

Another approach to the solution of the Schrödinger equation in the context of the theory of loges is based on the use of variational principles which are suitable for regional or loge localized discontinuous wavefunctions. We sketch here some of the results of the application of the variational principle of McCavert and Rudge [13] for obtaining independent particle model variational expressions for loge localized discontinuous orbitals [14].

Associating with each loge V_I the projection operator $P_I(x)$ which is defined as a three-dimensional Heaviside step function

$$P_I(x) = \begin{cases} 1 & x \in V_I \\ 1/2 & x \in S(V_I) \\ 0 & x \notin V_I \text{ or } S(V_I) \end{cases} \tag{62}$$

we can then define a loge localized discontinuous orbital $\phi_i^I(1)$ as the projection

$$P_I(x_1)\, \phi_i(1) = \phi_i^I(1) \tag{63}$$

where in general

$$\phi_i(1) = \varphi_i(x_1) \begin{cases} \alpha \\ \beta \end{cases}. \tag{64}$$

The problem we wish to consider is that of finding a wavefunction $\phi_i(1)$ which is made up of regionally discontinuous functions ϕ_i^I:

$$\phi_i(1) = \sum_{I=1}^{K} \phi_i^I(1). \tag{65}$$

We start from the variational functional

$$\mathscr{G} = \mathscr{H} + \mathscr{S} \tag{66}$$

where

$$\mathscr{H} = \iiint_{\mathbb{R}^3} d\Omega(1) \ldots \iiint_{\mathbb{R}^3} d\Omega(n) \, \Phi^*(1 \ldots n) \, [H - E] \, \Phi(1 \ldots n) -$$

$$- \sum_{i=1}^{n} \sum_{j \neq i}^{n} \sum_{I=1}^{K} \lambda_{ij} \iiint_{V_I} d\Omega(1) \, \phi_i^{*I}(1) \, \phi_j^I(1) \tag{67}$$

and

$$\mathscr{S} = \sum_{i=1}^{n} \sum_{I=1}^{K} \sum_{J>I}^{K} \frac{1}{2} \iint_{S_{IJ}} dS(1) \, \mathbf{n}_{IJ} [\phi_i^{*I}(1) \, \mathbf{V}_1 \phi_i^{*J}(1) - \phi_i^{*J}(1) \, \mathbf{V}_1 \phi_i^{*I}(1)]. \tag{68}$$

The $\{\lambda_{ij}\}$ are the lagrange multipliers which introduce the orthogonality conditions between the orbitals ϕ_i and ϕ_j.

The surfaces S_{IJ} are

$$S_{IJ} = S(V_I) \cap S(V_J) \tag{69}$$

and $\mathbf{n}_{IJ}$ is the vector normal to this surface. It is necessary to assume a particular form of the loge localized function in order to keep the variations within the domain of real functions [13]. We have taken

$$\phi_i^I(1) = \psi_i^I(\eta_1^I, \chi_1^I) \, e^{im_i \xi_1^I} \tag{70}$$

where η_1^I, χ_1^I and ξ_1^I are the coordinates of a particular system selected for V_I. Variation of the functional (66) leads to

$$2 \sum_{I=1}^{K} \iiint_{V_I} d\Omega(1) \, \delta\tilde{\phi}_i^I(1) \, [h_i(1) - \varepsilon_i] \, \tilde{\phi}_i^I(1) +$$

$$+ \sum_{I=1}^{K} \sum_{J>I}^{K} \frac{1}{2} \iint_{S_{IJ}} dS(1) \, \mathbf{n}_{IJ} [(\mathbf{V}_1 \delta\tilde{\phi}_i^I(1) + \mathbf{V}_1 \delta\tilde{\phi}_i^J(1)) \, (\tilde{\phi}_i^{*I}(1) - \phi_i^{*J}(1))$$

$$- (\delta\tilde{\phi}_i^I(1) + \delta\tilde{\phi}_i^J(1)) \, (\mathbf{V}_1 \tilde{\phi}_i^{*I}(1) - \mathbf{V}_1 \tilde{\phi}_i^{*J}(1))] = 0. \tag{71}$$

The orbitals $\tilde{\phi}_i$ are related to ϕ_i by a unitary transformation which diagonalizes the Lagrange multiplier matrix and the operator $h_i(1)$ is

$$h_i(1) = -\tfrac{1}{2}\mathbf{V}_1^2 + \sum_{p} \frac{Z_p}{r_{1p}} + \mathscr{V}_i(1) \tag{72}$$

where

$$\mathscr{V}_i(1) = \sum_{j \neq i}^{n} \sum_{J=1}^{K} \iiint_{V_J} d\Omega(2) \, \tilde{\phi}_j^J(2) \, \frac{(1-P_{12})}{r_{12}} \, \tilde{\phi}_j^J(2). \tag{73}$$

A matrix form of the variational expression (71) can be derived when the discontinuous loge localized functions in Equation (70) are expanded in terms of the sets $\{K_p^I\}$ of continuous functions suitably adapted to the particular region of the molecule. The discontinuity condition is implicitly taken into account by performing the integrations only within the respective loges. Equation (70) can be rewritten as

$$\phi_i^I(\eta_1^I, \chi_1^I, \xi_1^I) = \sum_{P=1}^{N_I} C_{ip}^I \, K_p^I(\eta_1^I, \chi_1^I) \, e^{im_i \xi_1^I}. \tag{74}$$

The coefficients $\{C_{ip}^I\}$ which define ϕ_i in each one of the loges V_I are found by solving the algebraic problem

$$\begin{bmatrix} \mathbb{M}_i^{11} & \cdots & \mathbb{M}_i^{1I} & \cdots & \mathbb{M}_i^{1K} \\ \vdots & & \vdots & & \vdots \\ \mathbb{M}_i^{I1} & \cdots & \mathbb{M}_i^{II} & \cdots & \mathbb{M}_i^{IK} \\ \vdots & & \vdots & & \vdots \\ \mathbb{M}_i^{K1} & \cdots & \mathbb{M}_i^{KI} & \cdots & \mathbb{M}_i^{KK} \end{bmatrix} \begin{pmatrix} \mathbb{C}_i^1 \\ \vdots \\ \mathbb{C}_i^I \\ \vdots \\ \mathbb{C}_i^K \end{pmatrix} = 0 \tag{75}$$

where the diagonal super matrix elements are

$$\mathbb{M}_i^{II} = \mathbb{H}_i^I + \mathbb{A}^{II} - (\mathbb{A}^{II})^T - \mathscr{E}_i \mathbb{S}^I \tag{76}$$

and the off-diagonal terms are

$$\mathbb{M}_i^{IJ} = \mathbb{A}^{IJ} - \mathbb{A}^{JI}. \tag{77}$$

The matrices appearing in these expressions are

$$\mathbb{H}_i^I = [(H_i^I)_{pq}] \tag{78a}$$

$$\mathbb{S}^I = [S_{pq}^I] \tag{78b}$$

and

$$\mathbb{A}^{IJ} = [A_{pq}^{IJ}]. \tag{78c}$$

The matrix elements are

$$(H_i^I)_{pq} = \iiint_{V_I} d\Omega(1) \, \mathscr{K}_p^I(1) \, h_i(1) \, \mathscr{K}_q^I(1) \tag{79a}$$

$$S_{pq}^I = \iiint_{V_I} d\Omega(1) \, \mathscr{K}_p^I(1) \, \mathscr{K}_q^I(1) \tag{79b}$$

and

$$A_{pq}^{IJ} = \iint_{S_{IJ}} dS(1) \, \mathbf{n}_{IJ} [\nabla_1 \mathscr{K}_p^I(1)] \, \mathscr{K}_q^J(1). \tag{79c}$$

The vector C_i^I is

$$C_i^I = \begin{pmatrix} C_{i1}^I \\ \vdots \\ C_{iN_I}^I \end{pmatrix}. \tag{80}$$

Let us notice that the off-diagonal blocks (77) are identically zero if the loges V_I and V_J have no common surface. In addition, let us remark that for the evaluation of the matrix elements in the diagonal blocks, the integrations are carried out only within each loge. Hence, the method advanced here leads to simplifications in the evaluation of integrals [14]. Because the operator $\mathscr{V}_i(1)$ in Equation (73) contains functions which are labelled by $j \neq i$, the solution of Equation (75) implies a self consistent field procedure.

4. Discussion

We have reviewed in the present work methods which have been advanced in the theory of loges for the ab-initio calculation of molecular wavefunctions. The first employs variational principles for continuous wavefunctions. The second has been devised to use discontinuous functions.

The basic premise of this work is that in molecular systems there is no need to treat in the same category all interactions. A molecule is very different from an electron gas in that there are regions in three dimensional space where there is a high probability of finding groups of electrons.

The most important effort in molecular wavefunctions calculation should be devoted to evaluating as exactly as possible the wavefunctions of these electron groups. Of course, it is also necessary to develop a rigorous method whereby the corrections to this first step should be successively calculated. This is precisely what we have attempted to do in Section 2. However, as the restriction that the loge functions drop to zero at the loge boundaries might prove to be limiting for a correct representation of the actual physical system, in Section 3 we have discussed a variational principle applicable to discontinuous loge functions. Because the surface terms guarantee that regional functions match at the loge boundaries, the total wavefunction need not have regions of zero density at these boundaries and this variational principle could be a more appropriate starting point for molecular wavefunction calculations. In Section 3 we have only discussed the case of an independent particle model wavefunction. It seems, however, that the extrapolation of this principle to pair equations (independent pair model) should prove fruitful in the treatment of large molecular systems.

References

1. a) Daudel, R., Odiot, S., and Brion, H.: *J. Chem. Phys.* **51**, 74 (1954); Daudel, R.: *The Fundamentals of Theoretical Chemistry* (Pergamon Press, 1968), p. 188; *Advan. Quantum Chem.* **1**, 115 (1964).

   b) Aslangul, C., Constanciel, R., Daudel, R., and Kottis, P.: *Advan. Quantum Chem.* **6**, 93 (1972); Aslangul, C.: *C.R. Acad. Sci., Ser. B.* **272**, 1 (1971).
   c) Daudel, R., Bader, R. F. W., Stephens, M. E., and Borret, D. S.: *Can. J. Chem.* **52**, 1310 (1974).
   d) Sperber, G.: *Int. J. Quantum Chem.* **5**, 177 (1971).
2. Eisenberger, P. and Marra, W. C.: *Phys. Rev. Lett.* **27**, 1413 (1971); Roux, M. and Epstein, I. R.: *Chem. Phys. Lett.* **18**, 18 (1973).
3. a) Ludeña, E. V. and Amzel, V.: *J. Chem. Phys.* **52**, 5923 (1970).
   b) Ludeña, E. V.: *Int. J. Quantum Chem.* **5**, 395 (1971).
   c) Ludeña, E. V.: *Int. J. Quantum Chem.* **6**, 1157 (1972).
4. Hirschfelder J. O., and Nazaroff, G. V.: *J. Chem. Phys.* **34**, 1666 (1961).
5. The name pseudo Hartree-Fock equations is used here in a sense which is different from the one given in Weber, T. A. and Parr, R. G.: *Phys. Rev.* **3A**, 81 (1971).
6. For the derivation of these equations in a density matrix formalism, see Aslangul, C., Constanciel, R., Daudel, R., Esnault, L., and Ludeña, E. V.: *Int. J. Quantum Chem.* **8**, 499 (1974).
7. Ludeña, E. V. and Ruette, F.: 'The Completely Localized Loge Model and the Solution of One-Electron Equations', (unpublished).
8. See section 5 of reference 3b.
9. Sinanoğlu, O.: *Adv. Chem. Phys.* **6**, 387 (1963); Nesbet, R. K.: *Adv. Chem. Phys.* **9**, 321 (1965); Freed, K.: *Phys. Rev.* **173**, 1, 24 (1968); *ibid., Chem. Phys. Lett.* **4**, 496 (1970).
10. See in particular, Sinanoğlu, O.: *J. Chem. Phys.* **36**, 706 (1962).
11. Sinanoğlu, O.: *Adv. Chem. Phys.* **14**, 237 (1968); *ibid., Proc. Roy. Soc.* **A260**, 379 (1961).
12. Hirschfelder, J. O., Byers Brown, W., and Epstein, S. T., *Adv. Quantum Chem.* **1**, 255 (1964).
13. McCavert, P. and Rudge, M. R.: *Proc. Roy. Soc. (London)* **A328**, 429 (1972).
14. Ludeña, E. V.: 'Variational Principles for Discontinuous Wavefunctions and the Independent Particle Model of Electronic Structure', *Int. J. Quantum Chem.* **9**, (1975).

DISCUSSION

The delocalization is introduced in the CLLM model by surface integrals. In the ZDO approximation it is introduced by monoelectronic integrals. In the two cases the kinetic energy is concerned.

The complete localized loge model gives a kind of generalization of the ZDO approximation.

EFFECTIVE MOLECULAR HAMILTONIANS, PSEUDOPOTENTIALS AND MOLECULAR APPLICATIONS

PH. DURAND and J. C. BARTHELAT

Laboratoire de Physique Quantique, Université Paul Sabatier,
31077 Toulouse Cedex, France

Abstract. The present paper is devoted to a general methodology to determine effective hamiltonians for various atomic or molecular fragments. These effective operators are derived from the previous knowledge of atomic or molecular wave functions and have an entirely theoretical status. When this formalism is applied to atomic fragments, the corresponding methods belong to the so-called pseudopotential methods, and lead to atomic core effective potentials. The efficiency of our pseudopotential method is shown through one-electron Hartree-Fock valence-only calculations on molecules containing various atoms of the periodic table including transition metal atoms. The calculated values of observables such as bond lengths, bond angles, vibrational force constants and internal rotation barriers are in good agreement with the experimental data when the latter are available.

1. Introduction

Most of the efforts in Quantum Chemistry are traditionally devoted to the resolution of the Schrödinger equation in order to obtain energies and wave functions for molecular systems. But as this equation cannot be exactly solved, one is frequently led to treat approximate effective molecular hamiltonians. Although the exact hamiltonian implicitly contains all of the information on the ground and excited states of the molecule, it is impossible to identify any natural constituent of the molecule (atom, bond, chemical functional group, etc...) through the basic operators such as kinetic energy operator, electron-nucleus and electron-electron energy operators. On the contrary, the familiar approximate effective hamiltonians of Quantum Chemistry provide effective operators directly connected to atomic or molecular contents (introduction of experimental data such as atomic electron affinities or ionization potentials, measure of the interaction between two chemically bonded atoms in Hückel methods, etc...). However in most cases these effective hamiltonians are dependent on minimal Slater basis sets, and have not a well-defined theoretical status. Up to now progress in the determination of theoretical effective hamiltonians has been slow, although the need for such hamiltonians well adapted for describing the great variety of chemical properties is obvious. With a view to clarifying the present situation, it is first useful to review briefly the methods of Quantum Chemistry and to divide them in three groups.

In the *first group* of methods, one attempts to obtain directly the best energy E and the wave function ψ of the system. Among these methods we find direct determinations of the best Slater determinant [1] within the one-electron approximation or more accurate configuration interaction calculations [2].

O. Chalvet et al. (eds.), Localization and Delocalization in Quantum Chemistry, Vol. II, 91–125. All Rights Reserved
Copyright © 1976 by D. Reidel Publishing Company, Dordrecht-Holland

The methods of the *second group* are more ambitious: they give simultaneously the total energy, the wave function and various effective hamiltonians or operators. This is the case of the classical closed-shell Hartree-Fock method that produces the Fock operator in its diagonal form. For open-shell systems various one-electron Fock-like operators have been proposed; they do not always have a clear physical meaning but play an important part in the convergence of the iterative process towards the solution [3]. Within the one-electron approximation, one can also obtain localized molecular orbitals and atomic potentials by solving equations of the type given by Adams and Gilbert [4, 5]. McWeeny's group function method also gives effective potentials and hamiltonians for the various groups of the molecule [6]. Although these methods introduce good true local potential, they usually suffer from mathematical difficulties in their applicability (bad convergence, oscillatory behavior, etc...); more generally these methods are too intricate for providing in only one calculation the wave function, the energy and various effective operators.

In the *third group* of methods, the wave functions and the energies of the molecule are supposed to be known. One only looks for effective hamiltonians and operators associated with constant and transferable atomic and molecular entities. The pseudo-potential method for example allows one to determine for each atom an effective atomic potential associated with the nucleus and the core electrons [7]. The theoretical parametrizations of semi-empirical methods [8] can also be considered as belonging to this third group of methods. Figure 1 summarizes the connection between these three groups of methods.

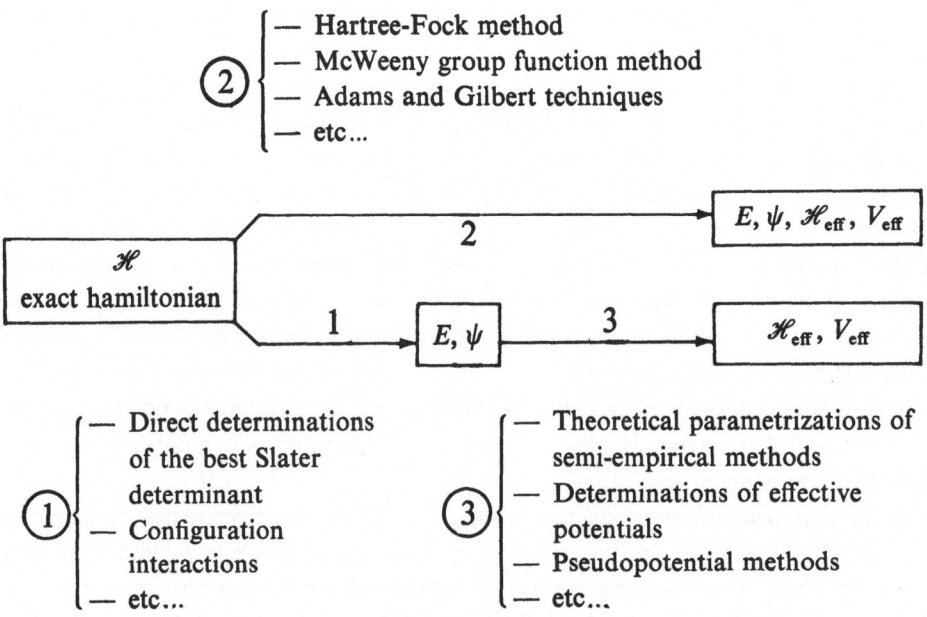

Fig. 1. A partitioning of the most fundamental methods of Quantum Chemistry.

It must be noticed that there is not a complete separation between the methods of groups 1 and 2. For instance once a group potential in a molecule has been determined by solving McWeeny's equations, it could be used for studying another molecule containing the same group of atoms. The methods of group 1 and 2 can also be distinguished by their mathematical techniques. In methods of group 1, E and ψ are generally determined by minimizing the mean value $\langle \psi | \mathcal{H} | \psi \rangle$ or the trace $\text{Tr}[\mathcal{H}]$ depending on whether one is looking for one or several stationary states. In methods of group 2, the good criterion for obtaining an effective hamiltonian $\mathcal{H}_{\text{eff}}$ once the wave function ψ is known, is to minimize the energy fluctuation or variance $\langle \psi | (\mathcal{H} - \mathcal{H}_{\text{eff}})^2 | \psi \rangle$ or $(\text{Tr}[\mathcal{H} - \mathcal{H}_{\text{eff}}]^2)$.

According to this partition of the quantum chemistry methods into three groups, the following general methodology is suggested:

(i) To determine systematically effective potentials by methods of group 3.

(ii) From these effective potentials to build up approximate effective hamiltonians.

(iii) To solve these effective hamiltonians by the most efficient mathematical methods of group 1.

The primary aim of this paper is to discuss and analyse these methods of group 3. It does not seem however that the originality or the importance of these methods of group 3 has been previously recognized and it is certainly unfortunate that the variance and least-squares techniques are still so little used in Quantum Chemistry (see however references 9 and 10). Among the various electronic, vibrational, vibronic, magnetic, ... effective operators that could be investigated, we choose to study only electronic effective potentials of atomic or molecular fragments. This paper is not a review of the various electronic effective potentials. We only report and analyse some methods closely related to our work. The partitioning of a molecule into atomic or molecular fragments is presented in Section 2. A general method then is developed to determine one-electron effective operators for these fragments. The practical determination of atomic fragment pseudopotentials forms the subject of Section 3 and our method is compared with other pseudopotential methods in this Section. Various molecular applications are presented in Section 4. They show the efficiency of our pseudopotential method to investigate such experimental quantities as bond lengths, bond angles, vibrational force constants and internal rotation barriers.

2. Partitioning of Molecules into Atomic or Molecular Fragments

The partitioning of a molecule into fragments can be performed at the experimental or theoretical level.

Unstable entities or molecular fragments frequently occur in chemical reactions. For instance the following intermediate entities are derived from methane [11]:

$$CH_3^+ \qquad :CH_3^- \qquad .CH_3 \qquad :CH_2$$

methyl cation   methyl anion   methyl radical   methylene radical

Various molecular fragments also happen in mass spectroscopy experiments (see reference 11, p. 867). Although these transient molecular species are directly deduced or conjectured from experiment, they cannot constitute a good general set of theoretical fragments because the geometries of these true *experimental fragments* are not the same than the geometries of *theoretical fragments* extracted from stable molecules in their ground state. Thus the free methyl radical is probably planar or pyramidal whereas it is tetrahedral in ethane.

To illustrate the definition of our theoretical fragments let us examine the hydrocarbon molecules. These molecules are all basically made of protons and electrons or of carbon and hydrogen atoms. If we introduce more structural information in the description of these molecules we can consider that they are also constituted from various molecular fragments fixed *in their standard geometry* (Figure 2).

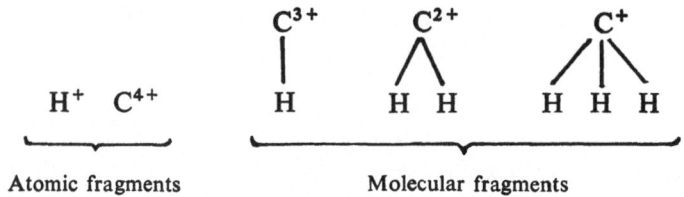

Fig. 2.   Examples of partitioning of hydrocarbon molecules into atomic and molecular fragments.

It must be noticed that the molecular fragments exhibited in Figure 2 are not the familiar neutral radicals of chemistry. We have chosen these electrically charged fragments for two reasons. First we want our fragments to have a closed-shell structure of nuclei and electron pairs. Secondly only the effective potentials of these fragments are easily transferable from one molecule to another without spin symmetry difficulties. However Vermeulin *et al.* [12] have proposed a general theory of molecular fragments including neutral fragments that generalizes the concept of atomic valence state. They have performed very accurate calculations of interatomic distances and force constants for the CF and NO fragments of fluorinated alkanes and saturated amine-oxides. Theoretical molecular fragments for further molecular calculations have also been proposed by Christoffersen [13] but these fragments are only defined

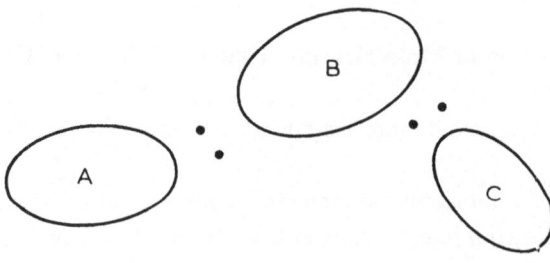

Fig. 3.   Molecule partitioned into fragments A, B, C,... and N electrons (represented by dots).

through sets of floating spherical gaussian orbitals and not by effective potentials. A general analysis of a molecular wave function into separated elements has been carried out by Constanciel [14] but this work also does not deal with the obtention of fragment effective potentials.

Generally any molecule can be partitioned into fragments and electrons. The effective hamiltonian for the molecule shown in Figure 3 is in atomic units

$$\mathscr{H}_{\text{eff}} = \sum_{i=1}^{N} [-\tfrac{1}{2}\Delta_i + \sum_A V_A(i)] + \sum_{i<j}^{N} \frac{1}{r_{ij}} \tag{1}$$

V_A : effective potential of fragment A.

The energy of the molecule is then given by

$$E = \sum_A E_A + \sum_{A<B} E_{AB} + E_{\text{eff}} \tag{2}$$

E_A : energy of fragment A.

E_{AB} : classical electrostatic interaction energy between the fixed charge distributions of fragments A and B.

E_{eff} : energy obtained by the resolution of the equation $\mathscr{H}_{\text{eff}}\Phi = E_{\text{eff}}\Phi$.

Φ : valence pseudo-wave function.

Obviously the expressions (1) and (2) are approximate ones but one can always construct a sequence of effective hamiltonians converging towards the exact molecular hamiltonian whereas the calculated energies converge towards the exact energy. Figure 4 shows two partitions into fragments of the ethane molecule.

The effective molecular hamiltonian corresponding to Figure 4a is given by

$$\mathscr{H}_{\text{eff}} = \sum_{i=1}^{2} [-\tfrac{1}{2}\Delta_i + V_A(i) + V_B(i)] + \frac{1}{r_{12}} \tag{3}$$

where V_A or V_B represents the interaction energy between one electron and the molecular fragment A or B.

In the case of Figure 4b, the molecule is made up of six protons and two C^{4+} atomic fragments denoted A and B. The effective molecular hamiltonian is then given by

$$\mathscr{H}_{\text{eff}} = \sum_{i=1}^{14} [-\tfrac{1}{2}\Delta_i + V(i)] + \sum_{i<j}^{14} \frac{1}{r_{ij}} \tag{4}$$

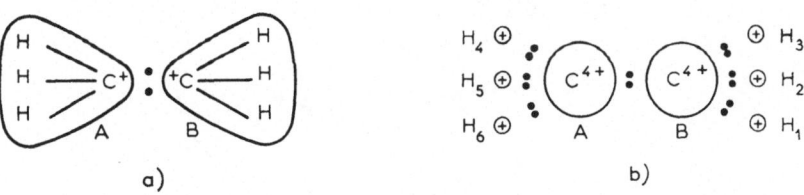

Fig. 4.   Two partitions of ethane in atomic or molecular fragments.

with

$$V = V_A + V_B - \sum_{k=1}^{6} \frac{1}{r_{H_k}}$$

where V_A or V_B represents the interaction energy between one electron and the atomic fragment A or B of Figure 4b. r_{H_k} is the distance from one electron to the proton H_k. The two effective hamiltonians (3) and (4) can be considered as belonging to a sequence of approximate hamiltonians converging towards the exact all-electron molecular hamiltonian defined by

$$\mathcal{H} = \sum_{i=1}^{18} \left[-\tfrac{1}{2} \Delta_i - 6 \left(\frac{1}{r_{Ai}} + \frac{1}{r_{Bi}} \right) - \sum_{k=1}^{6} \frac{1}{r_{H_k i}} \right] + \sum_{i<j}^{18} \frac{1}{r_{ij}} \tag{5}$$

where r_{Ai} and r_{Bi} are the distances of the electron i from the two carbon nuclei A and B.

The model a of Figure 4 allows the study of the C—C bond with the two-electron hamiltonian (3), and it could be easily used to calculate the electric or magnetic susceptibility of the C—C bond. But this model would be inadequate for a study of the internal rotation barrier that can be obtained with the model b of Figure 4 and the fourteen-electron hamiltonian (4) (see Section 4).

2.1. PRACTICAL DETERMINATION OF FRAGMENT POTENTIALS

To determine the effective potential of a fragment A, we first choose a molecule or an atom containing this fragment.
The effective hamiltonians for the two fragments A and B are given by

$$\mathcal{H}_A = \sum_{i=1}^{N_A} \left[\tfrac{1}{2} \Delta_i - \sum_{\alpha} \frac{Z_\alpha}{r_{\alpha i}} \right] + \sum_{i<j}^{N_A} \frac{1}{r_{ij}} \tag{6}$$

$$\mathcal{H}_B = \sum_{i=1}^{N_B} \left[-\tfrac{1}{2} \Delta_i - \sum_{\beta} \frac{Z_\beta}{r_{\beta i}} + V_A(i) \right] + \sum_{i<j}^{N_B} \frac{1}{r_{ij}} \tag{7}$$

Z_α and Z_β are the atomic numbers of the nuclei belonging to the fragments A and B respectively. N_A and N_B are the number of electrons of fragments A and B and the

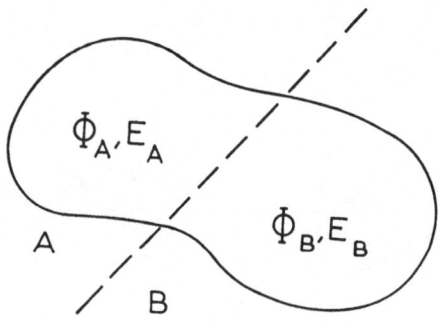

Fig. 5.   Partition of a molecule into two fragments A and B.

summations $\sum\limits_{\alpha}$ and $\sum\limits_{\beta}$ must be taken over the nuclei of fragments A and B respectively.

The first step is to extract from the molecular wave function ψ and energy E the two wave functions Φ_A, Φ_B and the energies E_A, E_B of the two fragments A and B. This is not an unambiguous mathematical operation and this can be carried out in various manners. The best one consists generally in using least-squares techniques [9–15]. One can for example partition the one-electron basis set $\{\chi_i\}$ of the entire molecule into two subsets $\{\chi_i^A\}$ and $\{\chi_i^B\}$ associated with the fragments A and B. The wave functions Φ_A and Φ_B are then constructed from these two subsets and determined by minimizing the fluctuations of the number of particles N_A and N_B in the states Φ_A and Φ_B,

$$\langle \hat{N}_A^2 - \langle \hat{N}_A \rangle^2 \rangle_{\mathrm{minimum}} \qquad \langle \hat{N}_B^2 - \langle \hat{N}_B \rangle^2 \rangle_{\mathrm{minimum}}. \tag{8}$$

The mean values are to be taken over the molecular wave function ψ. With Φ_A and Φ_B normalized to unity, the energies of the fragments A and B are given by the mean value of $\mathscr{H}_A$ and $\mathscr{H}_B$, respectively,

$$E_A = \langle \Phi_A | \mathscr{H}_A | \Phi_A \rangle \qquad E_B = \langle \Phi_B | \mathscr{H}_B | \Phi_B \rangle. \tag{9}$$

It must be noticed that the partition of the molecular energy E into E_A and E_B terms is not symmetrical since E_A is the energy of the fragment A and E_B is the energy of the fragment B plus the interaction energy between the fragments A and B through the term $\langle \Phi_B | \sum\limits_{i=1}^{N_B} V_A(i) | \Phi_B \rangle$ arising from (7).

The effective potential V_A is the superposition of a classical electrostatic potential $V_{\mathrm{elec},A}$ that predominates at large distances from the fragment A and a quantum potential $V_{\mathrm{quant},A}$ that arises principally from Pauli forces;

$$V_A = V_{\mathrm{elec},A} + V_{\mathrm{quant},A} \tag{10}$$

$$= V_{\mathrm{elec},A} + \sum_{\Gamma^{(\alpha)}} V_A^{(\alpha)}.$$

In the case of an atomic fragment the electrostatic term equals $-z_A/r$ where z_A is the net charge of the ion made up of the nucleus and the core electrons of atom A. The quantum effective potential is nonlocal and strongly repulsive in the neighbourhood of fragment A. With symmetrical fragments the effective quantum potential is the sum of terms associated with the various irreducible representations $\Gamma^{(\alpha)}$ of the molecular fragment [7]. Lastly the parameters of V_A are determined in such a way that the lowest solution of $\mathscr{H}_B$ is (Φ_B, E_B). Practically the quantity $\langle \Phi_B | (\mathscr{H}_B - E_B)^2 | \Phi_B \rangle$ is minimized.

The three principal steps in the determination of the fragment potential are summarized in the following scheme:

$$\psi, E \atop \text{Molecule} \qquad \longrightarrow \qquad \begin{cases} \Phi_A, E_A \\[2em] \Phi_B, E_B \\ \text{two fragments} \end{cases} \qquad \longrightarrow \qquad \begin{matrix} V_A \\ \text{Fragment potential} \end{matrix}$$

The determination of a fragment potential is worthwhile under two conditions:

(i) Both the fragment and its corresponding potential are more or less transferable from one molecule to another (this choice can be made from the chemist's intuitive knowledge).

(ii) The fragment is a universal component of chemistry.

Thus it appears that the atomic fragments or ions consisting of nuclei and core electrons are the most fundamental pieces of quantum chemistry. Their determination belongs to the general methodology of *pseudopotentials* (see Sections 3 and 4).

2.2. DISCUSSION

Our fragment potentials are *one-electron*, *ab-initio* and *variational*. One-electron effective potentials have been chosen for further computational convenience. Ab-initio means that their determination is entirely theoretical starting from ab-initio atomic or molecular wave functions. Variational means that when these fragment potentials are further used in an effective molecular hamiltonian (Equation (1)) the variational method can be used without any restriction (the orthogonality condition for example) to obtain the lowest solution of the effective molecular hamiltonian.

We now put our method among other methods which also introduce effective ab-initio fragment potentials.

First we investigate McWeeny group function theory [6] in the particular case of a molecule made up of two fragments A and B. The molecular wave function is the antisymmetrized product of the wave functions Φ_A and Φ_B of the two fragments.

The effective potential operator, V_A, for the nuclei and the electrons of fragment A is given by

$$V_A = - \sum_\alpha \frac{Z_\alpha}{r_\alpha} + J_\alpha - K_\alpha \qquad (11)$$

J_α and K_α are one-electron coulomb and exchange operators owing to the electrons of fragment A. One could try to identify the potential V_A with the potential previously given by Equation (10) but this cannot be easily done since our effective potential is variational and McWeeny's one is not. Practically our fragment potential can be used in an effective molecular hamiltonian without any restriction whereas with McWeeny's method the strong orthogonality condition must be maintained between the various groups of the molecule. More general effective valence shell hamiltonians have been derived by Westhaus *et al.* [16] but they introduce many-body potentials that should be rather unsuitable for current numerical applications.

To go beyond the strong orthogonality condition, Öhrn and McWeeny [17] have derived a one-body model for an electron outside a core. The effective potential 'seen' by the valence electron is rather intricate as it supposes the knowledge of the third order reduced density matrix of the core wave function. The generalizations of this model introduce generally many-body effective potentials [8] except when the core wave function is a fixed Slater determinant. This model for an atom is given in the

Appendix. It appears that the strong orthogonality condition that was previously required has now been transferred into the effective hamiltonian through projectors. Practically the same difficulties as before are to be met, i.e., mixing of the basis sets, entailing the calculation of all the one- and two-electron integrals, and difficulties to extract from this formalism effective one-electron fragment potentials. In conclusion the effective hamiltonians briefly reviewed in this discussion, although they are mathematically correct, are not a good starting point to derive fragment potentials and we think that this is more easily carried on by direct least-squares techniques presented from the outset of this Section.

3. Determination of Atomic Pseudopotentials

The fragments are the ions consisting of nuclei and core electrons of atoms. The effective potentials of these fragments are called pseudopotentials or model potentials. As we have already published a general method to determine these pseudopotentials [18] we only summarize the characteristic features of the method within the framework of the general methodology just presented in Section 2. The three steps of the method are detailed below and illustrated on the iron atom chosen as a typical example of a transition metal. At each step we compare our techniques with other similar techniques.

3.1. ATOMIC WAVE FUNCTIONS

The atomic wave functions are determined by the open-shell Hartree-Fock-Roothaan method [19]. When available we use the atomic orbitals determined by Clementi et al. [20–22]. Otherwise, especially for the lowest states arising from excited electronic configurations, we optimize the valence bases and solve the Hartree-Fock equations with the computer program of Roos et al. [21]. The $3d$ and $4s$ orbitals, occupied in the ground state of the atom, have been taken from ref. [22]. The $4p$ orbital has been calculated for the excited configuration $3d^6 4s^1 4p^1$ in the state ^7D.

3.2. DETERMINATION OF THE VALENCE PSEUDO-ORBITALS

The valence pseudo-orbital ϕ_v corresponding to a valence orbital φ_v is defined by the conditions that:

(i) the radial parts of ϕ_v and φ_v coincide at a distance r from the nucleus greater than a core radius R_c.

(ii) the radial part of ϕ_v decreases monotonically towards zero for $r < R_c$.

A good choice for R_c is the core loge radius as defined by Daudel et al. [23]. One must however note that the core loge radius of an atom depends both on the molecular environment and on the criterion chosen to define the core loge (cf. Table I).

TABLE I

Variation of the core loge radii (in Å) for lithium, boron and beryllium atoms

Atom or molecular fragment	R_c	Atom or molecular fragment	R_c	Atom or molecular fragment	R_c
LiH	1.42[a]	BeH	0.95[a]	BH	0.70[a]
LiH[+]	1.55[a]	BeH[+]	1.00[a]	–	–
LiH	1.34[b]	BeH	1.07[b]	BH	1.00[b]
LiH[+]	1.41[b]	BeH[+]	1.05[b]	–	–
Li	1.48[c]	Be	1.04[c]	B	0.79[c]

[a] Minimum of the missing information [23].
[b] Virial partitioning method [23].
[c] Intersections of the radial parts of the two outermost atomic orbitals [18].

Furthermore these authors have not determined the core loge radii for heavy atoms, so we have used the following definition: for a given orbital the value of R_c is obtained by intersecting the radial part of the valence orbital with the one of the outermost core orbital of the same symmetry. Thus we define a core radius for each symmetry s, p, d ... Table II gives the values that have been obtained from Li to Ar atoms.

TABLE II

Core radii, R_c, from lithium to argon obtained from atomic double-zeta orbitals [22].

Atom	R_c(Å) s symmetry	R_c(Å) p symmetry	Atom	R_c(Å) s symmetry	R_c(Å) p symmetry
Li	1.48	–	Na	1.70	–
Be	1.04	–	Mg	1.42	–
B	0.79	–	Al	1.22	1.38
C	0.64	–	Si	1.08	1.18
N	0.54	–	P	0.98	1.04
O	0.47	–	S	0.89	0.93
F	0.42	–	Cl	0.82	0.84
Ne	0.37	–	Ar	0.75	0.77

Once the core radius R_c is known, the pseudo-orbitals ϕ_v are determined by least-squares techniques as indicated in Reference [18]. As an example, we give in Table III the coefficients of the ϕ_{4s} and ϕ_{4p} pseudo-orbitals obtained with a double-zeta basis set of Slater-type orbitals. The radial part of the orbitals and of the corresponding pseudo-orbitals are represented in Figures 6 and 7.

We have not represented the pseudo-orbital ϕ_{3d} since it is identical to the nodeless φ_{3d} orbital, in the case of the iron atom.

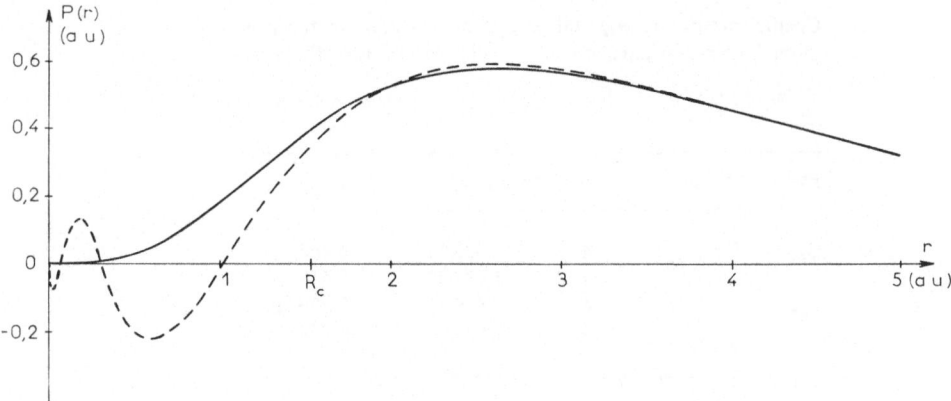

Fig. 6.   Radial parts of the iron φ_{4s} orbital (dotted line) and of the corresponding ϕ_{4s} pseudo-orbital (solid line).

Generally in pseudopotential techniques the valence pseudo-orbitals ϕ_v are linear combinations of the corresponding valence orbital and of the core orbitals φ_c of the same symmetry:

$$\phi_v = a_v \varphi_v + \sum_c a_c \varphi_c. \tag{12}$$

The mixing coefficients a_v and a_c are determined according to various criteria in order to obtain nodeless pseudo-orbitals. One can minimize the mean kinetic energy of the pseudo-orbital [24] or obtain the best fit with a Slater orbital [25] or de-orthogonalize variationally [26], etc... All these techniques give more or less the same results, but Equation (12) is an unnecessary constraint for building pseudo-orbitals [18–27]. We think that our definition of nodeless pseudo-orbitals that are identical to the true atomic Hartree-Fock orbital outside a core radius is the best adapted for further molecular calculations. Indeed it is expected that molecular bond properties primarily depend on the outermost part of the atomic valence orbitals.

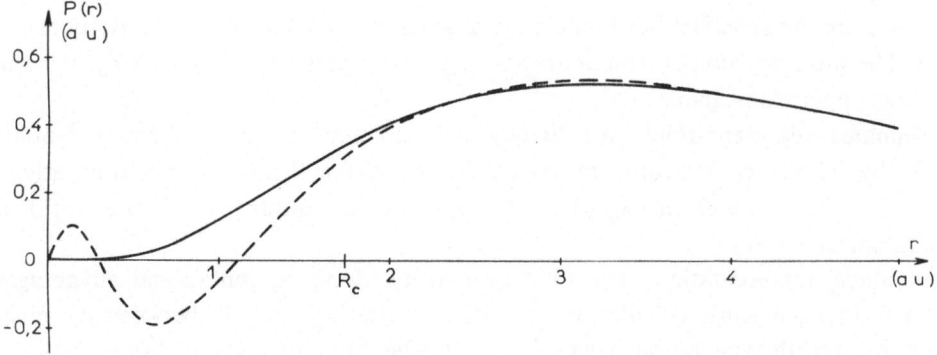

Fig. 7.   Radial parts of the iron φ_{4p} orbital (dotted line) and of the corresponding ϕ_{4p} pseudo-orbital (solid line).

TABLE III

Coefficients of the ϕ_{4s} and ϕ_{4p} pseudo-orbitals of the iron atom. The radial part of each pseudo -orbital is of the form

$$\sum_{i=1}^{2} N_i C_i \, r^{n_i-1} \, e^{-\zeta_i r} \text{ where } N_i \text{ is the normalization coefficient.}$$

Pseudo-orbital	n_i	ζ_i	c_i
ϕ_{4s}	4	1.925 17	0.468 33
	4	1.077 42	0.618 23
ϕ_{4p}	4	1.819 10	0.320 90
	4	0.981 16	0.759 81

3.3. PRACTICAL DETERMINATION OF THE ATOMIC PSEUDOPOTENTIALS

In this Section we use the notations of Ref. 18. The total valence pseudo-hamiltonian is given by

$$\mathcal{H}_{ps} = \sum_{i=1}^{N_v} \left[-\tfrac{1}{2} \Delta_i - \frac{z}{r_i} + W_R(i) \right] + \sum_{i<j}^{N_v} \frac{1}{r_{ij}} \tag{13}$$

where N_v is the number of valence electrons, and z which equals N_v is the net charge of the ion made up of the nucleus and of the core electrons of the atom. W_R is a nonlocal operator which is repulsive close to the nucleus.

With a view to further molecular calculations, the operator W_R is cast into simpler semilocal or nonlocal form:

$$W_R = \sum_l W_{R,l}(r) \, P_l; \qquad P_l = \sum_{m=-l}^{l} |Y_{lm}\rangle \langle Y_{lm}| \tag{14}$$

$$W_R = f(r) + \sum_l \sum_{ij} a_{ij,l} \sum_{m=-l}^{l} |\chi_{ilm}\rangle \langle \chi_{jlm}|. \tag{15}$$

The Y_{lm} are the spherical harmonics; the quantities $W_{R,l}$ and $f(r)$ are only functions of r. The wave function corresponding to $|\chi_{ilm}\rangle$ is represented by $f_{i,l}(r) \, Y_{lm}(\theta, \varphi)$ in spherical polar coordinates.

Semilocal representations give directly as a function of r the effective potentials 'seen' by the valence electrons. It is then possible to define theoretical electronegativity scales [28]. Thus the electronegativity concept finds a natural place in the world of Quantum Chemistry.

Nonlocal representations are more general and have computational advantages since further molecular calculations require only overlap integrals between the functions χ_{ilm} and the valence basis set [27, 29, 30]. One has, however, to choose carefully a sufficiently complete basis set in the atomic core region in order to have a truly variational pseudopotential (see Ref. [27] p. 969).

TABLE IV

Various analytical expressions for the $W_{R,l}(r)$ functions of the iron atom. All the quantities are in atomic units

$W_{R,0}$		$W_{R,1}$		$W_{R,2}$	
analytical expression	$\|\emptyset\|$	analytical expression	$\|\emptyset\|$	analytical expression	$\|\emptyset\|$
$\dfrac{1.715\,95}{r^2}$	9.51×10^{-2}	$\dfrac{1.245\,76}{r^2}$	4.91×10^{-2}	$-\dfrac{1.647\,07}{r}$	1.98×10^{-1}
$-\dfrac{0.908\,11}{r}+\dfrac{3.033\,38}{r^2}$	2.29×10^{-2}	$-\dfrac{0.667\,46}{r}+\dfrac{2.567\,99}{r^2}$	1.09×10^{-2}	$-\dfrac{1.365\,74}{r^2}$	8.87×10^{-2}
$8.906\,41\,\dfrac{e^{-0.954\,19\,r^2}}{r}$	3.08×10^{-3}	$6.705\,03\,\dfrac{e^{-0.731\,96\,r^2}}{r}$	1.58×10^{-3}	$-6.066\,54\,\dfrac{e^{-2.004\,51\,r^2}}{r}$	8.93×10^{-8}
$5.682\,35\,\dfrac{e^{-0.606\,88\,r^2}}{r^2}$	1.19×10^{-3}	$4.909\,72\,\dfrac{e^{-0.467\,75\,r^2}}{r^2}$	8.50×10^{-4}		

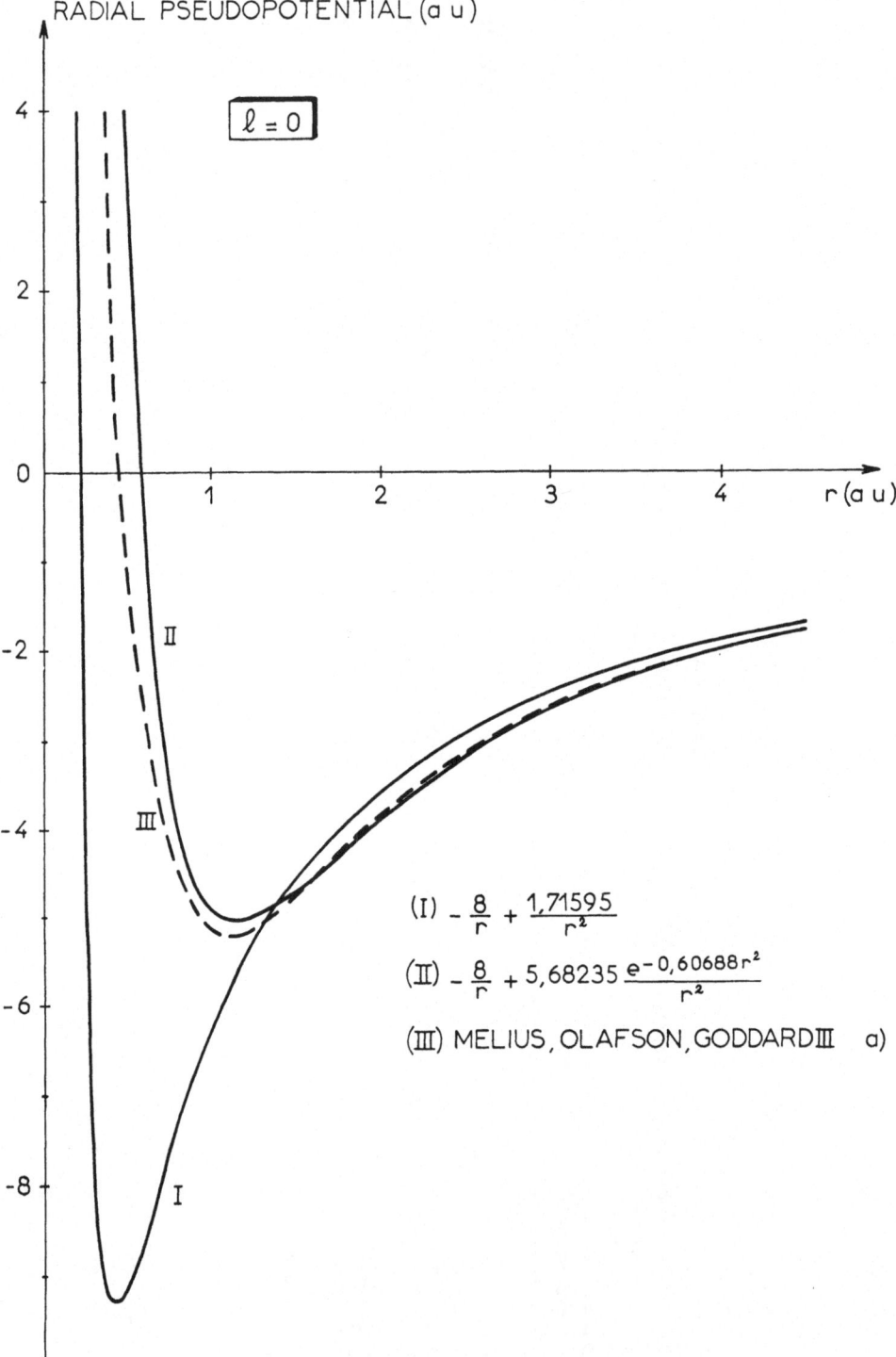

Fig. 8.   Radial pseudopotential $[-z/r + l(l+1)/2r^2 + W_{R,l}(r)]$ for the $4s$ electrons of the iron atom (see Table IV). (a) See Reference [31].

The variational resolution of Equation (13) by the open-shell Hartree-Fock-Root-haan equations leads to solving the following eigenvalue equation:

$$H_{ps}\phi'_v = \varepsilon'_v \phi'_v \qquad (16)$$

where H_{ps} is the corresponding monoelectronic pseudo-hamiltonian. The pseudo-

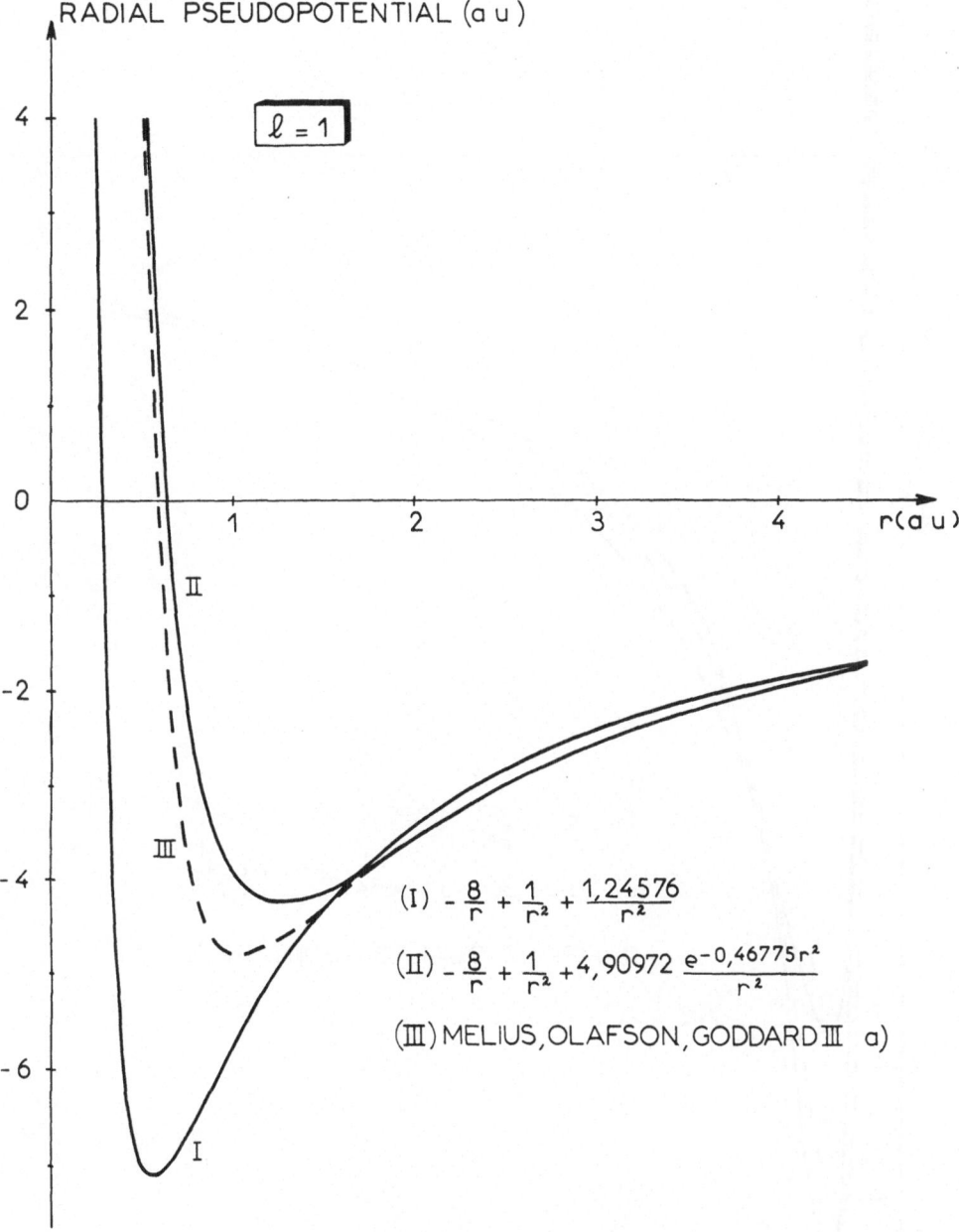

RADIAL PSEUDOPOTENTIAL (a u)

$\boxed{\ell = 1}$

(I) $-\dfrac{8}{r} + \dfrac{1}{r^2} + \dfrac{1,24576}{r^2}$

(II) $-\dfrac{8}{r} + \dfrac{1}{r^2} + 4,90972 \dfrac{e^{-0,46775 r^2}}{r^2}$

(III) MELIUS, OLAFSON, GODDARD III a)

Fig. 9. Radial pseudopotential $[-z/r + l(l+1)/2r^2 + W_{R,l}(r)]$ for the $4p$ electrons of the iron atom (see Table IV). (a) See Reference [31].

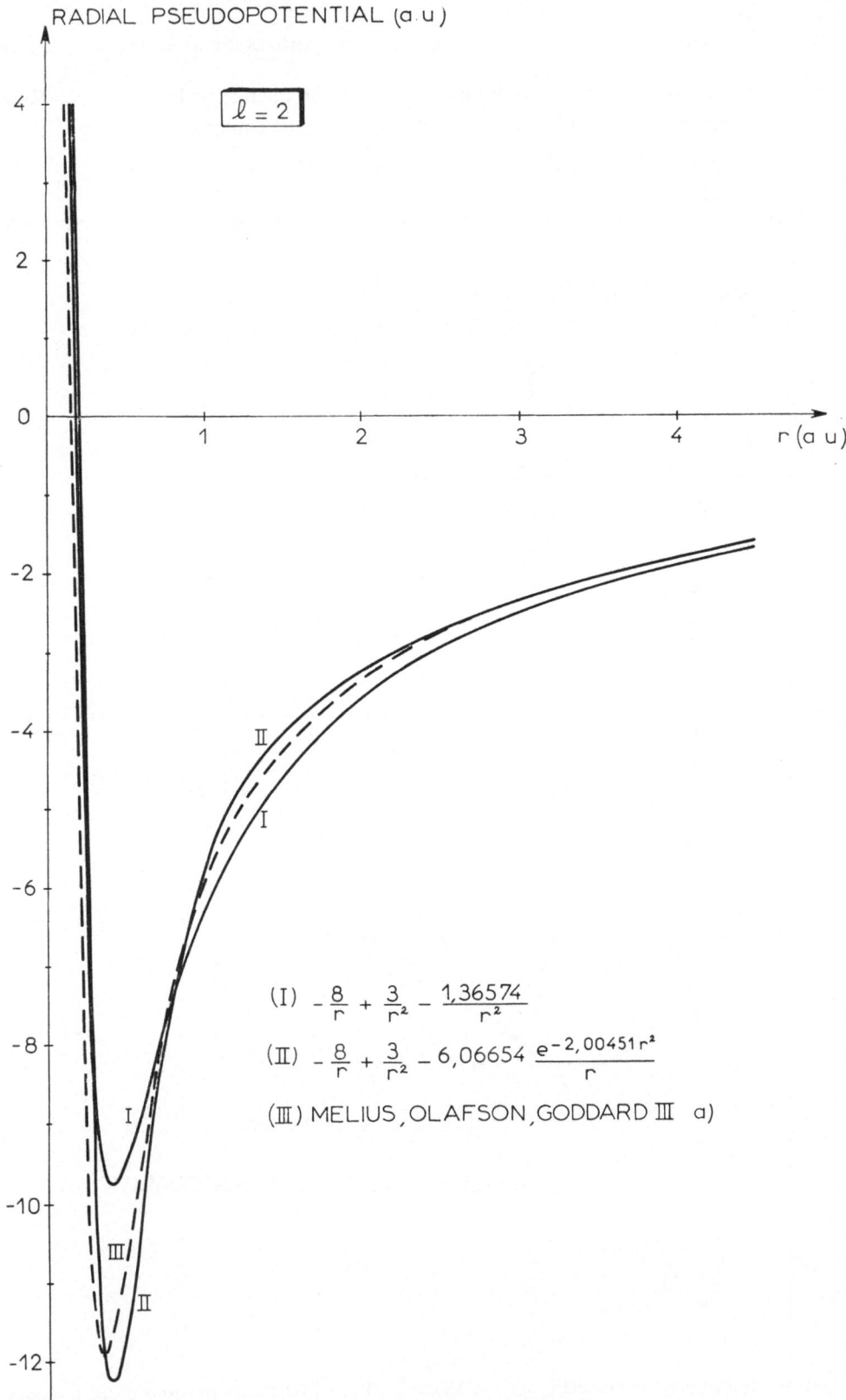

Fig. 10. Radial pseudopotential $[-z/r+l(l+1)/2r^2+W_{R,l}(r)]$ for the $3d$ electrons of the iron atom (see Table IV). (a) See Reference [31].

potential parameters are chosen in such a way that the lowest solution of Equation (16) $(\phi'_v, \varepsilon'_v)$ for each symmetry is exactly (ϕ_v, ε_v). ϕ_v is the valence pseudo-orbital obtained previously and ε_v is the energy of the corresponding orbital taken from an all-electron calculation. Practically, the parameters are best determined by a least-squares technique in minimizing the norm

$$\|\mathcal{O}\| = [\langle \phi_v | \mathcal{O}^2 | \phi_v \rangle]^{1/2} \tag{17}$$

of the operator defined by

$$\mathcal{O} = \varepsilon'_v | \phi'_v \rangle \langle \phi'_v | - \varepsilon_v | \phi_v \rangle \langle \phi_v |. \tag{18}$$

The smallness of the norm is a measure of the quality of the pseudopotential parametrization.

In this work we have calculated semilocal pseudopotentials with functions $W_{R,l}(r)$ of the following form

$$W_{R,l}(r) = \sum_i C_i \frac{e^{-\alpha_i r^2}}{r^{n_i}}. \tag{19}$$

Table IV gathers the various expressions calculated for these functions of symmetries $l = 0, 1, 2$ in the case of the iron atom.

In Table IV, $\|\mathcal{O}\|$ has been minimized with ϕ_{4s} and ϕ_{4p} from Table III. The $3d$ orbital and the valence energies ε_{3d} and ε_{4s} are from Reference [22]. The ε_{4p} energy has been obtained in the previously mentioned calculation of the Fe ^7D excited state. Figures 8, 9 and 10 give the shape of the effective potentials for the valence electrons of s, p and d symmetry.

It is to be noticed that we have incorporated the kinetic repulsive term $l(l+1)/2r^2$ in the effective potentials represented in Figures 8, 9 and 10. As expected the one-parameter pseudopotential (curve I of Figures 8, 9 and 10) badly reproduces the true pseudopotential which is better given by the two-parameter pseudopotential (curve II of Figures 8, 9 and 10). This latter compares very well with those obtained by Melius et al. [31]. This is not astonishing since these pseudopotentials have been obtained by similar techniques. These pseudopotentials are in agreement with the properties of the transition metal atoms: of the $3d$, $4s$ and $4p$ pseudopotentials, the $3d$ is the deepest and innermost. It is noticed that our $4p$ pseudopotential is slightly further out and shallower than the corresponding one of Melius et al. The comparison of these two $4p$ pseudopotentials with the corresponding $4s$ pseudopotentials of Figure 8 seems to indicate that our $4p$ pseudopotential agrees better with the fact that the excited $4p$ orbital is more diffuse than the $4s$ orbital.

4. Molecular Applications

The valence electronic pseudo-hamiltonian of a molecule is given by

$$\mathcal{H}_{ps} = \sum_{i=1}^{N_v} \left[-\tfrac{1}{2} \Delta_i - \sum_A \left(\frac{Z_A}{r_{Ai}} + W_{R,A}(i) \right) \right] + \sum_{i<j}^{N_v} \frac{1}{r_{ij}}. \tag{20}$$

For atom A of the molecule, z_A is the net charge of the ion made up of the nucleus and the core electrons and $W_{R,A}$ is the potential determined in Section 3 that is repulsive near the nucleus.

The molecular wave function is determined from Equation (20) with exactly the same method as from an ab-initio all-electron hamiltonian, i.e., one-electron methods or more elaborate configuration interaction methods taking into account the electronic correlation can be used.

In this work, we report calculations that have been performed at the restricted one-electron Hartree-Fock level. Our computer program uses a basis set of Slater orbitals expanded in one, two, three or four gaussian functions. These expansions* have been carried out by McWeeny's method [32]. The gaussian functions are then reproduced by lobe functions [33]. Thus we simulate s, p or d Slater orbitals that need

TABLE V

Pseudopotential parameters (in Å) for the following semilocal representation of $W_{R,l}(r)$: $W_{R,l} = \dfrac{C_1}{r} + \dfrac{C_2}{r^2}$

Atom		l	C_1	C_2
F	^2P	0	−1.081 50	1.272 83
		1	−	−0.034 32
Cl	^2P	0	−1.648 02	2.935 99
		1	−0.891 78	1.783 93
Br	^2P	0	−2.074 98	3.891 85
		1	−1.268 16	2.768 02
I	^2P	0	−2.088 19	4.820 36
		1	−1.368 91	3.709 51
O	^3P	0	−0.923 54	1.242 76
		1	−	−0.038 53
S	^3P	0	−1.418 16	2.770 39
		1	−0.701 02	1.602 56
Se	^3P	0	−1.588 42	3.301 05
		1	−1.098 52	2.541 11
C	^3P	0	−0.637 53	1.177 15
		1	−	−0.049 32
Si	^3P	0	−0.977 30	2.373 54
		1	−0.480 28	1.383 70
Ge	^3P	0	−1.050 96	2.519 64
		1	−0.615 45	1.745 07
N	^4S	0	−0.777 59	1.212 65
		1	−	−0.043 22
P	^4S	0	−1.231 99	2.637 94
		1	−0.606 88	1.527 63
B	^2P	0	−0.471 57	1.094 11
		1	−	−0.062 49

* These expansions of s, p, d Slater orbitals for the principal quantum number running from $n = 1$ to $n = 7$ (at intervals of 0.5) in $1s$, $2p$ and $3d$ gaussian functions can be obtained at the laboratory upon request.

not be centered on the nuclei and whose principal quantum number can take any real value from $n = 1$ to $n = 7$. All the matrix element calculations are those of classical ab-initio calculations except for those of operator $W_{R,A}$ that need indefinite integrals between modified spherical Bessel functions of the first order [25, 31, 34].

We present two sets of calculations. In the first set, our pseudopotential method is checked with some simple molecules for which we compare our results with similar ab-initio all-electron Hartree-Fock calculations. The molecules studied here are H_2O, HF, F_2 and ScH_3. In the second set, the pseudopotential method is applied to various series of molecules containing atoms of the second, third, fourth and fifth row of the periodic table. The calculated observables such as interatomic distances, bond angles, rotation barriers are directly compared with the experimental data.

All the following calculations have been carried out with minimal Slater basis sets. These Slater orbitals have been expanded in three gaussian functions (STO–3G

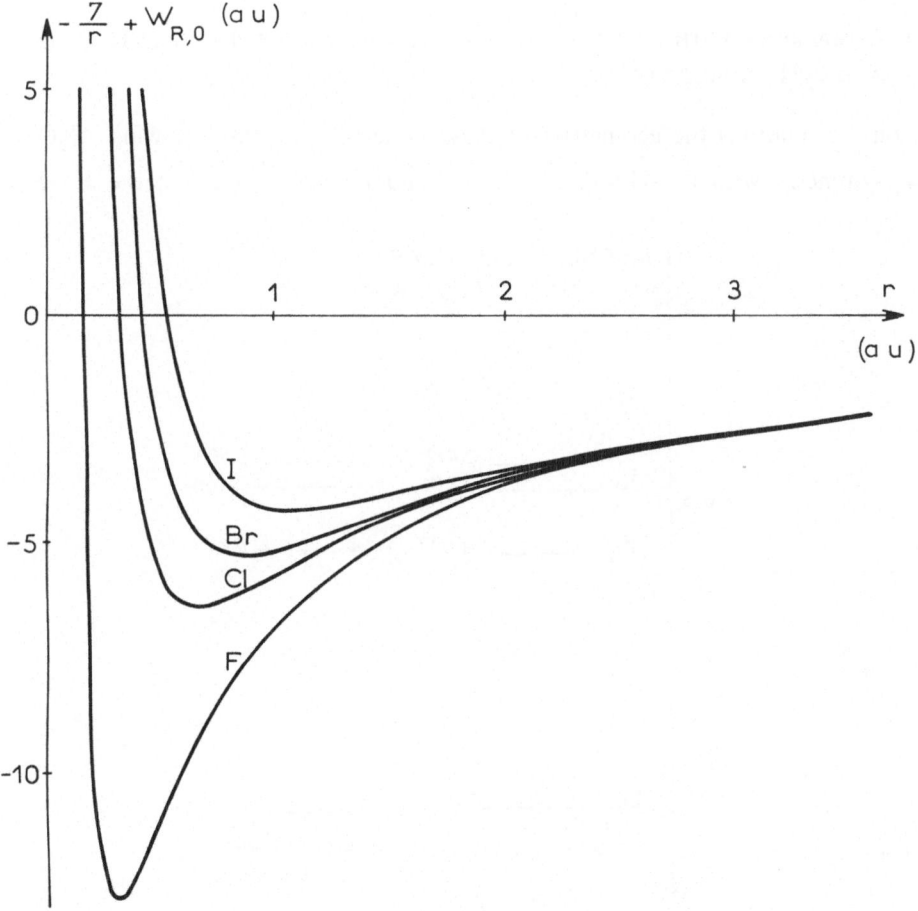

Fig. 11.   Radial s pseudopotential for the halogen atoms.

approximation level). The pseudopotential parameters used in this Section are listed in Table V except for Sc; for the latter, the following pseudopotentials have been used:

$$W_{R,0} = 4.411\,85\,\frac{e^{-0.315\,03\,r^2}}{r^2}\,; \qquad W_{R,1} = 4.406\,03\,\frac{e^{-0.250\,69\,r^2}}{r^2}\,;$$

$$W_{R,2} = -5.719\,87\,\frac{e^{-1.027\,03\,r^2}}{r}\,.$$

The pseudo-orbitals used to determine the pseudopotentials of Table V have been obtained from the valence double-zeta orbitals of Reference [20] except for I and Sc pseudo-orbitals that were obtained from Reference [22].

As an example, Figure 11 gives the aspect of the radial s pseudopotential (cf. Table V) for the halogen atoms; it is interesting to note that the shapes of these pseudopotentials are directly related to the electronegativities of these elements.

4.1. Comparison with Hartree-Fock all-electron calculations: H_2O, HF, F_2 and ScH_3 molecules

In our calculations the geometries of these molecules are the following: H_2O is of C_{2v} symmetry with O—H $= 0.9572$ Å and $\widehat{HOH} = 104.52$ degrees; the H—F and

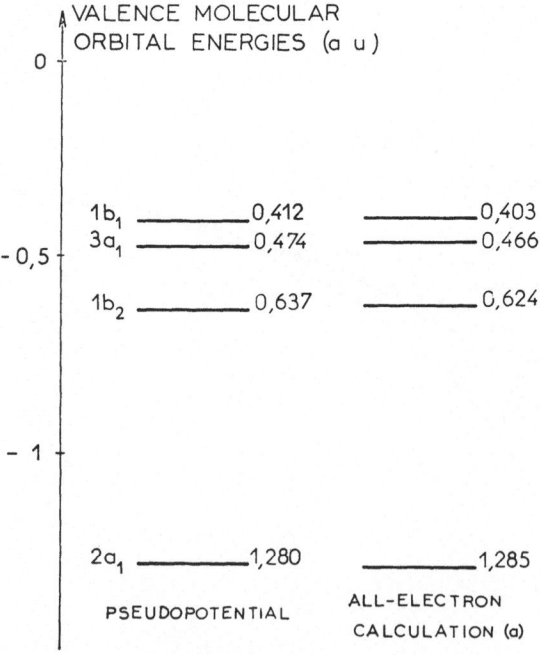

Fig. 12. Valence molecular orbital energies of the water molecule. (a) S. Aung, R. M. Pitzer and S. I. Chan: *J. Chem. Phys.* **49**, 2071 (1968).

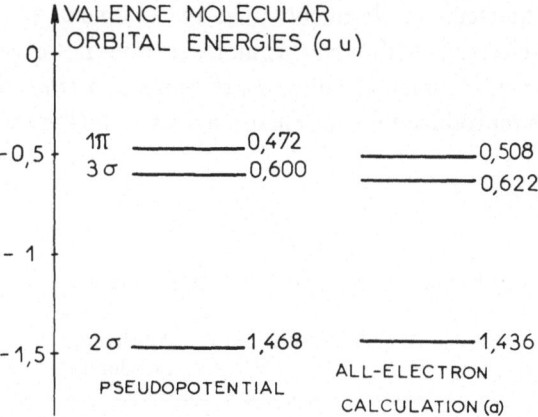

Fig. 13.   Valence molecular orbital energies of the HF molecule. (a) G. Nicolas and Ph. Durand: *J. Chim. Phys.* **71**, 496 (1974).

F—F bond distances are 0.915 Å and 1.418 Å respectively; the ScH_3 molecule is of C_{3v} symmetry i.e. all bond angles are tetrahedral and Sc—H = 1.743 Å [35]. The ζ Slater orbital exponents are the optimized values of Hehre *et al.* [36] for H_2O and HF, the best atom values of Clementi *et al.* [37] for F_2, and the same values as those used by Stevenson *et al.* [35] for ScH_3.

It can be seen in Figures 12 and 13 that for H_2O and HF molecules, very similar energies are obtained by the pseudopotential method and an ab-initio all-electron calculation which uses the same or very near minimal Slater orbital basis set.

Similar results are obtained for the molecular energy levels of the F_2 molecule (Figure 14).

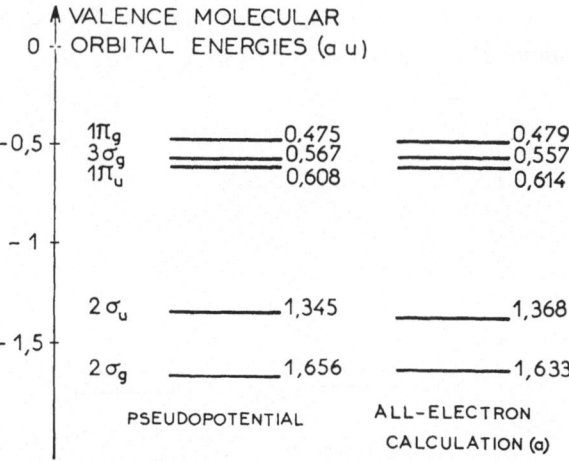

Fig. 14   Valence molecular orbital energies of the F_2 molecule. (a) B. J. Ransil, *Rev. Mod. Phys.* **32**, 239 (1960).

Finally, the ScH_3 molecule has been chosen as an example of a molecule containing an atom with one d electron. Although this molecule does not exist, we can compare our results with the very detailed calculations of Stevenson *et al.* on this same molecule [35]. Our results reproduce these calculations very well (Figure 15 and Table VI).

TABLE VI

Valence orbital populations (in Å) in the ScH_3 molecule.

Orbital		Pseudopotential	All-electron calculation[a]
Sc	$4s$	0.92	0.84
Sc	$4p_x$	0.38	0.39
Sc	$4p_y$	0.38	0.39
Sc	$4p_z$	0.02	0.03
Sc	$3d_{xy}$	0.21	0.18
Sc	$3d_{xz}$	0.10	0.09
Sc	$3d_{yz}$	0.10	0.09
Sc	$3d_{x2-y2}$	0.21	0.18
Sc	$3d_{z2}$	0.04	0.03
H_1	$1s$	1.22	1.27
H_2	$1s$	1.22	1.27
H_3	$1s$	1.22	1.27

[a] See Reference [35].

The calculations just presented, as well as previously published calculations on methane, ethylene, acetylene, $XH_4(X=C, Si, Ge, Sn)$ [38] and other pseudopotential studies by various authors [27, 29, 31, 39] show that valence-only ab-initio pseudopotential methods give valence molecular wave functions of the same quality as those obtained from ab-initio all-electron calculations.

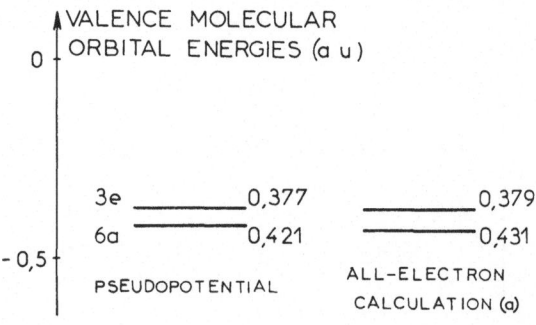

Fig. 15. Valence molecular orbital energies of the ScH_3 molecule. (a) See Reference [35].

4.2. INTERATOMIC DISTANCES AND FORCE CONSTANTS: SERIES HX (X = F, Cl, Br, I)

The calculations have been carried out with a Slater minimal basis set (see Table VII). The variation of valence molecular energy levels from HF to HI is given in Figure 16. The geometry of the molecule is the experimental one [40].

TABLE VII

Slater atomic orbital exponents

Atom	Orbital	Exponents
F	$2s$	2.55[a]
	$2p$	2.55
Cl	$3s$	2.3561[b]
	$3p$	2.0387
Br	$4s$	2.6382[b]
	$4p$	2.2570
I	$5s$	2.6807[b]
	$5p$	2.3223
H bonded to F	$1s$	1.32[a]
H bonded to Cl	$1s$	1.20
H bonded to Br	$1s$	1.18
H bonded to I	$1s$	1.13

[a] Reference [36]
[b] Reference [37].

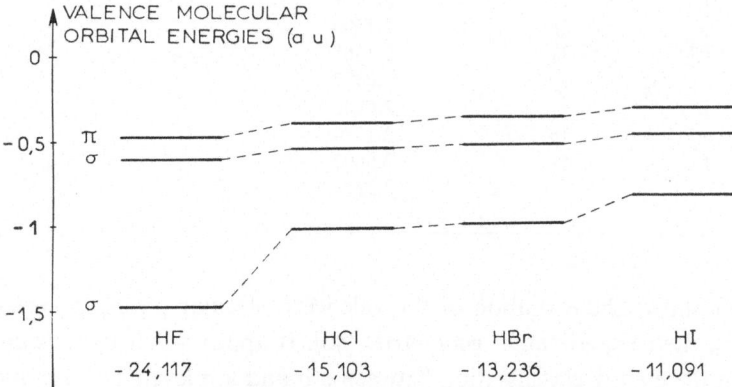

Fig. 16. Valence molecular orbital energies in the series HF, HCl, HBr, HI. The total valence energies (in Å) are reported at the foot of the figure.

The calculated interatomic distances and force constants are given in Table VIII. The largest error for the calculated interatomic distances is 9.4% for HBr. The force constants are obtained with an accuracy better than 21.3% which is the largest value for HI. All these results are very acceptable taking into account the use of a minimal Slater basis set and the limitation inherent to the one-electron model.

TABLE VIII

Interatomic distances and force constants for the molecules HF, HCl, HBr and HI.

| Molecule | Re(Å) | | k(mdyn/Å) | |
	Pseudo-potential	Experimental values[a]	Pseudo-potential	Experimental values[b]
HF	0.94	0.917	11.11	9.66
HCl	1.39	1.274	5.95	5.16
HBr	1.54	1.408	4.48	4.12
HI	1.75	1.609	3.81	3.14

[a] Reference [40].
[b] Reference [41].

4.3. INTERATOMIC DISTANCES AND FORCE CONSTANTS: SERIES F_2, Cl_2, Br_2, I_2

The Slater orbital exponents used in these calculations are listed in Table IX.

TABLE IX

Slater atomic orbital exponents. These values are from Reference [37]

Atom	Orbital	Exponents
F	2s	2.5638
	2p	2.5500
Cl	3s	2.3561
	3p	2.0387
Br	4s	2.6383
	4p	2.2570
I	5s	2.6807
	5p	2.3223

Figure 17 shows the evolution of the valence molecular orbital energies for these molecules in their experimental geometries [40]. It appears that there is an inversion from Cl_2 in the energy classification between a σ and a π level. This symmetry classification can be related to photoelectron spectroscopy experiments whereas a CNDO calculation leads to a wrong classification for Cl_2 [42].

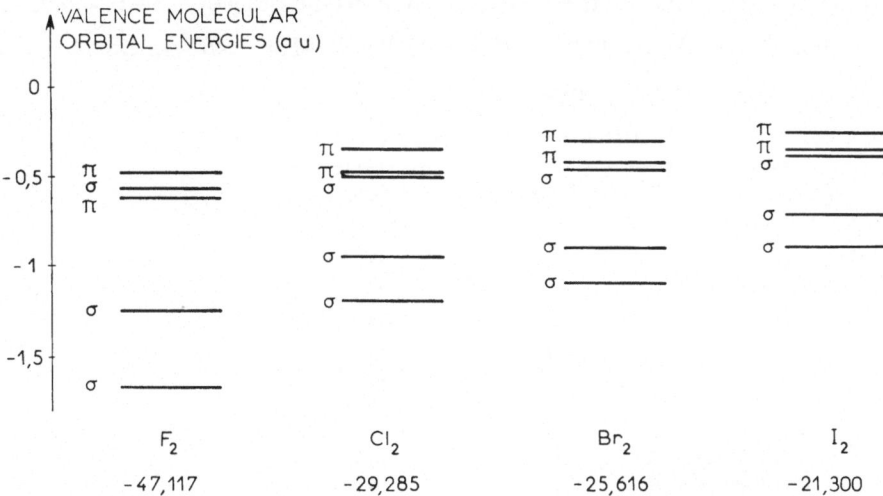

Fig. 17.   Valence molecular orbital energies in the series F_2, Cl_2, Br_2, I_2. The total valence energies (in Å) are reported at the foot of the figure.

The calculated interatomic distances and force constants are compared with the experimental data in Table X.

TABLE X

Interatomic distances and force constants in the halogen molecules.

Molecule	Re(Å)		k(mdyn/Å)	
	Pseudo-potential	Experimental values[a]	Pseudo-potential	Experimental values[b]
F_2	1.28	1.418	12.7	4.5
Cl_2	2.05	1.988	8.1	3.3
Br_2	2.28	2.284	6.3	2.5
I_2	2.68	2.666	4.2	1.7

[a] Reference [40].
[b] Reference [41].

Table X shows that the results are rather good for the interatomic distances but we can only obtain the right order of magnitude for the force constants. However it is well known that a one-electron Hartree-Fock calculation is quite rough for these molecules, particularly for F_2 [43]. It is however gratifying to observe that we have obtained approximatively the same discrepancy as that of Ransil's calculation for F_2.

4.4. BOND ANGLES AND BENDING FORCE CONSTANTS: SERIES H_2O, H_2S, H_2Se

The Slater orbital exponents used in these calculations are given in Table XI. Figure 18 shows the evolution of the valence molecular orbital energies from H_2O to H_2Se;

the molecules are here in their experimental geometries [40] (O—H = 0.9572 Å, $\widehat{HOH}$ = 104.52°; S—H = 1.328 Å, $\widehat{HSH}$ = 92.2°; Se—H = 1.460 Å, $\widehat{HSeH}$ = 91.0°).

TABLE XI

Slater atomic orbital exponents

Atom	Orbital	Exponents
O	2s	2.23 [a]
	2p	2.23
S	3s	1.94 [a]
	3p	1.94
Se	4s	2.4394 [b]
	4p	2.0718
H bonded to O	1s	1.26 [a]
H bonded to S	1s	1.15 [a]
H bonded to Se	1s	1.14

[a] Reference [36[.
[b] Reference [37].

This series has been chosen in order to check the possibilities of our pseudopotential method to reproduce bond angles. This rather low energetic observable is a good test for this type of method; it is well known for example that the X_α method met con-

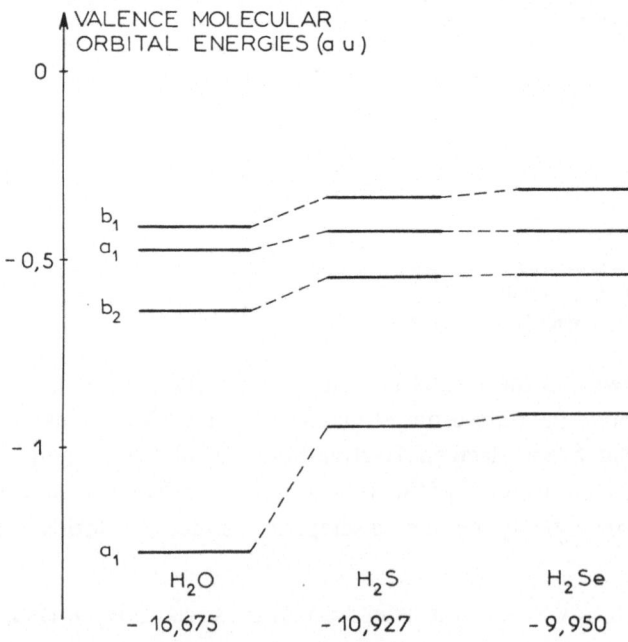

Fig. 18.   Valence molecular orbital energies in the series H_2O, H_2S, H_2Se. The total valence energies (in Å) are reported at the foot of the figure.

siderable difficulties with the water molecule that was predicted to be linear [44]. Table XII gathers the calculated and experimental values of the bond angle $\widehat{HXH}$ (X = O, S, Se) and the bending force constants. The error in the calculated values for these angles is lower than 4%.

TABLE XII

Bond angles and bending force constants in H_2O, H_2S and H_2Se molecules

Molecule H_2X	$\widehat{HXH}$ (degrees)		Bending force constant (mdyn/Å)	
	Pseudo-potential	Experimental values[a]	Pseudo-potential	Experimental values[b]
H_2O	101.0	104.52	1.33	0.69
H_2S	94.3	92.2	0.66	0.45
H_2Se	94.3	91.0	0.55	0.31

[a] Reference [40]
[b] Reference [41].

4.5. DONOR-ACCEPTOR BOND LENGTHS AND ROTATION BARRIER HEIGHTS OF THE MOLECULES BH_3NH_3 AND BH_3PH_3

These molecules have been chosen for their typical donor-acceptor N→B and P→B bonds. The length of this bond and the rotation barrier height in these two molecules have been calculated in our laboratory [45]. The Slater atomic orbital exponents used in these calculations are given in Table XIII. In all our rotation barrier calculations, bond angles and bond lengths are the same in the staggered and eclipsed molecular conformations (see Table XIV).

Table XV shows that the results agree very well with the experimental values.

TABLE XIII

Slater atomic orbital exponents

Molecule	Atom	Orbital	Exponents
BH_3NH_3[a]	B	2s	1.428
		2p	1.449
	N	2s	2.019
		2p	2.061
	H bonded to B	1s	1.080
	H bonded to N	1s	1.246
BH_3PH_3[b]	B	2s	1.50
		2p	1.50
	P	3s	1.90
		3p	1.90
	H bonded to B	1s	1.10
	H bonded to P	1s	1.10

[a] W. E. Palke: *J. Chem Phys.* **56**, 5308 (1972).
[b] Reference [36].

TABLE XIV

Geometries used in our calculations

Molecule	Bond lengths (Å)		Bond angles (degrees)	
BH₃NH₃[a]	B—N	1.560	HBH	109.47
	B—H	1.190	HNH	109.47
	N—H	1.008		
	B—P	1.937	HBH	114.6
BH₃PH₃[b]	B—H	1.212	HPH	101.3
	P—H	1.399		

[a] Reference [40].
[b] J. R. Durig, Y. S. Li, L. A. Carreira, and J. D. Odom: *J. Am. Chem. Soc.* **95**, 2491 (1973).

TABLE XV

The height of the rotation barriers and the length of the BN and BP bonds in the molecules BH_3NH_3 and BH_3PH_3

Molecule	Observable	Pseudo-potential	Hartree-Fock all-electron calculation	Experimental values
BH₃NH₃	bond length B—N (Å)	1.54	1.59[a]	1.56[c]
	rotation barrier (kcal/mole)	2.98	2.93[a]	–
BH₃PH₃	bond length B—P (Å)	1.92	2.08[b]	1.94[d]
	rotation barrier (kcal/mole)	2.00	1.72[b]	2.47[d]

[a] W. E. Palke: *J. Chem. Phys.* **56**, 5308 (1972).
[b] J. R. Sabin: *Chem. Phys. Letters* **20**, 212 (1973).
[c] Reference [40].
[d] J. R. Durig, Y. S. Li, L. A. Carreira and J. D. Odom; *J. Amer. Chem. Soc.* **95**, 2491 (1973).

4.6. ROTATION BARRIERS: SERIES XH_3YH_3 (X, Y = C, Si, Ge)

We have studied in this series:
(i) the evolution of the symmetry of the highest occupied orbital
(ii) the rotation barrier heights.

As all these investigations are going to be published elsewhere [46] we only report a comparison between the calculations and the experimental values – when the latter are known – of the rotation barrier heights. The calculated values agree well with the experimental ones and with previous all-electron calculations on ethane and methylsilane (Table XVI). This results are very satisfying inasmuch as these barrier heights are rather small.

TABLE XVI

Internal rotation barriers in kcal/mole for the molecules XH_3YH_3 (X, Y = C, Si, Ge)

Molecule	Pseudopotential	All-electron calculation	Experimental value
CH_3CH_3	3.19	3.3[a]	2.928[c]
CH_3SiH_3	1.40	1.44[b]	1.665[d]
CH_3GeH_3	1.22	–	1.239[e]
SiH_3SiH_3	1.05	–	~1.1[f]
SiH_3GeH_3	1.02	–	~1.1[f]
GeH_3GeH_3	1.09	–	1.49[g]

[a] R. M. Pitzer and W. N. Lipscomb: *J. Chem. Phys.* **39**, 1995 (1963).
[b] C. S. Ewig, W. E. Palke, and B. Kirtman: *J. Chem. Phys.* **60**, 2749 (1974).
[c] S. Weiss and G. E. Leroi: *J. Chem. Phys.* **48**, 962 (1968).
[d] D. R. Herschbach: *J. Chem. Phys.* **31**, 91 (1959).
[e] V. W. Laurie: *J. Chem. Phys.* **30**, 1210 (1959).
[f] A. P. Cox and R. Varma: *J. Chem. Phys.* **46**, 2007 (1967).
[g] J. E. Griffiths and G. E. Walfaren: *J. Chem. Phys.* **40**, 321 (1964).

5. Conclusion

We have presented in the first part of this paper a general methodology to determine effective potentials for various atomic or molecular fragments. These entirely theoretical effective potentials are obtained by various least-squares techniques from known atomic or molecular wave functions. We have then applied in the second part this formalism to determine atomic core effective potentials also called pseudopotentials. We have shown that when the latter are used in Hartree-Fock valence-only electron calculations with minimal Slater basis sets, it was possible, for various series of molecules containing atoms in the same column of the periodic table, to obtain a good agreement between the calculated values and the experimental data for observables such as interatomic distances, bond angles, vibrational force constants, internal rotation barriers, etc...

We think that this general methodology of determining effective potentials or hamiltonians at the various levels of the atomic or molecular construction must be thoroughly investigated. Indeed, it is a pity that more often than not in Quantum Chemistry the calculations are limited to some chemical properties whereas the corresponding wave functions remain unpublished and are consequently lost. A better attitude would be to extract systematically from the best calculated wave function high quality information such as effective potentials, coupling parameters, etc... that could be further re-employed. For computational convenience, these effective potentials should be put into simple forms, local or nonlocal representation as an example for atomic pseudopotentials. Thus it can be conjectured that too intricate models as for example many-body potentials should be gradually eliminated by 'natural selection'. Obviously, simple one-electron effective potentials that are constant and transferable from one molecule to another are approximations and therefore the wave functions obtained are also only approximate. What is needed in Quantum Chemistry is a well-defined hierarchy of effective approximate hamiltonians converging towards the exact ones, that could be easily and cheaply solved step by step and generally by perturbation techniques that keep the right level of description required to obtain useful results. All these models must be grounded on simple local analysis such as Daudel's loge concept which is closely related to our intuitive chemical knowledge.

Appendix: Determination of an Effective Hamiltonian for a System Containing N_v Valence Electrons

The monoelectronic functions χ_p of a complete basis set and the functions χ^p of the corresponding biorthogonal basis are connected by:

$$\chi^p = \sum_q \chi_q (S^{-1})_{qp} \tag{1}$$

with

$$\langle \chi_p | \chi_q \rangle = S_{pq} \quad \text{and} \quad \langle \chi^p | \chi^q \rangle = (S^{-1})_{pq}$$

$$\langle \chi_p | \chi^q \rangle = \langle \chi^p | \chi_q \rangle = \delta_{pq} \quad (\delta_{pq} = \text{Kronecker symbol}). \tag{2}$$

The closure relation may be written:

$$\sum_p |\chi_p\rangle \langle \chi^p| = \sum_p |\chi^p\rangle \langle \chi_p| = 1. \tag{3}$$

With these notations, the electronic hamiltonian of an atom is defined in second quantization by:

$$\mathscr{H} = \sum_{pq} \langle \chi^p | h | \chi^q \rangle \, C_p^+ C_q + \tfrac{1}{2} \sum_{pqrs} \left\langle \chi^p \chi^q \left| \frac{1}{r_{12}} \right| \chi^r \chi^s \right\rangle C_p^+ C_q^+ C_s C_r \tag{4}$$

where

$$\langle \chi^p | h | \chi^q \rangle = \int \chi^{*p}(1) \, h(1) \, \chi^q(1) \, d\tau_1 \quad \text{and} \quad h = -\tfrac{1}{2}\Delta - \frac{Z}{r} \tag{5}$$

$$\left\langle \chi^p \chi^q \left| \frac{1}{r_{12}} \right| \chi^r \chi^s \right\rangle = \int \chi^{*p}(1) \, \chi^{*q}(2) \, \frac{1}{r_{12}} \, \chi^r(1) \, \chi^s(2) \, d\tau_1 \, d\tau_2 \tag{6}$$

where C_p^+ and C_p are the creation and annihilation operators for the monoelectronic function χ_p. These χ_p functions form a complete and generally non-orthogonal basis set:

$$\langle \chi_p | \chi_q \rangle = S_{pq}. \tag{7}$$

The set of the χ_p functions is divided into two parts: the set of the core orbitals ϕ_α and the complementary set of the valence functions χ_i. From now on, the core orbitals will always be indexed by the Greek letters α, β, γ, etc... and the valence functions by the Latin letters i, j, k, etc... The state in which the N_c core levels of the atom are occupied is the reference state. In order that this reference state should be the vacuum state, $|0>$, we have to carry out a 'hole-particle type' transformation:

$$C_p^+ = a_p \quad \text{for} \quad p \in \alpha, \beta, \dots$$
$$C_p^+ = a_p^+ \quad \text{for} \quad p \in i, j, \dots. \tag{8}$$

After transformation (8) has been made, the hamiltonian $\mathcal{H}$ acts on N_v particle states denoted:

$$|i_1 i_2 \dots i_{N_v}\rangle = a_{N_v}^+ \dots a_{i_2}^+ a_{i_1}^+ |0\rangle. \tag{9}$$

By carrying out transformation (8) and by keeping in $\mathcal{H}$ only the terms which give a non-zero contribution when the matrix elements are taken between states defined by Equation (9), the effective hamiltonian is given by:

$$\mathcal{H}_{\text{eff}} = \text{constant term} + \sum_{i,j} \left\{ \langle \chi^i | h | \chi^j \rangle + \sum_{\alpha\beta} \left\langle \phi^\alpha \chi^i \left| \frac{1}{r_{12}} \right| \phi^\beta \chi^j \right\rangle_a S_{\beta\alpha} + \right.$$
$$+ \sum_{\alpha k} \left[\left\langle \phi^\alpha \chi^i \left| \frac{1}{r_{12}} \right| \chi^k \chi^j \right\rangle_a S_{k\alpha} + \left\langle \chi^k \chi^i \left| \frac{1}{r_{12}} \right| \phi^\alpha \chi^j \right\rangle_a S_{\alpha k} \right] \left. \right\} a_i^+ a_j +$$
$$+ \frac{1}{4} \sum_{ijkl} \left\langle \chi^i \chi^j \left| \frac{1}{r_{12}} \right| \chi^k \chi^l \right\rangle_a a_i^+ a_j^+ a_l a_k \tag{10}$$

with

$$\text{constant term} = \sum_{\alpha\beta} \langle \phi^\alpha | h | \phi^\beta \rangle S_{\beta\alpha} + \sum_{\alpha i} \left[\langle \phi^\alpha | h | \chi^i \rangle S_{i\alpha} + c.c. \right] +$$
$$+ \frac{1}{2} \sum_{\alpha\beta\gamma\delta} \left\langle \phi^\alpha \phi^\beta \left| \frac{1}{r_{12}} \right| \phi^\gamma \phi^\delta \right\rangle_a S_{\gamma\alpha} S_{\delta\beta} +$$
$$+ \sum_{\alpha\beta\gamma i} \left\langle \phi^\alpha \phi^\beta \left| \frac{1}{r_{12}} \right| \phi^\gamma \chi^i \right\rangle_a S_{\gamma\alpha} S_{i\beta} + c.c. +$$
$$+ \frac{1}{2} \sum_{\alpha\beta ij} \left\langle \phi^\alpha \phi^\beta \left| \frac{1}{r_{12}} \right| \chi^i \chi^j \right\rangle_a S_{i\alpha} S_{j\beta} + c.c. +$$
$$+ \sum_{\alpha\beta ij} \left\langle \phi^\alpha \chi^i \left| \frac{1}{r_{12}} \right| \chi^j \phi^\beta \right\rangle_a S_{j\alpha} S_{\beta i}. \tag{11}$$

In Equations (10) and (11), an antisymmetrized notation has been used for the bielectronic matrix elements:

$$\left\langle \chi^p \chi^q \left| \frac{1}{r_{12}} \right| \chi^r \chi^s \right\rangle_a = \left\langle \chi^p \chi^q \left| \frac{1}{r_{12}} \right| \chi^r \chi^s \right\rangle - \left\langle \chi^p \chi^q \left| \frac{1}{r_{12}} \right| \chi^s \chi^r \right\rangle. \tag{12}$$

If we consider an orthonormal basis set, the expression of $\mathscr{H}_{\text{eff}}$ is reduced to:

$$\mathscr{H}_{\text{eff}} = \sum_{\alpha} \langle \phi_{\alpha}|h|\phi_{\alpha}\rangle + \tfrac{1}{2}\sum_{\alpha\beta}\left[\left\langle \phi_{\alpha}\phi_{\beta}\left|\frac{1}{r_{12}}\right|\phi_{\alpha}\phi_{\beta}\right\rangle - \left\langle \phi_{\alpha}\phi_{\beta}\left|\frac{1}{r_{12}}\right|\phi_{\beta}\phi_{\alpha}\right\rangle\right]+$$

$$+ \sum_{ij}\left\{\langle \chi_i|h|\chi_j\rangle + \sum_{\alpha}\left[\left\langle \phi_{\alpha}\chi_i\left|\frac{1}{r_{12}}\right|\phi_{\alpha}\chi_j\right\rangle - \left\langle \phi_{\alpha}\chi_i\left|\frac{1}{r_{12}}\right|\chi_j\phi_{\alpha}\right\rangle\right]\right\} \times$$

$$\times a_i^+ a_j + \tfrac{1}{2}\sum_{ijkl}\left\langle \chi_i\chi_j\left|\frac{1}{r_{12}}\right|\chi_k\chi_l\right\rangle a_i^+ a_j^+ a_l a_k. \tag{13}$$

$$J_{\alpha} = \sum_{ij}\left\langle \phi_{\alpha}\chi_i\left|\frac{1}{r_{12}}\right|\phi_{\alpha}\chi_j\right\rangle a_i^+ a_j \quad \text{and} \quad K_{\alpha} = \sum_{ij}\left\langle \phi_{\alpha}\chi_i\left|\frac{1}{r_{12}}\right|\chi_j\phi_{\alpha}\right\rangle a_i^+ a_j$$

are the coulomb and exchange operators associated with the core orbitals. The total potential energy operator of the core orbitals has the expression:

$$V_{\phi_c} = \sum_{\alpha=1}^{N_c} (J_{\alpha} - K_{\alpha}). \tag{14}$$

In the conventional Schrödinger representation, the effective hamiltonian (13) may also be written:

$$\mathscr{H}_{\text{eff}} = \text{constant term} + \sum_{i=1}^{N_v} (1-P_i)\left[h(i) + V_{\phi_c}(i)\right](1-P_i) +$$

$$+ \sum_{i<j}^{N_v} (1-P_i)(1-P_j)\frac{1}{r_{ij}}(1-P_i)(1-P_j). \tag{15}$$

The constant term in Equation (15) is the same as in Equation (13). P_i is the projector on the N_c core states:

$$P_i = \sum_{\alpha=1}^{N_c} |\phi_{\alpha}(i)\rangle \langle \phi_{\alpha}(i)|. \tag{16}$$

References

1. Among many references, see for example:
   Levy, B.: *Chem. Phys. Letters* **4**, 17 (1969); *ibid.* **18**, 59 (1973).
   Barthelat, J. C. and Durand, Ph.: *Theoret. Chim. Acta (Berl.)* **27**, 109 (1972).
   Daudey, J. P.: *Chem. Phys. Letters* **24**, 574 (1974).
2. See for instance:
   Pipano, A. and Shavitt, I.: *Int. J. Quantum Chem.* **2**, 741 (1968).
   Whitten, J. L. and Hackenmayer, M.: *J. Chem. Phys.* **51**, 5584 (1969).
   Roos, B.: *Chem. Phys. Letters* **15**, 153 (1972).
   Huron, B., Malrieu, J. P., and Rancurel, P.: *J. Chem. Phys.* **58**, 5745 (1973).
   Downing, J. W., Michl, J., Jörgensen, P., and Thulstrup, E. W.: *Theoret. Chim. Acta (Berl.)* **32**, 203 (1974).
   Buenker, R. J. and Peyerimhoff, S. D.: *Theoret. Chim. Acta (Berl.)* **35**, 33 (1974).

3. Berthier, G.: in P. O. Löwdin and B. Pullman (eds.), *Molecular Orbitals in Chemistry, Physics, and Biology*, Academic Press, New York, 1964, p. 57.
   Caballol, R., Gallifa, R., Riera, J. M., and Carbó, R.: *Int. J. Quant. Chem.* **8**, 373 (1974).
4. Gilbert, T. L.: in P. O. Löwdin and B. Pullman (eds.), *Molecular Orbitals in Chemistry, Physics, and Biology*, Academic Press, New York, 1964, p. 405.
5. Millie, Ph., Lévy, B., and Berthier, G.: in O. Chalvet, R. Daudel, S. Diner and J. P. Malrieu (eds.), *Localization and Delocalization in Quantum Chemistry*, Vol. I, D. Reidel, Dordrecht-Holland, 1975, p. 59.
6. McWeeny, R. and Sutcliffe, B. T.: in D. P. Craig and R. McWeeny (eds.), *Methods of Molecular Quantum Mechanics*, Academic Press, New York, 1969, p. 180.
7. Weeks, J. D., Hazi, A., and Rice, S. A.: *Adv. Chem. Phys.* **16**, 283 (1969).
   Bardsley, J. N.: *Case Studies in Atomic Physics* **4**, 299 (1974).
8. Freed, K. F.: *Chem. Phys. Letters* **15**, 331 (1972).
   Freed, K. F.: *ibid.* **29**, 143 (1974).
9. Bader, R. F. W.: see Reference 5, p. 15.
10. Claverie, P., Malrieu, J. P., and Diner, S.: see Reference 5, p. 53.
11. Allinger, N. L., Cava, M. P., De Jongh, D. C., Johnson, C. R., Lebel, N. A., and Stevens, C. L.: in *Chimie Organique*, Ediscience/McGraw-Hill, Paris, 1975, p. 277.
12. Vermeulin, P., Lévy, B., and Berthier, G. : to be published.
13. Christoffersen, R. E.: *Adv. Quantum Chem.* **6**, 333 (1972).
14. Constanciel, R.: see Reference 5, p. 43.
15. Denis, A., Langlet, J., and Malrieu, J. P.: *Theoret. Chim. Acta (Berl.)* **29**, 117 (1973).
16. Westhaus, P.: *Int. J. Quant. Chem. Symp.* **7**, 463 (1973).
    Westhaus, P., Bradford, E. G., and Hall, D.: *J. Chem. Phys.* **62**, 1607 (1975).
17. Öhrn, Y. and McWeeny, R.: *Ark. för Fysik* **31**, 461 (1966).
18. Durand, Ph. and Barthelat, J. C.: *Chem. Phys. Letters* **27**, 191 (1974).
    Durand, Ph. and Barthelat, J. C.: *Theoret. Chim. Acta (Berl.)* **38**, 283 (1975).
19. Roothaan, C. C. J.: *Rev. Mod. Phys.* **32**, 179 (1960).
20. Clementi, E.: 'Tables of Atomic Functions', *IBM J. Res. Develop. Suppl.* **9**, 2 (1965).
    Clementi, E., Matcha, R., and Veillard, A.: *IBM Res. Note N J-117* (1967).
21. Roos, B., Salez, C., Veillard, A., and Clementi, E.: 'A General Program for Calculation of Atomic SCF Orbitals by the Expansion Method', *IBM Res. Note RJ-518* (1968).
22. Roetti, C. and Clementi, E.: *J. Chem. Phys.* **60**, 4725 (1974).
23. Daudel, R., Bader, R. F. W., Stephens, M. E., and Borrett, D. S.: *Can. J. Chem.* **52**, 1310 (1974).
24. Cusachs, L. C.: *Chem. Phys. Letters* **31**, 154 (1975).
25. Melius, C. F. and Goddard III, W. A.: *Phys. Rev.* **A10**, 1528 (1974).
26. Raffenetti, R. C. and Ruedenberg, K.: *J. Chem. Phys.* **59**, 5950 (1973).
27. Dixon, R. N. and Hugo, J. M. V.: *Mol. Phys.* **29**, 953 (1975).
28. Bloch, A. N. and Simons, G.: *J. Am. Chem. Soc.* **94**, 8611 (1972).
    St John, J. and Bloch, A. N.: *Phys. Rev. Letters* **33**, 1095 (1974).
29. Kleiner, M. and McWeeny, R.: *Chem. Phys. Letters* **19**, 476 (1973).
30. Bonifacic, V. and Huzinaga, S.: *J. Chem. Phys.* **60**, 2779 (1974); *ibid.* **62**, 1507 (1975); *ibid.* **62**, 1509 (1975).
31. Melius, C. F., Olafson, B. D., and Goddard III, W. A.: *Chem. Phys. Letters* **28**, 457 (1974).
32. McWeeny, R.: *Acta Cryst.* **6**, 631 (1953).
33. Preuss, H.: *Z. Naturforschg.* **11a**, 823 (1956).
    Whitten, J. L.: *J. Chem. Phys.* **44**, 359 (1966).
    Petke, J. D. and Whitten, J. L.: *J. Chem. Phys.* **56**, 830 (1972).
34. Chang, T. C., Habitz, P., Pittel, B., and Schwarz, W. H. E.: *Theoret. Chim. Acta (Berl.)* **34**, 263 (1974).
35. Stevenson, P. E. and Lipscomb, W. N.: *J. Chem. Phys.* **50**, 3306 (1969).
36. Hehre, W. J., Stewart, R. F., and Pople, J. A.: *J. Chem. Phys.* **51**, 2657 (1969).
    Hehre, W. J., Ditchfield, R., Stewart, R. F., and Pople, J. A.: *J. Chem. Phys.* **52**, 2769 (1970).
37. Clementi, E. and Raimondi, D. L.: *J. Chem. Phys.* **38**, 2686 (1963).
    Clementi, E., Raimondi, D. L., and Reinhardt, W. P.: *J. Chem. Phys.* **47**, 1300 (1967).
38. Barthelat, J. C. and Durand, Ph.: *Chem. Phys. Letters* **16**, 63 (1972).
    Barthelat, J. C. and Durand, Ph.: *J. Chim. Phys.* **71**, 505 (1974); *ibid.* **71**, 1105 (1974).

39. Switalski, J. D. and Schwartz, M. E.: *J. Chem. Phys.* **62**, 1521 (1975).
40. See *Interatomic Distances*, L. E. Sutton (ed.), the Chemical Society, London, Spec. Publ. **11–18**, 1958 and 1965.
41. Herzberg, G.: in *Molecular Spectra and Molecular Structure*, Vol. I, D. Van Nostrand, New York, 1950, p. 501.
42. Deb, B. M. and Coulson, C. A.: *J. Chem. Soc. A.*, 958 (1971).
43. Fraga, S. and Ransil, B. J.: *J. Chem. Phys.* **35**, 669 (1961).
44. Connolly, J. W. D. and Sabin, J. R.: *J. Chem. Phys.* **56**, 5529 (1972).
    Antoci, S., Mihich, L., and Nardelli, G. F.: *J. Chem. Phys.* **61**, 1245 (1974).
45. Nicolas, G.: unpublished results.
46. Nicolas, G., Barthelat, J. C., and Durand, Ph.: *J. Am. Chem. Soc.*, in press.

DISCUSSION

The results of the pseudopotential method do not change much when the radius of the core loge is changed. There are few differences between the pseudopotential of Si and that of Ge.

It would be interesting to introduce the configuration interaction procedure into the pseudo-potential method. It would be interesting to use such a process to study F_2 and various Rydberg and excited states.

The pseudopotential method is mainly justified when the fluctuation of the number of electrons (or of the property calculated) is small in the loges concerned.

THE CALCULATION OF INTERMOLECULAR
INTERACTION ENERGIES
IN TERMS OF LOCAL CONTRIBUTIONS*

PIERRE CLAVERIE

*Laboratoire de Chimie Quantique, Institut de Biologie Physico-chimique,
13 Rue P. et M. Curie, 75005 Paris, France*

Abstract. In the present paper, we examine the possibility of expressing in terms of local contributions four important parts of the interaction energy between two molecules, namely the three well-known long-range terms (electrostatic polarization, dispersion) and the short-range 1st order exchange term. Such approximate decompositions are found for all these four terms. For the electrostatic energy, this is done for any molecular function, but for the three other terms a very important role is played in the derivation by the possibility of representing the ground-state of the molecules by Slater determinants built from *localized* molecular orbitals. For the 2nd order terms (polarization and dispersion), we first show, by using also the 'excitonic' representation of excited states, that the decomposition of the total molecular polarizability into local (bond) contributions is theoretically approximately justified, and we may derive on this basis the decomposition into local contributions of the polarization and dispersion energies. The difference between the type of decomposition for these two energies is emphasized. Finally two problems not treated in the paper are mentioned: the problem of analysing the charge-transfer term, and the problem posed by the insufficient localizability of the π-orbitals of aromatic molecules.

1. Introduction

There exists in the domain of intermolecular interactions, a situation similar to the situation in the domain of the isolated molecule concerning the concept of the chemical bond. Namely, the concept of the molecule as a set of chemical bonds (and functional groups) is very strongly suggested by experimental practice, and extremely useful for the understanding (and foreseeing) of chemical facts; nevertheless, retrieving the concept of chemical bonds in the theoretical framework provided by present-day quantum mechanics is not a trivial task, as is perfectly exemplified by the contributions brought together in the present two volumes about 'Localization and Delocalization in Quantum Chemistry' (see in particular Daudey, Rojas and Malrieu in volume I [1]).

The analogous problem, in the domain of intermolecular interactions, is, of course, the decomposition of the interaction energy (or of other physical changes due to the interaction) in terms of local contributions. Such a decomposition, suggested by the possibility of retrieving elements like atoms and bonds in a molecule, was actually used in various domains of application, e.g. crystallography [2, 3], and even for the evaluation of *intra*-molecular interactions in conformational problems concerning large molecules [4] (a domain where the theoretical justification of such decomposition is certainly more difficult than in the case of the interaction between separate

---

* The content of this paper is also closely connected with the subject of Part III in Volume I (Expression of the Energy in Terms of Local Contributions').

O. Chalvet et al. (eds.), Localization and Delocalization in Quantum Chemistry, Vol. II, 127–152. All Rights Reserved

molecules: see [1]). Nevertheless, a direct calculation of the interaction energy according to the so-called 'supermolecule approach' (i.e. by simply taking the difference
between the energy of the complex (the supermolecule) and the sum of the energies
of the separate molecules, all these energies being calculated by one of the standard
techniques of quantum chemistry) displays only a single number, and the problem
remains of decomposing this number into various contributions in a physically
interesting and meaningful way: clearly, this problem is conceptually analogous to the
problem of analysing the quantum-mechanical computation of a molecule in terms
of bond contributions.

It is the purpose of the present paper to present such an analysis, based upon a
perturbation treatment of the interaction. Such a treatment allows a decomposition
of the energy into different terms (electrostatic, polarization, dispersion, first-order
'exchange' repulsion, charge transfer...), every one of them having a peculiar law of
variation as a function of the intermolecular distance. Then, the problem is to express
these terms as combinations of *local* contributions (atom-atom, atom-bond, bond-
bond contributions). As far as possible, these local contributions should be expressed
by using not only local, but transferable properties (e.g. the bond polarizabilities);
of course, it is likely that local, but non transferable, properties (e.g. atomic net
charges) also occur.

Of course, besides its conceptual interest, such a decomposition offers a sound
theoretical basis to the so-called 'empirical' formulae like the sums of atom-atom
terms used by numerous authors [2, 3, 4]. The simplicity (and the resulting short
computation time) of such formulae make them very attractive for all problems which
involve numerous molecules, and consequently minimization problems or statistical-
mechanical calculations: crystallographic problems [2, 3, 5, 6], theory of liquids
(see e.g. the recent developments in the theory of liquids of polyatomic molecules [7]).

In the present paper, we shall consider the three long-range terms (electrostatic,
polarisation, dispersion) and the first-order exchange term; the charge transfer term
(which may be classified as a second-order exchange term) will not be analyzed.

2. Basic Expressions for the Interaction Energy

A. GENERAL FRAMEWORK

Let us denote $H^{(m)}$ the hamiltonian of the molecule $m(m = 1, 2)$, H the total hamiltonian of the complex, $H_0 = H^{(1)} + H^{(2)}$ and

$$V = H - H_0 = H - (H^{(1)} + H^{(2)}) \tag{1}$$

the interaction hamiltonian, whose explicit expression is

$$V = \sum_{\mu^{(1)}} \sum_{\mu^{(2)}} \frac{Z_{\mu^{(1)}} Z_{\mu^{(2)}}}{r_{\mu^{(1)} \mu^{(2)}}} - \sum_{\mu^{(1)}} \sum_{j^{(2)}} \frac{Z_{\mu^{(1)}}}{r_{\mu^{(1)} j^{(2)}}} -$$

$$- \sum_{j^{(1)}} \sum_{\mu^{(2)}} \frac{Z_{\mu^{(2)}}}{r_{j^{(1)} \mu^{(2)}}} + \sum_{j^{(1)}} \sum_{j^{(2)}} \frac{1}{r_{j^{(1)} j^{(2)}}} \tag{2}$$

where r denotes the distance, μ the nuclei, j the electrons and Z the nuclear charges.

We also introduce the eigenfunctions $W_a^{(m)}$ and corresponding eigenvalues $E_a^{(m)}$ of the hamiltonian $H^{(m)}$. We shall also use the orthonormal set of products

$$\{W_0^{(1)} W_0^{(2)}, \ldots, W_a^{(1)} W_b^{(2)}, \ldots\}$$

denoted for brevity $\{\varphi_0, \ldots, \varphi_k, \ldots\}$ with a single index k replacing the double index (a, b), and the corresponding set of antisymmetrized products $\{\psi_0, \ldots, \psi_k, \ldots\}$, where ψ_k is defined by:

$$\psi_k = \mathfrak{a}\varphi_k = \mathfrak{a}_{(1, 2)}\varphi_k \tag{3}$$

where

$$\mathfrak{a} = [(n_1 + n_2)!]^{-1/2} \mathscr{A} = [(n_1 + n_2)!]^{-1/2} \sum_P (-1)^P P \tag{4}$$

is the usual antisymmetrizer for $(n_1 + n_2)$ particles, and

$$\mathfrak{a}_{(1, 2)} = Q^{-1/2} \mathscr{A}_{(1, 2)} \tag{5a}$$

with

$$\mathscr{A}_{(1, 2)} = 1 + \mathscr{A}'_{(1, 2)} = 1 - \mathscr{P}_{(1)} + \mathscr{P}_{(2)} - \ldots + (-1)^k \mathscr{P}_{(k)} + \ldots \tag{5b}$$

is the inter-molecular antisymmetrizer, which involves only the permutations *between* the two sets $(1, 2, \ldots, n_1)$ and $(n_1 + 1, \ldots, n_1 + n_2)$ of electron labels of the two molecules; $Q = (n_1 + n_2)!/n_1! n_2!$ is the number of these permutations which exchange exactly k labels between the two sets (therefore the maximum value of k is min (n_1, n_2), the smallest of the integers n_1 and n_2). These inter-molecular permutations suffice to transform φ_k into a totally antisymmetric function ψ_k since $\varphi_k = W_a^{(1)} W_b^{(2)}$ is already antisymmetric with respect to the 'intra-molecular permutations' acting inside one of the subsets $(1, \ldots n_1)$ and $(n_1 + 1, \ldots, n_1 + n_2)$ [8, 9]. The normalization factor $Q^{-1/2}$ of the intermolecular antisymmetrizer (5a) insures that $\langle \psi_k | \psi_k \rangle \to 1$ when the intermolecular distance $R \to \infty$.

The well-known treatment [10, 11 chap. 2] of the long-range part of the interaction energy consists in using $H_0 = H^{(1)} + H^{(2)}$ as unperturbed hamiltonian and performing a standard Rayleigh-Schrödinger (RS) perturbation expansion. But this treatment cannot give the correct short-range terms (exponentially decreasing with distance). The exact reason for this failure was made clear only rather recently [12, 13 chap. I]: the RS perturbation expansion connects the eigenfunction $W_0^{(1)} W_0^{(2)}$ of H_0 to an eigenfunction of H which is *not* physically admissible: it is orthogonal to the subspace of all antisymmetric functions of $(n_1 + n_2)$ electrons, and in fact has an eigenvalue lower than the lowest 'physical' eigenvalue which may be obtained in this 'physical' subspace. The difference between this too low 'mathematical' eigenvalue and the 'physical' desired one is exponentially small, and thus negligible, at long distance, but this is no more true at intermediate and short distances, where a modified treatment becomes necessary.

Many such attempts have been proposed since 1965 (the so-called 'exchange perturbation' treatments) (see e.g. the references in [13, chap. II] where some of these proposals are discussed). But almost all these treatments express the interaction energy under the form:

$$\Delta E = \mathscr{E}_1^{(HL)} + \sum_{r=2}^{+\infty} \mathscr{E}_r \tag{6}$$

where $\mathscr{E}_1^{(HL)}$ may be called the Heitler-London interaction energy:

$$\mathscr{E}_1^{(HL)} = \frac{\langle\psi_0|H|\psi_0\rangle}{\langle\psi_0|\psi_0\rangle} - \langle\varphi_0|H_0|\varphi_0\rangle \tag{7}$$

with $\psi_0 = \mathscr{A}\varphi_0$ and $\varphi_0 = W_0^{(1)}W_0^{(2)}$, so that

$$\langle\varphi_0|H_0|\varphi_0\rangle = \langle W_0^{(1)}|H^{(1)}|W_0^{(1)}\rangle + \langle W_0^{(2)}|H^{(2)}|W_0^{(2)}\rangle \tag{8}$$

is the sum of the ground-state energies $E_0^{(1)}$ and $E_0^{(2)}$ of the isolated molecules.

B. The first Order Term

If $W_0^{(m)}$ is an exact eigenfunction of $H^{(m)}$ (so that φ_0 is an exact eigenfunction of H_0), it is easily found [8, 14, 13 chap. II, Section 4B2] that (6) becomes

$$\mathscr{E}_1^{(HL)} = \frac{\langle\varphi_0|V|\psi_0\rangle}{\langle\varphi_0|\psi_0\rangle} = \frac{\langle\varphi_0|V|\mathscr{A}\varphi_0\rangle}{\langle\varphi_0|\mathscr{A}\varphi_0\rangle} = \frac{\langle\varphi_0|V|\mathscr{A}\varphi_0\rangle}{\langle\varphi_0|\mathscr{A}\varphi_0\rangle}. \tag{9}$$

If φ_0 is not an exact eigenfunction of H_0, then the expression (9) of $\mathscr{E}_1^{(HL)}$ is no more exact, but we still have the following property [13, Chapter III, Section 3A2]: if $W_0^{(m)}$ is the exact Hartree-Fock solution for $H^{(m)}$, then the difference between the expressions (7) and (9) is zero for the 1-exchange term, i.e. for $\mathscr{A}_{(1,2)}$ truncated to $1-\mathscr{P}_{(1)}$ (this result is a consequence of Brillouin's theorem).

Since the permutations $\mathscr{P}_{(k)}$ (with k exchanges) contribute by terms of order S^{2k}, where S denotes the order of magnitude of the intermolecular overlap integrals, this truncation of $\mathscr{A}_{(1,2)}$ to $1-\mathscr{P}_{(1)}$ is a good approximation at least in the region of the Van der Waals minimum.

Separating the exchange part $\mathscr{A}'$ in $\mathscr{A} = 1 + \mathscr{A}'$, Equation (9) becomes:

$$\mathscr{E}_1^{(HL)} = \frac{\langle\varphi_0|V|\varphi_0\rangle}{1 + \langle\varphi_0|\mathscr{A}'|\varphi_0\rangle} + \frac{\langle\varphi_0|V|\mathscr{A}'\varphi_0\rangle}{1 + \langle\varphi_0|\mathscr{A}'|\varphi_0\rangle} \tag{10}$$

where we find the first-order perturbation energy of the RS treatment

$$\mathscr{E}_1^{(RS)} = \langle\varphi_0|V|\varphi_0\rangle \tag{11}$$

so that, for large R,

$$\mathscr{E}_1^{(HL)} = \mathscr{E}_1^{(RS)} + O(S^2) \tag{12}$$

where the intermolecular overlap S actually decreases exponentially as a function of R.

In the minimum region, we may still truncate $\mathscr{A}_{(1,2)}$ to $1-\mathscr{P}_{(1)}$, i.e. $\mathscr{A}'_{(1,2)}$ to $-\mathscr{P}_{(1)}$, thus getting the 1-exchange term of order S^2:

$$\mathscr{E}_1^{(HL)} \# \mathscr{E}_1^{(RS)} + E_{1-\text{exch}} \tag{13a}$$

where

$$E_{1-\text{exch}} = -\langle\varphi_0|V\mathscr{P}_{(1)}|\varphi_0\rangle + \langle\varphi_0|V|\varphi_0\rangle\langle\varphi_0|\mathscr{P}_{(1)}|\varphi_0\rangle. \tag{13b}$$

In the case where $W_0^{(1)}$ and $W_0^{(2)}$ are approximated by Slater determinants (expressed in terms of spinorbitals $a_i^{(m)} = b_i^{(m)} \sigma_i^{(m)}$):

$$W_0^{(m)} = (n_m!)^{-1/2} \det |a_1^{(m)} \ldots a_i^{(m)} \ldots a_{n_m}^{(m)}|. \tag{14}$$

Williams *et al.* [15] have given for $E_{1-\text{exch}}$ the following expression (which we shall use in our analysis below):

$$E_{1-\text{exch}} = - \sum_{i=1}^{n_1} \sum_{j=1}^{n_2} \left[\iint \rho_{ij}^{(1)}(\alpha) \frac{1}{r_{\alpha\beta}} \rho_{ij}^{(2)}(\beta) \, d\alpha \, d\beta + \right.$$

$$\left. + S_{ij} \left(\int \rho_{ij}^{(1)}(\alpha) V^{(2)}(\mathbf{r}_\alpha) \, d\alpha + \int \rho_{ij}^{(2)}(\beta) V^{(1)}(\mathbf{r}_\beta) \, d\beta \right) \right] \tag{15}$$

where $\alpha = (\mathbf{r}_\alpha, \sigma_\alpha)$ and $\beta = (\mathbf{r}_\beta, \sigma_\beta)$ denote the sets of space and spin variables, $r_{\alpha\beta} = |\mathbf{r}_\alpha - \mathbf{r}_\beta|$, i labels the spinorbitals of molecule (1) and j those of molecule (2) as in formula (14), and

$$S_{ij} = \langle a_i^{(1)} | a_j^{(2)} \rangle \tag{16}$$

$$\rho_{ij}^{(1)} = a_i^{(1)} a_j'^{(2)} \quad \text{with} \quad a_j'^{(2)} = a_j^{(2)} - \sum_{k=1}^{n_1} S_{kj} a_k^{(1)} \tag{17a}$$

$$\rho_{ij}^{(2)} = a_i'^{(1)} a_j^{(2)} \quad \text{with} \quad a_i'^{(1)} = a_i^{(1)} - \sum_{l=1}^{n_2} S_{il} a_l^{(2)} \tag{17b}$$

($a_i'^{(1)}$ is the component of $a_i^{(1)}$ orthogonal to the subspace $\{a_i^{(2)}\}$ of the occupied orbitals defining $W_0^{(2)}$ and similarly for $a_j'^{(2)}$).

Finally, $V^{(m)}(\mathbf{r})$ denotes the negative of the electrostatic potential created at the point $\mathbf{r}$ by the unperturbed molecule (p):

$$V^{(m)}(\mathbf{r}) = - \sum_\mu^{(m)} \frac{Z_\mu}{|\mathbf{r} - \mathbf{r}_\mu|} + \sum_{k=1}^{n_m} \int \frac{|a_k^{(m)}(\beta)|^2}{|\mathbf{r} - \mathbf{r}_\beta|} \, d\beta. \tag{18}$$

$V^{(m)}(\mathbf{r})$ is defined as the negative of the electrostatic potential in order to compensate for the absence of the minus sign in the definition of the distributions $\rho_{ij}^{(m)}$. A detailed proof of (15) may be found in [13, Chapter III, Section 3A1).

C. THE HIGHER-ORDER TERMS

These terms are not identical in the various 'exchange perturbation' treatments (although their sum should always be the same, of course, provided these treatments are correct and their series expansion are convergent). But these terms should always appear under the form

$$\mathscr{E}_r = \mathscr{E}_r^{(\text{RS})}[1 + O(V)] + \mathscr{E}_r^{\text{exch}} \tag{19}$$

where $\mathscr{E}_r^{(\text{RS})}$ denotes the rth order perturbation term in the RS treatment, and $\mathscr{E}_r^{\text{exch}}$ is an 'exchange' term (exponentially decreasing at large intermolecular distance R). For example, the treatment proposed in [13, Chapter II, Section 4] gives

$$\mathscr{E}_2 = -\frac{1}{S_{00}} \sum_k{}' \frac{|H_{0k} - E_0 S_{0k}|^2}{H_{kk} - E_0 S_{kk}} \tag{20}$$

where $\sum_k{}'$ means that the value $k = 0$ is excluded from the summation, the matrix elements S_{kl}, H_{kl} are defined as $S_{kl} = \langle \psi_k | \psi_l \rangle$, $H_{kl} = \langle \psi_k | H \psi_l \rangle$ and $E_0 = E_0^{(1)} + E_0^{(2)}$. Since, as $R \to \infty$, $S_{kl} = \delta_{kl} + O(S^2) = 1 + O(e^{-CR})$ and $H_{kl} = E_k \delta_{kl} + V_{kl} + O(S^2)$, it is easily seen that

$$\mathscr{E}_2 = -\sum_k{}' \frac{|V_{0k}|^2}{(E_k - E_0) + (V_{kk} - V_{00})} = \mathscr{E}_2^{(RS)} + O(V), \tag{21}$$

since

$$\mathscr{E}_2^{(RS)} = -\sum_k{}' \frac{|V_{0k}|^2}{E_k - E_0}. \tag{22}$$

In the present paper, we shall deal only with $\mathscr{E}_2^{(RS)} . \mathscr{E}_2^{exch}$ and the higher-order terms $(r > 2)$ will not be analysed. This approximation

$$\Delta E \simeq \mathscr{E}_1^{(HL)} + \mathscr{E}_2^{(RS)} \tag{23}$$

has given good results for the He...He interaction [16]. Its main drawback is certainly the neglect of the charge-transfer term, which is contained in $\mathscr{E}_2^{(xch)}$; this term is non-negligible, for example, in the case of the hydrogen bond [19] and we have taken it into account in recent applications [5, 6] by an empirical reduction of the repulsive first-order exchange term.

As a conclusion for this section concerning basic theoretical concepts, it must be pointed out that an explicit expression like (6) for the interaction energy is easily obtained provided that $W_0^{(1)}$ (resp. $W_0^{(2)}$) is an exact eigenfunction of $H^{(1)}$ (resp. $H^{(2)}$); it is not necessary that the other basis functions $W_a^{(1)}$ (resp. $W_b^{(2)}$) are exact eigenfunctions of $H^{(1)}$ (resp. $H^{(2)}$). But, in practice, even the ground-state eigenfunctions of $H^{(1)}$ and $H^{(2)}$ are not known exactly, and this fact has unpleasant consequences on the expression (6): namely, the terms $\mathscr{E}_r (r \geq 2)$ are no more purely intermolecular when expressed in terms of the total hamiltonian H: they involve contributions corresponding to corrections of the intramolecular energies (which are not given exactly by the zeroth-order non-exact functions $W_0^{(1)}$ and $W_0^{(2)}$). This fact must be taken into account when very accurate ab initio calculation of the interaction energy are attempted in a necessarily finite basis, where $W_0^{(1)}$ and $W_0^{(2)}$ cannot be obtained exactly [13, Chapter II, Section 4D; 17; 18; 19]. Fortunately, this difficulty is irrelevant for our discussion of the decomposition of the energy into local contributions. A satisfactory procedure indeed consists in expressing the terms of (6) explicitly in terms of the interaction hamiltonian V (e.g. formula (9) instead of (7) or (21) instead of (20)): this may be easily done if $W_0^{(1)}$ and $W_0^{(2)}$ are exact eigenfunctions of $H^{(1)}$ and $H^{(2)}$ respectively. The resulting formulae may then be used even if $W_0^{(1)}$ and $W_0^{(2)}$ are no more supposed to be exact eigenfunctions; this implies a slight error, but it is not of intramolecular type and it may therefore be expected to remain negligible from the practical point of view.

3. The Rayleigh-Schrödinger (Long-Range) Terms and the
Matrix Elements of the Interaction Hamiltonian

By introducing explicitly the eigenfunctions $W_a^{(1)}$ and $W_b^{(2)}$ of the molecules, the RS terms (11) and (22) become:

$$\mathscr{E}_1^{(RS)} = \langle W_0^{(1)} W_0^{(2)} | V | W_0^{(1)} W_0^{(2)} \rangle \quad \text{(electrostatic energy)} \tag{24}$$

$$\mathscr{E}_2^{(RS)} = -\sum_a{}^{\prime(1)} \frac{|\langle W_0^{(1)} W_0^{(2)} | V | W_a^{(1)} W_0^{(2)} \rangle|^2}{E_a^{(1)} - E_0^{(1)}} \quad \begin{array}{l}\text{(polarization energy of (1)}\\ \text{by (2)}\end{array} \tag{25a}$$

$$-\sum_b{}^{\prime(2)} \frac{|\langle W_0^{(1)} W_0^{(2)} | V | W_0^{(1)} W_b^{(2)} \rangle|^2}{E_b^{(2)} - E_0^{(2)}} \quad \begin{array}{l}\text{(polarization energy of (2)}\\ \text{by (1)}\end{array} \tag{25b}$$

$$-\sum_a{}^{\prime(1)} \sum_b{}^{\prime(2)} \frac{|\langle W_0^{(1)} W_0^{(2)} | V | W_a^{(1)} W_b^{(2)} \rangle|^2}{(E_a^{(1)} - E_0^{(1)}) + (E_b^{(2)} - E_0^{(2)})} \quad \text{(dispersion energy)}. \tag{25c}$$

We therefore have to consider matrix elements of the general form

$$\langle W_a^{(1)} W_b^{(2)} | V | W_c^{(1)} W_d^{(2)} \rangle .$$

The expression of such matrix elements proposed by Longuet-Higgins [20] will be very useful for our purpose: these matrix elements may be considered as electrostatic interaction energies between appropriate charge distributions defined independently in molecules (1) and (2). This expression is obtained by using the 'charge density operators' associated with the two molecules ($m = 1, 2$):

$$\rho^{(m)}(\mathbf{r}) = \sum_\mu{}^{(m)} Z_\mu \, \delta(\mathbf{r} - \mathbf{r}_\mu) - \sum_i{}^{(m)} \delta(\mathbf{r} - \mathbf{r}_i) \tag{26}$$

and by remarking that the interaction operator V may be written as:

$$V = \iint \frac{\rho^{(1)}(\mathbf{r}^1) \, \rho^{(2)}(\mathbf{r}^2)}{|\mathbf{r}^1 - \mathbf{r}^2|} \, d\mathbf{r}^1 \, d\mathbf{r}^2 . \tag{27}$$

The operators $\rho^{(m)}(\mathbf{r})$ act on functions simply by multiplication. Then, using [27], we get:

$$\langle W_a^{(1)} W_b^{(2)} | V | W_c^{(1)} W_d^{(2)} \rangle$$

$$= \iint \frac{\langle W_a^{(1)} | \rho^{(1)}(\mathbf{r}^1) | W_c^{(1)} \rangle \langle W_b^{(2)} | \rho^{(2)}(\mathbf{r}^2) | W_d^{(2)} \rangle}{|\mathbf{r}^1 - \mathbf{r}^2|} \, d\mathbf{r}^1 \, d\mathbf{r}^2 \tag{28a}$$

$$= \iint \frac{f_{ac}^{(1)}(\mathbf{r}^1) \, f_{bd}^{(2)}(r^2)}{|\mathbf{r}^1 - \mathbf{r}^2|} \, d\mathbf{r}^1 \, d\mathbf{r}^2 \tag{28b}$$

where

$$f_{ac}^{(1)}(\mathbf{r}) = \langle W_a^{(1)} | \rho^{(1)}(\mathbf{r}) | W_c^{(1)} \rangle \tag{29a}$$

$$f_{bd}^{(2)}(\mathbf{r}) = \langle W_b^{(2)} | \rho^{(2)}(\mathbf{r}) | W_d^{(2)} \rangle \tag{29b}$$

are the charge distributions defined above. The advantage of this expression (28 b) of the matrix elements is twofold: first, the charge distributions involved are defined for each molecule independently of the other according to [29]; second, at intermediate (and, of course, large) intermolecular distances, the electrostatic interaction between these distributions will be obtained with sufficient accuracy by replacing the continuous (electronic) part of the charge distributions by suitable multipole expansions, and this procedure allows considerable conceptual and computational simplifications.

We shall end the present section by giving the explicit forms of the distributions [29] for the special case where the molecular functions W_a are Slater determinants (for the sake of simplicity, we drop the label (m) of the molecule). Since $\rho(\mathbf{r})$ is a mono-electronic operator (i.e. a sum of 1-electron terms), its matrix elements will be non-zero only if the two determinants W_a and W_c differ by at most one spin-orbital. Hence the two cases:

(1) $W_a \equiv W_c$: *state charge distribution*

$$f_{aa}(\mathbf{r}) = \langle W_a|\rho(\mathbf{r})|W_a\rangle = \sum_\mu Z_\mu \, \delta(\mathbf{r}-\mathbf{r}_\mu) - \sum_{i=1}^{n} |b_i(\mathbf{r})|^2 \qquad (30)$$

where the b_i's are the space parts of the occupied spin-orbitals from which the determinant W_a is built. Of course, if we have N orbitals b_I with occupation numbers $n_I^{occ}(n_I^{occ} = 1$ or $2)$, the sum $\sum\limits_{i=1}^{n}$ becomes a sum over the *orbitals*:

$$\sum_{I=1}^{N} n_I^{occ} |b_I(\mathbf{r})|^2 .$$

(2) W_a and W_c *differ by 1 spin-orbital: 'transition charge distributions'.*

Let $a_i = b_i \sigma_i$ and $a_j = b_j \sigma_j$ the different spin-orbitals involved in W_a, W_c respectively. Then

$$f_{ac}(\mathbf{r}) = \langle W_a|\rho(\mathbf{r})|W_c\rangle = -b_i^*(\mathbf{r})\, b_j(\mathbf{r})\, \delta_{ij}^\sigma \qquad (31)$$

where $\delta_{ij}^\sigma = \langle \sigma_i|\sigma_j\rangle$ (i.e. 1 if σ_i and σ_j are the same and 0 if they are different).

Let us remark that these 'transition charge distributions' are only mathematical aids for the computation of the matrix elements of V: *no direct physical* meaning should be ascribed to them. As concerns the state charge distributions, their physical meaning depends on the life-time of the state under consideration: there is no problem for the ground state, but, as concerns the excited states, some direct physical meaning may be attributed to their charge distribution only if they have a sufficiently long life-time. It must never be forgotten that the perturbation treatment uses the excited state functions only as a mathematically convenient basis, without any concern about their degree of physical meaning.

4. Analysis of the Short-range 1st Order Exchange Term in Terms of Local Contributions

We shall now, in the present section and in the following ones, use the fact that the ground-state function of many molecules may be quite well approximated by a single Slater determinant built from localized orbitals [21, 22].

The use of such orbitals is the logical basis for the analysis of the intermolecular interaction energy in terms of local distributions. Before performing such an analysis for the first order exchange energy, let us mention that Salem [23] evaluated the corresponding forces (by calculating the variations of charge density due to the interaction and applying the Hellman-Feynman theorem), and actually obtained the result that the localized orbitals displayed additive contributions to these forces (this conclusion remains valid for the interaction between conjugated systems like butadiene and ethylene [13, Chapter III, Section 3C2]).

Let us now consider the expression (15) of the first-order 1-exchange energy $E_{1-\text{exch}}$. We may decompose the charge distributions $\rho_{ij}^{(1)}$ and $\rho_{ij}^{(2)}$ as follows:

$$\rho_{ij}^{(1)} = \rho_{ij}'^{(1)} + \rho_{ij}''^{(1)} \tag{32a}$$

with

$$\rho_{ij}'^{(1)} = a_i^{(1)} a_j^{(2)} - S_{ij}(a_i^{(1)})^2 \tag{32b}$$

$$\rho_{ij}''^{(1)} = -\sum_{k \neq i}^{(1)} S_{kj} a_i^{(1)} a_k^{(1)} \tag{32c}$$

and similarly

$$\rho_{ij}^{(2)} = \rho_{ij}'^{(2)} + \rho_{ij}''^{(2)} \tag{33a}$$

with

$$\rho_{ij}'^{(2)} = a_i^{(1)} a_j^{(2)} - S_{ij}(a_j^{(2)})^2 \tag{33b}$$

$$\rho_{ij}''^{(2)} = -\sum_{l \neq j}^{(2)} S_{il} a_j^{(2)} a_l^{(2)}. \tag{33c}$$

The distributions $\rho_{ij}'^{(1)}$ and $\rho_{ij}'^{(2)}$ depend only on the pair of orbitals i and j, while $\rho_{ij}''^{(1)}$ and $\rho_{ij}''^{(2)}$ involve all orbitals of molecules 1 and 2 respectively. Then, since the orbitals are localized, $\rho_{ij}'^{(1)}$ and $\rho_{ij}'^{(2)}$ give, in the first integral of (15), an interaction between the regions (i) and (j) completely independent of the rest of the molecules:

$$-\iint \rho_{ij}'^{(1)}(\alpha) \frac{1}{r_{\alpha\beta}} \rho_{ij}'^{(2)}(\beta) \, d\alpha \, d\beta = -[\rho_{ij}'^{(1)}|\rho_{ij}'^{(2)}]$$

$$= -[a_i^{(1)} a_j^{(2)}|a_i^{(1)} a_j^{(2)}] + S_{ij}[a_i^{(1)} a_j^{(2)}|(a_i^{(1)})^2 + (a_j^{(2)})^2] -$$

$$- (S_{ij})^2 [(a_i^{(1)})^2|(a_j^{(2)})^2] \tag{34}$$

which is of the order of magnitude $(S_{ij})^2/R_{ij}$ (the hybrid term

$$S_{ij}[a_i^{(1)} a_j^{(2)}|(a_i^{(1)})^2 + (a_j^{(2)})^2]$$

being the most important one).

The other interactions involved in the first integral of (15), namely

$$[\rho_{ij}'^{(1)}|\rho_{ij}''^{(2)}], \quad [\rho_{ij}''^{(1)}|\rho_{ij}'^{(2)}], \quad [\rho_{ij}''^{(1)}|\rho_{ij}''^{(2)}]$$

are expected to give smaller values. Indeed, since the charge distributions $a_i^{(1)} a_k^{(1)}$ and

$a_j^{(2)} a_l^{(2)}$ are the product of *different* localized spinorbitals, they have much smaller values than the distributions $(a_i^{(m)})^2$ (this is evident if the corresponding orbitals b_i, b_k (or b_j, b_l) belong to distant parts of the molecule, but even if b_i^m, b_k^m associated with neighbour bonds with an atom in common, the atomic orbitals on this atom which contribute to $b_i^{(m)}$, $b_k^{(m)}$ respectively are not only orthogonal but localized in different directions). Moreover, the distributions $b_i^{(m)}$, $b_k^{(m)}$ are dipolar (total charge $\int b_i^{(m)} b_k^{(m)} \, \mathbf{dr} = 0$), so that the interactions $[\rho'^{(1)}|\rho''^{(2)}]$, $[\rho''^{(1)}|\rho'^{(2)}]$ are expected to vary like S^2/R^2 and $[\rho''^{(1)}|\rho''^{(2)}]$ like S^2/R^3 (if the orbitals are strictly orthogonal, they are not, in general, totally localized, but the orthogonalization 'tails' may be neglected in order to get simplified formulae).

Let us now consider the last two integrals with the factor S_{ij} in (15). We shall deal with the first one (the reasoning would be the same with the second one):

$$S_{ij} \int \rho_{ij}^{(1)}(\alpha) \, V^{(2)}(\mathbf{r}_\alpha) \, d\alpha = S_{ij}[\rho_{ij}^{(1)}|V^{(2)}]$$

$$= S_{ij}([\rho_{ij}'^{(1)}|V^{(2)}] + [\rho_{ij}''^{(1)}|V^{(2)}]). \tag{35}$$

Let us first consider the interaction $[\rho_{ij}'^{(1)}|V^{(2)}]$. We may distinguish in the potential $V^{(2)}$ the part $V_j^{(2)}$ which is due to the region j corresponding to the localized spinorbital $a_j^{(2)}$. Now, for the spinorbitals $a_i^{(1)}$ and $a_j^{(2)}$ which correspond to the regions of molecules (1) and (2) at the shortest distance (for given positions of the molecules), the part $V_j^{(2)}$ or $V^{(2)}$ will give a dominant contribution in the region of space where $a_i^{(1)} a_j^{(2)}$ is important, because the regions of molecule (2) contributing to $V^{(2)} - V_j^{(2)}$ lie outside this region $a_i^{(1)} a_j^{(2)}$. Consequently, these most important interactions obtained with the pairs $a_i^{(1)} a_j^{(2)}$ of spinorbitals at the closest distance may be expected to depend essentially on the part $V_j^{(2)}$ of the potential $V^{(2)}$ due to the region j.

For distributions $a_i^{(1)} a_j^{(2)}$ where $a_i^{(1)}$ and $a_j^{(2)}$ are not the closest possible, the previous reasoning does not hold, since there will be in molecule (2) some regions $l \neq j$ situated *between* regions $i^{(1)}$ and $j^{(2)}$, and the contributions $V_l^{(2)}$ of such regions to $V^{(2)}$ are as important as $V_j^{(2)}$. Consequently, the interaction $[\rho_{ij}'^{(1)}|V^{(2)}]$ cannot be expected to be expressible only in terms of the regions i and j *when these regions are not the closest possible*. But this does not matter from the practical point of view, because these interactions $[\rho_{ij}'^{(1)}|V^{(2)}]$ *decrease exponentially* as functions of the inter-region distance $R_{ij}(\rho_{ij}'^{(1)}$ is indeed of order S_{ij}, and $S_{ij}[\rho_{ij}'^{(1)}|V^{(2)}]$ decreases at least like S_{ij}^2): consequently, the total interaction $\sum_{i,j} S_{ij}[\rho_{ij}'^{(1)}|V^{(2)}]$ will practically reduce to the contribution of the (i, j) pair(s) with the shortest distance, for which a local-type formula (let us denote it by $e_{ij}'^{(1)}(R_{ij})$), may be assumed, and the use of the approximate formula $\sum_{i,j} e_{ij}'^{(1)}(R_{ij})$ will therefore be justified (it does not matter that the *negligible* contributions to $\sum^{(1)}_i \sum^{(2)}_j S_{ij}[\rho_{ij}'^{(1)}|V^{(2)}]$ coming from the non closest pairs of regions i, j are not accurately represented by the local-type formula $e_{ij}'^{(1)}(R_{ij})$).

It only remains to consider the second interaction term in (35), namely

$$S_{ij}[\rho''^{(1)}_{ij}|V^{(2)}] = -S_{ij}\sum_{k\neq i}^{(1)} S_{ik}[a_i^{(1)}a_k^{(1)}|V^{(2)}]. \tag{36}$$

Here the same arguments as those used above for the analysis of $[\rho^{(1)}_{ij}|\rho^{(1)}_{!j}]$ may be applied. The distribution $a_i^{(1)}a_k^{(1)}$ is a dipolar one; at large intermolecular distance, $V^{(2)}$ is at most a dipole potential (for a neutral molecule); at short distance, the atoms of molecule (2) which are the closest to the distribution $a_i^{(1)}a_k^{(2)}$ may contribute to $V^{(2)}$ by terms of order $-q/R$ (q: net charge), and hence by terms of the form $CqS_{ij}S_{ik}/R^2$ to (36). Since $a_{(1)}^i$ and $a_k^{(1)}$ are localized in different regions, $a_i^{(1)}a_k^{(1)}$ is a distribution with low values, therefore giving a small C coefficient. Moreover, for the pairs $a_i^{(1)}a_k^{(1)}$ which have a common atom μ, we shall have elementary charge distributions $\chi_{\mu(i)}^{(1)}\chi_{\mu(k)}^{(1)}$, where the χ's are orthogonal hybrid orbitals of atom μ; now, each one of these distributions has only a small dipole moment, and when the different products $\chi_{\mu(i)}^{(1)}\chi_{\mu(k)}^{(1)}$ pertaining to the same atom μ appear in the summation $\sum_i^{(1)}\sum_{k\neq i}^{(1)}$, the dipole moments of these products will add and tend to cancel each other (the cancellation would be exact for an atom with a symmetical environment like C in CH_4, so that an eventually non-negligible total dipole may be expected in the case of marked departure from spherical symmetry, like hetero-atoms bearing lone pairs).

We conclude from this discussion of the formula (15) in terms of localized orbitals that the first order 'exchange' energy between two molecules should appear, with a reasonable accuracy, as a sum of independent interaction terms between localized regions (bonds and eventually lone pairs). Since these terms involve S^2 (square of the overlap between such localized orbitals), it will be sufficient, in practise, to take into account the valence orbitals (since the inner shell orbitals, having larger Slater exponents, give much smaller overlap integrals).

It could be desired to go one step further, and to express E_{1-exch} as a sum of atom-atom interaction terms only (this is a very common practice [2, 4, 5, 6], owing to its computational simplicity). Now, bond-bond terms like S_{ij}^2 (where S_{ij} is a bond-bond overlap integral), cannot be reduced rigorously to a sum of atom-atom terms [23], owing to the occurrence of the power 2. Nevertheless, the error corresponds to a multiplicative factor which remains between $\frac{1}{2}$ and 2 [13, Chapter III, Section 3C2); this may seem at first sight prohibitive, but if we remember that the overlap integrals vary exponentially with the distance, it will be appreciated that a small variation of the distance suffices to compensate for a factor between $\frac{1}{2}$ and 2, and this explains that the use of atom-atom terms for evaluating E_{1-exch} may still give reasonable results. This theoretical analysis also justifies the use of exponential-type formulae for these atom-atom terms at intermediate (equilibrium region) and large distances (at very short distance the $1/R$ behaviour corresponding to the nuclear repulsion would dominate, so that the formulae based on the expression S^2/R first proposed by Murrel et al. [24] have a correct behaviour at all distances [13, Chapter III, Section 3B3)).

5. Analysis of the Electrostatic Interaction
in Terms of Local Contributions

The analysis of the electrostatic interaction is a straightforward application of the expressions derived in Section 3 above. We indeed have, according to (24) and (28):

$$E_{el} = \int\int \frac{f_{00}^{(1)}(\mathbf{r}^1)\, f_{00}^{(2)}(\mathbf{r}^2)}{|\mathbf{r}^1 - \mathbf{r}^2|}\, d\mathbf{r}^1\, d\mathbf{r}^2 \tag{37}$$

where

$$f_{00}^{(m)}(\mathbf{r}) = \langle W_0^{(m)} | \rho^{(m)}(\mathbf{r}) | W_0^{(m)} \rangle$$

is the (mean) charge distribution of molecule (m). If $W_0^{(m)}$ is a Slater determinant:

$$W_0^{(m)} = \frac{1}{\sqrt{n!}} \det |a_1^{(m)}(1) \ldots a_n^{(m)}(n)| \tag{38}$$

where the $a_i^{(m)} = b_i^{(m)} \sigma_i$ are the spin-orbitals, $f_{00}^{(m)}$ is given by the formula (30):

$$f_{00}^{(m)}(\mathbf{r}) = \sum_{\mu}^{(m)} Z_{\mu}^{(m)}\, \delta(\mathbf{r} - \mathbf{r}_{\mu}^{(m)}) - \sum_{i=1}^{n_m} |b_i^{(m)}(\mathbf{r})|^2 . \tag{39}$$

It has been a very frequent practice to introduce a multipole expansion of the distribution (39) around some center chosen in the molecule (m) in order to calculate the electrostatic interaction (37) as a sum of multipole-mutipole interactions (see e.g. [10], [11], [25], [26, part II]). But this procedure is satisfactory only if the multipole expansion for the interaction obtained in this way converges sufficiently, which is the case only for an intermolecular distance large enough with respect to the molecular sizes themselves. In the region of distance corresponding to the minimum of the energy, this condition is not realized except for atoms (which have a negligible electrostatic interaction anyway), and very small molecules; in other cases, a better procedure must be used [27]. Namely, we shall start from the exact expression (39) and suppose that the molecular orbitals b_i are expressed in terms of a basis of atomic orbitals $\{\chi_\alpha\}$, as it is almost always the case in practice (bond orbitals could also be used: the essential point is that the basis orbitals are strongly localized). Then inserting

$$b_i(\mathbf{r}) = \sum_{\alpha} c_{i\alpha}\, \chi_\alpha(\mathbf{r}) \tag{40}$$

into (39), we shall obtain $f_{00}(\mathbf{r})$ as a sum of nuclear point charges and electronic distributions among which the dominant ones will be strongly localized (for the sake of simplicity, we omit from now on the superscript (m) labeling the molecule: it is not necessary, since we analyse the charge distribution of each molecule quite independently from the other):

$$f_{00}(\mathbf{r}) = \sum_{\mu} Z_{\mu}\, \delta(\mathbf{r} - \mathbf{r}_{\mu}) - \sum_{\alpha}\sum_{\beta} P_{\alpha\beta}\, \chi_\alpha^*(\mathbf{r})\, \chi_\beta(\mathbf{r}) \tag{41}$$

where the $P_{\alpha\beta}$'s are the elements of the charge and bond order matrix:

$$P_{\alpha\beta} = \sum_{i(\text{spinorbitals})} c_{i\alpha}^*\, c_{i\beta} \tag{42}$$

(if we use orbitals b_I with occupation numbers n_I^{occ}, this formula, of course, becomes

$$P_{\alpha\beta} = \sum_{I(\text{orbitals})} n_I^{occ} c_{I\alpha}^* c_{I\beta} \, .$$

The elementary charge distributions $\chi_\alpha^*(\mathbf{r}) \chi_\beta(\mathbf{r})$ appearing in (41) have the states property: owing to the strongly localized character of the (atomic) orbitals $\chi_\alpha(\mathbf{r})$, these distributions are non negligible only if χ_α and χ_β belong to the same atom or to neighbour atoms. Then, the important elementary distributions, having a small size (size of an atom or of a bond), will be quite accurately approximated by a very limited multipole expansion (the first or the first two non-zero multipoles), for the purpose of calculating interactions at distance of the order of a few Angströms. Thus, instead of having a multipole expansion of $f_{00}(\mathbf{r})$ around a single center, we shall represent $f_{00}(\mathbf{r})$ as a sum of *several* multipole expansions (centered at the atoms, and eventually also at appropriate points of the bonds: see below), each one of these expansions containing very few terms (charge, dipole, quadrupole at most). As a consequence, the electrostatic interaction between two molecules will appear as a sum of interactions between these *local* multipoles.

The elementary charge distributions $\chi_\alpha^* \chi_\beta$ may be classified into three types [13, Chapter III, Section 4B]:

(1) *the 'diagonal' distributions* $|\chi_\alpha|^2$

They are, of course, one-center distributions and their main multipole is the charge $S_{\alpha\alpha} = \int |\chi_\alpha(\mathbf{r})|^2 \, d\mathbf{r} = 1$. In practice, it will be very rare to consider higher-order multipole moments (for an s orbital, all these moments are zero; for a p orbital, only the quadrupole is non-zero). Quite generally, for an orbital with azimuthal quantum number l, only multipoles between orders 0 and $2l$ may be non-zero (see e.g. [13, Chapter III, Section 4B]).

(2) *the 'non-diagonal' one-center distributions* $\chi_\alpha^* \chi_\alpha$

The orbitals χ_α and $\chi_{\alpha'}$ are different but belong to the same atom; they may be supposed orthogonal. Then, the total charge of such a distribution is

$$S_{\alpha\alpha'} = \int \chi_\alpha^*(\mathbf{r}) \, \chi_{\alpha'}(\mathbf{r}) \, d\mathbf{r} = 0 \, .$$

In practice, the distribution will therefore be approximated by its dipole moment. Anyway, only a finite number of multipole moments may be non-zero. If a basis of s and p orbitals is used, this dipole will be non-zero for the distributions $s - p$ (if a basis of hybrid orbitals was used, dipole moments corresponding to sp distributions would already appear in the 'diagonal' distributions).

In the CNDO-SCF method, these atomic dipole moments are well-known under the name of 'hybridization moments'. The total charge distribution of the molecule may be approximated by the set of *atomic net charges* (nuclear charge + total electronic charge given by the 'diagonal' distributions belonging to the atom), *and* of the *atomic dipoles*: the total dipole moment is actually computed by including the contribution of these atomic dipoles, and it would be *inconsistent* to compute electrostatic interactions between molecules studied by this method by using the net charges only (the interactions net charge-atomic dipoles and between the atomic dipoles themselves must also be included). In Hückel-type methods, on the contrary, it may be possible to consider net charges only.

If the inter-atomic overlap integrals are no more neglected (e.g. in an ab initio method), then the third type of distribution also occurs:

(3) *the 'non-diagonal' two-center distributions* $\chi_\alpha^* \chi_\beta$

The charge of such a distribution is the overlap integral $S_{\alpha\beta} = \int \chi_\alpha^*(\mathbf{r}) \chi_\beta(\mathbf{r}) \, d\mathbf{r}$.

In contrast with the two previous cases, the multipole expansion now contains an infinite number of non-zero terms. Here, too, it will be quite exceptional to consider multipoles beyond the charge and the dipole. There is now some arbitrariness in the approximate representation of such a distribution; four possibilities are listed below:

(a) use of the middle M of the segment $\alpha\beta^+$ as the center for the multipole expansion: charge $S_{\alpha\beta}$ at M and dipole of the distribution *calculated with respect to M* (the first non-zero multipole moment does not depend on the choice of the origin, but the following moments do depend on it).

(b) if $S_{\alpha\beta} \neq 0$, this charge is put at the point M' of the segment such that the longitudinal component of the dipole moment (calculated with M' as the origin) is zero. Then, we have at M' the charge $S_{\alpha\beta}$ and eventually a dipole perpendicular to the segment $\alpha\beta$.

(c) the charge $S_{\alpha\beta}$ is divided into two equal charges $S_{\alpha\beta}/2$, which are placed at the atoms α and β (Mulliken's approximation [28]). The purpose of this procedure is to avoid the introduction of new multipole centers besides the atoms, thus avoiding increasing the number of distances between such centers belonging to different molecules (distances which must be computed for getting the electrostatic interaction energy between the molecules). But the dipole moment of the distribution $\chi_\alpha^* \chi_\beta$ will *not* be reproduced, unless it is zero with respect to the middle M of the segment $\alpha\beta$. If this particular case is not realized, the non-zero dipole moment should be put at the point M; it is known as the 'homopolar' dipole [29].

(d) the previous discussion immediately suggests dividing the charge $S_{\alpha\beta}$ into two unequal parts q_α and q_β (placed at atoms α and β), such that the longitudinal component $\mu_{\alpha\beta}^L$ of the dipole moment is reproduced. If $\mu_{\alpha\beta}^L$ denotes the algebraic component on the axis $\boldsymbol{\alpha}\boldsymbol{\beta}$ (and consequently $\mu_{\beta\alpha}^L = -\mu_{\alpha\beta}^L$), it is easily found that

$$q_\alpha = \tfrac{1}{2} S_{\alpha\beta} - \frac{\mu_{\alpha\beta}^L}{R_{\alpha\beta}} \qquad q_\beta = \tfrac{1}{2} S_{\alpha\beta} - \frac{\mu_{\beta\alpha}^L}{R_{\alpha\beta}} \tag{43}$$

where $R_{\alpha\beta}$ is the distance between the atoms α and β.

This procedure is known as the Löwdin approximation [30, 31, 32]. In principle, the transversal component $\mu_{\alpha\beta}^T$ would remain (at the middle of $\alpha\beta$, say), but in practice the $\mu_{\alpha\beta}^T$'s corresponding to the different atomic orbitals belonging to the same pair

---

+ For brevity, we shall speak of 'atom α (resp. β)' instead of using the full expression 'atom bearing the orbital χ_α (resp. χ_β)'.

of atoms tend to cancel each other, so that it is a reasonable approximation to neglect them.

As a conclusion of this section, it must therefore be emphasized that using *only the Mulliken atomic net charges* (given, for example, by an ab initio program) for the calculation of an electrostatic interaction would be *incorrect for two reasons*: (1) the atomic dipoles must be taken into account (as we emphasized previously for the case of CNDO methods), and (2) the homopolar dipoles $\mu_{\alpha\beta}$ must be taken into account, or else atomic net charges obtained according to Löwdin's procedure should be used. An ab initio program which gives only Mulliken net charges is therefore quite misleading as concerns a reasonable representation of the molecular charge distribution in terms of charges and dipoles.

6. Decomposition of the Molecular Polarizability into a Sum of Local Contributions (Bond Polarizabilities)

The expressions (25) for the polarization and dispersion energies may be transformed in such a way as to involve the polarizability (static or dynamical), as is well-known in the long-range domain [25, 33, 34]. We shall extend this type of treatment in the medium-range, and for that purpose it is first necessary to decompose the total molecular polarizability as a sum of local contributions (for the sake of simplicity, we shall consider the case of the static polarizability, but the principle of the analysis is valid for the dynamical polarizability as well).

A not too strong static electric field ξ creates in a molecule an induced dipole μ_{ind} which depends linearly on ξ:

$$\mu_{ind} = A.\xi \tag{44}$$

where A is called the (static) polarizability tensor (the dynamical tensor $A(\hbar\omega)$ plays a similar role when the field is a sinusoidal function of time $\xi(t) = \xi_0 \cos \omega t$). Standard Rayleigh-Schrödinger perturbation theory gives for the tensor A the expression:

$$A = 2\operatorname{Re} \sum_I{}' \frac{\mu_{0I} \otimes \mu_{I0}}{E_I - E_0} \tag{45}$$

where $|I\rangle$ denotes the eigenfunctions of the molecule, E_I their eigenvalues, and μ_{0I} the 'transition dipole moment'

$$\mu_{0I} = \langle 0|\mu|I\rangle \tag{46}$$

where μ denotes the dipole moment operator:

$$\mu = \sum_v Z_v \mathbf{r}_v - \sum_i \mathbf{r}_i \tag{47}$$

and $\sum_I{}'$ means that the ground-state $|0\rangle$ is excluded from the summation. The symbol $\otimes$ denotes the dyadic or tensor product [35, Volume I, Chapter 1, p. 54]; if U and V are arbitrary vectors, $U \otimes V$ is the tensor defined by its action upon any vector X:

$(\mathbf{U} \otimes \mathbf{V})\mathbf{X} = \mathbf{U}(\mathbf{V} \cdot \mathbf{X})$; in an orthogonal basis, the matrix elements of $\mathbf{U} \otimes \mathbf{V}$ are $(\mathbf{U} \otimes \mathbf{V})_{ij} = U_i V_j$.

Now, we shall represent the excited states $|I\rangle$ in the framework of the so-called 'excitonic' treatment [36]: using *localized* orbitals (denoted by $u, v, w \ldots$), the ground state $|0\rangle$ will be represented by a Slater determinant W_0, and we shall introduce the '*localized*' monoexcited determinants $W_{u-w'}$ where an occupied orbital u of W_0 has been replaced by an 'excited' localized orbital w' (the prime will always indicate a virtual orbital); if it is necessary to indicate that w' belongs to the same region as some occupied orbital v, we shall use the notation $w'(v)$. These 'localized' determinants $W_{u-w'}$ will be denoted $|\alpha\rangle$ for brevity. Then the true monoexcited states $|I\rangle$ will be obtained, in a first approximation, by performing the Configuration Interaction of these basis states $|\alpha\rangle$:

$$|I\rangle = \sum_\alpha C_{I\alpha} |\alpha\rangle \tag{48}$$

where the matrix $C_{I\alpha}$ is unitary since both sets $\{|I\rangle\}$ and $\{|\alpha\rangle\}$ are orthonormal. States $|\alpha\rangle$ with the same mean energy (e.g. corresponding to analogous localized excitations on identical chemical bonds) will give a cluster of true eigenstates $|I\rangle$ under the effect of the Configuration Interaction.

Now, using (48), formula (45) becomes

$$\mathbf{A} = 2 \operatorname{Re} \sideset{}{'}\sum_\alpha \sideset{}{'}\sum_\beta \boldsymbol{\mu}_{0\alpha} \otimes \boldsymbol{\mu}_{\beta 0} \left(\sum_I \frac{C_{I\alpha} C_{I\beta}^*}{E_I - E_0} \right). \tag{49}$$

We shall simplify the bracketed factor

$$S_{\alpha\beta} = \sideset{}{'}\sum_I \frac{C_{I\alpha} C_{I\beta}^*}{E_I - E_0} \tag{50}$$

by using the above-mentioned fact that the eigenstates $|I\rangle$ may be classified into clusters corresponding to states $|\alpha\rangle$ with (almost) equal energies. We may distinguish two cases:

(1) $|\alpha\rangle$ and $|\beta\rangle$ have markedly different energies and therefore correspond to *different* clusters of eigenstates $|I\rangle$: then, there is no state $|I\rangle$ such that $C_{I\alpha}$ and $C_{I\beta}$ are *both* important, so that $s_{\alpha\beta}$ may be expected to be small.

(2) $|\alpha\rangle$ and $|\beta\rangle$ have (almost) equal energies and correspond to a single cluster of states $|I\rangle$; then $C_{I\alpha}$ and $C_{I\beta}$ will both be important for the states of this cluster, and we shall therefore use a kind of 'closure approximation' by replacing the energies E_I by the value E_α common to the 'localized' excited states $|\alpha\rangle$ generating the cluster; (50) then gives:

$$s_{\alpha\beta} \simeq \frac{1}{E_\alpha - E_0} \sum_I C_{I\alpha} C_{I\beta}^*,$$

i.e.

$$s_{\alpha\beta} \simeq \frac{\delta_{\alpha\beta}}{E_\alpha - E_0} \tag{51}$$

and this formula appears appropriate also for the first case.

Then (49) becomes

$$\mathbf{A} \simeq 2 \operatorname{Re} \sum_{\alpha}{}' \frac{\boldsymbol{\mu}_{0\alpha} \otimes \boldsymbol{\mu}_{\alpha 0}}{E_\alpha - E_0} \tag{52a}$$

$$\simeq 2 \operatorname{Re} \sum_u \sum_{w'} \frac{\boldsymbol{\mu}_{u-w'} \otimes (\boldsymbol{\mu}_{u-w'})^*}{\Delta E_{u-w'}} \tag{52b}$$

$$\simeq \sum_u \mathbf{A}_u \tag{52c}$$

$$\mathbf{A}_u = 2 \operatorname{Re} \sum_{w'} \frac{\boldsymbol{\mu}_{u-w'} \otimes (\boldsymbol{\mu}_{u-w'})^*}{\Delta E_{u-w'}} \tag{52d}$$

where

$$\boldsymbol{\mu}_{u-w'} = \langle W_0 | \boldsymbol{\mu} | W_{u-w'} \rangle \quad \text{and} \quad \Delta E_{u-w'} = E_{u-w'} - E_0.$$

$\mathbf{A}_u$ clearly appears as a polarizability associated with the localized orbital u. If there are several orbitals in the same region (σ and π orbitals for example), the corresponding tensors $\mathbf{A}_u$ may be added in order to define a polarizability for this region (in practice, the regions are the chemical bonds). We have therefore found a theoretical qualitative justification for the possibility of expressing the experimental polarizabilities of molecules as a sum of suitable bond polarizabilities [37]. From the practical point of view, an essential property of the bond polarizabilities is their *transferability* from a molecule to another (atomic polarizabilities may also be used, but they are much less transferable: thus in [5, 6] where they are used for computational reasons, they are calculated for each molecule by decomposing bond polarizabilities). It must be emphasized that the definition (52d) of $\mathbf{A}_u$ involves a summation over *all* virtual localized orbitals w' and not only over those of the region of u: in this context, the transferability of bond polarizability values would imply that the contribution of the virtual orbitals w' belonging to other regions (which cannot be neglected) is practically independent of the detailed nature of the bonds surrounding the bond u (this type of property has actually been found by Zwanzig [38] for an idealized model, namely a chain of coupled harmonic oscillators).

7. Analysis of the Polarization Energy in Terms of Local Polarizabilities

The results of the previous section may be easily applied to the polarization parts (25 a, b) of the second order long-range energy $\mathscr{E}_2^{(\mathrm{RS})}$. Let us consider the first term (25a) (the second term (25b) could be treated in exactly the same way); in order to agree with the notation of Section 6, we note $W_I^{(1)}$ instead of $W_a^{(1)}$ the excited eigenfunctions of the molecule (1):

$$E_{\mathrm{pol}}^{(1)} = - \sum_I{}'^{(1)} \frac{|\langle W_0^{(1)} W_0^{(2)} | V | W_I^{(1)} W_0^{(2)} \rangle|^2}{\Delta E_I^{(1)}}. \tag{53}$$

Inserting (48): $W_I^{(1)} = \sum'^{(1)}_\alpha C_{I\alpha} W_\alpha^{(1)}$, we get

$$\langle W_0^{(1)} W_0^{(2)}|V|W_I^{(1)} W_0^{(2)}\rangle = \sum^{(1)}_\alpha C_{I\alpha} V_\alpha^{(1)} \tag{54}$$

with

$$V_\alpha^{(1)} = \langle W_0^{(1)} W_0^{(2)}|V|W_\alpha^{(1)} W_0^{(2)}\rangle \tag{55a}$$

$$= \iint \frac{f_{0\alpha}^{(1)}(\mathbf{r}^1)\, f_{00}^{(2)}(\mathbf{r}^2)}{|\mathbf{r}^1 - \mathbf{r}^2|}\, d\mathbf{r}^1\, d\mathbf{r}^2 \tag{55b}$$

according to (28). Then, using (54), we transform (53):

$$E_{\text{pol}}^{(1)} = -\sum'^{(1)}_\alpha \sum'^{(1)}_\beta V_\alpha^{(1)} V_\beta^{(1)*} \left(\sum'_I \frac{C_{I\alpha} C_{I\beta}^*}{\Delta E_I^{(1)}}\right) \tag{56}$$

where the bracketed expression is nothing but the $s_{\alpha\beta}$ defined above by (50). Using its approximation (51), we get:

$$E_{\text{pol}}^{(1)} \simeq -\sum'^{(1)}_\alpha \frac{|V_\alpha^{(1)}|^2}{\Delta E_\alpha^{(1)}} = -\sum^{(1)}_u \sum^{(1)}_{w'} \frac{|V_{u-w'}^{(1)}|^2}{\Delta E_{u-w'}^{(1)}} \tag{57}$$

where u and w' have the same meaning as in the previous section (u: occupied localized orbital, w': virtual localized orbital).

Now we shall use the localized character of these orbitals in order to get a simple expression of the electrostatic energy

$$V_{u-w'}^{(1)} = \iint \frac{f_{u-w'}^{(1)}(\mathbf{r}^1)\, f_{00}^{(2)}(\mathbf{r}^2)}{|\mathbf{r}^1 - \mathbf{r}^2|}\, d\mathbf{r}^1\, d\mathbf{r}^2 \tag{58a}$$

$$\simeq -\boldsymbol{\xi}_u^{(2)} \cdot \boldsymbol{\mu}_{u-w'}^{(1)} \tag{58b}$$

where $\boldsymbol{\mu}_{u-w'}^{(1)}$ denotes the dipole moment of the transition distribution $f_{u-w'}^{(1)}(\mathbf{r}) = -u(\mathbf{r})\, w'(\mathbf{r})\, \delta_{uw'}^\sigma$ (see (31); for simplicity, we suppose that real orbitals are used), and $\boldsymbol{\xi}_u^{(2)}$ denotes the electric field created by the molecule (2) at some point of the region u where the transition dipoles $\boldsymbol{\mu}_{u-w'}^{(1)}$ (u fixed, w' variable) will be placed (for a bond, the middle may be chosen, but this is not compulsory and other choices could be tried for heteronuclear bonds). Of course, the approximation of (58a) by (58b) becomes less and less exact when w' is localized at larger and larger distances from u, but this poorer approximation will concern an increasingly smaller interaction, since the values of $f_{u-w'}^{(1)}(\mathbf{r})$ decrease exponentially when the distance between the *localized* orbitals u and w' increases: consequently, (58b) may be expected to approximate accurately enough the *important terms* among all the $V_{u-w'}^{(1)}$. Then

$$(V_{u-w'}^{(1)})^2 = (\boldsymbol{\xi}_u^{(2)} \boldsymbol{\mu}_{u-w'}^{(1)})^2 = \boldsymbol{\xi}_u^{(2)} (\boldsymbol{\mu}_{u-w'}^{(1)} \otimes \boldsymbol{\mu}_{u-w'}^{(1)})\, \boldsymbol{\xi}_u^{(2)} \tag{59}$$

according to the definition of the dyadic product $\otimes$ (see previous Section 6). Then (57) becomes:

$$E_{pol}^{(1)} \simeq - \sum_u^{(1)} \xi_u^{(2)} \left(\sum_{w'}^{(1)} \frac{\mu_{u-w'}^{(1)} \otimes \mu_{u-w'}^{(1)}}{\Delta E_{u-w'}^{(1)}} \right) \xi_u^{(2)} \qquad (60a)$$

$$\simeq - \frac{1}{2} \sum_u^{(1)} \xi_u^{(2)} \, \mathbf{A}_u^{(1)} \, \xi_u^{(2)} \qquad (60b)$$

where $\mathbf{A}_u^{(1)}$ actually appears as the local (bond) polarizability tensor according to (52d) (Section 6). We recognize here the classical expression of a polarization energy (hence the name given to the original expression (25a)), for the molecule (1), and this energy actually appears, within the approximations made, as the sum of polarization energies calculated independently for each region (bond) of the molecule corresponding to the localized orbitals. In practice, we use bond polarizabilities given by systematics deduced from experimental molecular polarizabilities [37].

The electric fields $\xi_u^{(2)}$ created by the molecule (2) are evaluated from the same representation of the electrostatic charge distribution which is used for calculating electrostatic interaction: appropriate simplified representations in terms of atomic charges; dipoles ... have been discussed above (Section 5).

The special form taken by $\xi \mathbf{A} \xi$ when $\mathbf{A}$ is axially symmetric (with respect to the bond axis), is easily found to be [13, Chapter III, Section 4E2]:

$$\xi \mathbf{A} \xi = \alpha^T (\xi)^2 + (\alpha^L - \alpha^T)(\xi \alpha^L)^2 \qquad (61)$$

where α^L and α^T are the longitudinal and transverse polarizabilities respectively, and α^L is a unit vector colinear to the longitudinal axis.

When the intermolecular distance is large (with respect to molecular sizes), all the fields $\xi_u^{(2)}$ become almost equal to a field $\xi^{(2)}$ created by the molecule (2) at some point (chosen as 'center'), in molecule (1). Then (60b) becomes:

$$E_{pol}^{(1)} \, (\text{long range}) = - \frac{1}{2} \xi^{(2)} \left(\sum_u^{(1)} \mathbf{A}_u^{(1)} \right) \xi^{(2)} \qquad (62a)$$

$$= - \frac{1}{2} \xi^{(2)} \, \mathbf{A}^{(1)} \, \xi^{(2)} \qquad (62b)$$

where $\mathbf{A}^{(1)}$ denotes the total polarizability tensor of molecule (1). It must be emphasized that (62b) is exact independently of the validity of the approximations which allowed us to obtain the decomposition (60b): (62b) may indeed be derived directly from (25a) and (45) in a rigorous way [25], [13, Chapter III, Section 4E3].

8. Analysis of the Dispersion Energy as a Sum of Local-Type Contributions (Bond-Bond Interactions)

We first transform the dispersion term (25c) rewritten as

$$E_{disp} = - \sum_I^{'(1)} \sum_J^{'(2)} \frac{|\langle W_0^{(1)} W_0^{(2)} | V | W_I^{(1)} W_J^{(2)} \rangle|^2}{\Delta E_I^{(1)} + \Delta E_J^{(2)}} \qquad (63)$$

as we did in the previous section for the polarization term, namely by introducing the locally excited determinants $W_\alpha^{(1)}$, $W_\beta^{(2)}$:

$$W_I^{(1)} = \sum_\alpha{}^{\prime(1)} C_{I\alpha}^{(1)} W_\alpha^{(1)} \qquad W_J^{(2)} = \sum_\beta{}^{\prime(2)} C_{J\beta}^{(2)} W_\beta^{(2)} \tag{64}$$

hence:

$$\langle W_0^{(1)} W_0^{(2)} | V | W_I^{(1)} W_J^{(2)} \rangle = \sum_\alpha{}^{\prime(1)} \sum_\beta{}^{\prime(2)} C_{I\alpha}^{(1)} C_{J\beta}^{(2)} V_{\alpha\beta} \tag{65}$$

where according to (28):

$$V_{\alpha\beta} = \langle W_0^{(1)} W_0^{(2)} | V | W_\alpha^{(1)} W_\beta^{(2)} \rangle = \iint \frac{f_{0\alpha}^{(1)}(\mathbf{r}^1) \, f_{0\beta}^{(2)}(\mathbf{r}^2)}{|\mathbf{r}^1 - \mathbf{r}^2|} \, d\mathbf{r}^1 \, d\mathbf{r}^2 . \tag{66}$$

Then inserting (65) into (63), we get:

$$E_{\text{disp}} = - \sum_\alpha{}^{\prime(1)} \sum_{\alpha'}{}^{\prime(1)} \sum_\beta{}^{\prime(2)} \sum_{\beta'}{}^{\prime(2)} V_{\alpha\beta} V_{\alpha'\beta'} \sum_I{}^{\prime(1)} \sum_J{}^{\prime(2)} \frac{C_{I\alpha}^{(1)} C_{I\alpha'}^{(1)} C_{J\beta}^{(2)} C_{J\beta'}^{(2)}}{\Delta E_I^{(1)} + \Delta E_J^{(2)}} \tag{67}$$

(for simplicity, we write this derivation in the case of real orbitals, so that all coefficients C's and matrix elements V's are real).

The essential difficulty with the expression (67) is that the bracketed factor is not explicitly separable into a product of quantities depending on one molecule only, a condition which is necessary if we want to express in a simple way the dispersion energy as a function of properties of the isolated molecule (on the contrary, in formula (56) for $E_{\text{pol}}^{(1)}$, the bracketed factor depended only on a single molecule). In order to realize such a separation, we shall mention two methods.

A. THE 'MEAN EXCITATION ENERGY' METHOD

This method is an approximate one: it consists in replacing the various denominators $\Delta E_I^{(1)} + \Delta E_J^{(2)}$ by some appropriate mean value, which is conveniently written as $\overline{U}^{(1)} + \overline{U}^{(2)}$ (this is the so-called 'closure' approximation [39] used very early by London [40] in his work on dispersion energy). Then $1/(\overline{U}^{(1)} + \overline{U}^{(2)})$ may be factored in (67), and the numerator of the bracketed expression actually separates into a product of two sums, each of them depending on a single molecule:

$$(\sum_I{}^{(1)} C_{I\alpha}^{(1)} C_{I\alpha'}^{(1)}) (\sum_J{}^{(2)} C_{J\beta}^{(2)} C_{J\beta'}^{(2)}) .$$

We shall not pursue here the development of this procedure (this has been done in [33], and more thoroughly in [13, Chapter III, Section 4F2]). The final result actually appears as a sum of independent bond-bond dispersion interactions:

$$E_{\text{disp}} \simeq - \frac{1}{4} \frac{\overline{U}^{(1)} \overline{U}^{(2)}}{\overline{U}^{(1)} + \overline{U}^{(2)}} \sum_u{}^{(1)} \sum_v{}^{(2)} \frac{1}{(R_{uv})^6} \text{Tr}[\mathbf{T}_{uv} \mathbf{A}_u^{(1)} \mathbf{T}_{uv} \mathbf{A}_v^{(2)}] \tag{68}$$

where Tr denotes the Trace (sum of the diagonal elements), u and v label the bonds (of molecules (1) and (2) respectively), $\mathbf{A}_u^{(1)}$ and $\mathbf{A}_v^{(2)}$ denote their polarizability tensors, R_{uv} their distance (calculated between their middles, for example), and $\mathbf{T}_{uv}$ the following tensor:

$$\mathbf{T}_{uv} = 3(\mathbf{r}_{uv} \otimes \mathbf{r}_{uv}) - \mathbf{1} \quad \text{with} \quad \mathbf{r}_{uv} = \frac{\mathbf{R}_{uv}}{R_{uv}} \tag{69}$$

this tensor expresses the electric field ξ created by a dipole μ, at a distance $\mathbf{R}$ from the dipole, by the formula

$$\xi = (\mathbf{T}\,\mu)/R^3 \tag{70}$$

and it is for this reason that it appears in the derivation of (68). If the polarizabilities are supposed to be isotropic:

$$\mathbf{A}_u^{(1)} = \alpha_u^{(1)}\mathbf{1} \qquad \mathbf{A}_v^{(2)} = \alpha_v^{(2)}\mathbf{1} \tag{71}$$

we have the simplified formula:

$$\mathrm{Tr}[\mathbf{T}_{uv}\alpha_u^{(1)}\mathbf{1}\,\mathbf{T}_{uv}\alpha_v^{(2)}\mathbf{1}] = \alpha_u^{(1)}\alpha_v^{(2)}\,\mathrm{Tr}(\mathbf{T})^2 = 6\alpha_u^{(1)}\alpha_v^{(2)} \tag{72}$$

(indeed $\quad (\mathbf{T})^2 = 9(\mathbf{r} \otimes \mathbf{r})^2 - 6\mathbf{r} \times \mathbf{r} + \mathbf{1} = 3(\mathbf{r} \otimes \mathbf{r}) + \mathbf{1} \quad$ and $\quad \mathrm{Tr}(\mathbf{r} \otimes \mathbf{r}) = 1, \quad \mathrm{Tr}\,\mathbf{1} = 3$). Therefore:

$$E_{\mathrm{disp}} \simeq -\frac{3}{2}\frac{\overline{U}^{(1)}\,\overline{U}^{(2)}}{\overline{U}^{(1)}+\overline{U}^{(2)}}\sum_u^{(1)}\sum_v^{(2)}\frac{\alpha_u^{(1)}\alpha_v^{(2)}}{(R_{uv})^6}\ . \tag{73}$$

At large intermolecular distance R, all $R_{uv} \neq R$, so that

$$E_{\mathrm{disp}} \simeq -\frac{3}{2}\frac{\overline{U}^{(1)}\,\overline{U}^{(2)}}{\overline{U}^{(1)}+\overline{U}^{(2)}}\frac{\alpha^{(1)}\alpha^{(2)}}{R^6}$$

(where $\alpha^{(1)} = \sum_u^{(1)}\alpha_u^{(1)}$ and $\alpha^{(2)} = \sum_v^{(2)}\alpha_v^{(2)}$ are the total polarizabilities of molecules (1) and (2) respectively): this is the well-known London formula [40].

The essential drawback of this method is the uncertainty in the choice of the mean excitation energies. The rule proposed by London [40], namely to take for $\overline{U}^{(m)}$ the ionization potential of (m), is satisfactory for the atoms H and He, but not for the heavier rare gas atoms, for which a value about twice the ionization potential would be more satisfactory [34, 41[+]]. Moreover, in the case of a molecule, the use of a single mean value $\overline{U}^{(m)}$ neglects possible differences between the bonds. The second method may avoid these defects.

B. THE 'DYNAMIC POLARIZABILITY' METHOD

For the principle of the method and its application to atoms, see [34, Section IIIC], [42] and references therein. The present treatment for molecules is taken from [13, Chapter III, Section 4F).

[+] The values given in [41] for the dispersion constant C_6 (corresponding to $E_{\mathrm{disp}} = -C_6/R^6$), are deduced from experimental virial coefficients; they appear slightly too large with respect to the theoretical accurate C_6 constants given in [34], probably because they take into account implicitly the contributions of further attractive terms $(-C_8/R^8, -C_{10}/R^{10})$ to the virial coefficient.

We apply the mathematical identity

$$\frac{1}{a+b} = \frac{2}{\pi} \int_0^{+\infty} \frac{ab}{(a^2+\xi^2)(b^2+\xi^2)} \, d\xi \quad (a, b>0) \tag{74}$$

to the factor $1/(\Delta E_I^{(1)} + \Delta E_J^{(2)})$ in (67). Then the separation of the bracketed factor becomes possible since the integrand in (74) is the *product* of two terms $a/(a^2+\xi^2)$ and $b/(b^2+\xi^2)$, each of them depending on a single molecule:

$$E_{\text{disp}} = -\sum_\alpha' \sum_{\alpha'}' \sum_\beta' \sum_{\beta'}' V_{\alpha\beta} V_{\alpha'\beta'} \, F(\alpha, \alpha'; \beta, \beta') \tag{75}$$

with

$$F(\alpha, \alpha'; \beta, \beta') = \frac{2}{\pi} \int_0^{+\infty} d\xi \, s_{\alpha\alpha'}^{(1)}(\xi) \, s_{\beta\beta'}^{(2)}(\xi) \tag{76}$$

where

$$s_{\alpha\alpha'}^{(1)}(\xi) = \sum_I'^{(1)} \frac{C_{I\alpha}^{(1)} C_{I\alpha'}^{(1)} \Delta E_I^{(1)}}{(\Delta E_I^{(1)})^2 + \xi^2} \qquad s_{\beta\beta'}^{(2)}(\xi) = \sum_J'^{(2)} \frac{C_{J\beta}^{(2)} C_{J\beta'}^{(2)} \Delta E_J^{(2)}}{(\Delta E_J^{(2)})^2 + \xi^2}. \tag{77}$$

Now, these quantities $s_{\alpha\alpha'}(\xi)$ are analogous to the $s_{\alpha\alpha'}$ encountered in Section 6 (Equation (50)), for the analysis of the static polarizability into local (bond) contributions (we see in fact that $s_{\alpha\alpha'}(0) = s_{\alpha\alpha'}$): they actually play the same role in the analysis of the 'polarizability at imaginary frequency' $\mathbf{A}(i\xi)$, which is obtained by analytic continuation of the dynamic polarizability:

$$\mathbf{A}(\hbar\omega) = 2 \operatorname{Re} \sum_I' \frac{(\boldsymbol{\mu}_{0I} \otimes \boldsymbol{\mu}_{0I}) \Delta E_I}{(\Delta E_I)^2 - (\hbar\omega)^2} \tag{78}$$

which describes the response (induced dipole) of the molecule to a periodic sinusoïdal electric field with angular frequency ω.

Then, the same argument (based on the consideration of localized orbitals) that was used in Section 6 for getting the approximate expression (51) for $s_{\alpha\beta}$ may be applied here to $s_{\alpha\alpha'}(\xi)$ thus giving:

$$s_{\alpha\alpha'}^{(1)}(\xi) \simeq \frac{\Delta E_\alpha^{(1)}}{(\Delta E_\alpha^{(1)})^2 + \xi^2} \delta_{\alpha\alpha'} \qquad s_{\beta\beta'}^{(2)}(\xi) \simeq \frac{\Delta E_\beta^{(2)}}{(\Delta E_\beta^{(2)})^2 + \xi^2} \delta_{\beta\beta'}. \tag{79}$$

Inserting (79) in (76) and (75), we get:

$$E_{\text{disp}} \simeq \sum_\alpha'^{(1)} \sum_\beta'^{(2)} (V_{\alpha\beta})^2 \frac{2}{\pi} \int_0^{+\infty} d\xi \, \frac{\Delta E_\alpha^{(1)}}{(\Delta E_\alpha^{(1)})^2 + \xi^2} \frac{\Delta E_\beta^{(2)}}{(\Delta E_\beta^{(2)})^2 + \xi^2}. \tag{80}$$

Then, denoting explicitly $W_\alpha^{(1)} = W_{u-w'}^{(1)}$ (localized mono-excitation $u \to w'$) and $W_\beta^{(2)} = W_{v-x'}^{(2)}$, the matrix element $V_{\alpha\beta}$ is:

$$V_{\alpha\beta} = V_{(u-w')(v-x')} = \iint \frac{f_{u-w'}^{(1)}(\mathbf{r}^1) f_{v-x'}^{(2)}(\mathbf{r}^2)}{|\mathbf{r}^1 - \mathbf{r}^2|} \, d\mathbf{r}^1 \, d\mathbf{r}^2 \tag{81}$$

and, for the same reason as previously (discussion in Section 7 between Equations (58) and (59)), this electrostatic interaction may be reduced to the interaction between the dipoles of the transition distributions $f_{u-w'}^{(1)}(\mathbf{r})$, $f_{v-x'}^{(2)}(\mathbf{r})$ (it must be recalled that the total charge of these distributions is zero, according to the definition (31) and the orthogonality of u, w' and v, x'). Then:

$$V_{(u-w')(v-x')} \simeq - \boldsymbol{\mu}_{u-w'}^{(1)} \frac{\mathbf{T}_{uv}}{(R_{uv})^3} \boldsymbol{\mu}_{v-x'}^{(2)} \tag{82}$$

and

$$(V_{(u-w')(v-x')})^2 \simeq \frac{1}{(R_{uv})^6} \, \mathrm{Tr}\left[\mathbf{T}_{uv}(\boldsymbol{\mu}_{u-w'}^{(1)} \otimes \boldsymbol{\mu}_{u-w'}^{(1)}) \, \mathbf{T}_{uv}(\boldsymbol{\mu}_{v-x'}^{(2)} \otimes \boldsymbol{\mu}_{v-x'}^{(2)})\right] \tag{83}$$

according to the tensor identity [33, appendix A], [43, appendix A II], [13, Chapter III, Appendix]:

$$(\mathbf{X}\,\mathbf{B}\,\mathbf{Y})(\mathbf{Y}\,\mathbf{B}\,\mathbf{X}) = \mathrm{Tr}[\mathbf{B}(\mathbf{Y}\otimes\mathbf{Y})\,\mathbf{B}(\mathbf{X}\otimes\mathbf{X})]. \tag{84}$$

Then, replacing in (80) $\sum_{\alpha}'^{(1)}$ by $\sum_{u}^{(1)}\sum_{w'}^{(1)}$ and $\sum_{\beta}'^{(2)}$ by $\sum_{v}^{(2)}\sum_{x'}^{(2)}$ and inserting (83), we easily get:

$$E_{\mathrm{disp}} \simeq - \frac{1}{2\pi} \sum_u^{(1)} \sum_v^{(2)} \frac{1}{(R_{uv})^6} \int_0^{+\infty} \mathrm{d}\xi \, \mathrm{Tr}\left[\mathbf{T}_{uv}\mathbf{A}_u^{(1)}(i\xi)\,\mathbf{T}_{uv}\mathbf{A}_v^{(2)}(i\xi)\right] \tag{85}$$

where

$$\mathbf{A}_u^{(1)}(i\xi) = 2\sum_{w'}^{(1)} \frac{\Delta E_{u-w'}^{(1)}}{(\Delta E_{u-w'}^{(1)})^2 + \xi^2} (\boldsymbol{\mu}_{u-w'}^{(1)} \otimes \boldsymbol{\mu}_{u-w'}^{(1)}) \tag{86a}$$

$$\mathbf{A}_v^{(2)}(i\xi) = 2\sum_{x'}^{(2)} \frac{\Delta E_{v-x'}^{(2)}}{(\Delta E_{v-x'}^{(2)})^2 + \xi^2} (\boldsymbol{\mu}_{v-x'}^{(2)} \otimes \boldsymbol{\mu}_{v-x'}^{(2)}) \tag{86b}$$

appear as the analytic continuation at imaginary frequency of *bond dynamic polarizabilities* $\mathbf{A}_u^{(1)}(\hbar\omega)$, $\mathbf{A}_v^{(2)}(\hbar\omega)$. Thus we have succeeded in expressing the total dispersion interaction between the two molecules as a sum of independent dispersion-like interactions between the regions (bonds) corresponding to the localized orbitals.

If the dynamic polarizabilities are isotropic:

$$\mathbf{A}_u^{(1)}(i\xi) = \alpha_u^{(1)}(i\xi)\,\mathbf{1} \qquad \mathbf{A}_v^{(2)}(i\xi) = \alpha_v^{(2)}(i\xi)\,\mathbf{1} \tag{87}$$

(85) becomes (see above Equations (71–73)):

$$E_{\mathrm{disp}} \simeq - \frac{3}{\pi} \sum_u^{(1)} \sum_v^{(2)} \frac{1}{(R_{uv})^6} \int_0^{+\infty} \mathrm{d}\xi \, \alpha_u^{(1)}(i\xi)\,\alpha_v^{(2)}(i\xi). \tag{88}$$

The polarizabilities $\alpha_u(i\xi)$ may be approximated in a convenient way by the following formula [34, Section IV]:

$$\alpha_u(i\xi) = \frac{\alpha_u}{1 + (b_u\,\xi)^2} \tag{89}$$

where α_u is the static polarizability and b_u is an appropriate constant for which various rules may be proposed [34], [44], [45] (e.g. according to [44] $b_u^2 = \alpha_u/n_u^{\text{eff}}$, where n_u^{eff} is an 'effective number' of electrons in the region u). With the formula (89), the integral in (88) is immediately given by (74):

$$E_{\text{disp}} \simeq -\frac{3}{2} \sum_u^{(1)} \sum_v^{(2)} \frac{1}{(R_{uv})^6} \frac{\alpha_u^{(1)} \alpha_v^{(2)}}{b_u^{(1)} + b_v^{(2)}}. \tag{90}$$

This is analogous to the formula (73) obtained by the 'mean excitation energy' method, with $b_u^{(1)}$ and $b_v^{(2)}$ corresponding to $1/\overline{U}^{(1)}$ and $1/\overline{U}^{(2)}$; but now we have different constants $b_u^{(m)}$ corresponding to the different bonds.

Further developments may be found in [13, Chapter III, Section 4F], especially the general expressions based on formulae analogous to (89) in the case of non-isotropic polarizabilities.

9. Conclusion

For the four contributions to the intermolecular interaction energy that we have studied in the present paper (namely: 1st order 'exchange' repulsion and the three long-range terms: electrostatic, polarization, dispersion), it has been possible to find a simple approximate expression in terms of local contributions. Except for the electrostatic contribution, the possibility of representing correctly the molecular ground-state function by a single Slater determinant built from *localized* molecular orbitals played a crucial role in the derivation of this result. Some differences between the different terms must be emphasized: for example, the dispersion term appears as a sum of independent bond-bond interactions, while the polarization energy appears as a sum of bond polarization energies, each of them being calculated with the *total* electric field of the other molecule (so that it is not possible to split further this energy as a sum of independent terms corresponding to regions of the polarizing molecule). Another difference concerns the *transferability* of the molecular properties involved: thus, bond polarizabilities seem rather transferable between different molecules, while atomic net charges are not (but bond dipole moments would perhaps be transferable to some extent).

Among the questions which have not been considered, we would mention the following two:

(1) the charge-transfer contribution. Owing to its exponential-like variation with distance, it is possible that this contribution has a local character, and that an analysis similar to the one performed for the 1st order 'exchange' repulsion (Section 4) would be successful.

(2) the aromatic molecules, where the π orbitals cannot be localized to a very high degree [21]. Deeper investigations would be necessary in order to see whether formulae based upon local contributions may be extended or generalized to handle this case, a possibility which seems supported by the practical successes of semi-empirical additive formulae [2, 3, 4].

References

1. Daudey, J. P., Malrieu, J. P., and Rojas, G.: in Chalvet, O., Daudel, R., Diner, S., and Malrieu, J. P. (eds.), *Localization and Delocalization in Quantum Chemistry*, Reidel, Dordrecht (1975), Vol. I, p. 155.
2a. Kitaigorodskii, A. I.: *Tetrahedron* **14**, 230 (1961); Some erroneous results of this paper are corrected in Kitaigorodskii, A. I. and Mirskaya, K. V.: *Soviet Phys. Crystallography* **9**, 137 (1964). See also Huron, M. J. and Claverie, P.: *J. Phys. Chem.* **76**, 2132 (1972) Section II.4.
2b. Kitaigorodskii, A. I.: *Acta Cryst.* **18**, 585 (1965).
2c. Kitaigorodskii, A. I. and Ahmed, N. A.: *Acta Cryst.* **A28**, 207 (1972).
3. Liquori, A. M. (a) *J. Polym. Sc.* **C12**, 209 (1966). (b) *Quart. Rev. Biophys.* **2**, 65 (1969).
4. Scheraga, H. A.: *Advances in Physical Organic Chemistry* **6**, 103 (1968), Academic Press, New York.
5. Caillet, J. and Claverie, P.: *Biopolymers* **13**, 601 (1974).
6. Caillet, J. and Claverie, P.: *Acta Cryst.* **A31**, 448 (1975),
7. Andersen, H. C. and Chandler, J. P.: *J. Chem. Phys.* **57**, 1930 (1972).
8. Murrell, J. N. and Shaw, G.: *J. Chem. Phys.* **46**, 1768 (1967).
9. Van Duijneveldt, F. B.: 'Intermolecular Forces and the Hydrogen Band', Thesis, Utrecht University (1969). Appendix A.
10. Hirschfelder, J. O. and Meath, W. J.: *Advances in Chemical Physics* **12**, 3 (1967).
11. Margenau, H. and Kestner, N. R.: *Theory of Intermolecular Forces*, Pergamon Press (1969).
12. Claverie, P.: *Intern. J. Quantum Chem.* **5**, 273 (1971).
13. Claverie, P.: 'Contribution à l'Étude des Interactions Moléculaires'. Thesis, Paris (1973). Registered at the C.N.R.S. library under number: A.O. 8214 (see also the chapter on 'Elaboration of Approximate Formulae for the Treatment of Large Molecules; Applications in Organic Chemistry' in *Intermolecular Interactions: from Diatomics to Biopolymers*, B. Pullman (ed.), Wiley, New York (scheduled for 1976 or 1977)).
14. Van der Avoird: *J. Chem. Phys.* **47**, 3649 (1967).
15. Williams, D. R., Schaad, L. J., and Murrell, J. N.: *J. Chem. Phys.* **47**, 4916 (1967).
16. Murrell, J. N. and Shaw, G.: *Mol. Phys.* (a) **12**, 475 (1967) (b) **13**, 325 (1968).
17. Daudey, J. P., Claverie, P., and Malrieu, J. P.: *Intern. J. Quantum Chem.* **8**, 1 (1974).
18. Daudey, J. P., Malrieu, J. P., and Rojas, O.: *Intern. J. Quantum Chem.* **8**, 17 (1974).
19. Daudey, J. P.: *Intern. J. Quantum Chem.* **8**, 29 (1974).
20. Longuet-Higgens, H. C.: *Proc. Roy. Soc. (London)* **A235**, 537 (1965).
21. Diner, S., Malrieu, J. P., Jordan, F., and Claverie, P.: *Theoret. Chim. Acta* **18**, 86 (1970).
22. Millie, Ph., Levy, B., and Berthier, G.: in Chalvet, O., Daudel, R., Diner, S. and Malrieu, J. P. (eds.), *Localization and Delocalization in Quantum Chemistry*, Reidel, Dordrecht (1975), Vol. I, p. 59.
23. Salem, L.: *Proc. Roy. Soc. (London)* **A264**, 379 (1961).
24. Murrell, J. N., Randic, M., and Williams, D. R.: *Proc. Roy. Soc. (London)* **A284**, 566 (1965).
25. Buckingham, A. D.: *Advances in Chemical Physics* **12**, 107 (1967).
26. Kihara, T.: *Suppl. Progr. Theoret. Phys.* **40**, 177 (1967).
27. Pollak, M. and Rein, R.: *J. Chem. Phys.* **47**, 2045 (1967).
28. Mulliken, R. S.: (a) *J. Chim. Phys.* **46**, 497 (1949) (see p. 521); (b) *J. Chem. Phys.* **23**, 1833 (1955).
29. Coulson, C. A. and Rogers, M. T.: *J. Chem. Phys.* **35**, 593 (1961).
30. Daudel, R., Lafforgue, A., and Vroelant, C.: *J. Chim. Phys.* **49**, 545 (1952).
31. Löwdin, P. O.: (a) *J. Chem. Phys.* **21**, 374 (1953); (b) *Adv. Phys.* **5**, 22 (1956).
32. Howard, B. B.: *J. Chem. Phys.* **39**, 2524 (1963).
33. Claverie, P. and Rein, R.: *Intern. J. Quantum Chem.* **3**, 537 (1969).
34. Dalgarno, A. and Davison, W. D.: *Advances in Atomic and Molecular Physics* **2**, 1 (1966).
35. Morse, P. M. and Feshbach, H.: *Methods of Theoretical Physics*, Mc Graw Hill, New York (1953).
36. Denis, A., Langlet, J., and Malrieu, J. P.: *Theoret. Chim. Acta* (a) **29**, 117 (1973); (b) **38**, 49 (1975).
37. Le Fevre, R. J. W.: *Advances in Physical Organic Chemistry* **3**, 1 (1965).
38. Zwanzig, R.: *J. Chem. Phys.* **39**, 2251 (1963).
39. Unsöld, A.: *Zeit. Physik.* **43**, 563 (1927).
40. London, F. (a): *Zeit. Physik. Chemie (B)* **11**, 222 (1930). (b) *Trans. Faraday Soc.* **33**, 8 (1937).

41. Pitzer, K. S.: *Advances in Chemical Physics* **2**, 59 (1959).
42. (a) Langhoff, P. W. and Karplus, M.: in *The Padé Approximant in Theoretical Physics*, Baker,
    G. A. and Gammel, J. L. (eds.), Academic Press, New York (1970), p. 41.
    (b) Langhoff, P. W. and Karplus, M.: *J. Chem. Phys.* **53**, 233 (1970).
    (c) Langhoff, P. W., Gordon, R. G., and Karplus, M.: *J. Chem. Phys.* **55**, 2126 (1971).
43. Rein, R., Claverie, P., and Pollak, M.: *Intern. J. Quantum Chem.* **2**, 129 (1968).
44. Mavroyannis, C. and Stephen, R. J.: *Mol. Phys.* **5**, 629 (1962).
45. Tang, K. T. and Karplus, M.: *Phys. Rev.* **171**, 70 (1968).

LOCALIZATION, BONDS, AND PHYSICAL MODELS
OF MOLECULAR REALITY

GIUSEPPE DEL RE

Cattedra di Chimica Teorica, Università di Napoli, 80134 Napoli, Italy

1. Bonds and Localized Quantum-Mechanical Models

A. THE CONCEPT OF BOND

Accurate *ab initio* calculations of purely electronic wave functions of molecules have given some grounds for doubting that privileged regions of molecules suggested by electron density maps and population analyses coincide with the traditional chemical bonds. This remark might lead to the surprising conclusion that the chemical bond is 'an arbitrary concept and can be defined in as many ways as one wishes' [1]; a conclusion resulting from having chosen to ignore that a chemical bond is not just a feature of the electron density at the equilibrium geometry. The concepts of bond and valency permit the prediction of the number of isomers of a molecule (say, a paraffin), its approximate geometry, dipole moment, frequencies of vibration, heats of formation. To a chemist, isomers differing by the topology of bonds are clearly distinct molecules, as is illustrated by the example of benzene I, Dewar's benzene II, prismane III.

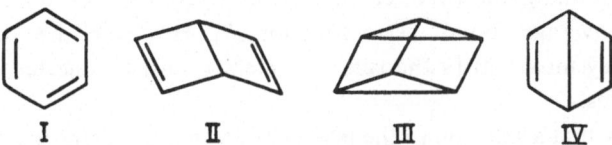

I   II   III   IV

A formula like IV denotes either the old (and obsolete) notion of a special resonant structure of the benzene aromatic system or an awkward way of writing a perfectly legitimate molecule; the awkwardness deriving from the fact that the 1–4 bond is formed by a tendentially tetrahedral carbon atom, and thus the molecule should not be drawn as if it were planar.

The concept of bond embodies the transferability of properties associated with certain pairs of atoms. Properties of certain parts or regions of molecules depend to first order only on the atoms present in those regions, so that all molecules having the same region or part will receive the same contribution from them. In a slightly broader sense, transferability also implies a rule specifying how the rest of a molecule affects the contribution of a given part. For instance, the transferability of the CC single bond means that the contributions of that bond to the total dipole moment, the heat of atomisation, etc., are very nearly the same in all the molecules containing that bond; and possible changes in those or other contributions arising from changes in molecular environment can be predicted by some simple rules.

Thus, transferability is required of most theoretical counterparts of chemical bonds [2, 3, 4]. However, it should be kept in mind that it accounts for just one aspect of the bond concept. Indeed, as is illustrated by the various additivity rules, it can sometimes apply to parts of molecules other than bonds. Group contributions to molecular properties like dipole moments are an elementary example, as well as the contributions of π systems to molecular refractions [5, 6]. Indeed, bonds like the CH bond are difficult to include in certain additive systematics [7, 8]; but, even then, geometric and above all valency considerations provide sufficient ground for speaking of the CH bond as a well defined entity; not to mention its spectroscopic characteristics.

B. BONDS AND LOCALIZATION

In conclusion, the concept of chemical bond does not seem to be in danger; electronic wave-functions for equilibrium geometries are just not sufficient to define unambiguously the bonds of a molecule. This is the reason why there has been a great interest in localization techniques and, to a lesser extent, in computations based on the preliminary assumption that the bond scheme somehow imposes the form of the wave function.

Actually, rigorous wave functions of bonds with no *a priori* restriction can be extracted from accurate *ab initio* calculations following exactly the way in which chemists arrived at the concept of chemical bond, *i.e.* by comparing a large number of molecules containing the same chemical bond, say a C—O single bond, and by constructing an average 'bond' wave function [9]. However, the idea that bonds exist and that they must satisfy the rules of valency must be injected even into such general procedures.

It is quite likely that a solution to the mystery of bonds will come from computations of potential energy surfaces. But even then we shall have the choice between *a priori* and *a posteriori* introduction of bonds in our interpretational if not computational schemes. This introduction can be realized either by more or less cautious attempts to isolate special regions in molecules, to be associated with electron pairs (the 'loges' [10]), or by referring to a physical model where 'functional localization' is admitted *a priori*, and discussing both the features it accounts for *and* those it does not account for.

In what follows we shall adopt the latter point of view, and proceed in accordance with remarks made long ago, among others by Coulson [11].

The translation of the bond concept into wave function language consists in the assignment of a convenient 'group function' in the sense of McWeeny [12]. Physically, a 'transferable' group of electrons (a bond, an aromatic sextet, etc.) can be treated as a quantum-mechanical system *formally* independent of the rest of the molecule. In the case of a bond, we may think of a fictitious diatomic molecule whose wave function represents the bond. In the independent particle model this reduces to the bond orbital concept; a notion which has often attracted the attention of physicists and chemists, from Hund's original work [13], to applications to diamond [14] and hydrocarbons [15], to a method proposed by the present author [16], to an excellent 'ab initio' investigation [17, 18], and to a very recent rediscovery by solid state physicists [19].

It is the analysis of localized models using bond orbitals that will be our main concern here.

C. INDUCTIVE AND MESOMERIC EFFECTS

The brilliant intuition of pre-quantistic theoretical chemists distinguished between two types of effects of the environment: inductive and mesomeric effects. Both are charge transfer effects: a perturbing group in the molecule – say, a chlorine atom – produces a charge shift or a polarization in the molecule. However, from the empirical point of view, they differ from each other mainly because the former appears both in saturated and in unsaturated systems, and seems to decrease exponentially with distance from the perturbing center; the latter is characteristic of conjugated systems, and propagates much further than the inductive effect, with marked preferences and special signs for certain sites (say, the o–p positions of substituted benzenes) [20, 21].

Qualitatively, the translation of these concepts into a localized theoretical scheme is straightforward, if not obvious, especially if one takes into account that saturated molecules are those for which the bond additivity rules hold best, conjugated molecules those for which they do not hold. In the case of inductive effects, the fictitious diatomic molecules that are the bonds undergo changes in field strengths (*viz.* the elements of the Hamiltonian matrix) depending on the environment, but remain formally independent; in the case of mesomeric effects, they are strongly coupled, and lose their identities (so that they must be incorporated in larger systems: e.g. the three π bonds of benzene become the benzene aromatic sextet). In a strictly perturbative model they are similar, but the mesomeric effect is inversely proportional to the energy differences between the substituent producing it and the π orbitals [22, 23].

2. A Priori Localization: a Simple Model

A. INTRODUCTION

We shall discuss here the localized model for bonds and related concepts – from hybridization to inductive effects – which is part of the MO-LCAO theory. It may be useful to emphasize that we are using the word *theory* as opposed to *method*; the former means that a well defined model replaces the molecule, loss of certain details being accepted as a price to pay for clarity in interpretation (think of the Hückel model of aromatic hydrocarbons), the latter is just a way to solve the essentially computational problem of a convenient choice of a one-electron basis for accurate determination of molecular wave functions. That a physical model – *i.e.* one to which a well defined Hamiltonian operator is associated – can be associated even to a minimal basis may not seems obvious; but it can be seen immediately in the second quantization formalism, by simply remarking that selection of a minimal basis just amounts to selection of a special coupling scheme between single particle 'propagators' [24], much as is currently done by solid-state physicists with their model Hamiltonians (*e.g.* Anderson's [25]; cf. also ref. 26).

The choice of the most significant basis in the simple MO-LCAO theory is a very delicate, not yet fully solved problem. Ideally, we should choose *modified valence atomic orbitals* (MVAO), as suggested by Mulliken [27]. A proper definition of MVAO's, however, involves definition of atoms *in situ* [28], which is still an unsolved problem, although some progress has been made toward it [24, 29]. In fact, for numerical estimates we shall have to be content with the adoption of Slater orbitals, although they are decidedly not an ideal choice – as was pointed out long ago by Mulliken [28]. We shall indicate, however, how they can be replaced by hopefully better AO's.

Having chosen a MVAO basis, the steps to reach a chemically significant description of a molecule can be schematized as follows: localization by hybridization, construction of the Hamiltonian for localized electron systems, introduction of non-bonded interactions, determination of geometries and conformations.

B. LOCALIZATION AND MODEL HAMILTONIANS

The whole procedure is based on the one-electron effective Hamiltonian which, in our model, determines the form of the molecular orbitals. If a block can be factorized out of the Hamiltonian matrix in question, there will be a certain number of orbitals which are formed with a subset of the given basis, and are formally completely independent of the rest of the molecular orbitals. For instance, in a molecule like CO_2, the MVAO basis $|\chi\rangle$ has the form

$$
\begin{aligned}
|\chi\rangle &\equiv (|\chi_1\rangle\ |\chi_2\rangle\ |\chi_3\rangle\ \cdots\ |\chi_{12}\rangle) \\
&= (|l_1\rangle\ |l_2\rangle\ |h_0\rangle\ |p\pi_0\rangle\ |h_c\rangle\ |h_c'\rangle\ |p\pi_c\rangle\ |p\pi_c'\rangle \\
&\quad |p\pi_0'\rangle\ |h_0'\rangle\ |l_1'\rangle\ |l_2'\rangle)
\end{aligned}
\tag{1}
$$

where the prime refers both to the right hand oxygen and to the plane perpendicular to that of the π orbitals; the letter l stands for the lone pairs, the letter h for the hybrids.

Suppose now that the effective Hamiltonian can be represented by a matrix **H** which consists of 1×1 and 2×2 blocks corresponding respectively to the lone pairs and to the four σ and π bonds of the molecule.

Then the whole system can be treated as if it consisted of eight independent electron pairs, to be described in terms of smaller bases of one or two elements (Equation (1)); the bonding and the lone pair orbitals being given by Equation (2).

In fact, the eigenvalue equation for the system in the above assumption is determined by the matrix

$$
\mathbf{H} = \begin{vmatrix}
H_{1,1} & 0 & 0 & 0 & 0 & 0 & 0 & 0 & 0 & 0 & 0 & 0 \\
0 & H_{2,2} & 0 & 0 & 0 & 0 & 0 & 0 & 0 & 0 & 0 & 0 \\
0 & 0 & H_{3,3} & 0 & H_{3,5} & 0 & 0 & 0 & 0 & 0 & 0 & 0 \\
0 & 0 & 0 & H_{4,4} & 0 & 0 & H_{4,7} & 0 & 0 & 0 & 0 & 0 \\
0 & 0 & H_{5,3} & 0 & H_{5,5} & 0 & 0 & 0 & 0 & 0 & 0 & 0 \\
0 & 0 & 0 & 0 & 0 & H_{6,6} & 0 & 0 & 0 & H_{10,6} & 0 & 0 \\
0 & 0 & 0 & H_{7,4} & 0 & 0 & H_{7,7} & 0 & 0 & 0 & 0 & 0 \\
0 & 0 & 0 & 0 & 0 & 0 & 0 & H_{8,8} & H_{8,9} & 0 & 0 & 0 \\
0 & 0 & 0 & 0 & 0 & 0 & 0 & H_{9,8} & H_{9,9} & 0 & 0 & 0 \\
0 & 0 & 0 & 0 & 0 & H_{6,10} & 0 & 0 & 0 & H_{10,10} & 0 & 0 \\
0 & 0 & 0 & 0 & 0 & 0 & 0 & 0 & 0 & 0 & H_{11,11} & 0 \\
0 & 0 & 0 & 0 & 0 & 0 & 0 & 0 & 0 & 0 & 0 & H_{12,12}
\end{vmatrix}
\tag{2}
$$

where, with four 1×1 blocks, there appear the following fully independent 2×2 blocks

$$
\mathbf{H}(\sigma) = \begin{vmatrix} H_{3,3} & H_{3,5} \\ H_{5,3} & H_{5,5} \end{vmatrix}
\quad \text{and} \quad
\mathbf{H}(\sigma') = \begin{vmatrix} H_{6,6} & H_{6,10} \\ H_{10,6} & H_{10,10} \end{vmatrix}
\quad \text{for the two } \sigma \text{ bonds;}
$$

$$
\mathbf{H}(\pi) = \begin{vmatrix} H_{4,4} & H_{4,7} \\ H_{7,4} & H_{7,7} \end{vmatrix}
\quad \text{and} \quad
\mathbf{H}(\pi') = \begin{vmatrix} H_{8,8} & H_{8,9} \\ H_{9,8} & H_{9,9} \end{vmatrix}
\quad \text{for the two } \pi \text{ bonds,}
$$

so that the lone pair molecular orbitals are just the corresponding basis elements, and the bonding orbitals have the general form

$$
|\phi_j\rangle = C_{\mu j} |\chi_\mu\rangle + C_{vj} |\chi_v\rangle .
\tag{3}
$$

We shall say that we have localized our system when we have factorized our effective Hamiltonian into a form of the type (2), possibly with blocks of order higher than 2, as will happen, for instance, in the case of benzene. Localization is introduced at the (model or effective) Hamiltonian level, not at the wavefunction level.

Several remarks are in order. First of all, the factorization under study is not imposed in principle, but is expected to result from the geometry of the given molecule and from the properties of the basis; which, for the purpose, can be built from spherical-harmonic orbitals by suitable hybridization [30, 31], as will be discussed later, but should not be delocalized *ad hoc* (*v. infra*).

Except in cases of different symmetries, the factorization will be but a first order property. Moreover, even if factorization held perfectly, it would not imply that the various subsystems into which the molecule is divided are actually independent of one another. The Hamiltonian *operator* which enters each of the matrix elements depends anyway on the whole molecule.

The kind of localization introduced here is not always a localization in the geometric sense, even though the fact that basis elements are supposed to be associated with individual nuclei implies some measure of spatial localization. In bond-orbital models, the latter is the consequence of approximate or rigorous separation into subsystems; and the word localization is used in a generalized sense. Two-center orbitals associated with bonds are often called localized orbitals, and so they are, in fact, if the AO's involved fall off very rapidly away from the corresponding nuclei; but orbitals formed by linear combination of $3s$ orbitals, say, would easily extend away from the nuclei over a large part of the whole molecule.

In short, it is important to keep in mind the distinction between the geometrical sense and the sense of 'separability' that can be assigned to the term localization [10].

As has been mentioned, the bond orbital concept has been repeatedly used by various authors [13, 19]; one of the most systematic attempts to application has been with molecular interactions and related topics [33]. However, contrary to the path traced by Mulliken [27, 32] the minimal basis problem and the resulting effective Hamiltonian concept do not seem to have been fully elaborated, even though model Hamiltonians were defined for typical π approximations [26].

C. LOCALIZATION AND INDUCTIVE EFFECTS: A 'NAIVE' APPROACH

Consider a saturated molecule like a paraffin, an alkylamine, a dialkylether. Such a system is made up of bonds, the additive rules hold quite well, and, due to the small 'size' of the atomic orbitals, the bond orbitals are even spatially localized. Now, we know that for the π systems of conjugated molecules a fairly deep understanding of the major features can be obtained by the simple Hückel method, not only in cases where topology is the only major factor (as with benzenoid hydrocarbons) but also when account of differences in electronegativity must be taken because of the presence of heteroatoms [34, 35].

In other words, empirical or semiempirical estimates of the matrix elements of the effective Hamiltonian for a subsystem of a molecular electron system is sufficient to provide at least semi-quantitative understanding of mesomeric effects. Is it possible to set up a localized scheme of the same level of sophistication to gain some insight into inductive effects in a completely saturated molecule?

Before entering the formal aspects of the problem, we shall describe a practical answer to the above question; the only preliminary remark needed is that – because of the definition of localization we are using here – the inductive effects must be introduced through some change in the matrix elements of the bond Hamiltonian, and not – as is the case with conjugation – through nonvanishing off-diagonal elements connecting blocks associated with different bonds.

The maximum of simplicity is obtained, first of all, by assuming that the hybrid orbitals entering a bond orbital are orthogonal to each other. This is somewhat strange at first sight, but is actually completely all right for the 2×2 problem, because the off-diagonal part of the Hamiltonian matrix is then always proportional to over-lap, and a simple transformation makes the orthogonalized and the non-orthogonal problems equivalent.

Next, we assume that the diagonal elements of the bond Hamiltonian $\mathbf{H}_{bond}$ are roughly a measure of the electronegativies of the atoms involved – an assumption already introduced in the standard Hückel method – whereas the off-diagonal elements measure at the same time the strength and the resistance to polarization of the bond. Calling X and Y the two atoms, and keeping in mind that in a saturated molecule we have one bond per atom pair, we write:

$$\mathbf{H}_{bond} = \begin{vmatrix} H_{XX} & H_{XY} \\ H_{YX} & H_{YY} \end{vmatrix} = \alpha \begin{vmatrix} 1 & 0 \\ 0 & 1 \end{vmatrix} + \beta \begin{vmatrix} \delta_X & \eta_{XY} \\ \eta_{XY} & \delta_Y \end{vmatrix} \tag{4}$$

which defines the dimensionless parameters δ_X, δ_Y, and η_{XY} in units β and with an energy zero-point α exactly as is customary in the Hückel method. As is well known, with that unit and that zero-point, $\mathbf{H}_{bond}$ can be replaced to all purposes by the matrix

$$\overline{\mathbf{H}}_{bond} = \begin{vmatrix} \delta_X & \eta_{XY} \\ \eta_{XY} & \delta_Y \end{vmatrix}, \tag{5}$$

whose eigenvectors give the coefficients indicated in Equation (2). With the necessary reservations [8], we can identify the atomic charges with the defect or excess electron populations

$$q_{X(Y)} = 1 - 2c_{X(Y)}^2; \quad q_{Y(X)} = -q_{X(Y)}, \tag{6}$$

where the partner of X is specified because atom X may form several bonds. In fact, its total net charge will be

$$q_X = \sum_Y q_{X(Y)} \tag{7}$$

where Y denotes all the atoms directly linked to X. For instance, in methane, for a CH bond one finds:

$$\overline{\mathbf{H}}_{CH} = \begin{vmatrix} 0.1343 & 1 \\ 1 & 0.0538 \end{vmatrix}$$

if the parameters are evaluated according to the method of Ref. 16.

Consider inductive effects. These were already allowed for to some extent in the Hückel method [36]. Following the same idea, but generalizing it to all atoms and taking into account the possibility of feedback (an atom increasing its electronegativity will cause an increase in the electronegativity of the very neighbor that induced the first increase), we proceed as follows.

Each parameters δ_X of atom X is assumed to consist of a contribution δ_X^0 which depends only on the nature of atom X and of a sum of contributions proportional to the δ's of the partners of X in the bonds X forms; the proportionality factors will be certain transfer coefficients characteristic of the bonds and of the directions in which they are taken:

$$\delta_X = \delta_X^0 + \sum_Y \gamma_{XY} \delta_Y. \quad \square \tag{8}$$

Since the δ's appearing on the right hand side of this equation are determined by equations of the same type, the set of the δ's is the solution of a linear system of equations which, in matrix formalism, can be written

$$\Gamma\Delta = \Delta^0 \tag{9}$$

Γ being a non-symmetric matrix which we may call the 'transfer matrix', with diagonal elements 1 and off-diagonal elements 0 for non bonded atom pairs. The numerical form of Equation (9) for methanol, with atoms taken in the order methyl-hydrogens—carbon—oxygen—hydrogen is

$$
\begin{array}{cccccc}
\text{H} & \text{H} & \text{H} & \text{C} & \text{O} & \text{H}
\end{array}
$$

$$
\begin{vmatrix}
1 & 0 & 0 & -.4 & 0 & 0 \\
0 & 1 & 0 & -.4 & 0 & 0 \\
0 & 0 & 1 & -.4 & 0 & 0 \\
-.3 & -.3 & -.3 & 1 & -.1 & 0 \\
0 & 0 & 0 & -.1 & 1 & -.3 \\
0 & 0 & 0 & 0 & -.4 & 1
\end{vmatrix}
\begin{vmatrix}
\delta_{H1} \\
\delta_{H2} \\
\delta_{H3} \\
\delta_C \\
\delta_O \\
\delta_H
\end{vmatrix}
=
\begin{vmatrix}
0 \\
0 \\
0 \\
0.07 \\
0.40 \\
0
\end{vmatrix}
\tag{10}
$$

The parameters used here are from the set originally chosen in order to improve the bond moment description of saturated molecules by allowing for inductive effects [16]. In the present context, only the fact that inductive effects appear in the net charges derived according to the above scheme is important; it is illustrated by the examples of methyl substituted water and amines (Table I).

The use of net populations in Mulliken's sense as measures of charge effects in molecules is open to criticism [8]. However, apart from surprising success in applications, the above method is presented here only as a simple model of localized treatment of a system of σ bonds, and the fact that it appears to take inductive effects into fairly correct account has little to do with the choice of Mulliken's populations or other indices. Anyway, it may be useful for the following to remark that use of those charges not only for inductive effects but also for estimating dipole moments implies that atomic and overlap moments cancel around a given atom in its normal valency: an assumption which strongly depends on the nature of the basis – or, in other words, is one of the many delicate conditions imposed on the MVAO's hidden behind the apparently simple model just presented.

TABLE I

Net charges q_C, q_N, q_O in ammonia and water derivatives according to Ref. 16

	q_C	q_N	q_O
Methylamine	−0.0585	−0.5258	–
Dimethylamine	−0.0664	−0.3520	–
Trimethylamine	−0.0726	−0.1978	–
Methylalcohol	−0.0133	–	−0.4539
Dimethylether	−0.0249	–	−0.2692
Ethylalcohol	+0.0355	–	−0.4582

In conclusion, a first-order chemical understanding of σ bonds and inductive effects can be obtained along the same lines as the Hückel π-model. Both models are extremely simplified schemes implying, for their approximate validity, at least a certain amount of separability and certain conditions on the basis. But, following the current practice of chemists and crystallographers, they may be invaluable in that they provide a conceptual and practical reference for the theoretical treatment of finer effects like for σ-π interactions, rotation barriers, etc.

This is the reason why global methods of computation [37], for all their important contribution to chemistry, cannot replace the 'naïve' Hückel methods with σ-π separation [38].

With these premises, the main question for theoreticians becomes the explicit formulation of the steps which connect a completely general treatment with the simplified schemes under study. In other words, the theoretician should indicate exactly what simplifications are made in order to obtain a scheme of the Hückel type, and what are the forms of the neglected contributions. If and when a complete answer to this becomes available, the above scheme will acquire entirely the dignity of a reference model because it will be possible to estimate neglected terms, and hence to assess its validity and limitations in any given application. To illustrate this point consider the well known example of correlation: it is not as important to include correlation in *ab-initio* accurate calculations as it is to know how much the results are affected by neglect of its variations from molecule to molecule [39].

Among other things hybridization obtained according to a particular criterion is an immediate result of an attempt to formalize and make rigorous the model presented in this section, as well as of attempts to obtain 'best-energy' bond orbitals [17, 18]; or to set up highly sophisticated localized orbital methods [40]. Before discussing questions concerning the atomic orbital basis, however, we shall concentrate on the effective Hamiltonian matrix. What partitioning of the total electronic Hamiltonian operator leads to an effective one-electron Hamiltonian of the type described in the present section?

3. Rigorous Definition of the Model Hamiltonian

A. GENERAL FORMALISM

The effective Hamiltonian which corresponds to Equations (1) and (6) can be defined starting from a closed shell determinantal wave function over bond orbitals. Some aspects of that partitioning have been the object of work by other quantum chemists [17, 40]. An attempt to define an effective Hamiltonian from the energy expression for that function has been recently published by us [41], extending to σ systems an analysis already made for π systems [42].

Here, we shall generalize those results using the second-quantization formalism, which provides a compact expression of the total Hamiltonian operator in terms of matrices over a given orbital basis, and thus makes consistent construction of a model Hamiltonian quite straightforward [24, 26].

Let $|i\rangle$ denote a bond orbital in the general sense – namely a two– or few-center σ or π LCAO orbital involving one AO per center. Let $|is\rangle$ and $|is'\rangle$ denote the corresponding spin-orbitals with spin components s, s'. Finally, let $\hat{a}_{is}^{+}$ and $\hat{a}_{is}$ denote the creation and annihilation operators for an electron in $|is\rangle$. The total molecular electronic Hamiltonian can be written

$$\mathscr{H} = \overline{\mathscr{R}} + \sum_{i,j}\sum_{s} H_{ij}^{(c)}\hat{a}_{is}^{+}\hat{a}_{js} + \tfrac{1}{2}\sum_{i,j}\sum_{k,l}\sum_{s,s'}(ij|kl)\,\hat{a}_{is}^{+}\hat{a}_{ks'}^{+}\hat{a}_{ls'}\hat{a}_{js} \tag{11}$$

where $\overline{\mathscr{R}}$ is the nuclear repulsion term, $H_{ij}^{(c)}$ is the core Hamiltonian matrix element associated with the bond orbitals $|i\rangle$ and $|j\rangle$, $(ij|kl)$ is the standard two-electron integral $(ij|1/r|kl)$.

Let $\hat{n}_{ks'}$ be the number operator for $|ks'\rangle$, and $\langle n_{ks'}\rangle$ its expectation value over the state taken as the reference one (here, of course, the state for which the chemical bonds are defined).

The standard procedure for obtaining the Fock-Hamiltonian [43] gives (in the spin-restricted scheme):

$$\mathscr{H} = \overline{\mathscr{R}} + \sum_{i,j}\sum_{s} H_{ij}\hat{a}_{is}^{+}\hat{a}_{js} + \varDelta\mathscr{H} \tag{12}$$

where

$$H_{ij} = H_{ij,s} = H_{ij}^{(c)} + \sum_{k}\sum_{s'}[ij|kk]_{ss'}\langle n_{js'}\rangle \tag{13}$$

is the ijth element of the Fock matrix H over the bond orbital basis;

$$[ij|kk]_{ss'} = (ij|kk) - (ik|kj)\,\delta(s,s') \tag{14}$$

where $\delta(s,s')$ is zero if s and s' are different, one otherwise;

$$\mathscr{R} = \overline{\mathscr{R}} - \tfrac{1}{2}\sum_{j,k}\sum_{s,s'}[jj|kk]_{ss'}\langle n_{js}\rangle\langle n_{ks}\rangle \tag{15}$$

is an effective nuclear repulsion. The term $\varDelta\mathscr{H}$ is a correction which contains not only correlation proper, but two additional contributions. One of them originates from the fact that, even if the standard SCF procedure is applied, the restrictions implicit in the bond orbital concept make the result different from the 'best' (Roothaan) orbital set [17]–for that, Equation (14) should contain also non-diagonal terms in the summation. In other words, unless the bond orbitals coincide with the Hartree-Fock orbitals, part of the screening to the nuclear repulsion and part of the one-electron term of the Hartree-Fock scheme are included in the correction term which must be analyzed when the scope and limitations of the bond orbital model are discussed. The other contribution arises from non-completeness of the basis, as discussed below.

B. THE BOND ORBITAL MODEL AND THE MVAO BASIS

We now turn to the (hybrid) MVAO basis $|\chi\rangle$. This is a truncated basis, and it has already been mentioned that truncation implies that we have separated out of the parts corresponding (i) to the couplings between the elements of $|\chi\rangle$ and those of the complementary set $|\chi'\rangle$, (ii) to the complementary set itself. Our complete model will neglect the couplings; its form such as we are studying here does not require explicit formulation of contribution (ii) either, unless certain excited states are to be studied.

The set $|\chi\rangle$ is also orthogonal as regards pairs of orbitals not belonging to the same bond or group: the overlap matrix S is block diagonal of the form (10). This condition can be at least reasonably well realized on a hybrid atomic orbital basis (*vide infra*) [50].

Denoting by **C** the matrix which transforms $|\chi\rangle$ into $|\psi\rangle = |\chi\rangle^C$, and by $\hat{a}_{\mu s'}^+$, α_{vs} the creation and annihilation operators for the dual AO basis, we can write, following Ref. 43:

$$\hat{\alpha}_{\mu s} = \sum_j C_{\mu j} \hat{a}_{js} \qquad \hat{\alpha}_{\mu s}^+ = \sum_j C_{\mu j}^* \hat{a}_{js}^+ \tag{16}$$

with

$$C^+ S C = I \tag{17}$$

whence

$$\langle \hat{\alpha}_{\mu s}^+ \hat{\alpha}_{vs} \rangle = \sum_{j,k} \langle \hat{a}_{js}^+ \hat{a}_{ks} \rangle C_{\mu j}^* C_{vk} = \sum_j C_{\mu j}^* C_{vj} \langle n_{js} \rangle \tag{18}$$

the latter equality holding only in a molecular state where the occupation number matrix of the $|k\rangle$'s is diagonal, i.e., where the bond orbitals have integral occupation numbers.

In the hybrid MVAO basis $|\chi\rangle$ the one-electron operator defined in Equation (12) is thus represented by the matrix whose general element is:

$$H_{\mu v} = H_{\mu v,\, s} = H_{\mu v}^{(c)} + \sum_{\kappa,\lambda} \sum_{s'} [(\mu v | \kappa \lambda) - (\mu \lambda | \kappa v)\, \delta(s, s')]\, C_{\kappa j}^* C_{vj} \langle n_{js'} \rangle \tag{19}$$

which holds for either value of the spin index s, so that, as we are not interested in 'magnetic' states, it is also the spinless Hamiltonian matrix of ordinary orbital theories.

In the closed-shell minimum-energy state (which we shall call the ground state even if it is not the HF one) we can write the well-known formula

$$H_{\mu\nu} = H_{\mu\nu}^{(c)} + \sum_{\kappa,\lambda} [2(\mu\nu|\kappa\lambda) - (\mu\lambda|\kappa\nu)] P_{\kappa\lambda} \tag{20}$$

with

$$P_{\kappa\lambda} = \sum_{j \, \text{occ}} C_{\kappa j}^* C_{\lambda j}. \tag{21}$$

The matrix (20) should be diagonalized in order to give the 'best' bond orbitals $|\varphi_0\rangle$ in the basis $|\chi\rangle$. However, this diagonalization need not give a block-factorized **C** matrix, as is necessary if the φ orbitals must be bond orbitals in the sense specified in Equation (3). Therefore, we need a further step, which consists in extracting from **H** the block-diagonal part which corresponds to the given bond scheme. How to find the latter will be indicated later.

Thus, we define three matrices (in the MVAO representation!):

(1) an effective Hamiltonian matrix $\mathbf{H}_{\text{eff}}$ for the bond orbital model, which is factorized into blocks (cf. Equation (2));

(2) a correction matrix $\mathbf{H}''$ which contains the coupling between blocks;

(3) a correction matrix $\mathbf{H}'$ containing terms neglected in $\mathbf{H}_{\text{eff}}$ because, for reasons explained below, approximations for integrals have been introduced in the latter.

With the above partitioning, we write

$$\mathbf{H} = \mathbf{H}_{\text{eff}} + \mathbf{H}' + \mathbf{H}''; \qquad \mathbf{H}_{\text{eff}} = \text{direct sum of } \mathbf{H}_{\text{eff}}^{(k)} \tag{22}$$

k indicating the kth bond or subsystem of the molecule, $\mathbf{H}_{\text{eff}}^{(k)}$ indicating the corresponding block of $\mathbf{H}_{\text{eff}}$ (cf. Equation (2)).

At this stage, with the standard occupation scheme of the Hückel model for the ground state, we have completely specified the Hamiltonian matrix (and the Hamiltonian operator) for our localized model of the electron distribution in a molecule. Of course, we have found once again the same old formulas; but, on one side, our purpose has been to introduce them so as to make evident that we have defined a physical model with a well specified Hamiltonian, derived in an unambiguous way from the general electronic Hamiltonian of a molecule; on the other side, we have prepared the ground for further discussion.

We shall consider explicitly only the ground state, but note that the dependence of $\mathbf{H}_{\text{eff}}$ on a particular choice of the occupation numbers of the bond orbitals is not a reason why it should not define a model for excited states. Indeed, a very interesting aspect of the latter is precisely provided by their differences in bond localization with respect to the ground state. Of course, in this case attention must be focused on the corrections appearing in Equations (22) and (12), and on the consequences of truncation of the basis. Distinctions like those between Rydberg and valence excited states [44], as well as the interpretation of certain properties of excited states originate from this kind of analysis.

C. PARTIAL SUMMARY

The results so far obtained in the analysis of the model under study are:

(1) Separation of the many-electron states into two complementary sets, one of which will be built by occupying in different ways appropriate one-electron states resulting from a (truncated) MVAO basis. The coupling between these two sets – arising from the fact that the states of either set cannot in general be exact eigenstates of the molecular electron system – is treated as a separate term of the total Hamiltonian, as is done for the coupling between singlet and triplet states or between the states of matter and those of radiation in standard quantum mechanical treatments.

(2) Definition of formally independent subsystems of the molecule (bonds, lone pairs, conjugated π-bond systems, etc.) to which a set $|\varphi\rangle$ of group orbitals, in particular bond orbitals, can be assigned. The elements of $|\varphi\rangle$, called here simply bond orbitals, are the one-electron states mentioned above. The requirement that they be the stationary states of the one-electron model under construction is equivalent to the requirement that the matrix representing the Hamiltonian operator of the model in a convenient (hybrid) MVAO basis be factorized into blocks corresponding to the various bonds or groups into which the molecule has been partitioned. The overlap matrix of the basis and the density matrices associated with the various bond orbitals must also be factorized in the same way.

(3) Definition of the reference state giving the occupation scheme to be used for defining the Fock-Hamiltonian for the bond orbitals. This will be taken as the closed shell state where the lowest-energy bond orbitals are doubly occupied for an ordinary molecule; and in general as the general chemist's ground state for the given species. Of course, this choice will make the aptness of the model to describe molecular 'reality' without corrections strongly dependent on the state to be studied and on the number of electrons of the group involved in possible excitation; but this is no drawback as long as the model is intended as a reference for explaining chemical facts. Indeed, the chemist always refers to the state represented by the chemical formula even in discussing excited states.

The above steps have led to a uniquely defined Hamiltonian matrix $\mathbf{H}_{\mathrm{eff}} + \mathbf{H}'$ (Equation (22), which differs from the complete SCF one-electron matrix of the molecule (over the MVAO basis) by a matrix $\mathbf{H}''$ containing the elements coupling the blocks of the other two. The next steps to be taken deal with the definition of $\mathbf{H}_{\mathrm{eff}}$ and the physical significance of its elements. They are

(4) Separation of the contributions to the kth block of $\mathbf{H}_{\mathrm{eff}} + \mathbf{H}'$ of, (a), cores and electrons of the kth bond; (b), electrons of other bonds formed by the atoms involved in the kth bond; (c), atoms and bonds other than those considered under (a) and (b).

(5) Introduction of the Mulliken approximation for orbital products. This is useful for simplifying the physical interpretation of the effect of neighboring atoms and bonds on a given bond and for the analysis of the major features of electrostatic interactions.

After these steps a problem remains, that of the definition of the MVAO basis.

We have often reminded the reader that the latter should be conveniently chosen and hybridized, but these warnings are not sufficient, of course. We shall briefly indicate how modification of the radial parts of free atom orbitals and hybridization can give a hybrid MVAO basis whose elements are reasonable from the intuitive point of view and at the same time satisfy certain conditions which are required by the model at hand.

Before discussing the basis problem, however, we shall complete the outline of the construction of our model.

D. EFFECTIVE BOND HAMILTONIAN

The effective bond Hamiltonian is constructed from the total electronic Hamiltonian following the steps just listed.

To show how the steps (4) and (5) can be carried out, we consider now the matrix H of Equations (20) and (23). We shall use the following notation:

$X, Y, \ldots$ or $X(k), Y(k), \ldots$ denote atoms participating in the kth bond (or sub system);

$A, B, \ldots$ denote atoms other than those involved in the kth bond;

$k, l, \ldots$ denote quantities, in particular atomic orbitals, belonging to the kth (lth, ...) bond; e.g. a hybrid orbital of X directly involved in the kth bond, which links X to Y;

$\bar{k}, \bar{l}, \ldots$ denote atomic orbitals and related quantities belonging to atoms participating in the kth (lth) bond, but not involved directly in it, e.g. a hybrid orbital of X involved in the lth bond ($l \neq k$);

$K, L, \ldots$ denote all the atomic orbitals and atoms not associated at all with the kth (lth, ...) bond or orbitals participating in it.

We now partition $\mathbf{H}^{(C)}$ into square blocks $\mathbf{H}^{(C)}(k)$, which form a block-diagonal matrix, and rectangular blocks $\mathbf{H}^{(C)}(k, K)$ which contain the coupling elements between k and the other bonds

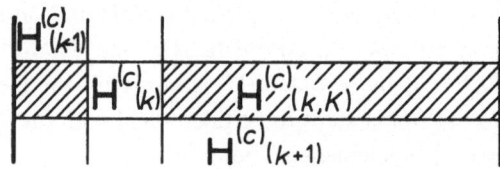

Next we partition in the same way the second term of Equation (20). This step is simple, because we just write:

$$H_{\mu\nu}(k) = H_{\mu\nu}^{(c)}(k) + \sum_{\kappa, \lambda} [2(\mu\nu|\kappa\lambda) - (\mu\lambda|\kappa\nu)] P_{\kappa\lambda} \qquad (24)$$

$$\mu, \nu \text{ in } k$$

$$H_{\mu\nu}(k) = 0 \qquad\qquad \mu, \nu \text{ not in } \kappa$$

and call $\mathbf{H}(k, K)$ the remaining rectangular block of $\mathbf{H}$. The matrix resulting from $\mathbf{H}$ when the blocks $\mathbf{H}(k)$ are suppressed is $\mathbf{H}''$ of Equation (22). Evaluation of its elements in view of assessing its importance can be obtained by well known approximation formulas [45].

Let us analyze a block $\mathbf{H}(k)$. Denoting by $\mathbf{T}$ the kinetic energy matrix, and by V_X^c the effective core potential of atom X, we can write

$$H_{\mu\nu}(k) = T_{\mu\nu} + \langle \mu | \hat{v} | \nu \rangle \tag{25}$$

$$\hat{v} = -\sum_A V_A^c - \sum_X V_X^c + \sum_{\kappa,\lambda \in k} \hat{\mathscr{L}}_{\kappa\lambda} P_{\kappa\lambda} + 2 \sum_{\kappa \in k} \sum_{\lambda \notin k} \hat{\mathscr{L}}_{\kappa\lambda} P_{\kappa\lambda} + \sum_{\kappa,\lambda \notin k} \times$$

$$\times \hat{\mathscr{L}}_{\kappa\lambda} P_{\kappa\lambda} \tag{26}$$

with

$$\hat{\mathscr{L}}_{\kappa\lambda} = \frac{1}{r_{12}} \left(2|\kappa\lambda| - \kappa|\lambda \right)$$

where $|\kappa\lambda|/r_{12}$ is the 'Coulomb operator', $\lambda|\kappa/r_{12}$ is the 'exchange operator'.

Next, we extract from $H_{\mu\nu}(k)$ a part $H_{\text{eff}\,\mu\nu}(k)$ which satisfies the following approximations

(a) the contribution of the term with $\kappa \in k$, $\lambda \notin k$ is zero;

(b) the contribution of the term corresponding to $\kappa \notin k$, $\lambda \notin k$ is just the Coulomb part. The latter is expressed by means of the generalized Mulliken approximation involving ad hoc parameters u [7, 46], i.e. by writing for the orbital product $(\kappa\lambda)$:

$$(\kappa\lambda) = \tfrac{1}{2} \left[S_{\kappa\lambda} u_{\kappa\lambda}(\kappa\kappa) + S_{\kappa\lambda} u_{\lambda\kappa}(\lambda\lambda) \right] \tag{27}$$

with $S_{\kappa\lambda} = S_{\lambda\kappa}$ the overlap integral and $u_{\kappa\lambda} + u_{\lambda\kappa} = 1$. With this approximation, the term in question becomes

$$\sum_{\kappa,\lambda \notin k} 2 \frac{|\kappa\lambda|}{r_{12}} \simeq 2 \sum_\kappa Q_\kappa \frac{|\kappa\kappa|}{r_{12}} \tag{28}$$

where use has been made of the equality

$$\sum_\kappa \sum_\lambda S_{\kappa\lambda} P_{\kappa\lambda} u_{\kappa\lambda}(\kappa\kappa) = \sum_\lambda \sum_\kappa S_{\kappa\lambda} P_{\kappa\lambda} u_{\lambda\kappa}(\lambda\lambda),$$

and

$$Q_\kappa = \sum_\lambda S_{\kappa\lambda} P_{\kappa\lambda} u_{\kappa\lambda} \tag{29}$$

is the (generalized) gross atomic population of the basis orbital $|\kappa\rangle$.

Note that (a) is just a special case of (b); because $|\kappa\rangle$ and $|\lambda\rangle$ are supposed to be orthogonal if they belong to different bonds.

With the above simplifications, and V_A^c replaced by Z_A/r_A we can write

$$-\sum_A \left[\frac{Z_A}{r_A} - \sum_{\kappa \in A} \sum_\lambda \hat{\mathscr{L}}_{\kappa\lambda} P_{\kappa''} \right] \simeq -\sum_A \left[\frac{Z_A}{r_A} - 2 \sum_{\kappa \in A} Q_\kappa \frac{|\kappa\kappa|}{r_{12}} \right] = \sum_A \hat{V}_A \tag{30}$$

where $\kappa \in A$ means: '$|\kappa\rangle$ is an atomic orbital of atom A'. Evidently, the complicated Coulomb-exchange potential acting on an electron in the SCF scheme as a result of the presence of atom A and of the occupation of its electrons has been replaced by an average potential V_A which depends only on the position vector r_A, of the electron with respect to A:

$$V_A = V(r_A) = -\frac{q_A}{r_A} + \text{dipole term} + \dots \tag{31}$$

where $q_A = Z_A - 2 \sum_{\kappa \in A} Q_\kappa$ is the 'net charge' of A. This potential will introduce the major (electrostatic) non-bonded contributions to the effective Hamiltonian and hence to the orbital energies and to the bond populations. The non-bonded contributions to the effective nuclear repulsion are a different story, and will be mentioned later.

With the above premises we define

$\mathbf{H}'(k)$ as the matrix whose elements are the contributions to (26) left out by approximation (a) and (b), and

$\mathbf{H}_{\text{eff}}(k)$ as the form taken by $\mathbf{H}(k)$ when the latter approximations are adopted, i.e. when (28) and hence (31) are introduced.

We emphasize that, in the philosophy of the present study, further approximations possibly made on $\mathbf{H}_{\text{eff}}$ (replacement of the SCF-type formula by a Hückel-type one, neglect of multipole terms in Equation (31), etc.) should always be understood as shifting certain contributions from $\mathbf{H}_{\text{eff}}$ to $\mathbf{H}'$. The latter should not be considered as a waste basket; on the contrary, it becomes vital to know its structure and to at least estimate its elements whenever the model represented by $\mathbf{H}_{\text{eff}}$ has to be critically examined; much in the same way as one may need to study the trimmings to reconstruct the original shape of an object if machining has modified it too much.

E. DISCUSSION OF $\mathbf{H}_{\text{eff}}$

The form of $\mathbf{H}_{\text{eff}}$ just obtained is

$$H_{\text{eff}, \mu\nu}(k) = T_{\mu\nu} + \langle \mu | \hat{V}(k) | \nu \rangle + \sum_A \langle \mu | \hat{V}_A | \nu \rangle \tag{32}$$

where $\hat{V}_A$ is given by (30) and

$$\hat{V}(k) = -\sum_X V_X^c + \sum_{\kappa, \lambda \in X} \hat{\mathscr{L}}_{\kappa\lambda} P_{\kappa\lambda}. \tag{33}$$

The contribution of the environment to the elements of $\mathbf{H}_{\text{eff}}$ is represented by the last term of (32). To briefly study it and to compare (32) with the simple model presented above, it is convenient to assume that also the basis elements entering the same bond or group are orthogonal to one another. This assumption requires no approximation or delocalization when the off-diagonal elements of $\mathbf{H}_{\text{eff}}(k)$ are proportional to the corresponding overlap integrals [51]; and this is always the case with bond orbitals obtained from a 2×2 Hamiltonian matrix. Nevertheless, it may complicate general analyses, so that we shall only adopt it for this particular discussion. The point is that, if $\langle\mu|v\rangle = \delta_{\mu v}$ and the Mulliken approximation (27) is applied everywhere, the contribution of $\hat{V}_A$ to $\mathbf{H}_{\mathrm{eff}}(k)$ involves only the diagonal terms, as is assumed in the naive picture presented in Section 2. Moreover, it is not difficult to devise some argument to show that, *to first order*, $\langle\mu|\hat{V}_A|\mu\rangle$ should be proportional to $\mathbf{H}_{\mathrm{eff},\kappa\kappa}(l_A)$, $|\kappa\rangle$ being an atomic orbital of A; for one thing, $\hat{V}_A$ depends on Q_A and Q_A depends on the density matrix elements of the orbitals with which A participates directly in bonds, and hence on the elements of the corresponding effective Hamiltonian. A relationship of the type (8) is thus not surprising; quite to the contrary, it appears to incorporate to some extent the fact that the various Hamiltonians (32) are coupled to one another through the $\hat{V}_A$'s. As regards strictly quantitative aspects, further work may be useful in this connection, following quite faithfully paths already explored with success [17]; we shall be content here with the conclusion that *a model Hamiltonian which considers the bonds as formally independent systems coupled to one another by transfer equations of the type (8) appears to be a physically acceptable picture combining localization with bond-bond interactions.*

As concerns inductive effects, it should be noted that if only nearest-neighbor (short-range) inductive effects are to be kept in the effective Hamiltonian to be diagonalized, only the atoms A which are directly linked to a given atom of the bond should be included in the matrix element (32) which corresponds to that atom. The remaining atoms will then provide a static contribution which must be included in the effective nuclear repulsion (vide infra).

F. ORBITAL ENERGIES, NUCLEAR REPULSIONS, BOND ENERGIES

For a discussion of bond energies it is convenient to adopt a simple (point-charge) electrostatic scheme. This, of course, will only serve to simplify the language in qualitative considerations. With this premise, we can list the various contributions to the orbital energies as follows:

(1) the kinetic energy (of the electrons) in the given bond orbital;

(2) the nuclear attractions $-Z_X/r_X$;

(3) the average repulsion of the other $(n-1)$ electrons of the bond or group under study on the electron in the given orbital;

(4) the repulsion of the electrons of X occupying orbitals not participating in k;

(5) the potential energy of the given electron in the field of the atoms (cores + electrons) not involved in k but linked to atoms participating in k.

This classification of contributions is useful to get some insight into the nature of the orbital energies; and also helps to understand how the shielding of nuclear repulsion comes about in the total energy expression. Take a point-charge classical model and calculate the corresponding potential energy of the molecule, E_{cl}:

$$E_{cl} = \tfrac{1}{2}\sum_{AB} \frac{q_A q_B}{R_{AB}} \quad (R_{AB} \equiv \text{distance } A \text{ to } B) \tag{34}$$

where the atoms have been assigned their net charges defined after Equation (31).

Now consider for simplicity one bond involving just two atoms X, Y; and call A, B the nearest neighbors of X, Y outside the bond, A', B', ... the other atoms. Then one can write (34) in the form

$$E_{cl} = \frac{Z_X Z_Y}{R_{XY}} - \frac{Q_X Q_Y}{2R_{XY}} - \left[\left(\sum_A \frac{q_A}{R_{AX}} + \frac{q_Y}{R_{XY}} \right) Q_X + \left(\sum_A \frac{q_A}{R_{AY}} + \frac{q_X}{R_{XY}} \right) Q_Y \right] +$$
$$+ \left[\sum_A \left(\frac{Z_X}{R_{AX}} + \frac{Z_Y}{R_{AY}} \right) q_A + \frac{1}{2} {\sum_{A'B'}}' \frac{q_{A'} q_{B'}}{R_{A'B'}} + \sum_{A,A'} \frac{q_A q_{A'}}{R_{AA'}} \right] \quad (35)$$

where Q_X stands for $2 \sum_{\lambda \in X} Q_\lambda$.

This formula shows that the potential energy of the system of electrons and atomic cores under consideration falls into four parts: the core-core repulsion, the electron cloud shielding, the 'dressed' atom-electron cloud interactions, the long-range electrostatic interaction between 'dressed' atoms. Now, the analysis of orbital energies just outlined indicates that the quantum-mechanical equivalent of the electrostatic energy E_{cl} is partly included in the orbital energies; in fact, the 'dressed' atom-electron cloud energy appears as the potential energy part of the effective Hamiltonian; whereas the other terms give rise to an effective nuclear repulsion.

The above intuitive considerations have been made quite rigorous in earlier papers [24, 41]. Here we shall briefly summarize those results in a form more consistent with the rest of the present study.

The total energy of a molecule, in the model where the wavefunctions are determined by H_{eff} and the non-bonded interactions are represented by electrostatic effects, can be partitioned into the following terms:

(1) effective bond nuclear repulsions (cf. Equation (15)):

$$E^0(k) = \frac{1}{2} \sum_{X,Y} \left[\frac{Z_X Z_Y}{R_{XY}} - \sum_{\mu, \nu \in k} (\mu\mu | \nu\nu) \bar{Q}_\mu \bar{Q}_\nu \right] \quad (36)$$

which represents, at the same level as (30), the core repulsions and the shielding effects of the electron clouds of the atomic orbitals of atoms involved in the kth bond. The bars above the Q's remind that, although they have essentially the form (29), they include corrections for exchange and for the fact that, one electron being treated by the effective Hamiltonian, it leaves in the atoms of the bond holes which repel each other;

(2) bond orbital energies:

$$E^{(1)}(k) = 2 \sum_{j \text{ occ} \in k} \varepsilon_j(k) \quad (37)$$

where $\varepsilon_j(k)$ is the orbital energy of jth orbital of the bond k;

(3) effective nuclear repulsion associated with the inductive effect:

$$E^{(2)}(k) = \sum_{X \in k} \sum_{A \text{ linked to } X} \left[\frac{Z_A Z_X}{R_{AX}} - \sum_{\kappa \in A, \mu \in X} \bar{Q}_\kappa \bar{Q}_\mu (\kappa\kappa | \mu\mu) \right]. \quad (38)$$

(4) long range electrostatic interactions:

$$E^{(3)} = \tfrac{1}{2} \sum_{A \neq B} \frac{q_A q_B}{R_{AB}}. \tag{39}$$

The energies $E^{(0)}(k) + E^{(1)}(k)$ may be treated as theoretical bond energies, provided care is taken to divide terms appearing in several bonds into partial contributions. An explicit form which applies to a bond proper, *i.e.* to a bond described by a two-center two-AO bond orbital, is

$$E^{(\text{bond})} = 2\varepsilon + V_{XY}^{cores} - \left[\left(\frac{1}{m_X} Q_X^2 - Q_{X(Y)}^2 \right) F_{XX} + \right.$$
$$\left. + \left(\frac{1}{m_Y} Q_Y^2 - Q_{Y(X)}^2 \right) F_{YY} + (Q_X Q_Y - Q_{Y(X)} Q_{X(Y)}) F_{XY} \right] \tag{40}$$

where F_{XX}, F_{YY}, F_{XY} are the *average* Coulomb integrals for the orbitals of X and Y; m_X, m_Y are the number of bonds formed by X and Y; $Q_{X(Y)}$ is the contribution to the population of X coming from a single electron on the XY bond orbital.

Equation (40) provides an example of the way in which properties can be associated to the localized model presented here; properties whose definition is unique within the adopted scheme, as is shown in the original papers [24, 41].

G. The basis problem

In order to conclude the present analysis, we come now to the question of the choice of the basis.

As has been mentioned, we can restrict our considerations to a 'Modified Valence Atomic Orbital' (MVAO) basis, hybridization being a supplementary feature to be discussed presently.

The fact that the choice of the basis is vital in the context of our model is made evident by the fact that we want some elements of the overlap and of the effective Hamiltonian matrices to be as small as possible, for otherwise the partitioning (23) becomes a purely formal thing. For the same reason, we want the Mulliken approximation (27) to hold as accurately as is compatible with the other requirements.

Two questions arise at this point: how can the required basis be constructed, and what is the price to pay for having chosen that particular basis.

Let us briefly consider the latter question. Of course, it is not sure at all that, among the infinite number of minimal bases – even AO minimal bases – that can be imagined and used for treating a molecular electronic problem, the basis which best block factorizes our **H** is that which will give the lowest ground state energy. Therefore, losses in accuracy of energy are unavoidable with the model at hand. However, in view of the purely theoretical interest of isolated-molecule finite-basis accurate energies, this does not seem to be a heavy price to pay; unless one wants to have a preliminary idea of the reliability of his calculated values. All reservations made, it stands to reason that an extended-basis ab-initio calculation has more chances of giving reasonable numerical results than any other procedure. On the other hand, as has been already pointed out, one may be interested in a quantum mechanical description of a molecule corresponding to a well defined model, the bond model. In that case, it is not so important to know that one will get the highest yield in numerical accuracy; it is important to know in every detail the translation that has been made of the chemical model into the mathematical quantum mechanical model. Moreover, for the reason presented in the introduction, we have to adopt practically as an experimental fact the possibility of using a localized model at least to describe the ground state of a normal molecule.

Thus we are left with the first question: how to prepare the basis. This question has been discussed at length in previous papers; therefore, here we just summarize the main results.

In the preceding considerations we have often spoken separately of MAO's and hybridization. The reason is the following. Start from a basis of Slater or hydrogenlike atomic orbitals, which we shall call 'pure' in the sense that their angular parts are the usual spherical harmonics. These are atomic orbitals because thay are one-center functions, the center being one of the atoms of the molecule. Keeping this characteristic unchanged, the radial parts of these AO's can be modified to meet various requirements. This is what we mean by 'modifying' the AO's; and procedures to determine the changes in the radial parts have been suggested [29, 47].

With pure MAO's hybrids can be constructed which are no longer symmetric with respect to the corresponding nucleus. In connection with our model, we can imagine that we have our VMAO basis and that we want to block factorize as far as possible by hybridization – i.e. mixing of MAO's of the same atom – the overlap matrix. This can be done by a rigorous mathematical technique [30]; and the hybrids thus obtained may be said to satisfy the *maximum localization criterion*, in the sense that they are so chosen that all overlap integrals with the other hybrids be as small as possible, except one. A number of reasons require that lone pairs should be treated here as special bonds [31].

Use of the maximum localization criterion for hybrids takes the overlap matrix into a quasi-block-factorized form, the size of the blocks being 2×2 for typical bonds like CC in saturated hydrocarbons [48], and becoming, of course, 6×6 in case of the π system of benzene. It is possible not to decide beforehand where the bonds are; indeed, it is not even necessary to impose the correct geometry, if the idea that bonds should be straight whenever possible is accepted [49]. Therefore, the maximum localization criterion seems to afford a method to determine the block form into which the overlap matrix and the effective Hamiltonian should finally be thrown.

Thus, the model under study ultimately rests on the choice of the pure MAO's. This is a very delicate point, which has been discussed at length elsewhere. We suggested that the hydrogenlike orbitals containing some parameters to adjust them to special conditions may be the best choice [47].

A special consideration must be made regarding the fact that in the above treatment the overlap matrix has been assumed to be rigorously factorized; whereas the maximum localization criterion does not destroy completely interblock elements. If some delocalization is admitted (1% of the other hybrids), it is not difficult to redefine hybrids which are very close to the quasi-orthogonal ones but have non-vanishing overlaps only within blocks corresponding to bonds or groups [41, 50]. Otherwise, use can be made of the flexibility introduced in the pure MAO's; in this case, an iterative scheme becomes necessary, because hybridization depends on overlaps, overlaps depend on the forms of MAO's.

4. Conclusion

The lines along which the physical bond model of a molecule can be developed and used to find mathematical counterparts of concepts currently used in chemistry have been our main concern in this paper. Our thesis is that localization, just as other concepts, need not be a feature of the exact wave-function associated with a molecular state, but can be a feature of the simplified model on which first-order interpretations are based.

In previous published work on this topic, we have insisted on the above points and attempted to show that this is a correct way to tackle the problems of theoretical chemistry (see references above). Here, we also wish to emphasize that, in our opinion, at the present stage of quantum chemical research, when the wave of interest in formidable 'accurate' calculations and the resulting controversies on the usefulness of simplified analyses and methods are dying away, the urgent task is to make applications to specific experimental problems. Only in that way will quantum chemistry retain its proper place in chemical and physical research.

References

1. Clementi, E.: *Chem. Rev.* **69**, 341 (1969).
2. Polak, R.: *Int. J. Quantum Chem.* **6**, 1077 (1972).
3. Polak, R.: *Coll. Czech. Chem. Comm.* **38**, 1450 (1972).
4. Bader, R. F. W. and Beddall, P. M.: *Chem. Phys. Letters* **8**, 29 (1971).
5. See, *e.g.*, von Auwers, K. and Eisenlohr, F.: *Chem. Ber.* **43**, 806 (1910) and ref. 6.
6. von Auwers, K.: *Liebigs Ann.* **437**, 63 (1924).
7. Coulson, C. A.: *Trans. Far. Soc.* **38**, 433 (1942).
8. Julg, A.: *Fortschr. Chem. Forsch.*, in press.
9. Nelander, B. and Del Re, G.: *J. Chem. Phys.* **52**, 5225 (1970).
10. Daudel, R. Bader, R. F., Stephens, M. E., and Borrett, D. S.: *Can. J. Chem.* **52**, 1310 (1974).
11. Coulson, C. A.: *Valence*, Oxford U. Press, 1953, p. 154ff.
12. McWeeny, R.: *Rev. Mod. Phys.* **32**, 335 (1960).
13. Hund, F.: *Zeits. f. Phys.* **73**, 24 (1931); **74**, 429 (1932).
14. Hall, G. G.: *Phil. Mag.* **43**, 338 (1952).
15. Sandorfy, C.: *Can. J. Chem.* **33**, 1337 (1955).
16. Del Re, G.: *J. Chem. Soc.* **1958**, 4031.
17. Petke, J. D. and Whitten, J. L.: *J. Chem. Phys.* **51**, 3166 (1969).
18. Wilhite, D. L. and Whitten, J. L.: *J. Chem. Phys.* **58**, 948 (1973).

19. Harrison, W. A.: *Phys. Rev.* **B8**, 4487 (1973).
20. Ingold, C. K.: *Chem. Revs.* **15**, 225 (1934).
21. Ingold, C. K.: *Structure and Mechanism in Organic Chemistry*, Cornell Univ. Press, Ithaca, New York (1953).
22. Narayan, B. and Murrell, J. N.: *Mol. Phys.* **26**, 1037 (1973).
23. Streets, D. G., and Ceasar, G. P.: *Mol. Phys.* **26**, 1037 (1973).
24. Del Re, G.: *Adv. Quantum Chem.* **8**, 95 (1974).
25. Anderson, P. W.: *Phys. Rev.* **124**, 41 (1961).
26. Koutecky, J.: *Chem. Phys. Letters* **1**, 249 (1967).
27. Mulliken, R. S.: *J. Am. Chem. Soc.* **88**, 1849 (1966).
28. Mulliken, R. S.: *J. Chimie Phys.* **46**, 497 (1949).
29. Del Re, G.: *Int. J. Quantum Chem.* **1**, 293 (1967).
30. Del Re, G.: *Theoret. Chim. Acta* **1**, 188 (1963).
31. Del Re, G., Esposito, U., and Carpentieri, M.: *Theoret. Chim. Acta* **6**, 36 (1966).
32. *E.g.*, Mulliken, R. S.: *Phys. Rev.* **40**, 55 (1932); **41**, 49 (1932); **43**, 279 (1933).
33. Magnasco, V. and Musso, G. F.: *J. Chem. Phys.* **60**, 10 (1974).
34. Orgel, L. E., Cottrell, T. L., Dick, P., and Sutton, L.E.: *Trans. Far. Soc.* **47**, 113 (1951).
35. Pullman, B. and Pullman, A.: *Les Théories électroniques de la chimie organique*, Masson, Paris (1952).
36. Groves, L. G. and Sugden, T. M.: *J. Chem. Soc.* **1937**, 1992.
37. Daudel, R. and Sandorfy, C.: *Semiempirical Wave-mechanical Calculations on Polyatomic Molecules*, Yale Un. Press, New Haven (1971).
38. Kutzelnigg, W. Del Re, G., and Berthier, G.: *Fortschr. Chem. Forsch.* **22**, 86 (1971).
39. Herigonte, P. V.: *Struct. Bonding (Berl.)* **12**, 1 (1972).
40. Malrieu, J. P., Claverie, P., and Diner, S.: *Theor. Chim. Acta* **13**, 1 (1969).
41. Del Re, G.: *Int. J. Quantum Chem.* **7S**, 193 (1973).
42. Del Re, G. and Parr, R. G.: *Rev. Mod. Phys.* **35**, 604 (1963).
43. Grimley, T. B.: *J. Phys. C. Solid St. Phys.* **3**, 1934 (1970).
44. Salahub, D. R. and Sandorfy, C.: *Theor. Chim. Acta* **20**, 227 (1971).
45. Cf. Klopman, G. and O'Leary, B. O.: *Fortschr. Chem. Forsch.* **15**, 445 (1970).
46. Daudel, R. and Laforgue, A.: *Compt. Rend.* **233**, 623 (1951).
47. Rastelli, A. and Del Re, G.: *Int. J. Quantum Chem.* **3**, 543 (1969).
48. Veillard, A. and Del Re, G.: *Theor. Chim. Acta* **2**, 55 (1964).
49. Rastelli, A. and Pozzoli, B.: *J. Chem. Soc. (Far. Trans.)* **69**, 256 (1973).
50. Lami, A. and Del Re, G.: to be published (available in preprint form).
51. Del Re, G.: *Nuovo Cim.* **17**, 644 (1960).

HOW DOES THE PCILO METHOD
TAKE ADVANTAGE OF LOCALIZATION?

J. P. DAUDEY

Institut de Biologie Physico-Chimique, Paris 75005, France

and

J. P. MALRIEU

Laboratoire de Physique Quantique, Université P. Sabatier, 31077 Toulouse, France

0. Introduction

The PCILO method was first introduced by Diner *et al* [1]. A general description of the main features of the method has been given in a recent review [2] as well as a detailed discussion of a new version of the method [3]. The purpose of the present contribution to the Localization/Delocalization problem is to recall briefly the central role played in the PCILO method by a localized picture of the molecular electronic structure. As it is shown in the first section, the use of localized molecular orbitals allows:

(α) the easy construction of a zeroth-order determinant which is a good approximation of the exact wave-function.

(β) a simplification (which is drastic if one uses a Zero Differential Overlap hypothesis) in the perturbative solution of the Configuration Interaction problem.

(γ) a decomposition of the total energy as a sum of localized transferable contributions attached to one bond, two bonds, three bonds, etc.

On the other hand, in the second section, the limitations, difficulties and possibilities of the PCILO method are discussed in connection with the likeliness of a localized description of the molecule.

1. Localized Orbitals in the PCILO Method

1.1. CONSTRUCTION OF THE ZEROTH ORDER WAVE-FUNCTION

Usually the molecular orbitals i are obtained (in the independant particle approximation) as the eigenvectors of a monoelectronic Hamiltonian operator h,

$$h \, |i\rangle = \varepsilon_i \, |i\rangle,$$

where h commutes with the symmetry operators of the molecule. As a consequence the eigenvectors $|i\rangle$ are delocalized over the whole molecule. It was first pointed out by Lennard-Jones *et al.* [4] that the N-electron wave-function $\Phi_0 = |1\bar{1}...i\bar{i}...n\bar{n}|$ built with the N-occupied molecular orbitals is invariant under any unitary transformation

O. Chalvet et al. (eds.), Localization and Delocalization in Quantum Chemistry, Vol. II, 175–183. All Rights Reserved

of the occupied molecular orbitals. Particularly it is possible to find *equivalent* sets of molecular orbitals which are localized on chemical bonds or correspond to lone pairs or inner shells. A lot of criteria for that *relocalization* process have been proposed. (For a review, see Millié *et al.* [5].)

On the other hand, it is possible to construct *a priori* localized molecular orbitals which are close to SCF molecular orbitals [6], such that the overlap between initially 'guessed' localized orbitals and final SCF orbitals is larger than 0.95 (and currently larger than 0.99 for σ-bonds and inner shells). The starting point of the PCILO method is to introduce the information contained in the chemical description of the molecule by constructing directly localized molecular orbitals corresponding to bonds, lone pairs and inner shells. The determinant built with these localized molecular orbitals is the zeroth order wave-function in the method.

1.2. IMPROVEMENT OF THE WAVE-FUNCTION

In a Configuration Interaction treatment, the weight of the excited configurations is related to the size of the molecular integrals. In a basis of delocalized molecular orbitals, all molecular integrals are roughly of the same size; it can be shown on a model problem that there are N^3 integrals varying as $1/N$, the other integrals varying as $1/N^2$ as N increases. On the contrary, in a basis of localized molecular orbitals, the molecular integrals $\langle ij|kl \rangle$ are divided into:

(α) local integrals where all molecular orbitals i, j, k and l are defined on the same bond (for instance $\langle i|i* \rangle$ where i is the occupied molecular orbital on bond i and $i*$ is one of the unoccupied orbitals on i). The size of these integrals is independent of the basis size and their number is proportional to N.

(β) integrals between local charge distributions (for instance $\langle ij|ij \rangle$ varying as $1/R$, $\langle ij|i*j* \rangle$ varying as $1/R^3$). The number of such integrals is proportional to N^2.

(γ) other integrals where i and k or j and l are defined on different bonds. Using Mulliken's approximation, $\langle ij|kl \rangle \propto S_{ik} S_{jl}$, where S_{ik} is the overlap between orbitals i and k; the size of these integrals decreases rapidly as a function of the distance between bonds.

In a Configuration Interaction treatment, only the local interactions will be important and they may be rationally selected. The advantage of using localized molecular orbitals has been suggested by Sinanoglu [7] and Nesbet [8] and verified both in semi-empirical and *ab-initio* [9, 10] calculations.

1.3. ZERO DIFFERENTIAL OVERLAP

The most used version of PCILO method is the semi-empirical one, in which the CNDO approximations [11] are used for the calculation of atomic integrals. The N-electronic Hamiltonian with the CNDO approximations can be written as

$$H_{\text{CNDO}} = \sum_{p,q} (a_p^+ a_q h_{pq} + a_p^+ a_q^+ a_q a_p g_{pq}),$$

in the second quantization formalism where h_{pq} is the matrix element of the monoelectronic operator in an atomic orbital basis and $g_{pq} = \langle pq|pq \rangle$ is the Coulomb bielectronic integral. It involves only N^2 terms (compared to N^4 terms for the ab-initio Hamiltonian) but in this form it is not possible to find a convenient approximation of the wave-function as an antisymmetrized product of *atomic orbitals*. In the CNDO SCF method, delocalized orbitals $\varphi_i = \sum_p c_{ip} \chi_p$ are introduced but in this basis, the N-electronic Hamiltonian has the same complexity as in an ab-initio treatment; all molecular integrals being different from zero except for symmetry considerations.

In the PCILO-CNDO method, totally localized molecular orbitals built on convenient hybrid atomic orbitals give an answer to the problem; as it was shown in 1.1, it is possible to define a good approximation Φ_0 in this basis, and the N-electronic Hamiltonian keeps a very simple form. The number of non-zero integrals is greatly reduced; in fact $\langle ij|kl \rangle$ is zero if i and k or (and) j and l are defined on different bonds. The N-electronic Hamiltonian in the PCILO CNDO method is

$$H_{\text{PCILO-CNDO}} = \sum_{i,j} a_i^+ a_j h_{ij} + \sum_{i,j} \langle ij|ij \rangle a_i^+ a_j^+ a_j a_i +$$

$$+ \sum_{i,j} \langle ij|i^*j \rangle (a_i^+ a_j^+ a_j a_{i*} + a_{i*}^+ a_j^+ a_j a_i) +$$

$$+ \sum_{i,j} \langle ij|i^*j^* \rangle (a_i^+ a_j^+ a_{j*} a_{i*} + a_i^+ a_{j*}^+ a_j a_{i*} +$$

$$+ a_{i*}^+ a_j^+ a_{j*} a_i + a_{i*}^+ a_{j*}^+ a_j a_i).$$

At this stage, it must be pointed out that this simplified form of the N-electronic Hamiltonian may be achieved with different approximations than CNDO ones. The important factor is the Zero Differential Overlap hypothesis between localized molecular orbitals on different bonds. For instance with INDO hypotheses [11], some additional terms have to be introduced but they are only local terms.

1.4. DECOMPOSITION OF THE TOTAL ENERGY INTO LOCAL AND TRANSFERABLE CONTRIBUTIONS

This point was examined in details from a general point of view in our contribution to the preceding volume [13] and we should only mention the results:

(α) a decomposition of the total nuclear field $\sum_A - Z_A/R_A$ into 'bond nuclear field' allows a partition of the total energy into contributions from one bond, two bonds and three bonds (if the perturbation expansion is limited to the third order). Such a decomposition was already made by many authors [14] for the energy of a single determinant,

(β) in a conformational study if one makes two additional hypotheses (the molecular orbitals are determined once and for all and the unperturbed Hamiltonian is fixed), the one-bond contributions are constant, the two-bond and three-bond contributions being a function of the relative geometry of the bonds. As a consequence the conformational energy (or the energy difference between two substituted molecules) is directly calculated (and possibly interpreted) in terms of *really varying* contributions. It is worthy to notice that the time consumed by the method is considerably reduced in this way. Further details will be found in reference 3.

2. Limitations, Applications and Extensions of the Method

2.1. LIMITATIONS

The method can be applied to all molecular systems for which a reasonable closed shell localized picture (for instance that of Lewis) may be proposed. Intermolecular problems may be treated, the method giving the various components of the usual intermolecular energies as derived from the perturbative formalism of London. Charged compounds may be treated as well. Atoms included in the program lie from H to Cl [14] (with the exception of He and Ne for which CNDO/2 parameters did not exist). The $3d$ atomic orbitals were not included for the third row elements since one does not see any evident criterion for their participation to the valence hybrids. They might be included as vacant molecular orbitals [15]. Therefore pentacoordinated compounds cannot be treated by the present program. This is a technical difficulty at the level of the construction of hybrids and bond orbitals but it might be solved without changing the architecture of the method since the chemical bonds of penta (or hexa) coordinated atoms are not *a priori* especially delocalized.

The main limitation of the method is, of course, the use of the CNDO N-electron Hamiltonian. It has been said that the CNDO hypotheses do not define an *approximate* Hamiltonian but a phenomenological model [16]. The neglect of the differential overlap should, in principle, result in a dramatic disappearance of the short range repulsion effects and the molecule should collapse as they actually do in the original CNDO/1 parametrization [17].

The repulsion is artificially reintroduced through a modification of the nuclear attraction integrals, resulting in an exaggerated repulsive exponential component of the penetration electrostatic terms. This phenomenological term mimics most of the short range effects but (α) it sometimes does not introduce a correct repulsion between non-hydrogen atoms resulting in excessively large intermolecular plane to plane complexes between conjugated systems [18, 19], (β) it does not give the rotational barrier around single bonds from overlap effects as was demonstrated by Sovers *et al.* [20] in *ab-initio* calculations. As was shown by England and Gordon [21] and by PCILO studies [22], the CNDO Hamiltonian introduces the barrier effect through delocalization contributions, which is erroneous.

Another limitation is the uncertainty about the (rate of) convergence of the perturbation expansion. The convergence cannot be guaranteed; one is compelled to check whether $|E_0| \gg |E_2| \gg |E_3|$ which is usually verified. The divergence may arise from excessive first order contributions in the wave-function. An intrapair double excitation may become too important if the corresponding bond is homolytically broken into two radicals, resulting in a near degeneracy with the doubly excited determinant. The delocalization single excitations may become too large if the localized picture is not realistic. For instance, in a very short range interaction between a lone pair and a bond, the delocalization from the lone pair might become too large to be treated as a perturbation. It is clear that the delocalization corrections will be dangerously large when the system is poorly localizable, such as in the intermediate stage of a chemical reaction or in aromatic systems. The series perhaps diverge from a kekulean structure of benzene, the most delocalized system. It is amazing to notice that highly delocalized systems may frequently be represented by *several* localized pictures; for instance the two kekulean structures of benzene, the two localized descriptions of a

COO⁻ group (—C(=O)(O⁻) or —C(O⁻)(=O)), the distorted initial and distorted final pictures of the transition complex of a chemical reaction (for the butadiene ⸺ cyclobutene isomerization, one may write the system as two butadiene double bonds or one bond on the central carbon atoms and a bond between the terminal carbon atoms [23]). *The (local) high delocalizations are the main difficulties of the PCILO algorithm. This difficulty frequently appears through an ambiguity in the choice of the localized zeroth order description.*

2.2. APPLICATIONS

The method is mostly applied as a purely numerical efficient tool. This success is linked to the way it handles the locality. The energy being summed up as one-bond, two-bond, three-bond transferable terms, is never calculated as a whole; it gives therefore a higher accuracy, avoiding cancellations between giant total repulsive and attractive terms and the possibility of a differential calculation of energy differences between conformers or interacting systems. Most of the methods of Quantum Chemistry consider the energy or other observables as a mysterious collective phenomenon, while PCILO remembers both the local character of the operators and the strong experimental evidence of the chemical bond as a constitutive element of molecular structure. It takes benefit of this preliminary knowledge as a starting point and makes feasible calculations of the energy (or energy differences) of very large systems, for which the variational calculations *a priori* allow a completely new organization of the density, which in fact never occurs.

From a numerical point of view the method generally compares with CNDO SCF calculations. In a few cases it gives better results: the rotational barriers around a single bond between two conjugated systems correctly appear in PCILO calculations [24] while CNDO SCF calculations give perpendicular structures as the stable conformations [25]. The TCNE-benzene complex and analogous plane to plane molecular interactions between conjugated systems are about 10 kcal mole^{-1} in PCILO method [26] instead of 300 kcal mole^{-1} in CNDO SCF approaches. As an example

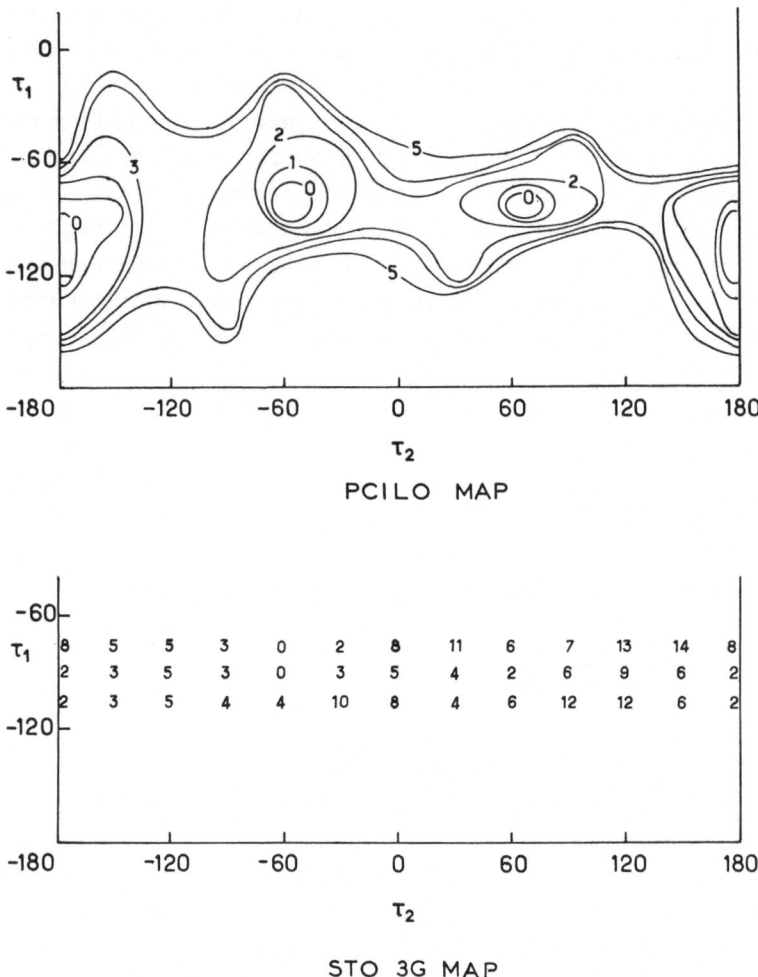

Fig. 1.   Comparison of energy maps for amphetamine molecule obtained with PCILO method
and STO 3G method.

of accuracy of the method one gives in Figure 1 the comparison of energy maps for
amphetamine molecule as calculated by PCILO and an *ab-initio* STO-3G SCF
calculation [27]. The deviation appears to be about 2 kcal mole^{-1} (which is not far
from the limit of accuracy of the *ab-initio* technique!). The method has been applied to
hundreds of compounds (essentially large (bio)molecules). For suggestive examples
see references [2, 28].

But the main conceptual interest certainly is interpretative since the method allows
one to seek the local origin of phenomena and their physical components. We already
gave in Volume I an example concerning rotation barriers [12], the proportionality
of the delocalization energy and the number of electrons [29] and in Volume II the
origin of the stereospecificity of electrocyclic reactions [23].

2.3. Extensions of the Method

The method has been already extended to density matrix calculations in various bases (bond molecular orbitals, hybrid atomic orbitals) and to the dipole moment in a double perturbation expansion performed to the second order. As concerns the ground state properties of closed shell systems one may quote the calculation of nuclear spin-spin NMR coupling constants (contact terms only) through a multiple perturbation expansion [30]. The method allows a study of the relative roles of hybridization, bond polarity, direct through space delocalization and step by step processes through intermediate bonds as already illustrated in Volume I [31]. The method may be extended to non-equilibrium compounds or non-closed shell structures. Localized free radicals have been studied a long time ago [32], ESR coupling constants have been calculated showing the stereospecific origin of the 'W rule' for coupling constants [31].

Delocalized excitations or ionizations strongly modify the structure of the molecule so that no localized description with only one determinant appears to be likely enough for a zeroth order description (except for the $n\pi^*$ excited states of (conjugated) carbonyl groups [32]). A multiconfigurational zeroth order description becomes necessary, obtained through a variational first step, namely the diagonalization of a restricted configuration interaction matrix. The method meets there the excitonic methods [34]. In many cases the zeroth order description cannot include singly excited determinants only as do usual excitonic methods, and one is compelled to build a more general procedure, joining the PCILO CNDO algorithm to the CIPSI procedure [35] which selects and treats variationally the most important interactions in the Configuration Interaction matrix, weaker interactions being included by a perturbation expansion. This program already exists and allows the treatment of strong $\sigma\pi$ mixing in distorted excited states [36], or the transition states in chemical reactions [37].

PCILO did not give yet all its possibilities. Among the various possible improvements, one may quote

(α) a more rational choice of bond hybrids. Bond polarity is determined variationally through the achievement of a local Brillouin theorem. One may imagine a rapid choice of the best hybridization of each atom, eventually coupled with the polarity determination which would really give the best fully localized determinant.

(β) a rational inclusion of $3d$ atomic orbitals in the fully localized description.

(γ) the inclusion of INDO or MINDO approximations. The INDO parameters already imply a considerable increase of the number of contributions in the series expansion but the exchange processes remain local between adjacent bonds and the computation time does not increase significantly [38].

(δ) a rational study of substituent effects on chemical reactions, for instance a bond cleavage, by introducing a local multiconfigurational zeroth order wave-function and calculating the substituent inductive influence through changes in bond polarities and the influence of delocalization through space.

But the main effort is now applied to the achievement of an *ab-initio* algorithm including overlap effects which are so difficult to treat explicitly but which play such a dominant role in Quantum Chemistry. Section IV of our preceding contribution [12] gave an idea about the strategy developed along that direction.

References

1. Diner, S., Malrieu, J. P., Claverie, P., and Jordan, F.: *Chem. Phys. Letters* **2**, 319 (1968); Diner, S., Malrieu, J. P., and Claverie, P.: *Theoret. Chim. Acta* **13**, 1 (1969); Malrieu, J. P., Claverie, P., and Diner, S.: *Theoret. Chim. Acta* **13**, 18 (1969); Diner, S., Malrieu, J. P., Jordan, F., and Gilbert, M.: *Theoret. Chim. Acta* **15**, 100 (1969); Jordan, F., Gilbert, M., Malrieu, J. P., and Pincelli, U.: *Theoret. Chim. Acta* **15**, 211 (1969).
2. Malrieu, J. P.: to appear in *Modern Quantum Chemistry* (ed. by G. A. Segal).
3. Daudey, J. P.: to be published.
4. Lennard-Jones, J. E.: *Proc. Roy. Soc. (London)* **A198**, 114 (1949); Hall, G. G. and Lennard-Jones, J. E.: *Proc. Roy. Soc. (London)* **A202**, 155 (1950); **A205**, 367 (1951); Hall, G. G.: *Proc. Roy. Soc. (London)* **A202**, 166 (1950); **A213**, 102 (1952).
5. Millié, Ph., Levy, B., and Berthier, G.: in *Localization and Delocalization in Quantum Chemistry*, Vol. I (ed. by Chalvet, O., Daudel, R., Diner, S., and Malrieu, J. P.), Reidel, Dordrecht, Holland (1975), p. 59.
6. Daudey, J. P.: *Chem. Phys. Letters* **24**, 574 (1974).
7. Sinanoglu, O.: in *Adv. Chem. Phys.*, Volume VI (ed. by I. Prigogine), Interscience, New-York (1964), p. 315.
8. Nesbet, R. K.: in *Adv. Chem. Phys.*, Volume IX (ed. by I. Prigogine), Interscience, New-York (1965), p. 321.
9. Staemmler, V. and Kutzelnigg, W.: *Theoret. Chim. Acta* **7**, 67 (1967); Diner, S., Malrieu, J. P., Jordan, F., and Claverie, P.: *Theoret. Chim. Acta* **18**, 86 (1970).
10. Wilhite, D. L. and Whitten, J. C.: *J. Chem. Phys.* **58**, 948 (1973); Masson, A., Levy, B., and Malrieu, J. P.: *Theoret. Chim. Acta* **18**, 197 (1970).
11. Pople, J. A. and Beveridge, D. L.: *Approximate Molecular Orbital Theory*, Mc Graw-Hill Book Company (1970).
12. Daudey, J. P., Malrieu, J. P., and Rojas, O.: in *Localization and Delocalization in Quantum Chemistry* (ed. by Chalvet, O., Daudel, R., Diner, S. and Malrieu, J. P.), Reidel, Dordrecht, Holland (1975), Volume I, p. 175.
13. Musso, G. F. and Magnasco, V.: *Chem. Phys. Letters* **23**, 79 (1973); van Duijneveldt, F. B. and Murrell, J. N.: *J. Chem. Phys.* **46**, 1759 (1967).
14. Q.C.P.E. program number 220.
15. Giessner-Prettre, Cl.: private communication.
16. Malrieu, J. P.: *J. Chem. Phys.*, in press.
17. Pople, J. A., Santry, D. P., and Segal, G. A.: *J. Chem. Phys.* **43**, 5129 (1965).
18. Chesnut, D. B. and Wormer, P. E.: *Theoret. Chim. Acta* **20**, 250 (1971).
19. Mo, O., Yanez, M., and Fernandez-Alonso, J. I.: *J. Phys. Chem.* **79**, 137 (1975).
20. Sovers, O. J., Kern, C. W., Pitzer, R. M., and Karplus, M.: *J. Chem. Phys.* **49**, 2592 (1968); Kern, C. W., Pitzer, R. M., and Sovers, O. J.: *J. Chem. Phys.* **60**, 3583 (1974).
21. England, W. and Gordon, M. S.: *J. Amer. Chem. Soc.* **93**, 4649 (1971).
22. Diner, S., Malrieu, J. P., Jordan, F., and Gilbert, M.: *Theoret. Chim. Acta* **15**, 100 (1969).
23. Langlet, J. and Malrieu, J. P.: *J. Amer. Chem. Soc.* **94**, 7254 (1973) and the contribution to the present volume.
24. Perahia, D. and Pullman, A.: *Chem. Phys. Letters* **19**, 73 (1973).
25. Gropen, G. and Seip, H. M.: *Chem. Phys. Letters* **11**, 445 (1971).
26. Faramond-Baud, D.: Thèse de 3ᵉ Cycle, Université de Grenoble (1973); Arnaud, R., Faramond-Baud, D., and Gelus, M.: *Theoret. Chim. Acta* **31**, 335 (1973).
27. Pullman, B., Berthod, H., and Courriere, Ph.: *Int. J. Quant. Chem., Quantum Biology Symp.* **1**, 93 (1974).

28. Pullman, B. and Pullman, A.: *Adv. in Prot. Chemistry* **28**, 347 (1974); Pullman, B. and Saran, A.: *Prog. in Nucleic Ac. Research and Mol. Biology*, in press; Pullman, B.: in *Molecular and Quantum Pharmacology* (ed. by Bergmann, E. and Pullman, B.), Reidel, Dordrecht-Holland (1974): Pullman, A. and Pullman, B.: *Quarterly Review of Biophysics* **7**, 505 (1975).
29. Malrieu J. P.: see reference 12, page 335.
30. Denis, A. and Malrieu, J. P.: *Mol. Phys.* **23**, 581 (1972).
31. Ellinger, Y., Levy, B., Millie, Ph., and Subra, R.: see reference 12, page 283.
32. Langlet, J., Gilbert, M., and Malrieu, J. P.: *Theoret. Chim. Acta* **22**, 80 (1971).
33. Langlet, J.: *Theoret. Chim. Acta* **27**, 223 (1972).
34. For a discussion, see Langlet, J. and Malrieu, J. P.: contribution to this volume.
35. Huron, B., Malrieu, J. P., and Rancurel, P.: *J. Chem. Phys.* **58**, 5745 (1973).
36. Langlet, J. and Malrieu, J. P.: *Theor. Chim. Acta* **33**, 307 (1974); Mommichioli, F., Bruni, C., and Langlet, J.: *Chem. Phys. Letters*, in press.
37. Daudey, J. P., Langlet, J., and Malrieu, J. P.: *J. Amer. Chem. Soc.* **96**, 3393 (1974).
38. Douady, J., Ellinger, Y., and Subra, R.: private communication.

PART III

EXCITONS AND LOCALIZATION

EXCITONS AND ELECTRONIC COLLECTIVE
EXCITATIONS IN MOLECULAR ORGANIC SOLIDS

RENÉ VOLTZ

Laboratoire de Physique des Rayonnements et d'Electronique Nucléaire, Centre de Recherches Nucléaires et Université Louis Pasteur, 67200 Strasbourg-Cronenbourg, France

and

PHILEMON KOTTIS

Université de Bordeaux I et E.R. 134 C.N.R.S., 33405 Talence, et Centre de Mécanique Ondulatoire Appliquée du C.N.R.S., rue du Maroc, 75019 Paris, France

1. Introduction

In order to provide a general background for the following discussions on the localizability and motion of electronic excitation, we attempt to review some of the basic concepts related to excitons in molecular crystals [1–7]. One of our purposes is to define rigorously, in the limit of a weak matter-radiation interaction, the light absorbing entities and the subsequent processes of electronic excitation energy transformation. We know that individual molecules can only be taken as the light absorbing entities under very special conditions, which must be discussed in each case.

Excitons are collective excitations corresponding to normal electronic oscillation modes pertaining to the crystal as a whole [8]; ideally, they are typical particle-wave entities characterized by definite expectation values of momentum (wave vector) and energy (frequency). The properties of these electronic elementary excitations are however critically determined by the interaction with vibrational modes in the crystal. In molecular crystals, these modes separate into low-energy lattice modes (torsional, optical, acoustic) and high-energy modes closely related to the intramolecular vibrations of the separate molecules. In general, the interaction between intramolecular excitations (vibronic states) and lattice phonon modes is sufficiently weak that the former can be treated using a rigid lattice model for the crystal; the approximate results hence obtained are subsequently corrected by regarding the coupling to the lattice phonons as a perturbation.

In Section 2 of the present report, we begin with the rigid lattice approximation to define the excitons in a molecular material. For simplicity the treatment is restricted to an idealized model of a crystal, consisting of a cubic array of identical molecules. In the rigid lattice approximation, two limiting cases must usually be distinguished, depending upon the coupling of the electronic excitation to the intramolecular vibrations in the crystal:

(a) the weak coupling limit: if spacing ΔE between vibronic levels of the free molecule is larger than the electronic resonance interaction U between neighbours ($\Delta E > U$),

O. Chalvet et al. (eds.), Localization and Delocalization in Quantum Chemistry, Vol. II, 187–208. All Rights Reserved

the functions associated to the motion of electrons and nuclei do not factorize. Therefore, the states in this case will be called 'vibronic excitons', for obvious reasons.

(b) the strong coupling limit: $U > \Delta E$; the functions associated to the motion respectively of electrons and nuclei factorize. These exciton states are called 'electronic excitons [9].

In the weak coupling limit, the exciton states are related to the molecular vibronic levels: each vibronic state of the free molecule gives rise to an exciton band; typical examples can be found among the lowest excited triplet and singlet states of aromatic hydrocarbon crystals. In the following, only the weak coupling case ('vibronic excitons') will be discussed in some detail, since it is the limit which is considered in most of the accompanying contributions [13–16].

In Section 3, the necessary corrections to the rigid lattice model of the vibronic excitons are introduced by analyzing the influence of the crystal lattice vibrations. Interaction with phonons essentially determines the energy shifts and widths, as well as finite lifetimes which characterize the excitons as they must be invoked for the interpretation of experimentally observed spectral line-shapes and transport properties. A connection is established of the present treatment with the other more phenomenological approaches that are widely used in current literature [cf. refs. 13–16 and references therein]. All along the discussion, the point of view adopted is that generally used to describe coherence and properties of elementary excitations in condensed media [17–20, 37]. Based on the many-body theoretical methods and concepts, it considers the excitons as carefully defined 'quasiparticles' and allows one to concentrate the attention to the most essential properties of the excitations in the solid, with a minimum of complications arising from the very large number of degrees of freedom.

2. Vibronic Frenkel Excitons

In the rigid lattice approximation, molecular excitons are described in the Frenkel exciton model, in terms of excited configurations with the positive hole and the excited electron located at the same site. Hence, the treatment usually begins by choosing the set of locally excited states as a basis for the representation.

2.1. Locally Excited State Representation

For the rigid lattice cubic crystal of N identical molecules, the hamiltonian is of the form:

$$H_e = \sum_i^N H(\mathbf{i}) + \tfrac{1}{2}\sum_i \sum_{j \neq i} V(\mathbf{i}, \mathbf{j}) \tag{1}$$

where $H(\mathbf{i})$ is the energy operator of the molecule $\mathbf{i}$ at site $\mathbf{R}_i$, and $V(\mathbf{i}, \mathbf{j}) = V(|\mathbf{R}_i - \mathbf{R}_j|)$ represents the Coulomb interaction between the molecules i and j. Periodic boundary conditions are assumed. Since the energy operator of the crystal is invariant under space translations, its eigenstates are also eigenstates of the momentum operator of the crystal with eigenvalues k. For the ground state of the crystal, with no molecules excited, one writes:

$$H_e \,|g\rangle = E_g \,|g\rangle. \tag{2}$$

In the weak coupling limit where the crystalline excited states are related to definite vibronic states of the separate molecules, it is natural to introduce, as a convenient basis of description, the set of locally excited states [2]:

$$|n, \mathbf{i}\rangle = B_n^+ (\mathbf{i}) \,|g\rangle \tag{3}$$

in which molecule $\mathbf{i}$ is excited in the free molecule vibronic configuration $|n\rangle$, with all the other molecules in the ground state; $B_n^+ (\mathbf{i})$ is a one-particle creation operator at site $\mathbf{R}_i$, which satisfies the Boson commutation rules:

$$[B_n(\mathbf{i}), B_m^+ (\mathbf{j})] = \delta_{nm} \delta_{\mathbf{ij}}.$$

In the occupation number representation thus defined, the exciton hamiltonian can be taken in the form:

$$H_e = E_g + \sum_n \left[\sum_{\mathbf{i}} E_n B_n^+ (\mathbf{i}) B_n(\mathbf{i}) + \sum_{\mathbf{i}, \mathbf{j} \neq \mathbf{i}} U_n(\mathbf{i}, \mathbf{j}) B_n^+ (\mathbf{i}) B_n(\mathbf{j}) \right] \tag{4}$$

with $E_n = \langle ni|H_e| ni \rangle$ and $U_n(\mathbf{i}, \mathbf{j}) = \langle ni| V(\mathbf{i}, \mathbf{j})| nj \rangle$.

E_n is the excitation energy of the free molecule corrected by the change, upon excitation, of the interaction energy with $N-1$ non-excited molecules. $U_n(\mathbf{i}, \mathbf{j})$ is a matrix element accounting for the excitation exchange between sites $\mathbf{j}$ and $\mathbf{i}$; it usually contains a Coulomb and an exchange term, the former being important for singlet configurations (multipole-multipole resonance interaction) and the latter for triplet states [21].

Since excitons correspond to normal modes of electronic oscillations in the crystal of interacting molecules, they may be introduced following the formal treatment of elementary excitations in manybody systems [17].

2.2. FREE EXCITONS

The elementary excitations of the crystal, each related to a molecular vibronic level, are conveniently defined by looking for an operator $B_n^+ (\mathbf{k})$ which satisfies the equation

$$[H_e, B_n^+ (\mathbf{k})] \,|g\rangle = E_n(k) B_n^+ (\mathbf{k}) \,|g\rangle \tag{5}$$

where $E_n(k)$ is a real number; bearing in mind Equation (2), one verifies that Equation (5) is equivalent to:

$$H_e B_n^+ (\mathbf{k}) \,|g\rangle = [E_g + E_n(k)] B_n^+ (\mathbf{k}) \,|g\rangle \tag{5'}$$

which shows that

$$|n\mathbf{k}\rangle = B_n^+ (\mathbf{k}) \,|g\rangle \tag{6}$$

is an eigenfunction of the crystal hamiltonian, with an excitation energy $E(k)$ above the ground state; i.e., $B_n^+(\mathbf{k})$ is the creation operator for the elementary excitations of the crystal. The suitable form satisfying (5) is easily verified to be

$$B_n^+(\mathbf{k}) = N^{-1/2} \sum_i \exp(i\mathbf{k}\mathbf{R}_i) \, B_n^+(\mathbf{i}) \tag{7}$$

with

$$E_n(k) = E_n + L_n(k), \quad \text{and} \quad L_n(k) = \sum_{j \neq i} U_n(R_{ij}) \exp(-i\mathbf{k}\mathbf{R}_{ij}) \tag{8}$$

where $\mathbf{R}_{ij} = \mathbf{R}_j - \mathbf{R}_i$. Usually, the interaction $U_n(R_{ij})$ is sufficiently short-ranged that the summation can be limited to the nearest neighbors at the distance R ('restricted' Frenkel model): one then gets:

$$L_n(k) = 2U_n \sum_{v=1}^{3} \cos(k_v R) \tag{9}$$

where $U_n = U_n(R)$ and k_v represents the three components of the crystal 'momentum' vector $\mathbf{k}$, the possible values of k_v being $(2\pi/R) \cdot (n_v/N_v)$, with $n_v = 1, \ldots N_v$; one has $N_1 \cdot N_2 \cdot N_3 = N$. Since in Equation (7), the creation operators $B_n^+(\mathbf{i})$ satisfy Boson commutation rules, the same applies for the creation and annihilation operators $B_n^+(\mathbf{k})$ and $B_n(\mathbf{k})$. In this momentum space representation, the crystal hamiltonian becomes:

$$H_e = E_g + \sum_n \sum_k E_n(k) \, B_n^+(\mathbf{k}) \, B_n(\mathbf{k}). \tag{10}$$

The elementary excitations thus defined, generated by operator $B_n^+(\mathbf{k})$, (see Equation 7), characterized by definite values of momentum $\mathbf{k}$ and energies $E_n(k)$, [Equations (8) and (9)], are termed (vibronic) *excitons*. For the given vibronic molecular state $|n\rangle$, the corresponding energies lie in an energy band which – in the presently examined cubic case – is centered on E_n and has a total width $12U$, under the validity conditions of expression (9).

The excitons are collective electronic normal excitation modes which must be created by light; the creation operator $B_n^+(\mathbf{k})$ defined in a formal way in this section has accordingly a physical significance that will now be analyzed.

2.3. CREATION OF EXCITONS BY LIGHT

We have here to consider the total hamiltonian:

$$H = H_m + H_r + H_{mr} \tag{11}$$

where H_m describes the material subsystem, as approximated hitherto by (1) or (10): $H_m \sim H_e$. Operator H_r characterizes the free electromagnetic field in the conventional Fock space representation. $H_m + H_r = H^0$ defines the zero order operator; $H_{mr}(=H_{er})$ is the weak interaction coupling radiation and matter; it will be taken in the electric dipole interaction approximation under the form:

$$H_{mr} = - \sum_i \mathbf{d}(\mathbf{i}) \, \mathbf{E}(\mathbf{i}) \tag{12}$$

where $\mathbf{d}(\mathbf{i})$ is the electric dipole moment operator for molecule $(\mathbf{i})$. $\mathbf{E}(\mathbf{i})$ is the electric field operator at site $\mathbf{R}_i$ given by:

$$\mathbf{E}(\mathbf{i}) = \sum_k \mathbf{E}(\mathbf{k}, \mathbf{i}) \tag{13}$$

$$\mathbf{E}(\mathbf{k}, \mathbf{i}) = i(2\pi \hbar \omega_k / V)^{1/2} \, \hat{\varepsilon}(\mathbf{k}) \, [a(\mathbf{k}) \exp(i\mathbf{k}\mathbf{R}_i) - a^+(\mathbf{k}) \exp(-i\mathbf{k}\mathbf{R}_i)]$$

V is the quantization volume: $\hbar\omega$, $\mathbf{k}$ and $\hat{\varepsilon}(\mathbf{k})$ represent energy, wave vector and polarization of a photon whose creation and annihilation operators are $a^+(\mathbf{k})$, $a(\mathbf{k})$. To obtain the proper form of the operator $\mathbf{d}(\mathbf{i})$, operating in the material subspace [22], we note that, in terms of vibronic state vectors of molecule $(\mathbf{i})$, one can write

$$\mathbf{d}(\mathbf{i}) = \sum_{n,m} |n, \mathbf{i}\rangle \langle n, \mathbf{i}|\mathbf{d}|m, \mathbf{i}\rangle \langle m, \mathbf{i}|$$

$$= \sum_n |o, \mathbf{i}\rangle \langle o, \mathbf{i}|\mathbf{d}|n, \mathbf{i}\rangle \langle n, \mathbf{i}| + \sum_n |n, \mathbf{i}\rangle \langle n, \mathbf{i}|\mathbf{d}|o, \mathbf{i}\rangle \langle o, \mathbf{i}| + \ldots$$

Restricting the analysis to linear effects that involve only transitions from (or to) the molecular ground state $|o, \mathbf{i}\rangle$, we need to keep only the two written terms in the last equation; the two dipolar transition matrix elements are taken real: $\langle o|\mathbf{d}|n\rangle = \langle n|\mathbf{d}|o\rangle = \mathbf{d}_n$; the transition operators $|o, \mathbf{i}\rangle \langle n, \mathbf{i}|$ and $|n, \mathbf{i}\rangle \langle o, \mathbf{i}|$ are respectively equal to the $B_n(\mathbf{i})$ and $B_n^+(\mathbf{i})$ operators defined before, hence:

$$\mathbf{d}(\mathbf{i}) = \sum_n \mathbf{d}_n [B_n(\mathbf{i}) + B_n^+(\mathbf{i})]. \tag{14}$$

If we insert the expression (13) and (14) in operator (12), one obtains:

$$H_{mr} = i(N/V)^{1/2} \sum_n d_n \sum_k (2\pi \hbar \omega_k)^{1/2} \{[B_n^+(\mathbf{k}) + B_n(-\mathbf{k})] \, a(\mathbf{k}) -$$

$$- [B_n^+(-\mathbf{k}) + B_n(\mathbf{k})] \, a^+(\mathbf{k})\} \tag{15}$$

with $d_n = \mathbf{d}_n \hat{\varepsilon}$ (assuming light with given polarization $\hat{\varepsilon}(k) = \hat{\varepsilon}$) and remembering definition (7) for $B_n(\mathbf{k})$ and $B_n^+(\mathbf{k})$.

Exciton formation in the crystal by absorption of photons is accounted for by the first term of this interaction operator. One starts with the crystal in the ground state $|g\rangle$ and with one photon $a^+(\mathbf{q}) |vac\rangle$ present, i.e. with an eigenstate of $H^0 = H_m + H_r$ of the form:

$$a^+(\mathbf{q}) |o\rangle$$

where $|o\rangle$ represents the ground state $|g\rangle |vac\rangle$ of the total system. Upon application of radiation-matter coupling one obtains states with excitons and no photons:

$$B_n^+(\mathbf{q}) |o\rangle.$$

More precisely it is seen that:

$$H_{mr} \, a^+(\mathbf{q}) \, |o\rangle = \sum_n g_n(q) \, B_n^+(\mathbf{q}) \, |o\rangle \tag{16}$$

where $g_n(q)$ is the coupling constant:

$$g_n(q) = \langle o|B_n(\mathbf{q}) \, H_{mr} \, a^+(\mathbf{q})|o\rangle = i \, (N/V)^{1/2} \, (2\hbar\omega_q)^{1/2} \, d_n.$$

Equation (16) shows that the dipole interaction operator (12) is indeed a physical realization of the formal creation operator $B_n^+(\mathbf{k})$ of Subsection 2.2. To derive Equation (16) account was taken of $a(\mathbf{k})a^+(\mathbf{q}) \, |vac\rangle = \delta_{\mathbf{k,q}} \, |vac\rangle$, which means that momentum is conserved. Due to this momentum conservation condition, the exciton wave numbers are equal to those of the absorbed photons. Now if the light has a spectral width Δv which characterizes the photons coherence, this leads to a momentum distribution Δq, with $\Delta q = (h/c)\Delta v$, hence to the preparation of a coherent superposition (packet) of excitons with $\Delta k = \Delta q$ (coherence transfer from radiation to matter) [20]. In the spectral regions of interest (visible or ultraviolet) one generally has $q \sim 10^5 \mathrm{cm}^{-1}$, so that $qR \ll 1$ is the rule for any optical exciton. As a consequence, the optical properties can be described in the limit $k \to 0$; in particular, the energy is given by (cf. Equations (8) and (9)):

$$E_n(o) = E_n + L_n(o)$$

$$L_n(o) = \sum_{\mathbf{j} \neq \mathbf{i}} U_n(R_{i,j}) \tag{17}$$

$$= 6U_n(R) \quad \text{(restricted Frenkel approximation)}.$$

Having defined the excitons, we must next consider their properties; these are indeed needed to describe the actual physical situations as observed in experiments of spectral or temporal characteristics of the crystals.

2.4. Properties of Vibronic Exciton

The exciton properties may be conveniently expressed in terms of Green functions, the definitions of which will be first summarized [cf. for example refs. 17, 18, 19].

(a) *Definitions*: for simplicity, we take the origin of the energy scale at the ground level of the system and take units so that $\hbar = 1$. The basic operator to be used is 'the propagator in energy space':

$$G(E^+) = \lim_{\varepsilon \to 0^+} (E - H + i\varepsilon)^{-1}. \tag{18}$$

It is related to the 'time dependent Green operator'

$$U^+(t) = \theta(t) \, U(t), \quad \text{with} \quad U(t) = \exp(-iHt) \tag{19}$$

($\theta(t)$ is Heavyside's step function) by the transformation

$$U^+(t) = \frac{i}{2\pi} \int_{-\infty}^{+\infty} dE \exp(-iEt) \, G(E^+).$$

(20)

Exciton properties may be described by means of 'single-particle propagators' given by matrix elements of the operators (18)–(20). For the *spectral* characteristics, one hence defines

$$G_n(\mathbf{k}, E) = \langle o|B_n(\mathbf{k}) \, G(E^+) \, B_n^+(\mathbf{k})|o\rangle$$

(21)

The *temporal* evolution properties are discussed in terms of the Fourier transform of (21):

$$G_n(\mathbf{k}, t) = \frac{1}{2\pi} \int_{-\infty}^{+\infty} dE \exp(-iEt) \, G_n(\mathbf{k}, E).$$

(22)

Using Equation (20), this becomes,

$$G_n(\mathbf{k}, t) = -i\theta(t) \, \langle o|B_n(\mathbf{k}, t) \, B_n^+(\mathbf{k})|o\rangle$$

(22′)

where state vectors and operators are given in the Heisenberg representation: $B_n(\mathbf{k}, t) = \exp(iHt) \, B_n(\mathbf{k}) \exp(-iHt)$.

The single-particle propagator $G_n(\mathbf{k}, t)$ is in fact a probability amplitude that the system remains, at time $t>0$, in the initially prepared state $B_n^+(\mathbf{k})|o\rangle$, [cf. expression (16)]; $|G_n(\mathbf{k}, t)|^2$ represents the decay law of the elementary excitation. The energy transport properties may likewise be expressed by the propagator in the locally excited representation:

$$G_n(\mathbf{R}_{ij}, t) = N^{-1} \sum_{\mathbf{k}} \exp(-i\mathbf{k}\mathbf{R}_{ij}) \, G_n(\mathbf{k}, t)$$

$$= -i\theta(t) \, \langle o|B_n(\mathbf{R}_j, t) \, B_n^+(\mathbf{R}_i)|o\rangle$$

(23)

with the same definition of $B_n(\mathbf{R}_j, t)$ as for $B_n(\mathbf{k}, t)$; the second line of expression (23) follows by using transformation (7); $G_n(\mathbf{R}_{ij}, t)$ is the amplitude of probability to find at $t>0$, the excitation at site $\mathbf{R}_j$, if initially it was located at $\mathbf{R}_i$.

Through the single-particle propagators, [Equations (21–23)], we focus our attention on the physical subspace where only matter is excited, i.e. on the subspace spanned by the $B_n^+(\mathbf{k})|o\rangle$ vectors and selected by a projection operator of the form:

$$M = \sum_n \sum_{\mathbf{k}} B_n^+(\mathbf{k})|o\rangle \, \langle o| \, B_n(\mathbf{k}).$$

(24′)

For the interacting exciton-photon system introduced in Subsection 2.3, this set of states is not complete and must be supplemented by the set of states $a^+(\mathbf{q})|o\rangle$ where the radiation field is excited. Let R label the corresponding projection operator; it reads

$$R = \sum_{\mathbf{q}} a^+(\mathbf{q}) \, |o\rangle \, \langle o| \, a(\mathbf{q})$$

(24″)

where the sum is defined in Equation (13). In addition to their characteristic Hermitian and idempotent properties, these projectors satisfy the relations:

$$MR = RM = 0; \quad M+R = 1 \tag{25}$$

With the partition of the physical space thus introduced, the propagator accounting for the exciton properties – in the presence of photons – is the 'reduced' operator $MG(E^+)M$, which by standard methods is shown to be of the form [23–25]:

$$MG(E^+) M = M(E - \mathscr{H}_M + i\varepsilon)^{-1} M \tag{25'}$$

In Equation (25'),

$$\mathscr{H}_M = MHM + MH_{mr} R(E - RHR + i\varepsilon)^{-1} RH_{mr} M \tag{25''}$$

is an 'effective hamiltonian', in which the second term is a complex 'level shift operator', of the form:

$$W_M = \sum_q MH_{mr} \frac{a^+(\mathbf{q}) |o\rangle \langle o| a(\mathbf{q})}{(E - E_q + i\varepsilon)} H_{mr} M . \tag{25'''}$$

Bearing in mind that:

$$(E - E_q + i\varepsilon)^{-1} = Pp(E - E_q)^{-1} - i\pi\delta(E - E_q)$$

(P_p for principal part), it is seen that the real part of W_M (P_p of the integral $\sum\limits_q$) describes radiative level shifts, that are generally small; the imaginary part (δ-function of the integral) is responsible for the level widths and lifetime properties of the excitons in the rigid lattice, due to coupling to the radiation field as it will now be seen more in detail.

(b) *Quasiparticle properties*: according to expression (25''') of the level shift operator, the corresponding matrix elements are given by:

$$\langle o|B_n(\mathbf{k}) W_M B_n^+(\mathbf{k})|o\rangle = \Delta_n(k) - i\Gamma_n(k)/2$$

so that one gets for the 'effective hamiltonian' $\mathscr{H}_M$, the complex energy:

$$\langle o|B_n(\mathbf{k}) \mathscr{H}_M B_n^+(\mathbf{k})|o\rangle = E_n(k) + \Delta_n(k) - i\Gamma_n(k)/2$$

and for the single-particle propagator (21):

$$\begin{aligned} G_n(k, E) &= \langle o|B_n(\mathbf{k}) MG(E^+) MB_n^+(\mathbf{k})|o\rangle \\ &= \{E - [E_n(k) + \Delta_n(k)] + i\Gamma_n(k)/2\}^{-1} . \end{aligned} \tag{26}$$

Following the common procedure to determine the quasiparticle properties from this function, it may be stated that the real part of the pole of $G_n(k, E)$ gives the energy, and the imaginary part the lifetime of the exciton [18].

To establish a more practical connection with experimental observables, we may note that the light absorption cross sections are proportional to the imaginary part of the matrix element $\langle o|a(\mathbf{k})\, T(E)\, a^+(\mathbf{k})|o\rangle$ of the transition operator, (cf. for example ref. 26):

$$T(E) = H_{mr} + H_{mr}\, G(E^+)\, H_{mr}.$$

For the cross section of absorption of photons with energy E_k, one then gets:

$$\sigma(E_k) \propto \mathrm{Im}\ \langle o|a(\mathbf{k})\, H_{mr}\, G(E_k^+)\, H_{mr}\, a^+(\mathbf{k})|o\rangle$$

or, bearing in mind Equation (16):

$$\sigma(E_k) \propto \mathrm{Im}\ \sum_n \langle o|B_n(\mathbf{k})\, G(E_k^+)\, B_n^+(\mathbf{k})|o\rangle$$

$$= \mathrm{Im}\ \sum_n G_n(\mathbf{k}, E_k).$$

The spectral line-shape for a vibronic exciton $B_n^+(\mathbf{k})|o\rangle$, is hence of the form:

$$\mathrm{Im}\ G_n(\mathbf{k}, E_k) = \frac{\Gamma_n(k)/2}{[E_k - E_n(k) - \Delta_n(k)]^2 + [\Gamma_n(k)/2]^2} \tag{27}$$

which is a Lorentzian centered at the energy $E_n(k) + \Delta_n(k)$, with a width $\Gamma_n(k)$. For the reason given in Subsection 2.3, all these quantities must be taken in the limit $k \to 0$ for the optical spectra.

The time dependent single-particle propagator $G_n(\mathbf{k}, t)$, as given by Equation (22), is the Fourier transform of function (26):

$$G_n(\mathbf{k}, t) = -i \exp(-\Gamma_n(k)\, t/2) \exp[-i(E_n(k) + \Delta_n(k))\, t]\, \theta(t). \tag{28}$$

From the square of this probability amplitude, one gets the decay law (spontaneous emission):

$$P_n(k, t) = \exp[-\Gamma_n(k)\, t] \tag{29}$$

where $\Gamma_n(k)^{-1}$ is the characteristic lifetime of the exciton $B^+(\mathbf{k})|o\rangle$.

Excitation transfer from site $\mathbf{R}_i$ to $\mathbf{R}_j$ is described by the propagator (23), which is here of the form:

$$G_n(\mathbf{R}_j - \mathbf{R}_i,\, t) = -i\theta(t)\, N^{-1} \sum_k \exp(i\mathbf{k}\mathbf{R}_i) \exp(-i[\mathbf{k}\mathbf{R}_j + E_n(k)\, t +$$

$$+ \Delta_n(k)\, t])\, \exp(-\Gamma_n(k)\, t/2). \tag{30}$$

In practical instances, one starts at $t = 0^+$, with a more or less localized, optically prepared exciton packet, described by an amplitude distribution function $A(\mathbf{R}_i, t=0) = A(\mathbf{R}_i)$; the probability amplitude to find site $\mathbf{R}_j$ excited at time $t > 0$ is then given by:

$$\Phi_n(\mathbf{R}_j,\, t) = \sum_{\mathbf{R}_i} G_n(\mathbf{R}_j - \mathbf{R}_i,\, t)\, A(\mathbf{R}_i) \tag{31}$$

or:

$$\Phi_n(\mathbf{R}_j,\, t) = -i\theta(t)\, N^{-1} \sum_k A(\mathbf{k}) \exp(-i[\mathbf{k}\mathbf{R}_j + E_n(k)\, t + \Delta_n(k)\, t]) \times$$

$$\times \exp(-\Gamma_n(k)\, t/2) \tag{32}$$

if $A(\mathbf{k})$ denotes the transform in momentum space of the spectral excitation amplitude distribution function $A(\mathbf{R}_i)$, i.e.,

$$A(\mathbf{k}) = \sum_{\mathbf{R}_i} \exp(i\mathbf{k}\mathbf{R}_i)\, A(\mathbf{R}_i) \tag{33}$$

$A(\mathbf{k})$ characterizes the coherence transfer from radiation to matter: it is centered at a value $k_0(=q_0)$ with a distribution width $\Delta k(\sim \Delta q) > \Delta R^{-1}$, ΔR being the width of $A(\mathbf{R}_i)$, covering a large number of sites. Employing the common stationary phase method arguments [27] to discuss Equation (32), it is verified that the excitation wave packet moves with the group velocity:

$$\mathbf{v}_n = \nabla_{\mathbf{k}}[E_n(k) + \Delta_n(k)]_{\mathbf{k}=\mathbf{k}_0}.$$

Disregarding the generally negligible radiative energy shift $\Delta_n(k)$, one obtains in the restricted Frenkel approximation (Equations (8–9)), the velocity components ($v = 1, 2, 3$):

$$(v_n)_v = 2\, U_n\, R \sin k_0 R$$

$$\sim 2\, U_n\, R^2 k_0, \quad \text{since} \quad k_0\, R \ll 1. \tag{34}$$

The wave packet also undergoes spreading: for an initial width $\Delta R \sim (\Delta k)^{-1}$, one has, at time t, a width equal to:

$$\Delta R(t) = \Delta R [1 + (t/t_0)^2]^{1/2} \tag{35}$$

where $t_0 = (\Delta R)^2/|\nabla_{\mathbf{k}}^2 E_n(k_0)| \sim (\Delta R)^2/4\, U_n R^2$.

The time t_0 is of the order of the intermolecular excitation exchange time ($\sim U_n^{-1}$), if the initial excitation was localized on one site, i.e. $\Delta R \sim R$ (this problem is discussed in detail for dimers [22] and for chains [15]). Finally the excitation wave packet dissipates in time according to a decay law, approximately given by:

$$P_n(t) = \exp(-\Gamma_n(k_0)\, t). \tag{36}$$

We may conclude this section by noting that the vibronic excitons, defined in the rigid lattice approximation, have the general particle-wave attributes of 'quasi-particles' in condensed materials: they are characterized by definite values of momentum (wave vector) $\mathbf{k}$ and energy (frequency) $E_n(k)$; as any particle or elementary excitation, they may be created and annihilated; they have a finite lifetime $\Gamma_n(k)$. Creation by light leads to excitons with wave numbers fixed around $\mathbf{k} = 0$. The good knowledge of momentum of such excitons is not compatible with any localization: they must be regarded as distributed over spatial regions much larger than the intersite distance R in the crystal. In principle, no initially localized excitation can therefore be prepared by a light pulse in an ideal (translationally invariant) rigid lattice crystal (this statement remains valid even in 'parallel' finite chains, see ref. 15). Such localized excitations could however be prepared by other means: the inelastic scattering processes of charged particles (electrons, protons …) for example, give rise to excited crystal states, localized within a spatial distance $\Delta R = \hbar v/E_n$), with v for the particle velocity and

E_n the exciton energy [40]. These states are wave packets $A(\mathbf{k})$ of the form (33), in which excitons are superposed with a range of momenta such as $\Delta k \sim (\Delta R)^{-1}$. They have the characteristic dynamical properties given by Equations (34–36).

In practice, however, the rigid lattice description of excitons can only be considered to account for the major energy level distribution, relative to those of the isolated molecules. It cannot interpret the optical line-shapes and the transport properties in rate processes which are both essentially determined by the interaction of the excitons with the lattice vibrations and which be will analyzed in the next section.

3. Interaction of Excitons with Phonons

In Section 2, the vibronic excitons were defined by assuming all the molecules fixed in space so that the material subsystem was represented by the model hamiltonian H_e (cf. Equations (1), (4) and (10)). Here, we begin to discuss the form of the energy operator H_m of the crystal, which will be taken to describe the influence of small lattice vibrations; as in the past section, the locally excited representation is first considered.

3.1. COUPLING OF VIBRONIC MOLECULAR STATES TO LATTICE VIBRATIONS

The general form of the material hamiltonian will be:

$$H_m = H_e + H_p + H_{ep} \tag{37}$$

where H_e is the vibronic exciton energy operator discussed in Section 2. H_p represents the energy of the lattice phonons the spectrum of which can usually be taken as quasicontinuous. The zero-order hamiltonian in the following treatment will be given by

$$H_m^0 = H_e + H_p.$$

The basis vectors corresponding to the locally excited states (3) are accordingly products of localized vibronic states, $|ni\rangle$ and lattice phonon, $|\mathbf{p}\rangle$, factors:

$$|ni; \mathbf{p}\rangle = |ni\rangle |\mathbf{p}\rangle$$

($\mathbf{p}$ is understood to represent phonon wave vector together with polarization). For the ground state of the crystal, an equation like (2) applies, where E_g and $|g\rangle$ also include the phonon quasi vacuum. The coupling operator H_{ep} in Equation (37) includes the perturbation through which the molecular vibronic and the lattice phonon excitations interact: it may be written as a sum of two types of contributions:

$$H_{ep} = \sum_i H_m'(\mathbf{i}) + \tfrac{1}{2} \sum_{i, j \neq i} H_m'(\mathbf{i}, \mathbf{j}). \tag{38}$$

Operator $H_m'(\mathbf{i})$ may be regarded as a correction to the $H(\mathbf{i})$ terms in the rigid lattice expression (1). It is termed 'local' perturbation and it essentially accounts for the influence of the small molecular vibrations $\delta\mathbf{R}_i$ around the equilibrium positions $\mathbf{R}_i$: to first order one has:

$$H'_m(\mathbf{i}) = [\mathbf{V}_{\mathbf{R}_i} H(\mathbf{i})]_{\mathbf{R}_i} \delta\mathbf{R}_i \tag{39}$$

where the first factor operates in the subspace of vibronic excitations ($|n, \mathbf{i}\rangle$), while $\delta\mathbf{R}_i$ is a lattice phonon operator that should be expanded in terms of the normal lattice vibrational modes. This perturbation is responsible for single phonon transitions which are seldom accompanied by intramolecular transitions between different vibronic levels. Usually the latter transitions (vibrational relaxation) involve many phonon processes described in higher-order approximations, and have accordingly relatively lower probabilities [30].

In a similar way, the second terms in expression (38) are first order corrections to the $V(\mathbf{i}, \mathbf{j})$ interactions in the rigid lattice hamiltonian (1) (they are termed 'non-local' perturbations). They account for intermolecular electronic resonance interactions by the lattice vibrations; to first order one has:

$$H'_m(\mathbf{i}, \mathbf{j}) = [\mathbf{V}_{\mathbf{R}_{ij}} V(\mathbf{i}, \mathbf{j})]_{\mathbf{R}_{ij}} \delta\mathbf{R}_{ij}. \tag{40}$$

This perturbation is responsible for intermolecular vibronic energy exchange with simultaneous single phonon transitions. For later references, we denote the typical matrix elements thus defined by:

$$W_{p'p}(\mathbf{i}) = \langle n\mathbf{i}; \mathbf{p}'| H'_m(\mathbf{i}) |n\mathbf{i}; \mathbf{p}\rangle \tag{41}$$

$$W_{p'p}(\mathbf{i}, \mathbf{j}) = \langle n\mathbf{i}; \mathbf{p}'|H'_m(\mathbf{i}, \mathbf{j})|n\mathbf{j}; \mathbf{p}\rangle.$$

We may now extend the treatment of Subsection 2.1 to take into account the influence of phonons; this leads us to define an 'effective' hamiltonian for the locally excited states in the crystal.

3.2. Effective Hamiltonian for the Locally Excited Configurations

In the present case, the localized vibronic excited state $|n, \mathbf{i}\rangle$, introduced in Section 2 to build up the exciton states generated by light, are represented by state vectors of the form $|n, \mathbf{i}; \mathbf{p}=0\rangle$ [31]. To describe the spectral and temporal modifications due to the interaction of these excited states with the lattice vibrations, we may define an effective hamiltonian acting in the subspace of the (interesting) $|n\mathbf{i}; \mathbf{p}=0\rangle$ state vectors.

This is done by partitioning the physical space; let

$$P = \sum_{n, \mathbf{i}} |n\mathbf{i}; O\rangle \langle O; n\mathbf{i}| \tag{42}$$

denote the projection operator into the interesting subspace, so that for an arbitrary state $|\psi\rangle$, satisfying the Schrödinger equation of the material system:

$$(E - H_m)|\psi\rangle = 0. \tag{43}$$

$P|\psi\rangle$ is that portion of $|\psi\rangle$ for which the crystal is in the excited configurations of special interest. The complementary projection $Q|\psi\rangle$ ($Q = 1 - P$) is on all the other configurations, including those with phonons excited.

Inserting

$$|\psi\rangle = P|\psi\rangle + Q|\psi\rangle$$

in Equation (43), one obtains the two separate groups of equations:

$$[E - PH_m P] P|\psi\rangle - (PH_m Q) Q|\psi\rangle = 0$$

$$[E - QH_m Q] Q|\psi\rangle - (QH_m P) P|\psi\rangle = 0$$

the solution of which leads to:

$$[E - \mathcal{H}_e(E^+)] P|\psi\rangle = 0 \qquad (44)$$

Here $\mathcal{H}_e(E^+)$ is the effective hamiltonian

$$\mathcal{H}_e = PH_m P + PH_m Q [E^+ - QH_m Q]^{-1} QH_m P \qquad (45)$$

which determines motion in the exciton subspace. It fully accounts for the influence of the lattice vibrations on the locally excited vibronic states entering in the definition of the collective exciton states. As in operator (25″) the real part of (45) describes energies with level shifts; but now these shifts are not necessarily very small [32]. The imaginary parts are responsible for the level width and finite lifetime properties due to interactions with phonons.

The effective hamiltonian, $\mathcal{H}_e$, thus defined, generalizes the H_e operator of Section 2. The locally excited state occupation number representation also involves creation operators $B_n^+(\mathbf{i})$ defined as:

$$|n\mathbf{i}; O\rangle = B_n^+(\mathbf{i}) |g\rangle. \qquad (46)$$

If one denotes the corresponding projection operators by P_i, the total projector on the discrete subspace is given by:

$$P = \sum_i P_i, \quad \text{with} \quad P_i = |n\mathbf{i}; O\rangle \langle O; n\mathbf{i}|. \qquad (47)$$

The complementary projectors are likewise displayed as the sum:

$$Q = \sum_i Q_i, \quad \text{with} \quad Q_i = \sum_{\mathbf{p}'} |n\mathbf{i}; \mathbf{p}'\rangle \langle \mathbf{p}'; n\mathbf{i}| \qquad (48)$$

where $\sum_{\mathbf{p}'}$ represents a summation over wavevector and polarization of the lattice phonons, and may eventually be replaced by an integral with a phonon state density.

The physical properties of the excited states (46) in the presence of the phonons are now seen to be determined by the hamiltonian $\mathcal{H}_e$, which can be written as:

$$\mathcal{H}_e = E_g + \sum_n \left[\sum_i \mathcal{E}_n B_n^+(\mathbf{i}) B_n(\mathbf{i}) + \sum_{\mathbf{i}, \mathbf{j} \neq \mathbf{i}} \mathcal{U}_n(\mathbf{i}, \mathbf{j}) B_n^+(\mathbf{i}) B_n(\mathbf{j}) \right]. \qquad (49)$$

This means that we recover the same form as (4), except that the matrix elements $\mathcal{E}_n$

and $\mathcal{U}_n(\mathbf{i}, \mathbf{j})$ pertain to the effective operator (45), and contain, in addition to the real parts, imaginary parts as well.

The complex terms $\mathcal{E}_n$ are more precisely given by:

$$\mathcal{E}_n = \langle ni; O|\mathcal{H}_e|ni; O\rangle = \langle ni; O|P_i \mathcal{H}_e P_i|ni; O\rangle \tag{50}$$

with

$$P_i \mathcal{H}_e P_i = P_i H_e P_i + P_i H_{ep} Q[E^+ - QH_m Q]^{-1} QH_{ep} P_i$$

$$= P_i H_e P_i + P_i H'_m(i) Q_i[E^+ - Q_i H_m Q_i]^{-1} Q_i H'_m(i) P_i +$$

$$+ \sum_{i, j \neq i} P_i H'_m(i,j) Q_j[E^+ - Q_j H_m Q_j]^{-1} Q_j H'_m(i,j) P_i.$$

The elements (50) thus involve three terms

$$\mathcal{E}_n = E_n + \sum_{p'} \frac{\langle ni; O|H'_m(i)|ni; \mathbf{p}'\rangle \langle ni; \mathbf{p}'|H'_m(i)|ni; O\rangle}{E - (E_n + E_{p'}) + i\varepsilon}$$

$$+ \sum_{j \neq i} \sum_{\mathbf{p}'} \frac{\langle ni; O|H'_m(i,j)|nj; \mathbf{p}'\rangle \langle nj; \mathbf{p}'|H'_m(i,j)|ni; O\rangle}{E - (E_n + E_{p'}) + i\varepsilon}. \tag{50'}$$

E_n is the rigid lattice approximation energy (Section 2). The two other terms are seen to describe real energy shifts (principal part of the integrals), $\delta_n(\mathbf{i}) + \sum_{j \neq i} \delta_n(\mathbf{i}, \mathbf{j})$, together with imaginary energy widths $(-i\pi \delta(E - E_n - E_{p'})$ distributions in the integrals), $[\hbar \gamma_n(\mathbf{i})/2] + \sum_{j \neq i} [\hbar \gamma_n(\mathbf{i}, \mathbf{j})/2]$, with:

$$\delta_n(\mathbf{i}) = Pp \sum_{\mathbf{p}'} \frac{|W_{p'}(i)|^2}{E - (E_n + E_{p'})}$$

$$\delta_n(\mathbf{i}, \mathbf{j}) = Pp \sum_{\mathbf{p}'} \frac{|W_{p'}(i,j)|^2}{E - (E_n + E_{p'})}$$

$$\hbar \gamma_n(\mathbf{i}) = 2\pi \sum_{\mathbf{p}'} W_{\mathbf{p}'}(i) \delta(E - E_n - E_{p'}) W_{\mathbf{p}'}^*(i)$$

$$\hbar \gamma_n(\mathbf{i}, \mathbf{j}) = 2\pi \sum_{\mathbf{p}'} W_{\mathbf{p}'}(i,j) \delta(E - E_n - E_{p'}) W_{\mathbf{p}'}^*(i,j) \tag{51}$$

using the notations given by Equation (41). Collecting all the terms we thus get:

$$\mathcal{E}_n = E_n + [\delta_n(\mathbf{i}) + \sum_{j \neq i} \delta_n(\mathbf{i}, \mathbf{j})] - (i\hbar/2) [\gamma_n(\mathbf{i}) + \sum_{j \neq i} \gamma_n(\mathbf{i}, \mathbf{j})]. \tag{52}$$

The non-diagonal elements $\mathcal{U}_n(\mathbf{i}, \mathbf{j})$ in expression (49) are given in a similar way by

$$\mathcal{U}_n(\mathbf{i}, \mathbf{j}) = \langle ni; O|\mathcal{H}_e|nj; O\rangle = \langle ni; O|P_i \mathcal{H}_e P_j|nj; O\rangle \tag{53}$$

with

$$P_i \mathcal{H}_e P_j = P_i H_e P_j + P_i H'_m(i,j) Q_j[E^+ - Q_j H_m Q_j]^{-1} Q_j H'_m(j) P_j +$$

$$+ P_i H'_m(i) Q_i[E^+ - Q_i H_m Q_i]^{-1} Q_i H'_m(i,j) P_j.$$

From this, one easily gets:

$$\mathcal{U}_n(\mathbf{i,j}) = U_n(\mathbf{i,j}) + \sum_{\mathbf{p'}} \frac{\langle ni; O|H'_m(\mathbf{i,j})|nj; \mathbf{p'}\rangle \, \langle nj; \mathbf{p'}|H'_m(\mathbf{j})|nj; O\rangle}{E - (E_n + E_{p'}) + i\varepsilon} +$$

$$+ \sum_{\mathbf{p'}} \frac{\langle ni; O|H'_m(\mathbf{i})|ni; \mathbf{p'}\rangle \, \langle ni; \mathbf{p'}|H'_m(\mathbf{i,j})|nj; O\rangle}{E - (E_n + E_{p'}) + i\varepsilon}. \quad (54)$$

As before the real and imaginary parts of the two latter terms correspond to energy shifts (exciton band narrowing) $\delta'_n(\mathbf{i,j})$ and linewidths $\hbar\gamma'_n(\mathbf{i,j})/2$; these quantities are given by

$$\delta'_n(\mathbf{i,j}) = Pp \sum_{\mathbf{p'}} \frac{W_{\mathbf{p'}}(\mathbf{i,j}) \, W_{\mathbf{p'}}^*(\mathbf{j}) + W_{\mathbf{p'}}(\mathbf{i}) \, W_{\mathbf{p'}}^*(\mathbf{i,j})}{E - (E_n + E_{p'})} \quad (55)$$

$$\hbar\gamma'_n(\mathbf{i,j}) = 2\pi \sum_{\mathbf{p'}} [W_{\mathbf{p'}}(\mathbf{i,j}) \, W_{\mathbf{p'}}^*(\mathbf{j}) + W_{\mathbf{p'}}(\mathbf{i}) \, W_{\mathbf{p'}}^*(\mathbf{i,j})] \, \delta(E - E_n - E_{p'}).$$

Thus, we have in expression (49):

$$\mathcal{U}_n(\mathbf{i,j}) = U_n(\mathbf{i,j}) + \delta'_n(\mathbf{i,j}) - (ih/2)\,\gamma'_n(\mathbf{i,j}). \quad (56)$$

3.3. RELATION WITH TIME DEPENDENT FORMULATIONS

At this point, it might be useful to establish a connection with the widely used treatments of the problem based on time dependent stochastic models [33, 34]. The starting expressions for the energy operator, in these formulations, are also taken in the locally excited vibronic basis, but the influence of phonons is taken into account by letting fluctuate the diagonal (E_n) and non-diagonal $U_n(\mathbf{i,j})$ energy matrix elements; this introduces time dependent stochastic perturbations assumed to be Gaussian Markov processes with $\delta(t)$-like correlation functions (white noise). The relation with the foregoing discussion of the same problem, by the less phenomenological effective hamiltonian method, clearly appears by noting that the various line-widths, (51) and (55), can be formally expressed in the form of a Fourier transform with respect to a time dependent energy correlation function [35]. Consider, for example, in Equation (51):

$$\hbar\gamma_n(\mathbf{i}) = 2\pi \sum_{\mathbf{p'}} \langle ni; O|H'_m(\mathbf{i})|ni; \mathbf{p'}\rangle \, \langle ni; \mathbf{p'}|H'_m(\mathbf{i})|ni; O\rangle \, \delta(E - E_n - E_{p'})$$

and introduce the representation:

$$\delta(x) = (2\pi h)^{-1} \int_{-\infty}^{+\infty} dt \, \exp(-ixt/h).$$

If, we further define the operator $H'_m(\mathbf{i}; t)$ in the Heisenberg picture:

$$H'_m(\mathbf{i}; t) = \exp(iH_m^0 t/\hbar) \, H'_m(\mathbf{i}) \, \exp(-iH_m^0 t/\hbar)$$

we can successively write $\hbar\gamma_n(\mathbf{i})$ as:

$$\hbar\gamma_n(\mathbf{i}) = (h)^{-1} \int_{-\infty}^{+\infty} dt \, \exp[-i(E-E_n)\,t/\hbar] \times$$

$$\times \left\{ \sum_{p'} \langle ni; 0|H'_m(\mathbf{i}; t)|ni; \mathbf{p}'\rangle \langle ni; \mathbf{p}'|H'(\mathbf{i}; 0)|ni; 0\rangle \right\}$$

$$= (h)^{-1} \int_{-\infty}^{+\infty} dt \, \exp[-i(E-E_n)\,t/\hbar] \, \langle ni; 0|H'_m(\mathbf{i}; t) \times$$

$$\times H'_m(\mathbf{i}; 0)|ni; 0\rangle. \qquad (57)$$

The matrix element of the product of the two Heisenberg operators at different times may be regarded as a correlation function. In the present case, with the system initially in the pure quantum state $|ni; \mathbf{p}=0\rangle$, the correlation function is merely a quantum mechanical expectation value [36]. In the model of Haken et al. [14. 34], it is prescribed that:

$$\langle ni; 0|H'_m(\mathbf{i}, t)\, H'_m(\mathbf{i}; 0)|ni; 0\rangle = 2\hbar\gamma_0 \, \delta(t)$$

where γ_0 is a measure of the strength of local fluctuations: a comparison with (57) shows that Haken's γ_0 parameter equals $(\hbar\gamma_n/2)$ as defined by Equation (51).

In the same way, expression (51) of $(\hbar\gamma_n(\mathbf{i}, \mathbf{j})/2)$ can be displayed as a Fourier integral of a correlation function related to the parameter γ_1 defined in the previously mentioned work [34] to measure the non-local fluctuations between the molecules $\mathbf{i}$ and $\mathbf{j}$:

$$\langle ni; 0|H'_m(\mathbf{i},\mathbf{j}; t)\, H'_m(\mathbf{i},\mathbf{j};0)|ni; 0\rangle = 2\hbar\gamma_1 \, \delta(t)$$

γ_1 is hence identified with $(\hbar\gamma_n(\mathbf{i}, \mathbf{j})/2)$ in the present work.

The last energy width $(\hbar\gamma'_n(\mathbf{i}, \mathbf{j})/2)$, Equation (55), of special importance for intermolecular energy transfer, can likewise be related to the correlation function:

$$\langle ni; 0|H'_m(\mathbf{i},\mathbf{j}; t)\, H'_m(\mathbf{j};0)|ni; 0\rangle + \langle ni; 0|H'_m(\mathbf{i}; t)\, H'_m(\mathbf{i},\mathbf{j};0)|ni; 0\rangle.$$

These terms, which describe mechanisms resulting from interference of local and non-local fluctuations via phonons, are neglected in most of the current studies [49]. There does not seem to be an apparent reason that the corresponding contributions are necessarily smaller than those described by γ_0 and γ_1, when both of the latter are operative. This point deserves further study, especially when specific contributions to the diffusion tensor $\mathbf{D}$ can be related to γ_0 and γ_1, respectively [38].

3.4. EFFECTIVE EXCITON HAMILTONIAN

Having defined the effective hamiltonian in the locally excited state representation, (Subsection 3.2), we may turn to the ultimate step of our analysis and determine the vibronic exciton state of the crystal (the actual resonances of the crystal) in the same way as in Section 2, by introducing a creation operator:

$$B_n^+(\mathbf{k}) = N^{-1/2} \sum_{\mathbf{i}} \exp(i\mathbf{k}\mathbf{R}_i) \, B_n^+(\mathbf{i})$$

assumed to satisfy now the equation of motion:

$$[\mathcal{H}_e, \, B_n^+(\mathbf{k})] = \mathscr{E}_n(k) \, B_n^+(\mathbf{k}) \tag{58}$$

similar to (5). But the 'energy' of the elementary excitation is here complex; it is indeed verified to be equal to

$$\mathscr{E}_n(k) = \mathscr{E}_n + \mathscr{L}_n(k), \quad \text{with} \quad \mathscr{L}_n(k) = \sum_{j \neq i} \mathscr{U}_n(\mathbf{R}_{ij}) \exp(-i\mathbf{k}\mathbf{R}_{ij}) \tag{59}$$

with $\mathscr{E}_n$ and $\mathscr{U}_n(\mathbf{R}_{ij})$ given by Equations (52) and (56). Separating explicitly the real and imaginary parts of expression (59):

$$\mathscr{E}_n(k) = \mathscr{E}_n'(k) - i\mathscr{E}_n''(k) \tag{60}$$

one gets for the real term an expression similar to (8):

$$\mathscr{E}_n'(k) = \bar{E}_n + \sum_{j \neq i} \bar{U}_n(\mathbf{R}_{ij}) \exp(-i\mathbf{k}\mathbf{R}_{ij}) \tag{61}$$

or, in the restricted Frenkel approximation, (cf. Equation (9)),

$$\mathscr{E}_n'(k) = \bar{E}_n + 2\bar{U}_n(R) \sum_{v=1}^{3} \cos(k_v R) \tag{61'}$$

where $\bar{E}_n$ and $\bar{U}_n$ are now renormalized energies:

$$\bar{E}_n = E_n + \delta_n(\mathbf{i}) + \sum_{j \neq i} \delta_n(\mathbf{i},\mathbf{j})$$

$$\bar{U}_n = U_n(\mathbf{R}_{ij}) + \delta_n'(\mathbf{i},\mathbf{j}), \quad \text{cf. Equations (51) and (56)}. \tag{62}$$

The imaginary term is similarly given by:

$$\mathscr{E}_n''(k) = (\hbar/2) \left[\gamma_n(\mathbf{i}) + \sum_{j \neq i} \gamma_n(\mathbf{i},\mathbf{j}) + \sum_{j \neq i} \gamma_n'(\mathbf{i},\mathbf{j}) \exp(-i\mathbf{k}\mathbf{R}_{ij}) \right] \tag{63}$$

with definitions (51) and (55). When only interactions between nearest neighbors are notable, we get

$$\mathscr{E}_n''(k) = (\hbar/2) \left[\gamma_n + 6\gamma_n(\mathbf{R}) + 2\gamma_n'(\mathbf{R}) \sum_{v=1}^{3} \cos(k_v R) \right] \tag{63'}$$

where we have set $\gamma_n(\mathbf{R}) = \gamma_n(\mathbf{i},\mathbf{j})$ and $\gamma_n'(\mathbf{R}) = \gamma_n'(\mathbf{i},\mathbf{j})$ for $\mathbf{R} = \mathbf{R}_{ij}$.

In terms of the quantities thus defined, the effective hamiltonian for the 'dressed' vibronic excitons may be written in a form similar to Equation (10):

$$\mathcal{H}_e = E_g + \sum_n \sum_{\mathbf{k}} \mathscr{E}_n(k) \, B_n^+(\mathbf{k}) \, B_n(\mathbf{k}). \tag{64}$$

From this hamiltonian, the new energy-time properties of the excitons interacting with phonons are easily deduced, by means of the same quantities as in Subsection 2.4.

3.5. Spectral and Transport Properties

The *spectral* vibronic exciton properties are discussed in terms of hamiltonian (11), where the material term is now given by Equation (37). Account must be taken of the partition of the material excited states in the two subspaces selected by the projection operators P and $Q = M - P$: the excitons are defined, as in Subsection 2.3, as generated by the radiation-matter interaction term, $H_{mr} = H_{er}$, and are selected by projector P, (Equation (42)); the other material states represented by Q are not excited by light [39]. Here, the spectral properties then depend upon the reduced propagator $PG(E^+)P$ rather than on $MG(E^+)M$ as before. By partitioning $M = P + Q$ in the material subspace, it is easy to verify that:

$$PG(E^+)\,P = P[E^+ - \mathscr{H}_p(E)]^{-1}\,P \tag{65}$$

where $\mathscr{H}_p$ is the effective hamiltonian for the excitons interacting with both the phonon and photon fields:

$$\mathscr{H}_p = PH_eP + PH_{er}R\,[E^+ - RHR]^{-1}\,RH_{er}P +$$
$$+ PH_{ep}Q\,[E^+ - QHQ]^{-1}\,QH_{ep}P \tag{65'}$$

where the two last terms are the complex level shift operators which respectively account for the influence of photons and phonons in the way discussed before, (cf. Equations (25) and (45)).

The single-particle propagator (21) for the excitons hence becomes:

$$G_n(k, E) = \langle o|B_n(\mathbf{k})\,PG(E^+)\,PB_n^+(\mathbf{k})|o\rangle$$
$$= [E - \mathscr{E}_n(k) - \varDelta_n(k) + i\varGamma_n(k)/2]^{-1}.$$

Quite generally, the radiative quantities $\varDelta_n$ and $\varGamma_n$ are found negligible compared to the real and imaginary parts of $\mathscr{E}_n(k)$, so that we may take:

$$G_n(\mathbf{k}, E) = [E - \mathscr{E}_n'(k) + i\mathscr{E}_n''(k)]^{-1} \tag{66}$$

with $\mathscr{E}_n'(k)$ and $\mathscr{E}_n''(k)$ given by Equations (59) and (63). Using the same arguments as in Section 2.4, we note that the spectral line-shape for the exciton $B_n^+(k)|o\rangle$ is given by the imaginary part of expression (66)

$$\text{Im } G_n(\mathbf{k}, E) = \mathscr{E}_n''(k)\,\{[E - \mathscr{E}_n'(k)]^2 + [\mathscr{E}_n''(k)]^2\}^{-1} \tag{67}$$

where the energy $\mathscr{E}_n'$ and the width $\mathscr{E}_n''$ are to be taken in the limit of $k \to q \sim 0$ (cf. Equation (61'), (63')) e.g.,

$$\mathscr{E}_n'(o) = \bar{E}_n + 6\bar{U}_n(R)$$

$$\mathscr{E}_n''(o) = [\gamma_n + 6\gamma_n(R) + 6\gamma_n'(R)]/2.$$

The time dependent single-particle propagator (22') is here of the form:

$$G_n(\mathbf{k}, t) = -i\exp[-\mathscr{E}_n''(k)\,t]\,\exp[i\mathscr{E}_n'(k)\,t]\,\theta(t).$$

The probability of survival of the exciton $B_n^+(\mathbf{k})|o\rangle$ at time t is hence

$$P_n(k, t) = \exp[-2\mathscr{E}_n''(k)\, t] \tag{68}$$

which shows that its lifetime is given by $\tau_n(k) = [2\,\mathscr{E}_n''(k)]^{-1}$, where $\mathscr{E}_n''(k)$ is obtained from Equations (63) or (63′).

To discuss the *transport* properties, we may begin in the same way as in Section 2.4. The appropriate single-particle propagator (23) becomes:

$$G_n(\mathbf{R}_j - \mathbf{R}_i; t) = -i\theta(t)\, N^{-1} \sum_{\mathbf{k}} \exp(i\mathbf{k}\mathbf{R}_i)\, \exp\{-i[\mathbf{k}\mathbf{R}_j + \mathscr{E}_n(k)\, t]\} \times$$

$$\times \exp[-\mathscr{E}_n''(k)\, t]. \tag{69}$$

Also, taking wave packets constructed around a mean value $\mathbf{k}_0$ of wave vector, we can define a quasi-particle velocity $\mathbf{v}_n(k_0)$ and lifetime $\tau_n(k_0)$, (cf. Equations (31)–(36)):

$$\mathbf{v}_n(k_0) = [\mathbf{\nabla}_{\mathbf{k}}\, \mathscr{E}_n'(k)]_{\mathbf{k}=\mathbf{k}_0} \tag{70}$$

$$(v_n)_\nu(k_0) = 2\,\overline{U}_n\, R^2\, \sin k_0 R \quad \text{with} \quad \nu = 1, 2, 3$$

where $\overline{U}_n$ is given by Equation (62); $\tau_n(k_0)$ is given by $[2\,\mathscr{E}_n''(k_0)]^{-1}$.

4. Concluding Remarks

In practice, exciton transport is invoked for the interpretation of the experimental results on various rate processes, involving for example excitation transfer to an impurity [40], exciton-exciton interaction [41, 42] modification of triplet exciton ESR lines [43], etc... A direct description of such actual situations by means of the simple quantities (69) and (70) is not possible, and in any case more or less complicated 'master equation' or relaxation matrices must be derived and applied [44, 45]. In the present context only very general remarks will be made.

We first note that the exciton rate processes are generally followed by experimental techniques with resolution times notably larger than the mean life time $\tau_n(k_0)$ (for usual low temperatures $\gtrsim 10^{-10}$ s) of the coherent exciton wave packets; before any observation is possible, the non-diagonal elements $\rho_n(\mathbf{k}, \mathbf{k}') = \langle o|B_n(\mathbf{k})\,\rho B_n^+(\mathbf{k}')|o\rangle$ of the exciton density matrix ρ_n in the energy momentum representation have therefore decayed. The exciton states are then best represented by a diagonal density matrix in energy space

$$\rho_n = \sum_{\mathbf{k}} B_n^+(\mathbf{k})\, |o\rangle\, P(\mathbf{k})\, \langle o|\, B_n(\mathbf{k}) \tag{71}$$

where $P_n(\mathbf{k})$ may usually be taken as the Boltzmann distribution function for the excitons in thermal equilibrium with phonons. The mixed state (71) describes an ensemble of $B_n^+(\mathbf{k})|o\rangle$ excitons, each of which being present with a statistical weight $P_n(\mathbf{k})$. Taking advantage of the quasi-particle picture of these excitons, we may use the common Fokker-Planck theoretical arguments [46] to derive an expression of the diffusion coefficient D_n for the ensemble. According to Zwanzig [47] we write:

$$D_n = (1/3) \int_0^T d\tau \langle \mathbf{v}_n(0) \, \mathbf{v}_n(\tau) \rangle \tag{72}$$

where $\mathbf{v}_n$ denotes the particle velocity $\mathbf{v}_n = \nabla_{\mathbf{k}}[\mathscr{E}'_n(k)]$. The velocity time correlation function is given by

$$\langle \mathbf{v}_n(0) \, \mathbf{v}_n(\tau) \rangle = \sum_{\mathbf{k}} P_n(\mathbf{k}) \, [\nabla_{\mathbf{k}} \, \mathscr{E}'_n(k)]^2 \, \exp[-\mathscr{E}''_n(k) \, \tau].$$

The integration limit T in Equation (72) is a macroscopic time, with $T \gg [2\mathscr{E}''_n(k)]^{-1}$, so that D_n becomes

$$D_n = (1/3) \sum_{\mathbf{k}} P_n(\mathbf{k}) \, [\nabla_{\mathbf{k}} \, \mathscr{E}'_n(k)]^2 / (\mathscr{E}''_n(k))$$

$$\simeq (1/3)|\mathbf{v}_n(k_0)|^2 / (\mathscr{E}''_n(k)) \tag{73}$$

where k_0 is a mean value determined by the distribution function $P_n(\mathbf{k})$; if $k_0 R \ll 1$, the various factors in expression (73) can be related to spectral properties; see for example ref. 13.

The fundamental derivations of the master equations in Quantum Statistical Physics [44–47] show that the simple kinetic equations like the diffusion equation are obtained under the condition that the initial density matrix is diagonal. As shown by Equation (71), this is the case in the collective exciton representation but not in the representation of the locally excited configuration. In the latter basis, the density matrix includes indeed non-diagonal elements, since

$$\rho_n = N^{-1} \sum_{i, j} B_n^+ (\mathbf{i}) \, |o\rangle \, P(\mathbf{R}_i - \mathbf{R}_j) \, \langle o| \, B_{nj} \, (\mathbf{j}) \tag{74}$$

$P(\mathbf{R}_i - \mathbf{R}_j)$ is the transform of $P(\mathbf{k})$:

$$P(\mathbf{R}_i - \mathbf{R}_j) = N^{-1} \sum_{\mathbf{k}} \exp[i k(\mathbf{R}_i - \mathbf{R}_j)] \, P(\mathbf{k}). \tag{75}$$

A diagonal form can be recovered if the distribution function $P(\mathbf{k})$ is broad enough so that its width Δk is larger than R^{-1}; the distribution $P(\mathbf{R}_i - \mathbf{R}_j)$ then becomes very sharp and we are left with

$$\rho_n \sim N^{-1} \sum_i B_n^+ (i) \, |o\rangle \, \langle o| \, B_n \, (i) \tag{76}$$

which describes an incoherent mixure of all the locally excited configurations. The validity condition $\Delta k R \gg 1$, together with the initial representation (76), as a mixed state, should be born in mind when the currently most frequent transport equations in the locally excited basis are used; i.e. when 'random walk' motion, 'hopping excitons', 'jumping' site spin hamiltonians [48]; etc., are used as models to describe observed properties such as trap emissions, line-shape broadening or narrowing. delayed fluorescence etc. (see in particular refs. 13–16).

References

1. Davydov, A. S.: *Theory of Molecular Excitons*, Mc Graw-Hill, New York, 1962.
2. Davydov, A. S.: *Theory of Molecular Excitons*, Plenum Press, New York, 1971. This reference is different in view-point from ref. 1.
3. Craig, D. P. and Walmsley, S. H.: *Excitons in Molecular Crystals*, Benjamin, New York, 1968.
4. Rice, S. A. and Jortner, J.: in *Physics and Chemistry of the Organic Solid State* (D. Fox, M. Labes, and A. Weissberger, eds.), Wiley-Interscience, New York, 1967, Vol. III, p. 199.
5. Robinson, G. W.: *Ann. Rev. Phys. Chem.* **21**, 429 (1970).
6. Sheka, E. F.: *Usp. Fiz. Nauk.* **104**, 593 (1971).
7. Phillpott, M.R.: *Adv. Chem. Phys.* **23**, 227 (1973).
8. These modes may be seen in the following way: a molecular excitation to level $\ln|n\rangle$ may be represented by an electronic oscillator $\mu_{on} \exp\left[i(E_n - E_o)t/\hbar\right]$, where μ_{on} is a matrix element of an electronic dipole operator. The resonant electronic oscillators, associated with each site, couple weakly and form modes of electronic oscillations: $\mu_{ok}^n \exp\left[i(E_k - E_o)t/\hbar\right]$.
9. These situations may be illustrated as follows [10, 11] (noting by ΔE_{ss} the Stokes shift): in the weak coupling case the excitation relaxes in the molecular potential ($\Delta E_{ss} > U$) and vibrates many times before transfer to the neighbours occurs ($\Delta E \gg U$), i.e. the nuclei of the molecule feel mainly the electronic potential of a free excited molecule [12]. In the strong coupling case by contrast, the excitation transfers to the neighbours before relaxing ($U > \Delta E_{ss}$) and before any vibation takes place ($U > \Delta E$), i.e. the nuclei of each molecule vibrate in the same potential of a de-localized electronic state. These 'pictures' involving relaxation and transfer between identical molecules can be invoked only when an external perturbation of strength $\gamma_0 \gg U$ destroys the excitonic coherence between the identical molecules (see conclusion of the present paper).
10. Simpson, W. T. and Peterson, D. L.: *J. Chem. Phys.* **26**, 588 (1967).
11. Forster, Th.: in *Modern Quantum Chemistry* (O. Sinanoğlu, ed.), Academic Press, New York (1965), Vol. III, p. 93.
12. Witkowski, A. and Moffitt, M.: *J. Chem. Phys.* **33**, 872 (1960).
13. Ern, V. and Schott, M.: p. 249 in this volume.
14. Reineker, P. and Haken, H.: p. 285 in this volume.
15. Aslangul, Cl., Lemaistre, J.-P., and Kottis, Ph.: p. 209 in this volume.
16. Lemaistre, J.-P., Aslangul, Cl., and Kottis, Ph.: p. 239 in this volume.
17. Lipkin, H. J.: *Quantum Mechanics, New Approaches to Selected Topics*, North Holland Publ. Co., Amsterdam, 1973, Chap. 9 and 11.
18. Pines, D.: *The Many Body Problem*, Benjamin, New York, 1961.
19. Ziman, J. H.: *Elements of Advanced Quantum Theory*, Cambridge University Press, 1969, Chap. 4.
20. Cohen-Tannoudji, C.: *Ann. Phys.* **7**, 423 (1962); Novikov, L. N., Pokazanev, V. G., and Strotskii, G. V.: *Usp. Fiz. Nauk.* **101**, 273 (1970).
21. $U_n(i, j)$ has no more or no less significance for motion of the excitation, than the integrals K_{ij}, in the molecular orbital theory, have significance for motion of the electrons between orbitals i and j.
22. Hopfield, J. J.: *Phys. Rev.* **112**, 1555 (1958).
23. Messiah, A.: *Mécanique Quantique*, Dunod, Paris, 1965, Chap. 21. Cohen-Tannoudji, C., Diu, B., and Laloë, F.: *Mécanique Quantique*, Hermann, Paris 1973.
24. Levine, R. D.: *Quantum Mechanics of Molecular Rate Processes*, Clarendon Press, Oxford, 1969, Part 3.
25. Voltz, R.: in *Organic Molecular Photophysics* (J. B. Birks, ed.), John Wiley, New York, 1975, Chap. 5.
26. Shore, B. W.: *Rev. Mod. Phys.* **39**, 439 (1967).
27. Messiah, A.: *Mécanique Quantique*, Dunod, Paris, 1965, Chap. 6.
28. Aslangul, C. and Kottis, Ph.: *Phys. Rev.* **B10**, 4364 (1974).
29. Klein, G.: Thèse, Strasbourg, 1975 and references therein.
30. These transitions are better described by the non-adiabatic operator, i.e. the energy term neglected in the adiabatic approximation separating the intramolecular motions from the lower energy lattice motions [25].

31. The assumption $\mathbf{p} = 0$ implies that we take a zero temperature assumption. This is sufficient in the present context and may be extended without essential difficulties to finite temperature conditions, cf. ref. 2, Chap. IV.

32. See in particular Fischer, J. and Rice, S. A.: *J. Chem. Phys.* **52**, 2089 (1970); Grover, M. K. and Silbey, R.: *J. Chem. Phys.* **52**, 2099 (1970).

33. See for instance models for small polarons Holstein, T.: *Ann. Phys. (N.Y.)* **8**, 343 (1959), part II; Sewell, G. L.: *Phys. Rev.* **129**, 597 (1963).

34. Haken, H. and Reineker, P.: in *Excitons, Phonons and Magnons* (A. Zahlan, ed.), Cambridge U.P., London, 1968.

35. See, for example, ref. 17, Chap. 11.

36. In a more general case, applying for instance at finite temperatures, the significance of the bracket $\langle\ \rangle$ must be generalized [37] as $\langle A \rangle = \mathrm{Tr}\ \rho A$, where ρ represent the canonical distribution over the possible initial states $\{|ni; \mathbf{p}\rangle\}$. Another way to introduce finite temperature, through the amplitude of the parameter γ_0, and its limitations, is discussed in refs. 15 and 28.

37. Maradudin, A. A.: in *Elementary Excitations in Solids* (A. A. Maradudin and G. F. Nardelli, eds.), Plenum Press, New York, 1969, p. 455.

38. Ern, V., Suna, A., Tomkiewicz, Y., Avakian, P., and Groff, R. P.: *Phys. Rev.* **B5**, 3222 (1972).

39. This means, in particular, that indirect radiative transitions with simultaneous formation of excitations and phonons, are ignored.

40. See for example, Voltz, R.: *Radiation Res. Rev.* **1**, 301 (1968).

41. Suna, A.: *Phys. Rev.* **B1**, 1716 (1970).

42. Trlifaj, M.: *Pure and Applied Chem.* **37**, 197 (1974).

43. See for instance, Lemaistre, J.-P. and Kottis, Ph.: in *Electron Spin Relaxation in Liquids* (L. T. Muus and P. W. Atkins, eds.), Plenum Press, New York, 1972, p. 492.

44. See, for instance, Fujita, S.: *Introduction to Non-Equilibrium Quantum Statistical Mechanics*, Saunders, Philadelphia, 1966.

45. Zwanzig, R.: in *Quantum Statistical Mechanics* (P. H. E. Meijer, ed.), Gordon and Breach, New York, 1966, p. 139.

46. See for example, Kittel, C.: *Elementary Statistical Physics*, Wiley, New York, 1958, Chap. 32.

47. Zwanzig, R.: *Ann. Rev. Phys. Chem.* **16**, 67 (1965).

48. Abragam, A.: *The Principles of Nuclear Magnetism*, Oxford (1961) Chap. 10.

49. Note however that this crossed relaxation term disappears by symmetry for the dimers. (Cl. Aslangul, private communication).

ELECTRONIC COLLECTIVE EXCITATIONS
IN MOLECULAR ONE-DIMENSIONAL SYSTEMS

I: *Optical Response, Localization and*
Motion of Linear Excitons

CLAUDE ASLANGUL, JEAN-PIERRE LEMAISTRE and PHILÉMON KOTTIS

1. Introduction

The existence and properties of electronic collective excitation [1] will be discussed in the domain of finite collections of weakly interacting identical molecules: the molecular aggregates. For the latter, excitonic coherence states are formed in general when the local fluctuation of the energy is much weaker than the quantum resonance interactions.

The attention will be focused on the optical response of an ensemble of molecular aggregates oriented in a crystalline non-resonant matrix and related properties and phenomena such as selection rules, which replace the 'momentum' conservation rule in crystals [1], excitation localization and subsequent transfer in the aggregate [4], in conjunction with complementary investigations in infinite systems [5, 6]. As a type of such aggregate, we may consider 'guest polymers A—A—...—A' of light naphthalene in a deuterated crystal, guests of finite multichains of α-dibromonaphthalene and tetrachlorobenzene [5], or more complex systems of helix-molecular chains encountered in biological systems such as polynucleotides, for instance [6].

For these systems, we consider the lower excited electronic states which have the essential peculiarity to be very closely spaced; vibrations of the chain as a whole, as well as those of the matrix (phonons) are treated semiclassically as random fluctuations of the electronic matrix elements and whose main effect is to destroy the excitonic coherence [2].

Apart of the definition of electronic collective excitation (ECE) stationary states, we will have to define an energy density operator in order to describe in a rigorous manner, and otherwise than intuitively, observable properties such as localization and excitation transfer. The discussion of the optical preparation of a state (the initial state) needs to analyse selection rules, proper to collective excitations for which, as excitonic coherence states, we have to *add the excitation amplitudes of the different sites*, interfering with phases defined from the invariance of the aggregate hamiltonian. This discussion will be made with a semi-classical description of the matter-radiation interaction.

Two sections, 2 and 3, will cover this contribution: in Section 2, states of excitonic coherence are derived for a one-dimensional chain from a model hamiltonian. An

O. Chalvet et al. (eds.), Localization and Delocalization in Quantum Chemistry, Vol. II, 209–237. All Rights Reserved
Copyright © 1976 by D. Reidel Publishing Company, Dordrecht-Holland

isomorphism is established between collective excitations and a finite Lie algebra in which operations analogous [1] to creation and annihilation are defined. The perturbations are described with a special class of superoperators whose compact formalism has been shown to allow straightforward calculations [2]: an optical susceptibility function, a macroscopic evolution equation and excitation moments will be examples of such calculations for one-dimensional chains performed in the present contribution.

One of the main conclusions of this section is to show that in the interaction with the radiation, the individual molecule excitation loses completely its independence, i.e. its resonance frequency, polarization, lifetime (or excitation residence time), for a physical measure and that this is due to the formation of excitonic coherence states which do not imply any motion. The molecule is coupled coherently to its neighbours in the same way a $2p$ orbital state of the carbon atom loses completely its individuality when it contributes to form a π-state in a benzene molecule [7]; the latter is no more described in terms of motion of 'the π electrons'. The fact that the resonance interaction in the π states (the β's) are much larger than the resonance interactions in the excitons (the V's) makes conceptually absolutely no difference at low temperatures. Section 3, devoted to the description of the excitation motion, is divided in two parts. In one, we discuss the possibility of an optical preparation of a localized state $|\psi(0^+)\rangle$ for which a function is defined to measure the degree of its localization. In the second part, we define an energy density operator through which the first two moments of the excitation are calculated for one-dimensional systems. These moments allow a rigorous description of the time evolution of an excitation singularity prepared at one end, or in the middle, of the molecular chain, with subsequent reflections of the exciton at the ends of the chain. For an infinite chain, the moments coincide with the usual quantities describing the excitation transport and its nature [4, 8]. In conclusion, it is pointed out that the coherent transport is not always amenable to simple pictures such as 'site-to-site motion with conservation of energy' as is the case in a dimer-aggregate [2].

2. The Static Problem

2.1. THE MODEL HAMILTONIAN FOR ONE-DIMENSIONAL CHAINS

The degrees of freedom of the system are divided into two classes, according to the density and frequencies of the states:

(a) A class of degrees of freedom of low density and high energies: these are the lower electronic excitations and intrasite vibrations with quantum numbers v. In fact, in a low temperature approximation we consider excitations to vibrationless ($v=0$) first electronic states and, hence, we have to consider a truncated space reduced to the ground and first excited electronic state for each site [9]. For an aggregate containing N identical, weakly interacting sites, we note by $|M_p^0\rangle$ and $|M_p^x\rangle$ the ground and first excited state of the site p. A complete electronic basis is formed with the following configurations.

$$|0\rangle = \hat{\mathscr{A}} \, |M_1^0 \, ... \, M_N^0\rangle,$$

$$\{|M_p\rangle = \hat{\mathscr{A}} \, |M_1^0 \, ... \, M_{p-1}^0, \, M_p^x, \, M_{p+1}^0, \, ... \, M_N^0\rangle\}_{1 \leqslant p \leqslant N}. \tag{2.1}$$

Utilizing the next-nearest-neighbour approximation, the truncated hamiltonian, which describes the ground state and the first exciton band, one writes:

$$H_0 = \hbar\omega_0 \sum_{p=1}^{N} |M_p\rangle \langle M_p| + \hbar\Delta\omega \sum_{p=1}^{N-1} \left[|M_p\rangle \langle M_{p+1}| + |M_{p+1}\rangle \langle M_p| \right] \tag{2.2}$$

where ω_0 is the resonance frequency of a site, corrected by change, upon excitation, of the interaction energy with the $N-1$ non-excited molecules. $\Delta\omega$ is the matrix element of the excitation exchange between two neighbour sites. $\hat{\mathscr{A}}$ is the operator antisymmetrizing the electrons in each configuration.

The frequencies ω_k and the eigenfunctions $|A_k\rangle$ of the collective stationary excitations are obtained by a straightforward diagonalization of H_0; then we get:

$$|A_k\rangle = \frac{\sqrt{2}}{\sqrt{N+1}} \sum_{p=1}^{N} \sin kp\theta \, |M_p\rangle; \qquad \omega_k = \omega_0 + 2\Delta\omega \cos k\theta \tag{2.3}$$

$$\hat{H}_0 = \sum_{k=1}^{N} \hbar\omega_k \, |A_k\rangle \langle A_k| \tag{2.4}$$

with $\theta = \pi/(N+1)$ and $1 \leqslant k \leqslant N$.

H_0 is the hamiltonian of a rigid perfect chain and describes N electronic collective stationary excitations. This is in the sense that it results from N electronic resonant oscillators $\mu_n e^{i\omega_0 t}$, with $n = 1...N$, weakly coupled and which form N modes of electronic oscillation $\mu_k e^{i\omega_k t}$, with $k = 1...N$, or N excitonic coherence states.

In many papers, the molecular exciton state is presented as really independent molecules transfering very rapidly the excitation from one to another. What the molecules are rather exchanging is coherence, or information, and that makes that the molecules: (i) have not arbitrary phases, (ii) at each instant all the sites vibrate at the same frequency ω_k with constant comparable weights; this picture excludes any motion or migration for molecules prepared in a state $|A_k\rangle$. Indeed, it is contradictory to use exciton selection rules which imply that the sites are indistinguishable and in phase and to use a migration picture which implies transitions from site to site and random phase; this not a question of semantics but of observation results: see in ref. 10 the EPR response of a linear parallel chain according to the two languages.

(b) A class of degrees of freedom of high density and low energies: these are collective vibrations in the molecular chain and in the crystalline matrix (translational and torsional modes of phonons) for which we may distinguish site-site vibrations in the chain and site-matrix vibrations. From the model definition of the hamiltonian the former and the latter vibrations have, respectively, zero-diagonal and zero non-diagonal matrix elements on the electronic basis $\{|M_p\rangle\}_p$, cf. (2.1). The essential advantage of the stochastic model, used for small polarons [11] and for excitons [12], is to allow an exact derivation of the gross features of the effect of these vibrations on the electronic properties, from a stochastic semi-classical field $\hat{\varepsilon}(t)$, instead of considering explicitly a huge amount of phonon states and their spectral properties (for a comparative discussion see ref. 1). However, in order to represent quantized vibration energies by such a field $\hat{\varepsilon}(t)$, a high temperature approximation must be valid; this amounts here to the condition that $kT \gg \Delta E_{ph}$, where ΔE_{ph} is the band width of the phonons. This condition implies that the local vibrations do not correlate to form phonons and that we can describe the local vibration itself by a brownian motion which gives to the local electronic oscillator a random phase when expressed in the interaction representation $\hat{\mu}_I(t) = \exp(i\hat{H}_0 t/\hbar)\, \hat{\mu}(t)\, \exp(-i\hat{H}_0 t/\hbar)$.

2.2. FORMALISM AND TECHNIQUE OF SUPEROPERATORS

The exciton stochastic hamiltonian is developed in a Lie algebra of operators where the superoperators technique allows exact solutions for evolution equations of complex aggregates such as simple or double one-dimensional linear chains.

Therefore, the formalism is developed for the statistical treatment in terms of the density operator of a special class of quantum systems, namely those whose stochastic (incoherent) interactions are present together with non-stochastic (coherent) interactions, with which they do not commute. We focus our attention on a statistical ensemble of oriented non-interacting, one-dimensional systems, composed of identical subunits whose resonance quantum interactions can be described with a single parameter $(\Delta\omega)$ in the next nearest neighbour approximation. As usual one such aggregate is called the microsystem, the statistical ensemble being called the system. From a rigorous quantum-statistical point of view, our system should be enlarged as to include an external reservoir which would define the macroscopic thermodynamic properties such as temperature (and populations). In this first approach, we merely take the reservoir into account through the magnitude of the mean-square deviations of the semi-classical stochastic terms of $\hat{\varepsilon}(t)$; we are then left with the state space pertaining to a single microsystem.

Because of stochastic interactions introduced in the system through the 'Brownian' local motion of sites, our system cannot be described by a vector $|\psi\rangle$, but only through a statistical operator $\hat{\rho}$. This operator is not giving us only the physical response of the sample, its resonances for instance, but gives us additional informations such as the coherence of the system which is a crucial concept for the comprehension of the excitation migration in the system.

The starting point is then the Liouville equation and the calculation of the density matrix $\hat{\rho}(t)$. For a microsystem, the density operator obeys the Liouville equation:

$$i\hbar \frac{d\hat{\rho}}{dt} = [\hat{H}, \hat{\rho}] \tag{2.5}$$

where $\hat{\rho}$ is written in the space of states. In this section we show that if we expand $\hat{\rho}$

in a vectorial space of linear operators $\{\hat{\Omega}_n\}_n$, then Equation (2.5) takes the form:

$$ih \frac{d}{dt}|\rho(t)) = \hat{H}_-|\rho(t)), \tag{2.5'}$$

where $|\rho(t))$ is a vector, defined in space $\{\hat{\Omega}_n\}_n$ and $\hat{H}_-$ is a superoperator acting in the same space (as $|\psi\rangle$, in the Schrödinger equation, is defined in the Hilbert space of the eigenkets $|m\rangle$ and $\hat{H}$ acts in the same space). The integration of (2.5') and the derivation of its observables become a straightforward operation without mathematical complicated intermediates such as repeated commutators, functional derivations, or generalized density operators. Such a procedure has been first proposed by Fano [13]. Now, if we are able to find a particular basis of linear operators, say basis $\{\hat{T}_n\}_n$, that commutes with $\hat{H}$, or with the unperturbed part of it, the matrix of $\hat{H}_-$ is diagonalized or block diagonalized. Each block corresponds to a specific part of the perturbation, coupling specific states of the unperturbed Hamiltonian. Therefore, if we are interested in a specific property of the system, for instance, its absorption line shape or its excitation motion by resonance transfer, we will have to integrate the components of $|\rho(t))$ corresponding only to one block of the matrix of $\hat{H}_-$, which is of considerably reduced dimension. Below, we develop the formalism which allows, given $\hat{H}$, choosing the privileged basis $\{\hat{T}_n\}_n$ and to derive the supermatrix $\hat{H}_-$, cf. Equation (2.9), in its 'physically irreducible form', or more generally, given $\hat{\Xi}$, $\hat{B}$, and $\hat{A}$ related by $\hat{A} = [\hat{\Xi}, \hat{B}]$, to find the supermatrix $\hat{\Xi}_-$ so that $\hat{\Xi}_-|B) = |A)$, the relation of which (2.5') is a particular case. The reader may, for the first time, skip the derivation of the supermatrix $(\Omega_n|\hat{H}_-|\Omega_m)$ given below and come back later after seeing simple examples of its application handled for the dimer in Section 5 of ref. 4.

Let us call $\mathcal{H}$ the domain of a complete set of linear operators $\{\hat{\Omega}_n\}_n$ and $\mathcal{A}(\mathcal{H})$ the vectorial space they define. The set $\mathcal{A}(\mathcal{H})$ is obviously an algebra but we essentially need its Lie-Algebra substructure; to every operator $\hat{A}$ acting in $\mathcal{H}$ and represented by a matrix $\langle m|\hat{A}|n\rangle$ on some complete basis $\{|m\rangle\}_m$, we associate a vector $|A)$ in $\mathcal{A}(\mathcal{H})$ (we use the notation of Lynden-Bell [14]). A trivial complete basis in that space is made up by $S_{mn} = |m\rangle\langle n|$, but usually it is not convenient since it does not include any mathematical or physical symmetry properties of the system under consideration; some rules concerning the choice of a complete basis in $\mathcal{A}(\mathcal{H})$ are given below.

If $\mathcal{H}$ is a finite-dimension space, so is $\mathcal{A}(\mathcal{H})$ and the scalar product $(|A), |B)) = \text{Tr}(\hat{B}^+\hat{A})$ gives a Hilbertian structure to $\mathcal{A}(\mathcal{H})$; the dual space $\mathcal{A}'(\mathcal{H})$ has elements $(A|$, so that the preceding scalar product can be noted $(|A), |B)) \equiv (B|A)$. We shall especially be concerned with a special class of linear operators whose domain is $\mathcal{A}(\mathcal{H})$ and which we will now define. In $\mathcal{H}$, the operator equation, the so-called internal derivative (id) is

$$[\Xi, \hat{B}] = \hat{A}. \tag{2.6}$$

For any operator $\hat{L}$, using an orthonormal basis set such that $\text{Tr}(\hat{\Omega}_\lambda^+ \hat{\Omega}_\mu) = (\Omega_\lambda|\Omega_\mu) = \delta_{\lambda\mu}$, we can write

$$\hat{L} = \sum_\lambda L_\lambda \hat{\Omega}_\lambda$$

with

$$L_\lambda = \mathrm{Tr}(\hat{\Omega}_\lambda^+ \hat{L}) \tag{2.7}$$

being a complex number. Equation (2.6) can then be rewritten as

$$\sum_\mu B_\mu \, \mathrm{Tr}(\hat{\Omega}_\nu^+ [\hat{\Xi}, \hat{\Omega}_\mu]) = A_\nu.$$

By using the cyclic properties of the trace we get

$$\sum_\mu B_\mu \, \mathrm{Tr}([\hat{\Omega}_\nu, \hat{\Omega}_\mu^+]^+ \hat{\Xi}) = A_\nu.$$

Since $\{\hat{\Omega}_\lambda\}_\lambda$ is complete, there must exist a unique set of complex numbers $\sigma_{\nu\mu}^\tau$ such that

$$[\hat{\Omega}_\nu, \hat{\Omega}_\mu^+] = \sum_\tau \sigma_{\nu\mu}^\tau \hat{\Omega}_\tau.$$

The constants $\sigma_{\nu\mu}^\tau$ are the constants of structure of the Lie subalgebra [15]. We then finally get:

$$\sum_\mu B_\mu \, \mathrm{Tr}([\hat{\Omega}_\nu, \hat{\Omega}_\mu^+]^+ \, \hat{\Xi}) = \sum_\mu B_\mu \sum_\tau \sigma_{\nu\mu}^{\tau*} \, \mathrm{Tr}(\hat{\Omega}_\tau^+ \, \Xi) = A_\nu.$$

This equation may be rewritten in space $\mathscr{A}(\mathscr{H})$ as

$$\sum_\mu (\Omega_\mu|B) \, ([\hat{\Omega}_\nu, \hat{\Omega}_\mu^+]| \, \Xi)$$
$$= \sum_\mu (\Omega_\mu|B) \, (\sum_\tau \sigma_{\nu\mu}^{\tau*}(\Omega_\tau^+|\Xi)) = (\Omega_\nu|A). \tag{2.8}$$

We then define the superoperator $\hat{\hat{\Xi}}_-$, acting in $\mathscr{A}(\mathscr{H})$, by its matrix elements on the basis $\{|\Omega_\lambda)\}_\lambda$

$$(\Omega_\nu|\hat{\hat{\Xi}}|\Omega_\nu) = ([\hat{\Omega}_\nu, \hat{\Omega}_\mu^+]|\Xi) = \sum_\tau \sigma_{\nu\mu}^{\tau*}(\Omega_\tau|\Xi) = \sum_\tau \sigma_{\nu\mu}^{\tau*} \Xi_\tau. \tag{2.9}$$

Then (2.8) is written

$$\sum_\mu (\Omega_\nu|\hat{\hat{\Xi}}_-|\Omega_\mu) \, (\Omega_\mu|B) = (\Omega_\nu|A)$$

and leads to the linear transformations $\hat{\hat{\Xi}}_-$ in the vectorial space $\mathscr{A}(\mathscr{H})$:

$$\hat{\hat{\Xi}}_-|B) = |A). \tag{2.10}$$

The operator $\hat{\hat{\Xi}}_-$ will be called the id operator associated to $\hat{\Xi}$; we see that the initial Equation (2.6) has now the usual form of a simple linear transformation in $\mathscr{A}(\mathscr{H})$ space; we do not claim that every difficulty is avoided in this way, but we can directly use all the procedures involved by conventional eigenvector equations such as perturbation methods, variational methods and so on; anyhow every formal expression will be made simpler by the use of (2.10) instead of (2.6).

The construction of the matrix representing any id operator is summed up by Eq. (2.9); first, compute the table of the commutators $[\hat{\Omega}_\nu, \hat{\Omega}_\mu^+]$, second, replace the resulting operator by the component of $|\Xi)$ on that operator; so the essential numbers are obviously the constants of structure $\sigma_{\mu\nu}^\tau$ fixed as soon as the basis is known.

The choice of the latter is a matter of convenience, but may be decisive for an actual full analytic determination: because of the usual high dimensionality of the matrices $(\hat{\Xi}_-)_{\lambda\mu}$, one must select a basis which leads to the best block diagonalization of the involved operators.

Particularly, we want to solve a Liouville equation (2.5) and one of the best choices for the basis is to take the 'eigenoperators' of the constants of motion; let us examine that point further. Let be $\hat{K}$ such that $\partial\hat{K}/\partial t = 0$, $[\hat{H}, \hat{K}] = \hat{O}$. It is easy to show that this implies $[\hat{H}_-, \hat{K}_-] = \hat{O}$.

In effect for any $|X)$ belonging to $\mathscr{A}(\mathscr{H})$ we have

$$[\hat{A}_-, \hat{B}_-]|X) = \hat{A}_-(\hat{B}_-|X)) - \hat{B}_-(\hat{A}_-|X))$$
$$= \hat{A}_-|[\hat{B}, \hat{X}]) - \hat{B}_-|[\hat{A}, \hat{X}])$$
$$= [\hat{A}, [\hat{B}, \hat{X}]] - [\hat{B}, [\hat{A}, \hat{X}]]$$
$$= -([[\hat{B}, \hat{X}], \hat{A}] - [[\hat{X}, \hat{A}], \hat{B}])$$
$$= [[A, B], X] \equiv ([A, B])_-|X);$$

this latter equation resulting from the Jacobi identity

$$[\hat{A}, [\hat{B}, \hat{C}]] + [[\hat{B}, \hat{C}], \hat{A}] + [[\hat{C}, \hat{A}], \hat{B}] = \hat{O}.$$

We then have

$$[\hat{A}, \hat{B}] = \hat{C} \Rightarrow [\hat{A}_-, \hat{B}_-] = \hat{C}_-.$$

We see that if one can find a constant of motion $\hat{K}$, and if we take as members of the basis the eigenvectors of K_-, $|\chi_i)$,

$$\hat{K}_-|\chi_i) = \chi_i|\chi_i) \Leftrightarrow [\hat{K}, \hat{\chi}_i] = \chi_i\hat{\chi}_i$$

the matrix of $\hat{H}_-$ will be diagonal on the basis $|\chi_i)$ except for degeneracies in the spectrum of $\hat{K}_-$; we note that, for every $\hat{K}$, $|K)$ is itself an eigenvector of $\hat{K}$ with the eigenvalue zero:

$$[\hat{K}, \hat{K}] = 0 \Leftrightarrow \hat{K}_-|K) = 0.$$

Let us sum up the choice of this particular basis, called from now on $\{|T_n)\}_n$: we first look for the constants of motion, $\hat{K}_i$, and among them, we take those which generate a set of mutually commuting operators; the basis is made up with their common eigenvectors (if the set of operators is complete, then $\hat{H}_-$ will be fully diagonal). This procedure is fully illustrated in Section 5 of ref. 2.

We finally note that the components of $\hat{\rho}$ on the $\hat{T}_n$'s represent the expectation values of these latter operators in the system described by $\hat{\rho}$; particularly the stationary states of the system, $|\rho_0)$, will belong to the subspace included in $\mathscr{A}(\mathscr{H})$, generated by these constants of motion.

2.3. MACROSCOPIC EVOLUTION EQUATION FOR A MODEL HAMILTONIAN

As a first application of the id-operators technique explained above, we shall derive here a Schrödinger-type evolution equation for a statistical ensemble density operator $\hat{\rho}$. The total time-dependent Hamiltonian for the ith microsystem is partitioned into an unperturbed common part $\hat{H}_0$ and a stochastic part $\hat{\varepsilon}^i(t)$,

$$\hat{H}^i(t) = \hat{H}_0 + \hat{\varepsilon}^i(t).$$ (2.11)

The random operator $\hat{\varepsilon}^i(t)$ represents fluctuations due to interaction with a large number of degrees of freedom; generally speaking these may be external (electrostatic interactions with the lattice) and internal (intersite vibrations and Coulombic interactions for instance). For an aggregate, the physical conditions have been discussed in Section 1. The diagonal and nondiagonal elements of $\hat{\varepsilon}^i(t)$, on the basis of localized states, represent the so-called local and nonlocal fluctuations. Recent measurements [17] of widths for monomer and dimer isotopic traps in crystals suggest that the main contribution to $\hat{\varepsilon}^i(t)$ originates indeed from internal degrees of freedom, as has been predicted [18] and justifies our assumption that $\hat{\varepsilon}^i(t)$ contains only effects from intersite vibrations.

The Liouville equation for the ith aggregate is written in the space of states $\mathscr{H}$,

$$ih \frac{\mathrm{d}}{\mathrm{d}t} \hat{\rho}^i(t) = [\hat{H}^i(t), \hat{\rho}^i(t)].$$

According to Equation (2.10) and through (2.9), this equation can be written in a Schrödinger form:

$$ih \frac{\mathrm{d}}{\mathrm{d}t} |\rho^i(t)) = \hat{H}^i_-(t) |\rho^i(t)).$$ (2.12)

Equation (2.12) may be formally integrated

$$|\rho'(t)) = \exp\left(\frac{i}{ih} \int_{t_0}^{t} \hat{H}^i_{0-}(t') \, \mathrm{d}t'\right) \exp\left(\frac{1}{ih} \int_{t_0}^{t} \hat{\varepsilon}^i_-(t') \, \mathrm{d}t'\right) |\rho^i(t_0)).$$

Here we have used Feynman's convention [19], t' being the ordering parameter. The density operator for the system will then be given by

$$|\rho(t)) = \exp\left(\frac{1}{ih} \int_{t_0}^{t} \hat{H}_{0-}(t') \, \mathrm{d}t'\right) \left\langle \exp\left(\frac{1}{ih} \int_{t_0}^{t} \hat{\varepsilon}^i_-(t') \, \mathrm{d}t'\right) \right\rangle |\rho(t_0)),$$

assuming a pure preparation and where $\langle \, \rangle$ stands for the statistical average $(1/\mathscr{N}) \Sigma_{i=1}^{N}$. The last time-evolution operator can be evaluated with the cumulants defined by Kubo [20]

$$\left\langle \exp\left(\frac{1}{ih} \int_{t_0}^{t} \hat{\varepsilon}^i_-(t') \, \mathrm{d}t'\right) \right\rangle =$$

$$= \exp\left(\sum_{n=0}^{\infty} (i\hbar)^{-n} \int_{t_0}^{t} dt_1 \int_{t_0}^{t_1} dt_2 \cdots \int_{t_0}^{t_{n-1}} dt_n \, \langle \hat{\varepsilon}^i_-(t_1) \, \hat{\varepsilon}^i_-(t_2) \ldots \hat{\varepsilon}^i_-(t_n) \rangle_c \right).$$

Assuming that $\hat{\varepsilon}^{(i)}_-(t)$ is a centered normal (Gaussian) operator, which physically means that it is the result of a great number of independent perturbations (central-limit theorem), the preceding expression reduces to

$$\left\langle \exp\left(\frac{1}{i\hbar} \int_{t_0}^{t} \hat{\varepsilon}^i_-(t') \, dt' \right) \right\rangle = \exp\left(-\frac{1}{\hbar^2} \int_{t_0}^{t} dt_1 \int_{t_0}^{t_1} dt_2 \, \langle \hat{\varepsilon}^i_-(t_1) \, \hat{\varepsilon}^i_-(t_2) \rangle \right).$$

Now, Feynman's time ordering will be convenient only if the time evolution operators depend on one ordering parameter; such a situation is produced if we assume that $\hat{\varepsilon}^i_-(t)$ has a unique correlation time which is beyond the time scale of the system. Then, we can safely assume that [22]

$$\langle \hat{\varepsilon}^i_-(t_1) \, \hat{\varepsilon}^i_-(t_2) \rangle = \langle \hat{\varepsilon}^2 \rangle \, \delta(t_1 - t_2),$$

where $\langle \hat{\varepsilon}^2 \rangle$ is a time-independent operator. Note that this assumption is consistent with the fact that $\hat{\varepsilon}^i(t)$ is due to fluctuations of site coordinates whose correlations may be neglected. This is a high-temperature approximation of a first kind introduced in this theory. Indeed, at very low temperatures, correlations between site vibrations become important and the evolution Equation (3.4) will be of integrodifferential form. The Dirac correlation function assumed for $\hat{\varepsilon}^i_-(t)$ implies

$$\left\langle \exp\left(\frac{1}{i\hbar} \int_{t_0}^{t} \hat{\varepsilon}^i_-(t') \, dt' \right) \right\rangle = \exp\left(-\frac{1}{2\hbar^2} \int_{t_0}^{t} \langle \hat{\varepsilon}^2 \rangle \, (t') \, dt' \right),$$

which leads to a straightforward integration for the vector $|\rho(t))$,

$$|\rho(t)) = \exp\left(\frac{1}{i\hbar} \int_{t_0}^{t} \hat{H}_{0-}(t') \, dt' \right) \times$$

$$\times \exp\left(-\frac{1}{2\hbar^2} \int_{t_0}^{t} \langle \hat{\varepsilon}^2 \rangle \, (t') \, dt' \right) |\rho(t_0))$$

$$= \exp\left\{ \frac{1}{i\hbar} \left(\hat{H}_{0-} - \frac{i}{2\hbar} \langle \hat{\varepsilon}^2 \rangle \right) (t - t_0) \right\} |\rho(t_0)).$$

In the last equation, we have reestablished the usual way of writing. Putting now $\langle \hat{\varepsilon}^2 \rangle = 2\hbar^2 \hat{\Gamma}$ and differentiating we get a macroscopic time-evolution equation for the statistical-ensemble density operator [24]

$$i\hbar \frac{d}{dt} |\rho(t)) = (\hat{H}_{0-} - i\hbar\hat{\Gamma}) |\rho(t)) \equiv \hat{H}| \, \rho(t), \qquad (2.13)$$

with

$$\langle \hat{\varepsilon}^i_-(t_1)\, \hat{\varepsilon}^i_-(t_2) \rangle = 2h^2\, \hat{\Gamma} \delta(t_1 - t_2). \tag{2.14}$$

We note the macroscopic character of the operator $\hat{\Gamma}$: there does not exist an operator $\hat{\Gamma}'$ acting in $\mathcal{H}$ generating an id operator Γ'_- acting in $\mathcal{A}(\mathcal{H})$ such that $\hat{\Gamma}'_- \equiv \hat{\Gamma}$ (that is the reason why we have suppressed the subscript $-$ in the macroscopic operator $\hat{H}$ defined by Equation (2.13); this results from having taken the statistical mean, an operation under which the Lie subalgebra is not closed; as every statistical property (temperature, pressure, ...), $\hat{\Gamma}$ cannot be defined at a microscopic scale; this point is exhibited in the detailed treatment of Section 5 in ref. 2; we see that the relaxation operator $\hat{\Gamma}$ has nonzero matrix elements where every id operator has zero matrix elements, and vice-versa; indeed the mathematical structure of $\hat{\Gamma}$ is quite different from that of an id operator. This is in connection with the discussion made by Fano on the Ritz combination principle [21].

2.4. The Linear Optical Response of an Electronic Collective Excitation

In this paragraph, we use Equation (2.13) and we derive the imaginary part of the electric susceptibility of the system, noted $\chi''(\omega)$, which gives the resonances and the selection rules (line-shape) proper to electronic collective excitations.

(a) The generalized optical susceptibility.

Let $\hat{W}(t)$ be a weak semiclassical radiation-matter coupling with $\hat{W}(t) = f(t)\hat{W}$, $f(t)$ being a classical function and $\hat{W}$ a time-independent operator. $\hat{W}(t)$ constitutes the external perturbation of the system and $\hat{W}_-(t)$ is its corresponding id operator. The magnitude of $\hat{W}_-(t)$ defines the first order of the perturbation series and the linear time response [26, 27] $|\Delta\rho(t))$ of the system is obtained from Equation (2.13) where $\hat{W}_-(t)$ was added to $\hat{H}$,

$$i\hbar \frac{d}{dt} |\Delta\rho(t)) = \hat{H} |\Delta\rho(t)) + \hat{W}_-(t)\, \rho(t)), \tag{2.14}$$

where $|\rho(t))$ is the density operator in the absence of external perturbation. Equation (2.14) integrates easily to give

$$|\Delta\rho(t)) = i\hbar \int_{-\infty}^{+\infty} dt'\, \hat{K}(t-t')\, \hat{W}_-(t')\, \hat{K}(t'-t_0)\, |\rho(t_0)). \tag{2.15}$$

$\hat{K}(\tau)$ is the propagator of the system, with $\hat{K}(\tau) = (i\hbar)^{-1}\theta(\tau)\, e^{\hat{H}\tau/i\hbar}$ with $\theta(\tau) = 1$ (respectively, 0) for $\tau > 0$ (respectively, $\tau < 0$). $\hat{H}$ is the Hamiltonian defined in Equation (2.13). If, at the beginning t_0 of perturbation by $\hat{W}(t)$, the system is in a stationary state, then we note $|\rho(t_0)) = |\rho_0)$ and we obtain $i\hbar \hat{K}(t'-t_0)|\rho_0) = |\rho_0)$. Hence:

$$|\Delta\rho(t)) = \int_{-\infty}^{+\infty} dt'\, f(t')\, \hat{K}(t-t')\, \hat{W}_-|\rho_0).$$

The Fourier transform of $|\Delta\rho(t))$, is noted $|\Delta R(\omega))$ and, using the convolution theorem, the latter becomes

$$|\Delta R(\omega)) = \hat{\mathscr{K}}(\omega) \, F(\omega) \, \hat{W}_-|\rho_0), \tag{2.16}$$

where $\hat{\mathscr{K}}(\omega)$ and $F(\omega)$ are the Fourier transforms of $K(t)$ and $f(t)$, respectively, with [28] $\hat{\mathscr{K}}(\omega) = (\hbar\omega \, \hat{I} - \hat{H})^{-1}$.

The total absorbed energy is written

$$\Delta E = \int_{-\infty}^{+\infty} dt \, \mathrm{Tr} \left(\Delta\hat{\rho}(t) \frac{\partial \hat{W}}{\partial t} \right) = \int_{-\infty}^{+\infty} dt \, \frac{\partial f}{\partial t} \, (\hat{W}|\Delta\rho(t)). \tag{2.17}$$

Replacing $f(t)$ and $|\Delta\rho(t))$ by their Fourier transforms, the absorbed energy is expressed as follows:

$$\Delta E = \frac{i}{2\pi} \int_{0}^{+\infty} d\omega \, \omega |F(\omega)|^2 \, (W|[\hat{\mathscr{K}}(\omega) - \hat{\mathscr{K}}(-\omega)] \, \hat{W}_-|\rho_0).$$

Defining the complex susceptibility of the system by the expression [29, 27]

$$\chi(\omega) = \chi'(\omega) - i\chi''(\omega) = (W|\hat{\mathscr{K}}(\omega) \, \hat{W}_-|\rho_0) \tag{2.18}$$

the absorbed energy takes its usual form

$$\Delta E = \frac{1}{\pi} \int_{0}^{+\infty} d\omega \, \omega |F(\omega)|^2 \, \chi''(\omega) \equiv \int_{0}^{+\infty} I(\omega) \, d\omega, \tag{2.19}$$

with

$$\chi''(\omega) = - \, \mathrm{Im} \, (W|\hat{\mathscr{K}}(\omega) \, \hat{W}_-|\rho_0), \tag{2.20}$$

which shows that the absorption line shape is directly related to the imaginary part of the complex susceptibility and to the spectral density of the perturbation, a well-known result [30].

(b) The line shape of one-dimensional systems.

It is easy to show that the space $\mathscr{A}(\mathscr{H})$ of linear operators, on which operate the superoperators, can be divided in five independent orthogonal subspaces [31]; each one of these spaces sustains a reduced evolution Equation (2.13) and describes specific physical phenomena, such as exciton transfer ($\mathscr{A}^x$) absorption ($\mathscr{A}^a$), emission $\mathscr{A}^e$) etc... Thus we have

$$\mathscr{A} = \mathscr{A}_I \oplus \mathscr{A}_K \oplus \mathscr{A}^x \oplus \mathscr{A}^a \oplus \mathscr{A}^e \tag{2.21}$$

generated as follows (for more details see ref. (2.31): $-\mathscr{A}_I$ and $\mathscr{A}_K$, of dimension one, are generated respectively by $|I)$ and $|K)$, with

$$\hat{I} = \frac{1}{\sqrt{N+1}} \sum_{k=0}^{N} |A_k\rangle \langle A_k| \quad \text{and} \quad \hat{K} = \frac{1}{\sqrt{N(N+1)}} \sum_{k=1}^{N} \times$$
$$\times (|A_k\rangle \langle A_k| - N \, |A_0\rangle \langle A_0|)$$

which are the two constants of motion for one-dimensional systems in the adiabatic approximation (spontaneous radiations and radiationless transitions to the ground state are neglected; the total population $\hat{I}$ and the exciton band population $\hat{K}$ are constants). $-\mathscr{A}^x$ describes the exciton motion. It is of dimension N^2-1, with $N-1$ diagonal operators:

$$|a_l^x) = |\sum_{k=1}^{N} \alpha_{k,l} \, |A_k\rangle \, \langle A_k|), \quad \text{with} \ \ 1 \leqslant l \leqslant N-1 \ \text{and} \ \alpha_{0l}=0 \ \text{for any} \ l.$$

N^2-N non-diagonal operators:

$$|b_m^x) = |\sum_{k \neq l}^{N} \beta_{kl,m} \, |A_k\rangle \, \langle A_l|), \quad \text{with} \ \ 1 \leqslant k, l \leqslant N$$

$-\mathscr{A}^a$ and $\mathscr{A}^e$ are subspaces, each one of dimension N, generated respectively by vectors $|a_k^+)$ and $|a_k)$, with

$$|a_k^+) = \||A_k\rangle \, \langle A_0|) \quad \text{and} \quad |a_k) = \||A_0\rangle \, \langle A_k|).$$

From the formula of the generalized susceptibility (2.20) and the independence of the subspaces $\mathscr{A}^i$, we have $\chi(\omega) = (W|\hat{\mathscr{K}}^a(\omega) \, \hat{W}_-|\rho_0)$, where $\mathscr{K}^a(\omega)$ is the resolvant of the restriction of $\hat{H}$ in the subspace $\mathscr{A}^a$, cf. (2.21).

Using the long-wavelength approximation the semiclassical matter-radiation interaction is written:

$$\hat{W}(t) = (iA(t)/c) \sum_k \omega_k \, \mathbf{u}\boldsymbol{\mu}_k (\hat{a}_k - \hat{a}_k^+) \equiv f(t) \, \hat{W}$$

with $f(t) \equiv (iA(t)/c)$ being a classical function which carries the time dependence of the perturbation, $\mathbf{u}$ is the radiation polarization with $\mathbf{A}(t) = \mathbf{u}A(t),$ $\boldsymbol{\mu}_k$ is the transition moment of the stationary state $|A_k\rangle$, with

$$\boldsymbol{\mu}_k = \frac{\sqrt{2}}{\sqrt{(N+1)}} \sum_{p=1}^{N} (\sin kp\theta) \, \boldsymbol{\mu}_p, \quad \text{and} \quad \boldsymbol{\mu}_p = \langle M_0|\sum_i er_i|M_p\rangle$$

being a transition moment slightly different from that of the isolated molecule p.

In space $\mathscr{A}(\mathscr{H})$, we may write:

$$|W) = \sum_{k=1}^{N} W_k \, \mathbf{u}\boldsymbol{\mu}_k (|a_k) - |a_k^+)). \tag{2.22}$$

Since operators $\hat{I}$ and $\hat{K}$ are the only constants of motion we write the stationary state of the sample, $|\rho_0)$ in the form $|\rho_0) = (I|\rho_0) \, |I) + (K|\rho_0) \, |K)$, which is explicitly:

$$|\rho_0) = \frac{1}{\sqrt{N+1}} \, |I) - r \, \frac{1}{\sqrt{N(N+1)}} \, |K) \tag{2.23}$$

with $r = \sum\limits_{k=1}^{N} (\rho_{oo} - \rho_{kk})$ indicating the difference of population between the excited space (the first exciton band) and the ground state. Then, using the compact form of $\chi(\omega)$, we derive the expression:

$$\chi(\omega) = (W|\hat{\mathcal{X}}^a(\omega)\,\hat{W}_-|\rho_0) = \sum_{k_1,k'}^{N} (W|a_{k'}^+)\,(a_{k'}^+|\hat{\mathcal{X}}^a(\omega)|a_k^+)\,(a_k^+|\hat{W}_-|\rho_0)$$

(2.24)

with the help of the following results:

$$W_-|K) = [\hat{W},\hat{K}] = \frac{\sqrt{N+1}}{\sqrt{N}} \sum_k \omega_k\,\mathbf{u}\boldsymbol{\mu}_k(|a_k) - (a_k^+))$$

$$(a_k^+|W_-|K) = \frac{N+1}{\sqrt{N}}\,\omega_k\mathbf{u}\boldsymbol{\mu}_k$$

and approximating in the numerator all the ω_k's by ω_0, we obtain:

$$\chi(\omega) = \frac{2r\omega_0^2}{N(N+1)}\left[\sum_{k=1}^{N}\left(\sum_{p=1}^{N}(\mathbf{u}\boldsymbol{\mu}_p)\,\sin kp\theta^2\right)^2 (a_k^+|\hat{\mathcal{X}}^a(\omega)|a_k^+) + \right.$$

$$\left. + \sum_{k\neq k'}^{N}\sum_{p,p'}^{N}(\mathbf{u}\boldsymbol{\mu}_p)\,(\mathbf{u}\boldsymbol{\mu}_{p'})\,\sin kp\theta\,\sin k'p'\theta\,(a_k^+|\hat{\mathcal{X}}^a(\omega)|a_{k'}^k)\right]$$

(2.25)

after some elaboration we find:

$$(a_k^+|\hat{\mathcal{X}}^a(\omega)|a_{k'}^+) = \frac{\delta_{kk'}}{\hbar(\omega-\omega_k+i\Gamma)} +$$

$$- \frac{4\gamma_1}{\hbar(N+1)}\{\xi_0(k,k')\,F_0(\omega)+\xi_1(k,k')\,F_1(\omega)\}\,\frac{\sin k\theta\,\sin k'\theta}{(\omega-\omega_k+i\Gamma)(\omega-\omega_{k'}+i\Gamma)}$$

with $\xi_0(k,k')$ (respectively $\xi_1(k,k')$) is zero when k and k' are not both even (respectively odd) and equals one when k and k' are both even (respectively odd). Furthermore $F_i(\omega)$ are special functions given in ref. 31 where line shapes for one-dimensional systems are derived in detail.

(c) Optical selection rules.

Taking the imaginary part of expression (2.24) for the absorption, we see that the first term gives the absorption from N independent Lorentzian lines which are the excitonic resonances with intensities proportional to $|\mu_k|^2$. The second term, non definite positive, is an interference contribution from the very closely spaced states $|A_k\rangle$ with homogeneous width γ_1 and non-zero transition amplitudes due to the finite number of sites in the chain. This term is small; it depends on the preparation of the excited state and on the size of the chain [31] and may be neglected for chains with $N \gtrsim 20$. Consequently we may safely discuss the optical selection rules from considering the first term of (2.24):

$$\chi''(\omega) \propto \sum_{k=1}^{N}\left(\sum_{p=1}^{N}\mathbf{u}\boldsymbol{\mu}_p\,\sin kp\theta\right)^2 \frac{\Gamma}{(\omega-\omega_k)^2+\Gamma^2}\,(\Gamma=\gamma_0+2\gamma_1).$$

(2.26)

Note that Γ is not dependent on k, which is a consequence of the stochastic model utilized in this work [32]. The weighting factors of the Lorentziansal low establishing selection rules for collective excitations (notice that we are squaring the sum of individual site transition amplitudes, introducing steps of phase $k\theta$). In particular, when all sites have parallel μ_p's, the square of the sum equals to $[1 - (-1)^k] \cotg^2 k (\theta/2)$ and allows us to draw the following conclusions:

(*i*) the lines of pair states are forbidden.

(ii) the states with small k values have the strongest transition, which is a reminiscent of a $k = 0$ selection rule (remember we are working in the long-wavelength approximation) or accounts for the fact that small phase steps, k, between sites, help constructive interference in the absorption process.

(iii) when N becomes very large only states with $k = 0$ absorb; then a collective excitation is a quasi-particle created by light [1], with conservation of the momentum $q = 2\pi/\lambda \sim 0$.

3. The Dynamical Problem

3.1. PREPARATION OF LOCALIZED EXCITATIONS

(a) Definition of a measure of the localization: From conclusion (iii) of the last paragraph, it is obvious that for configurations with parallel sites we cannot produce localized states with optical excitation.

In order to discuss this problem more systematically, let us measure the degree of localization of an arbitrary excited prepared state $|\psi^x\rangle$, with the help of the following function:

$$S(|\psi^x\rangle) = -\frac{1}{\log N} \sum_{p=1}^{N} |\langle M_p|\psi^x\rangle|^2 \log |\langle M_p|\psi^x\rangle|^2$$

$$\text{with} \quad \sum_p |\langle M_p|\psi^x\rangle|^2 = 1 \tag{3.1}$$

and whose defining properties are

$$0 \leqslant S(|\psi^x\rangle) \leqslant 1$$

$$S(|\psi^x\rangle) = 0 \quad \text{when} \quad \langle M_p|\psi^x\rangle = \delta_{pp'}, \; p' \text{ being one of the } N \text{ sites}.$$

$$S(|\psi^x\rangle) = 1 \quad \text{when} \quad \langle M_p|\psi^x\rangle = \frac{1}{\sqrt{N}}, \text{ for any one of the } N \text{ sites}.$$

Let us consider now the two limiting cases of an optical excitation; (i) the excitation is monochromatic (ω): then we can reach only one of the excitonic resonances $|A_k\rangle$, with $\omega_k = \omega$. Calculation of $S(|A_k\rangle)$ shows that its value is always near 1, for any value of k. Consequently for any value of the number N of sites, the excitation is strongly delocalized, see Table I.

TABLE I

Variation of the degree of delocalization of a state $|A_k\rangle$ according to the number N of site, as given by function S. For any value of $N \gtrsim 10$, the function $S(|A_k\rangle)$ remains larger than 0.80 for any k

(N)	1	2	3	4	5	6	7	8	9	10
(k)										
1	1.	1.	.95	.93	.92	.91	.91	.91	.91	.91
2		1.	.63	.93	.86	.91	.89	.91	.90	.91
3			.95	.93	.68	.91	.90	.86	.91	.91
4				.93	.86	.91	.71	.91	.90	.91
5					.92	.91	.90	.91	.73	.91
6						.91	.89	.86	.90	.91
7							.91	.91	.91	.91
8								.91	.90	.91
9									.91	.91
10										.91

(ii) the excitation is obtained with a pulse of very short duration τ. In a first order perturbation theory, one obtains:

$$|\psi^x\rangle \propto \sum_k |A_k\rangle \langle A_k|\hat{W}|0\rangle = \sum_k \sum_p |A_k\rangle \langle A_k|M_p\rangle \langle M_p|\hat{W}|0\rangle .$$

Since we are considering a parallel configuration of sites, $\langle M_p|\hat{W}|0\rangle$ is independent of p, hence we are left with:

$$|\psi^x\rangle \propto \sum_k \sum_p |A_k\rangle \langle A_k|M_p\rangle = \sum_p |M_p\rangle .$$

After normalization, we may write

$$|\psi^x\rangle = \frac{1}{\sqrt{N}} \sum_p |M_p\rangle \tag{3.2}$$

which is the most delocalized state ($S(|\psi^x\rangle) = 1$).

From (i) and (ii), it follows that, for any kind of optical excitation, it is impossible to create a localized excitation in a linear parallel chain. It is obvious that in order to create an observable excitation migration (exciton transport), we must prepare a non-stationary state. Intuitively we know that the more the initial state is localized the more the transport becomes sensitive. Radicalizing, we consider in what follows the possibility to prepare a most localized state, i.e. $S(|\psi^x\rangle) = 0$.

(b) Preparation of a localized state: Let us consider the preparation

$$|\psi^x\rangle \propto \sum_k c_k |A_k\rangle . \tag{3.3}$$

The experimental problem is to know in which conditions we may get

$$c_k = \sin kp\theta \quad \text{for any value of } k. \tag{3.4}$$

Since in the development of any state $|M_p\rangle$, we find nearly all the resonances $|A_k\rangle$, see Equation (2.3), one of the conditions of realizing (3.2) is that nearly all the resonances are accessible by excitation; the proportionality condition (3.4) will be satisfied if the two following conditions are fulfilled.

(1) the spectral width of the radiation must be uniform in a range $4\Delta\omega$ around ω_0. This may be realized when the duration τ of the excitation pulse is $\tau \ll 2\pi/4\Delta\omega$. On the other hand, to measure a resonance near ω_0 we must have $\tau \gg (2\pi/\omega_0)$. Typical values of ω_0 and $4\Delta\omega$, respectively $\sim 20\,000$ and $\sim 20\,\text{cm}^{-1}$, lead to the limits:

$$10^{-15}\,\text{s} \ll \tau \ll 10^{-12}\,\text{s}. \tag{3.5}$$

relation (3.5) concerns the energy of the photons.

(2) In conjunction with relation (3.5), the transition amplitudes of the resonances, $|o\rangle \rightarrow |A_k\rangle$, must be comparable; this condition concerns the selection rules of the system.

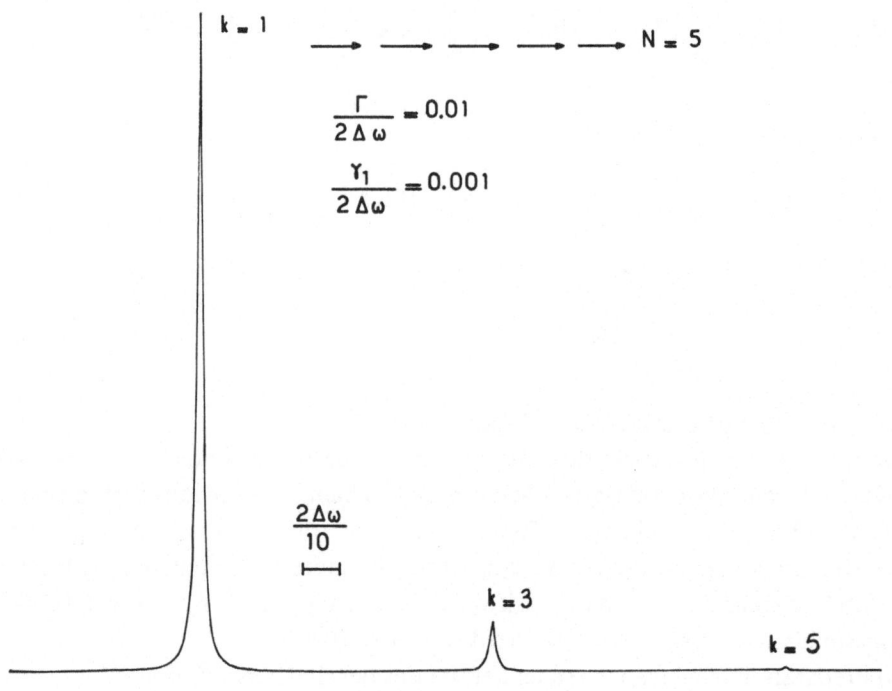

Fig. 1.   Calculated absorption spectrum for a chain of five parallel sites and for a radiation polarization **u** parallel to the μ_p's.

From part II, it was concluded that for parallel chains this condition is not satisfied; for pair lines are missing and states with small values of k have predominant transitions. The impossibility of preparing localized excitations appears obvious when remembering that the photon is not punctual and the uncertainty relation $\Delta q \Delta p \sim \hbar$, with $E = pc$ and $4\hbar\Delta\omega \ll \Delta E \ll \hbar\omega_0$, leads to a localization $10^3 \, \text{Å} \ll \Delta q \ll 10^6 \, \text{Å}$. Thus, since a photon has an extension much larger than a site, it is difficult to see why it would relax to a particular site among a collection of strictly equivalent ones. However, in order to privilege a site, we can play on the polarization property of the radiation and find out whether some 'orientational trap' might not pump the photon from the radiation field.

In the first order we obtain for the coefficient (3.4):

$$c_k \propto \langle A_k | \hat{W} | O \rangle = \sum_{p'} \mathbf{u}\boldsymbol{\mu}_{p'} \sin kp'\theta .$$

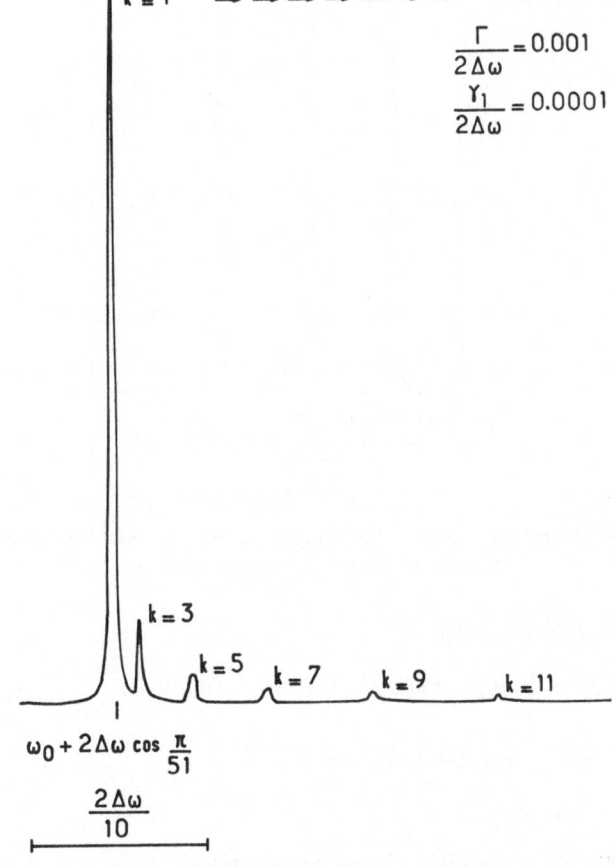

Fig. 2.   Calculated absorption spectrum for a chain of fifty parallel sites and for a radiation polarization u parallel to the $\boldsymbol{\mu}_p$'s.

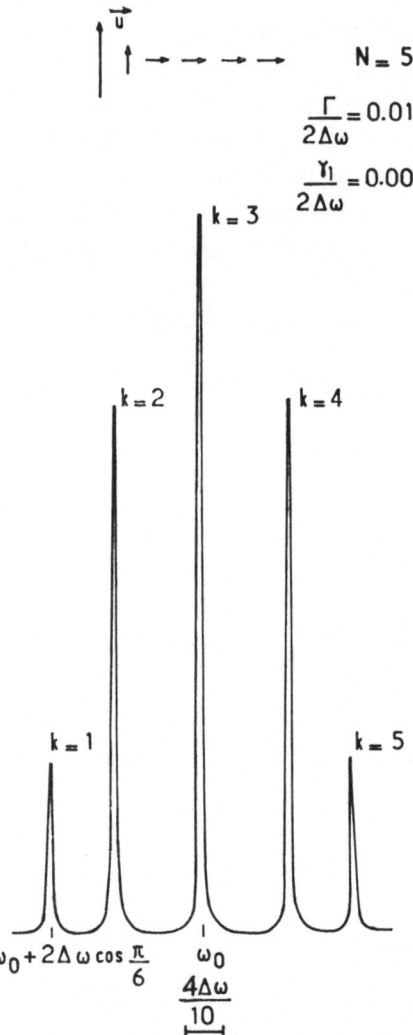

Fig. 3.   Calculated absorption spectrum for a chain of five sites with an 'orientational trap' μ_1 parallel to the radiation polarization u.

Condition (3.4) is realized with:

$$\sin kp\theta = \sum_{p'} \mathbf{u\mu}_{p'} \sin kp'\theta \quad \text{for any value of } k. \tag{3.6}$$

In turn, relation (3.6) is possible only when

$$\mathbf{u\mu}_{p'} = \mathbf{u\mu}_p \delta_{pp'}. \tag{3.7}$$

Therefore, all sites have transition moments perpendicular to the radiation polarization, except one of them. Then the photon will 'relax' in this site and the excitation is localized in $|M_p\rangle$ for $t = \tau \equiv o^+$; this result was a priori obvious. However, as an

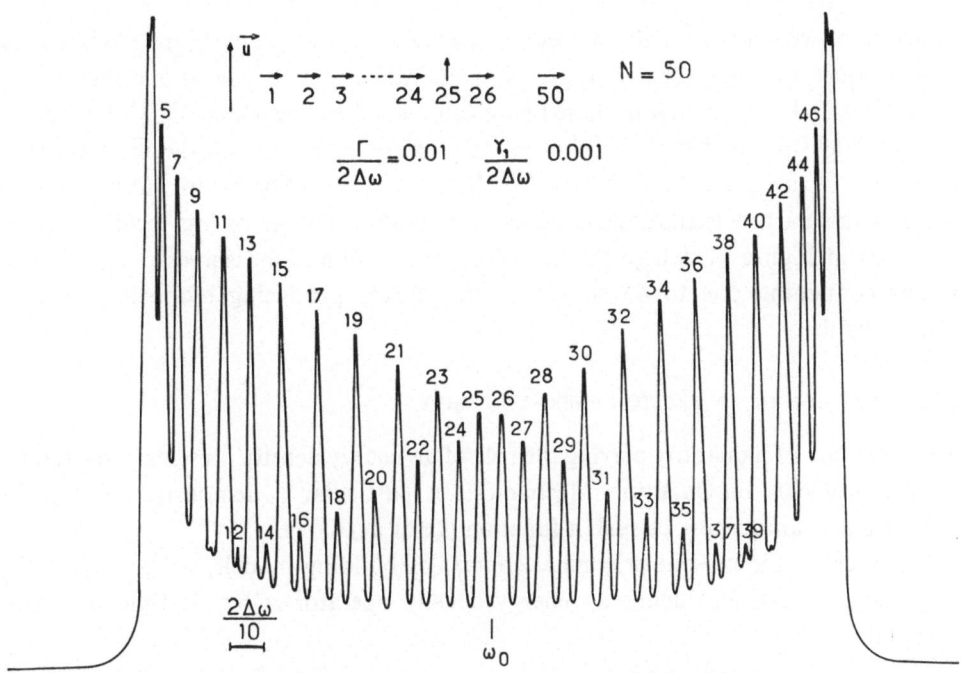

Figs. 4a-b. Calculated absorption spectra for parallel chains containing one orientation trap. (a) In the middle of the chain, with $\mu_n \| \mathbf{u}$. (b) At one end of the chain, with $\mu_n \| \mathbf{u}$.

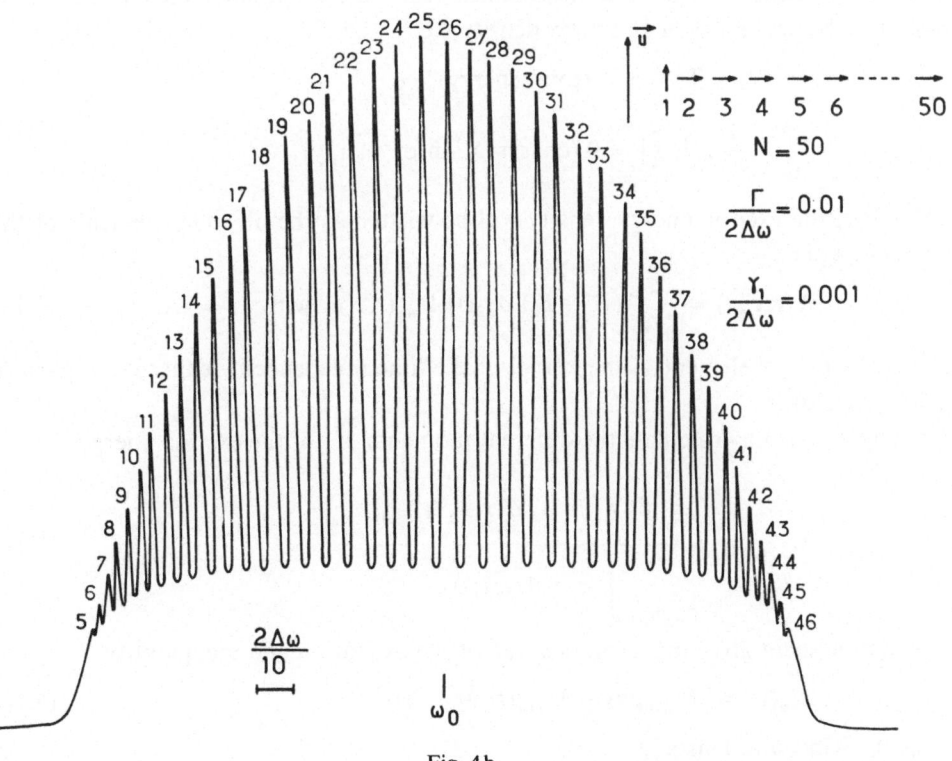

Fig. 4b.

illustration, we show how the preceding spectra (Figures 1 and 2) are modified for small (Figure 3) or large chains (Figure 4) containing a singular site at one end or near the middle [33]. Notice that in these figures almost all the resonances $|A_k\rangle$ have comparable intensities and that the photon must excite coherently all the resonances in order to prepare a localized state $|M_p\rangle$. If the last condition is not fulfilled, i.e. if more or less narrow excitations are used, we obtain the more or less localized states indicated in Figure 11, where the degree of localization is measured by $S(|\psi^x\rangle)$ for chains containing one to 45 sites and for sources producing monochromatic to chaotic light.

3.2. DEFINITION OF AN EXCITON ENERGY DENSITY

The definition of a quantity playing the role of an energy density presents a few fundamental problems in Quantum Mechanics. In particular a demonstration for the existence and uniqueness of such a function is not available.

However, in the present pragmatic context of building models, we generalize the proposal of Jauch and define an energy density operator $\hat{\mathscr{E}}$ by the following relation [34]

$$\hat{\mathscr{E}} = \tfrac{1}{2}(\hat{H}\hat{\rho} + \hat{\rho}\hat{H}) \tag{3.8}$$

where $\hat{\rho}$ is the usual density operator obeying Equation (2.13). The symmetrization in Equation (3.8) is a usual procedure when functions of products of non-commuting operators appear in Quantum Mechanics from the correspondance principle. $\hat{\mathscr{E}}$ possesses the properties of an energy density, i.e.

$$\mathrm{Tr}\,\hat{\mathscr{E}}\hat{\rho} = \mathrm{Tr}\,\hat{H}\hat{\rho} = (\text{total energy})$$

$$\frac{\mathrm{d}\hat{\mathscr{E}}}{\mathrm{d}t} + \frac{1}{i\hbar}\,[\hat{H},\hat{\mathscr{E}}] = 0 \ (\text{conservation of energy}). \tag{3.9}$$

Therefore, the exciton energy density will be written as the diagonal elements of the kernel of $\hat{\mathscr{E}}$, i.e.

$$\mathscr{E}(\mathbf{r},\mathbf{r}',t) = \sum_{m,n} \tfrac{1}{2}(E_m + E_n)\,\rho_{mn}(t)\,D_{mn}(\mathbf{r},\mathbf{r}') \quad \text{with} \quad \mathbf{r} = \mathbf{r}'. \tag{3.10}$$

D_{mn} is the matrix element of the kernel of the time independent reduced-one-particle density operator.

Using $\hat{\mathscr{E}}$, we may calculate the different moments M_λ of the exciton energy:

$$M_{\lambda;\,u,\,v\,\ldots} = \frac{\displaystyle\int \mathrm{d}^3 r\,\{\hat{u}\hat{v}\,\ldots\,\hat{\mathscr{E}}(\mathbf{r},\mathbf{r}',t)\}_{r=r'}}{\displaystyle\int \mathrm{d}^3 r\,\hat{\mathscr{E}}(\mathbf{r},\mathbf{r},t)}. \tag{3.11}$$

The first moment gives the mean position of the excitation and the quantity

$$\Delta_{uv}(t) = [M_{2;\,uv}(t) - M_{1;\,u}(t)\,M_{1;\,v}(t)]^{1/2} \tag{3.12}$$

gives its dispersion tensor.

3.3. Motion and delocalization in one-dimensional systems

(a) The dimer: In Figures 5a and b, we present the variation of $M_1(t)$ and $\Delta(t)$, for a statistical dimer, see ref. 2, p. 4378. For the situation $\gamma_0 \ll \Delta\omega$ termed quasi-coherent, the exciton looks like an energy ball, fluctuating between the two sites: it contracts, and is stationary in time, when visiting a site. On the contrary for the situation $\gamma_0 \gg \Delta\omega$, for which the excitonic coherence does not survive, the excitation prepared in one site decays and delocalizes in a monotonic way; for $t > \gamma_0^{-1}$ the excitation is completely delocalized with $M_1(t)$ and $\Delta(t)$ having reached their maximum value, $\sim \frac{1}{2} R$, see Figure 5.

(b) The one-dimensional chain: In the present work we present a few preliminary results on the excitation transfer in linear systems containing an arbitrary number of sites.

In order to present this discussion in a very general manner, we determine first the

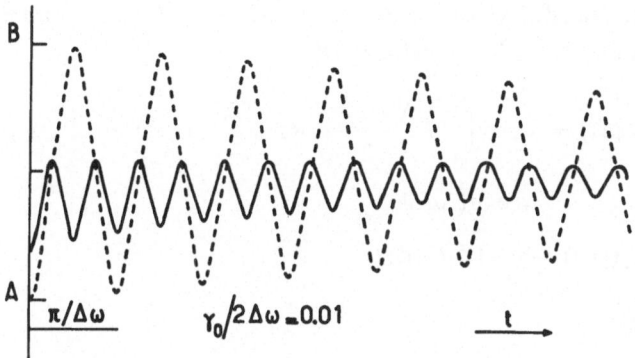

Figs. 5a-b.   Motion of the excitation ($M_1(t)$) and its volume ($\Delta(t)$) for a statistical dimer and for a pure initial state $\rho(o^+) = |A\rangle\langle A|$. ($AB = R$ is the distance between the two sites). (a) variation of $M_1(t)$ and $\Delta(t)$ for a quasi-coherent case. (b) variation of $M_1(t)$ and $\Delta(t)$ for a quasi-coherent case.

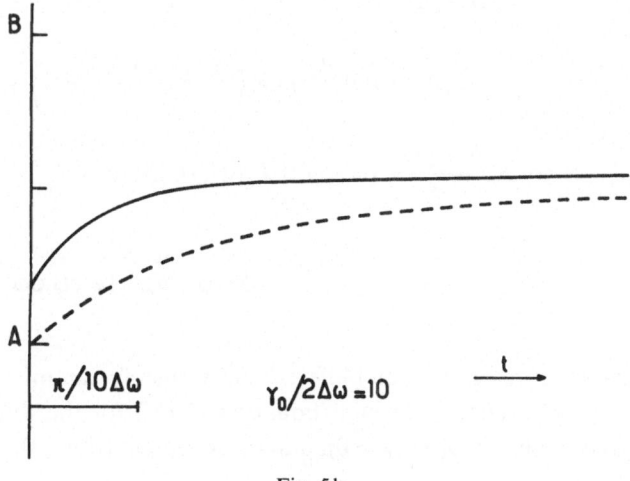

Fig. 5b.

temporal evolution of a generating function of spatial moments, $F(\phi, t)$ [35] in the approximation of weak overlap and narrow exciton band proper to vibronic excitons, see ref. 1.

Thus, such a function depends only on the elements $\langle M_p | \hat{\rho}(t) | M_p \rangle \equiv \rho_{pp}(t)$. Then, we can choose for function $F_0(\phi, t)$, in one dimensional systems, the following expression (R is the distance between two adjacent sites):

$$F_0(\phi, t) = \exp\left(-i \frac{N+1}{2} R\phi\right) \sum_{p=1}^{N} \exp(ipR\phi) \, \rho_{pp}(t) \tag{3.13}$$

with subsequent expressions for the moments:

$$M_n(t) = i^{-n} \left\{ \frac{\partial^n}{\partial \phi^n} F_0(\phi, t) \right\}_{\phi=0}. \tag{3.14}$$

The time evolution of F_0 is easily deduced from the differential equation for $|\rho\rangle$, projected in the transfer subspace $\mathcal{A}^x$; it is shown that F_0 is coupled to the $N-1$ functions $F_r(\phi, t)$, $1 \leqslant r \leqslant N-1$, defined as:

$$F_r(\phi, t) = \tfrac{1}{2} \exp\left(-i \frac{N-r+1}{2} R\phi\right) \sum_{p=1}^{N-r} \{(-1)^r \rho_{p+r,p} + \rho_{p,p+r}\} \times$$

$$\times \exp(ipR\phi)$$

$$F_{-r}(\phi, t) = (-1)^r F_r(\phi, t).$$

We then obtain:

$$\frac{\partial F_r}{\partial t} = -2\{(1-\delta_{r0}) \gamma_0 + (2 - 2\delta_{r0} \cos R\phi + \delta_{r1}) \gamma_1\} F_r(\phi, t) +$$

$$+ 2\Delta\omega \sin R \frac{\phi}{2} \{F_{r+1}(\phi, t) - F_{r-1}(\phi, t)\} +$$

$$+ \frac{1}{\pi} \int_0^{2\pi} \left\{ 4\gamma_1 \delta_{r0} \sin R \frac{\phi}{2} \sin R \left(\frac{N}{2} \phi - \frac{N-1}{2} \phi' \right) + \gamma_1 (1 - \delta_{r0}) \times \right.$$

$$\left. \times \cos R \frac{N-r-1}{2} (\phi - \phi') \right\} F_r(\phi', t) \, d\phi' +$$

$$+ \frac{1}{\pi} \int_0^{2\pi} i\Delta\omega \sin R \frac{N-r+1}{2} (\phi - \phi') F_{r-1}(\phi', t) \, d\phi'. \tag{3.13'}$$

The integral terms represent the 'end effects' and vanish for an infinite chain.

The motion may be discussed through Equation (3.13') for the three usual situations, namely purely coherent ($\gamma_1 = \gamma_0 = 0$), purely incoherent ($\Delta\omega = 0$) and the general case ($\Delta\omega, \gamma_0, \gamma_1 \neq 0$).

(a) The purely coherent case: with the assumption of an initial state $\hat{\rho}(O^+) = |M_q\rangle \langle M_q|$, we get

$$F_0(\phi, t) = \left(\frac{2}{N+1}\right)^2 \sum_{p=1}^{N} \sum_{l,m}^{N} \sin lp\theta \, \sin mp\theta \, \sin lq\theta \, \sin mq\theta \, e^{ipR\phi} \times$$

$$\times \, e^{-i[(N+1)/2] \, R\phi} \, e^{2i\Delta\omega t(\cos l\theta - \cos m\theta)}. \tag{3.14}$$

$F_0(\phi, t)$ is a pseudo periodic function, with a pseudo period increasing dramatically with N, see Figure 6; its value is $\sim 10^4 (2\pi/2\Delta\omega)$ for $N = 10$. For a singlet excited state, with a lifetime τ not of the order of 10^{-8} s, the pseudoperiod never becomes apparent (the inverse would be true for triplet excited states).

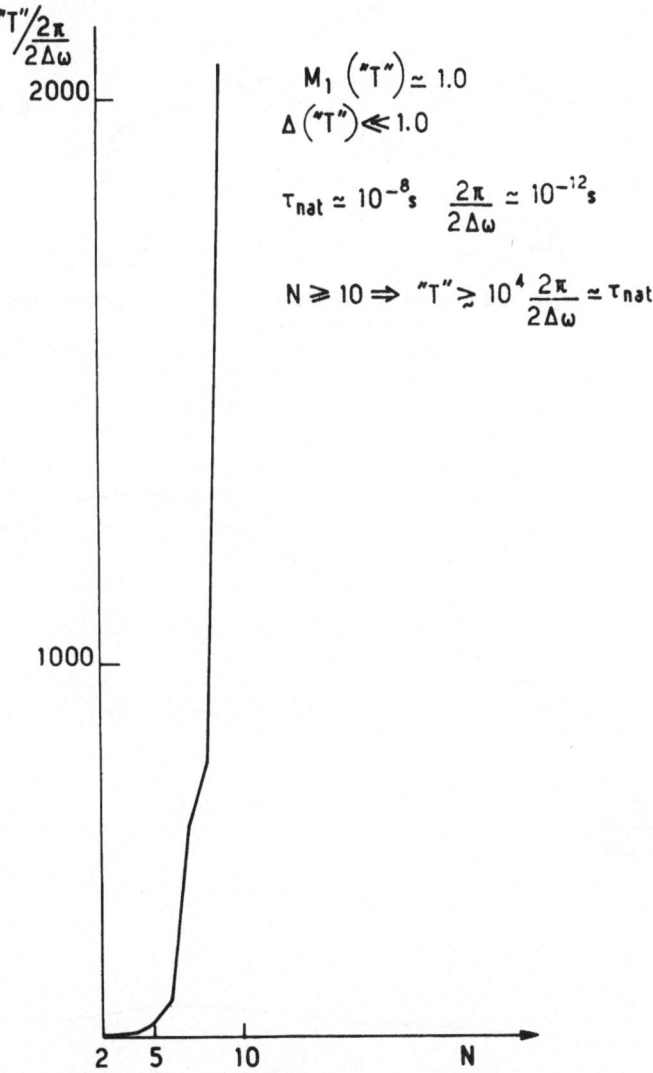

Fig. 6.   Approximate variation of the pseudoperiod 'T' as a function of the number N of sites, for a purely coherent case. 'T' is measured in $(2\pi/2\Delta\omega)$ units.

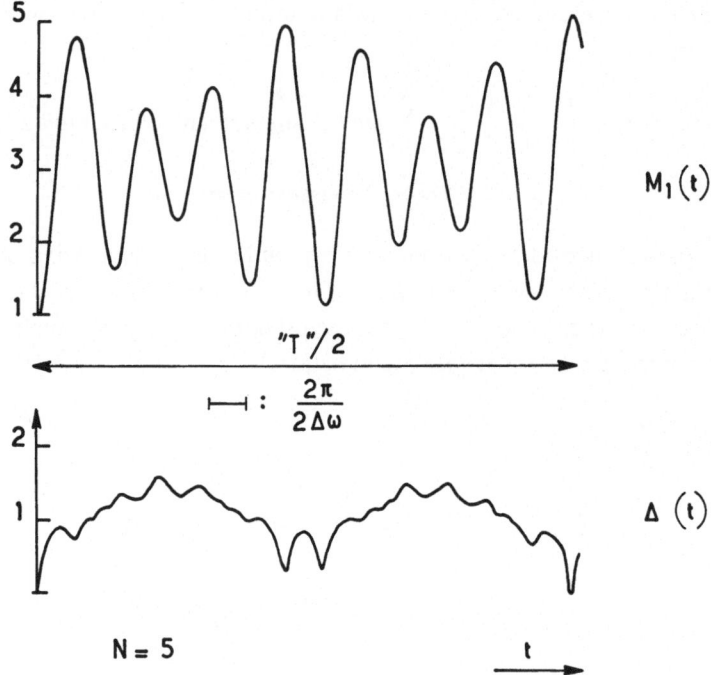

Fig. 7.   Variation of $M_1(t)$ and of $\Delta(t)$ during the first half pseudoperiod when $N = 5$ and for a purely coherent case.

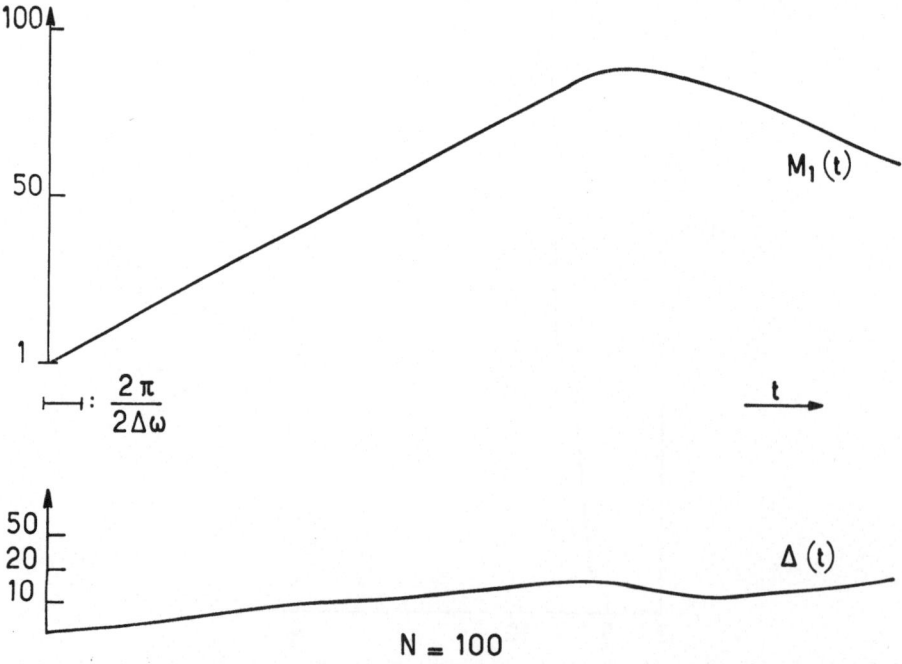

Fig. 8.   Variation of $M_1(t)$ and $\Delta(t)$ when $N = 100$, for a purely coherent case.

In Figures 7 and 8, we present the variation of $M_1(t)$ and $\Delta(t)$ for chains with $N = 5$ and $N = 100$. An essential difference appears when comparing with the motion in the dimer: there is no contraction of the energy ball, or stationarity, when the excitation crosses a site, except for the ends of the chain. In the intermediate phase, between localization and complete delocalization, the excitation covers a large number of sites ($\sim 40\%$). The localization of the excitation, which is the consequence of interferences between the resonances induced by the radiation, appears as a very complex phenomenon and its description is not reduced to a model of a site-to-site motion of the excitation with conservation of energy as could suggest the extrapolation of the analysis of the quantities $M_1(t)$ and $\Delta(t)$ for a dimer.

(b) The purely incoherent case: using the initial state $\hat{\rho}(O^+) = |M_q\rangle \langle M_q|$ the generating function becomes (with $\psi = \pi/N$):

$$F_0(\phi, t) = \frac{\sin \dfrac{NR\phi}{2}}{N \sin R \dfrac{\phi}{2}} - \frac{4i}{N} \sin R \frac{\phi}{2} \sum_{k=1}^{N-1} \frac{\{e^{-i(N/2)R\phi} - (-1)^k e^{i(N/2)R\phi}\}}{1 - 2e^{iR\phi} \cos k\psi + e^{2iR\phi}} \times$$

$$\times e^{iR\phi} \cos k(\psi/2) \cos k(q - \tfrac{1}{2}) \psi \, e^{-8\gamma_1 t \sin^2 k(\psi/2)}. \tag{3.15}$$

On Figure 9 we present the variation of $M_1(t)$ and $\Delta(t)$. Their variation is analogous to that obtained for the dimer: $M_1(t)$ and $\Delta(t)$ increase monotonically towards their

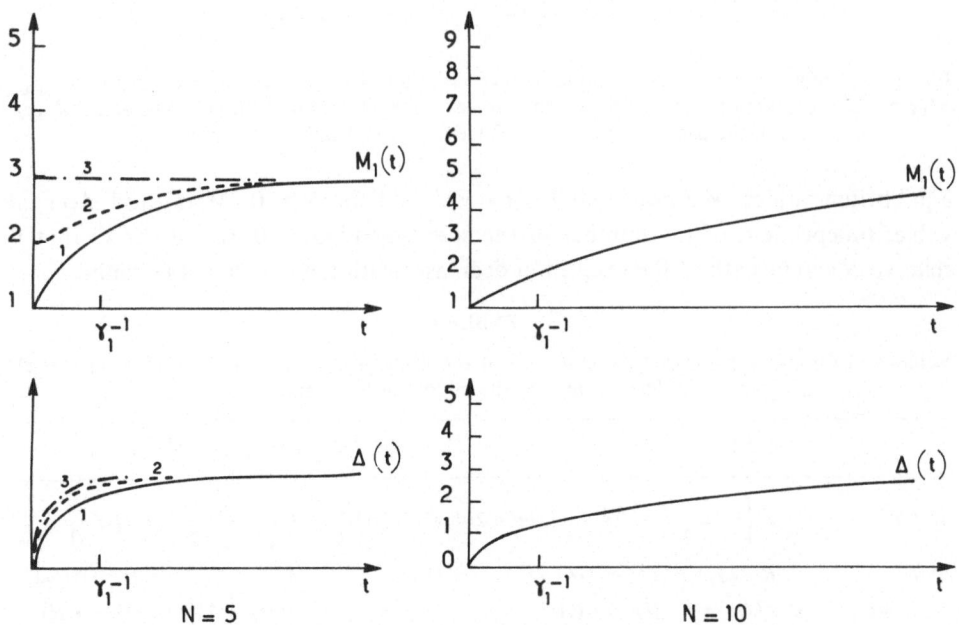

Fig. 9.   Variation of $M_1(t)$ and of $\Delta(t)$ when $N = 5, 10$, for a purely incoherent case. For $N = 5$, the index of the curves indicate the initial excited site (denoted as q in the text).

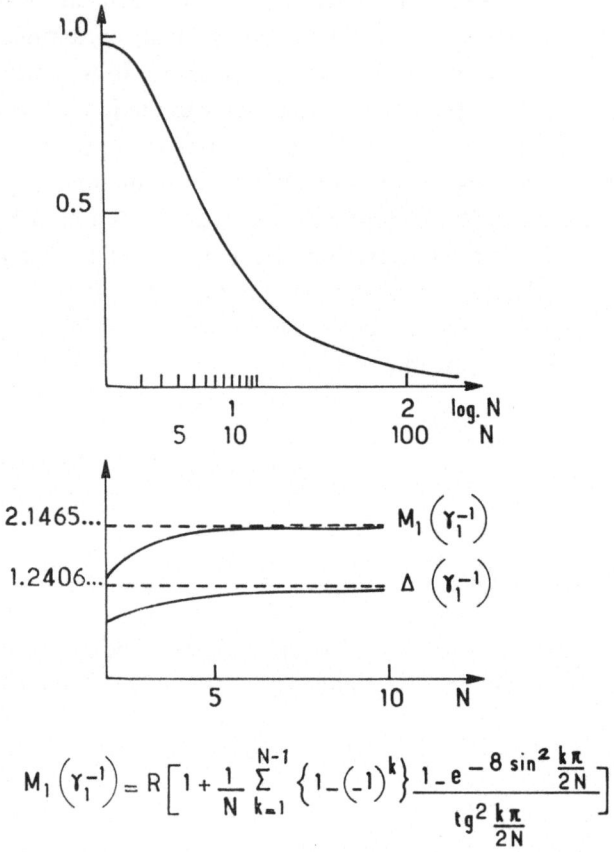

$$M_1\left(\gamma_1^{-1}\right) = R\left[1 + \frac{1}{N}\sum_{k=1}^{N-1}\left\{1-(-1)^k\right\}\frac{1-e^{-8\sin^2\frac{k\pi}{2N}}}{tg^2\frac{k\pi}{2N}}\right]$$

Fig. 10.   Purely incoherent case: – upper curve: variation of the relative coherent path $(M_1(\gamma_1^{-1})/M_1(\infty))$ as a function of length of the chain. – lower curve: variation of the coherent path, $M_1(\gamma_1^{-1})$ and of the packet $\Delta(\gamma_1^{-1})$, as a function of the length of the chain (N).

equilibrium values. We note that for $t = \gamma_1^{-1}$ and for $N \gtrsim 10$, $M_1(t)$ and $\Delta(t)$ take values independent of the number of the sites, see Figure 10, which shows that the relative coherent path of the excitation decreases with the length of the chain.

TABLE II

Values and variations of the first moment and of the dispersion, just after the excitation $(t = O^+)$. The origin is at the middle of the chain.

	$q = 1$	$q \neq 1, 2, N, N-1$
$M_1(t)$	$R\left\{-\dfrac{N-1}{2} + 2\gamma_1 t + (\Delta\omega^2 - 2\gamma_1^2)\,t^2 + O(t^2)\right\}$	$R\left\{q - \dfrac{N+1}{2} + O(t^2)\right\}$
$\Delta^2(t)$	$R^2\left\{2\gamma_1 t + (\Delta\omega^2 - (N+2)\,\gamma_1^2)\,t^2 + O(t^2)\right\}$	$R^2\left\{4\gamma_1 t + 2\Delta\omega^2 t^2 + O(t^2)\right\}$
Wavelike	$\Delta^2(t) = R^2\Delta\omega^2 t^2 + O(t^2)$	$\Delta^2(t) = 2R^2\Delta\omega^2 t^2 + O(t^2)$
'Diffusional'	$\Delta^2(t) = 2R^2\gamma_1 t - (N+2)\,R^2\gamma_1^2 t^2 + O(t^2)$	$\Delta^2(t) = 4R^2\gamma_1 t + O(t^2)$

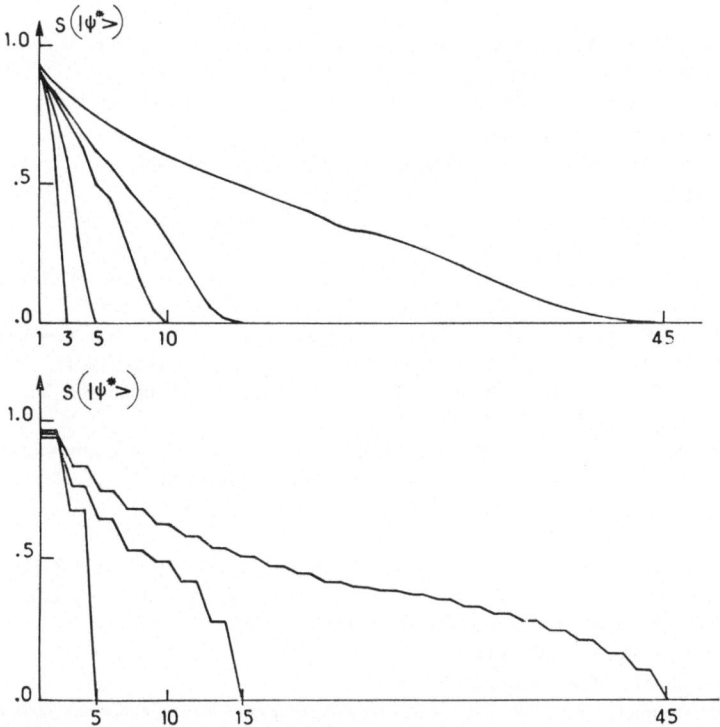

Fig. 11.   Calculation of the localization of the initial prepared state, as a function of the position and the spectral width $\delta\omega(k)$ of the exciting light. $\delta\omega(k) = (\omega_k + \tfrac{1}{2}\delta\omega) - (\omega_1 - \tfrac{1}{2}\delta\omega)$, with $\delta\omega \ll (\omega_k - \omega_{k-1})$ for any k. The abscissa numbers are the various values of k in the preceding formula. Upper curve: the end site is parallel to the radiation field, the other ones being perpendicular. Lower curve: the middle site is parallel to the radiation field, the other ones are perpendicular.

(c) The general case: comparisons with similar works on exciton transport in infinite chains [4, 8], can be made only for times near the origin, $t \sim 0$ (indeed in finite chains, the moments have bounded values). In particular, with $\rho(O^+) = |M_q\rangle \langle M_q|$ we obtain from Table II:

$$\dot{M}_1(0) = 2R\gamma_1(\delta_{q1} - \delta_{qN}); \qquad \dot{\Delta}^2(0) = 2R^2\gamma_1(2 - \delta_{q1} - \delta_{qN}).$$

For diffusion-like motion and for $t \sim 0$, we obtain for the diffusion constant: (Inasmuch as speaking of diffusion in finite chains is meaningful).

$$D = R^2\gamma_1(2 - \delta_{q1} - \delta_{qN}). \tag{3.16}$$

D depends on the nature of the prepared state: if the excitation is prepared at one end or elsewhere we get respectively the values $D = R^2\gamma_1$ and $D = 2R^2\gamma_1$. The latter result is identical to that obtained for an infinite chain [4, 8b]: this is quite normal for $t \sim 0$; later reflections of the exciton by the ends of the finite chain or by impurities in an infinite chain add complexity to the problem of the motion as show the different figures.

References

1. Voltz, R. and Kottis, Ph.: this volume, p. 187, and all the references on collective and elementary excitation therein.

2. Aslangul, Cl. and Kottis, Ph.: *Phys. Rev.* **B10**, 4364 (1974); *Comptes Rendus* **B278**, 735 (1974).

3. Ern, V. and Schott, M.: this volume, p. 249.

4. Reineker, P. and Haken, H.: this volume, p. 285, and references therein.

5. Port, H., Vogel, D., and Wolf, H. C.: *Chem. Phys.*, in press (1975); Vogel, D.: *Diplomarbeit*, Univ. of Stuttgart (1974); Whiteman, J. D.: Thesis, Univ. of Pennsylvania, 1971.

6. Vigny, P. and Duquesne, M.: in *Excited States of Biological Molecules*, Lisbon 1974; Eisinger, J., Lamola, A. A., Longworth, J. W., and Gratzer, W. B.: *Nature* **226**, 113 (1970).

7. Craig, D. P. and Walmsley, S. H.: *Excitons in Molecular Crystals*, Benjamin, New York, 1968.

8. Grover, M. and Silbey, R.: *J. Chem. Phys.* **52**, 2099 (1970); **54**, 4843 (1971); Fischer, J. and Rice, S. A.: *J. Chem. Phys.* **52**, 2089 (1970); Ern, V., Suna, A., Tomkiewicz, Y., Avakian, P., and Groff, R. P.: *Phys. Rev.* **B5**, 3222 (1972).

9. To consider transitions, from many vibrations of the ground state to many vibrations of the excited electronic state, complicates, but not essentially, the problem.

10. Lemaistre, J. P., Aslangul, Cl., and Kottis, Ph.: this volume, p. 239.

11. Sewell, G. L.: *Phys. Rev.* **129**, 597 (1963).

12. Haken, H. and Strobl: in *The Triplet State* (ed. by Zahlan A.), (Cambridge U.P. London, 1968); Haken and Reineker: in *Excitons, Phonons and Magnons* (ed. by Zahlan A.), (Cambridge U.P. London 1968); Reineker, P.: Thesis, Stuttgart, 1971.

13. Fano, U.: *Rev. Mod. Phys.* **29**, 74 (1957).

14. Lynden-Bell, R. M.: *Molec. Phys.* **22**, 837 (1971).

15. See for instance Bacry, H.: *Lecons sur la théorie des groupes et les symétries des particules élémentaires*, Gordon and Breach, Chap. 5.

16. id. stands for internal derivative (see ref. 15. where the genesis of these operators is recalled).

17. Mauser, K. E., Port, H., and Wolf, H. C.: *Chem. Phys.* **1**, 74 (1973).

18. On the strong effect of the degenerate site vibrations on the exciton fluctuation see Sternlicht, H. and McConnell, H. M.: *J. Chem. Phys.* **35**, 1793 (1961).

19. Feynman, R. P.: *Phys. Rev.* **89**, 108 (1951).

20. Kubo, R.: *J. Phys. Soc. Jap.* **17**, 1100 (1962).

21. Wang, M. C. and Uhlenbeck, G.: *Rev. Mod. Phys.* **17**, 323 (1945).

22. We must mention that such an assumption is stronger than the Markovian one, for which the correlation function are exponentials (Doob's theorem, see ref. 20, 21 and 23).

23. Doob, J. L.: *Ann Math.* **43**, 351 (1942).

24. Haken and Reineker arrived at a similar equation for the density operator using a method of functional derivation. However, they conserved the Liouville equation form which prevents the use of standard techniques for a direct derivation of the observables. We present our method here, first for its compact form which allows extension of calculations to large systems and, second, because the use of a privileged basis in space $\mathscr{A}(\mathscr{H})$ takes a priori into account the symmetry and the physics of the problem and produces a supermatrix 'physically irreducible'; this allows straightforward calculations. For instance the excitation moments and the electric susceptibility are derived without using such intermediates as a dipole autocorrelation function which needs in turn the definition of a generalized density operator, cf. Ref. 12. Although the work of these authors has greatly motivated us, we think that the understanding and the theory of transport in organic systems are still in a pioneering stage and that any improvement of the theoretical tool helps to control the approximation of the models and to provide a better comprehension of the physics of the problem.

25. Fano, U.: *Phys. Rev.* **131**, 259 (1963).

26. Kubo, R.: *J. Phys. Soc. Jap.* **12**, 570 (1957); Kubo, R. and Tomita, T.: *J. Phys. Soc. Japan* **9**, 888 (1954).

27. Des Cloiseaux, D.: in *Theory of Condensed Matter*, International Atomic Energy Agency, Vienna (1968).

28. As usual, $\mathcal{K}(\omega)$ is more precisely defined as $\lim_{\varepsilon \to 0^+}$ of:

$$\int_{-\infty}^{-\infty} dt \, \frac{e^{i(\omega + i\varepsilon)t} \, \hat{K}(t)}{(\hbar\omega + i\varepsilon) \, \hat{I} - \hat{H}}.$$

29. $\chi(\omega)$ is the Fourier transform of a real quantity which implies
$\chi(\omega) = \chi'(-\omega) \quad \chi''(\omega) = -\chi''(-\omega)$

30. See for instance, Slichter, C. P.: *Principles of Magnetic Resonance*, Harper and Row, New York, 1963, sec. 29.

31. Aslangul, Cl. and Kottis, Ph.: *Comptes Rendus* **B279**, 523 (1974);
Aslangul, Cl and Kottis, Ph.: submitted to *Physical Review*.

32. Expression (2.25) reduces to expression (5.7) of ref. 2 obtained for the dimer, with $\Gamma = \gamma_0 + \gamma_1$.

33. We assume that the resonance couplings are not modified when introducing a singular site in the calculations.

34. Jauch, J. M.: *Foundations of Quantum Mechanics*, Addison-Wesley, Mass., 1968.

35. Similar to the characteristic functions of Probability Theories, see for instance Renyi, A.: in *Calcul des Probabilités*, Dunod, Paris, 1966, p. 284.

ELECTRONIC COLLECTIVE EXCITATIONS IN ONE-DIMENSIONAL MOLECULAR SYSTEMS

II: *Motional Effects and EPR Response of Linear Triplet Excitons*

JEAN-PIERRE LEMAISTRE, CLAUDE ASLANGUL and PHILÉMON KOTTIS

Centre de Mécanique Ondulatoire Appliquée, Paris

1. Introduction

Since in these discussions we are mainly concerned by the existence and the properties of electronic collective excitations, a preliminary remark must be made on the interpretation of ESR line-shapes of triplet excitons. This is that the stochastic theories of line-shapes, often utilized by molecular spectroscopists, have been built up for paramagnetic systems for which intersite couplings are usually weak, or randomized, with respect to local fluctuation parameters, hence in the absence of electronic collective excitations, see for instance refs. 1, 2 and references therein.

Therefore, all these theories embody in the calculations the random phase approximation: the sample is considered as an assembly of independent molecules, excited or not, the intermolecular interactions manifest themselves by an excitation transfer which adds to each molecule supplementary routes of relaxation by transfer. Then, in EPR problems, the description of the electronic excitation motion by successive random transfers – the electronic excitation carries the spin polarization with it – is satisfactorily described by semi-classical theories, based on two stochastic models, the Markovian and the Gaussian Markovian, which are mathematically tractable [3]. It is obvious, therefore, that these theories cannot be used, without careful examination, for excitons in solid phase where, to spin dipolar interactions, we have to add strong intersite couplings from quantum electronic resonance interactions $\hat{V}$ which dominate the local fluctuations, are by no means randomized, and lead to collective excitation stationary states [2]. Subsequently, the random phase approximation (on the time phase of the sites) which implies randomized $\hat{V}$ interactions, has no validity and has to be expelled from our calculations, when not explicitly justified.

For molecular crystals or aggregates, it is easy to show that if the fluctuation parameters γ_0 – see ref. 1, Equations (51–56) – due to exciton-phonon, zero average, couplings are large compared to the quantum resonance coupling $\hat{V}$, the electronic collective excitations created by light do not survive after a short time τ. After this time and for all properties with characteristic times (such as T_1, T_2, since we are measuring ESR properties) notably larger than τ, it is possible to avoid the use of rigorous but difficult formulations [1] and utilize a time dependent stochastic model,

O. Chalvet et al. (eds.), Localization and Delocalization in Quantum Chemistry, Vol. II, 239–248. All Rights Reserved

i.e.: (i) describe the excitation motion in the random phase approximation, (ii) forget about interactions $\hat{V}$ in the calculation of the unperturbed states; (iii) define a transfer rate $k \sim |V|^2/\gamma_0$, between unperturbed localized states.

This rate gives the 'jumping probability' used to mix memories of sites and modify the individual molecule ESR responses, causing broadening, shifts and narrowing by classical exchange [3].

Let us point out, however, that for molecular excitons, we expect seldom to see broadening of individual site resonances, say ω_p, since the condition $\gamma_0 \gg V$, which allows us to use the transfer concept, leads generally also to $k \gg |\omega_p - \omega_q|$, which is the narrowing coalescence case [4]. Therefore, the theories which extend to molecular excitons the time dependent stochastic model, and which do not take into account the interaction $\hat{V}$ for the calculation of the unperturbed states, are valid in high temperatures but, when extrapolated to low temperatures, may lead to two contradictory answers, besides both of them incorrect. The first answer is obtained when reasoning in the following way: the exchange matrix element of $\hat{V}$, say V_{pq}, measures sort of a velocity of the electronic excited state – and of the spin carried with it – with which it migrates among the sites, sometimes V_{pq} is related to the residence time of the excitation through the uncertainty principle $V_{pq}\tau \sim 1$ [5]. Then for $V_{pq} \gg |\omega_p - \omega_q|$, we expect to see, in the absence of saturation effects, a *unique narrowed line*, with a width $\sim \beta|\omega_p - \omega_q|^2/V_{pq}(\beta = 1$ or $= 10/3)$. This narrow line originates from a temporal averaging on the sites absorption moment amplitudes $\langle \mu(t) \rangle$. The second answer comes when utilizing a density matrix [6] equation, defined on the electronic spin variables:

$$i\hbar\dot{\rho} = [H(\sigma), \rho] + \left(\frac{d\rho}{dt}\right)_{tr} \tag{1}$$

ρ is a density matrix defined on spin variables, only, with $H(\sigma) = \sum_p H_p(\sigma_p)$, where H_p^s is the spin hamiltonian of the pth site. $[d\rho/dt]_{tr}$ is a phenomenological classical contribution to $\dot{\rho}$ from the transfer; it accounts for the temperature effects and cancels at low temperatures [6a].

Equation (1) implies: (a) the validity of the random phase approximation; (b) the possibility to localize, as a result of relaxation, the excitation and to define spin hamiltonians for sites; (c) the existence of an excitation transfer from site-to-site, accompanied by phonon absorption and emission in the lattice.

All these concepts are valid at high temperatures and Equation (1) gives satisfactory results. When extrapolated to low temperatures, where the concepts (a–c) are meaningless because of collective excitations, see refs. 1 and 2, the results obtained from Equation (1) are incorrect. Indeed for $[d\rho/dt]_{tr} = 0$, Equation (1) becomes separable with respect to site variables: we obtain a *broad absorption from independent sites*, with static (large) hyperfine structures.

Therefore, we see that using naively the time dependent model we obtain results which are incorrect; this confirms what we know that quantum answers cannot be described as a fast classical motion and this for all degrees of freedom.

In the following contribution, we calculate ESR line-shapes of finite collections of weakly interacting identical molecules, without using assumptions such as independent sites, transfer etc..., which may be valid only at certain temperatures. We start from the total time independent hamiltonian and derive the ESR line-shape – fine and hyperfine structures – of collective excitations. Then on two examples, the dimer [7] A–B (or A–A) and a parallel finite chain $\{A$–$A...A\}_N$, we show that when we increase the temperature and create a Boltzmann population on $|k\rangle$ states, we encounter in a natural way the following situations:

(a) a narrow ESR resonance is observed at a very low temperature: only the $|k=0\rangle$ state is populated, the hyperfine structure is narrowed not because of any kind of motion, but because of the normalization of the electron spin density on the sites of the chain; the hyperfine structure width diminishes as $\sim 1/\sqrt{N}$, where N is the number of sites. The resonance $\omega(k=o)$ reflects the symmetry of the system (dimer or chain).

(b) At a higher temperature both homogeneous and inhomogeneous broadenings appear: the former is due to the lifetime shortening of the states $|k\rangle$, because of thermal transitions $|k\rangle \rightarrow |k'\rangle$; the latter is due to thermally populated $|k\rangle$ states which absorb independently and at different frequencies $\omega(k)$ when the inverse of their lifetime τ_k is not shorter than their static widths $\lambda_k (\lambda_k \sim T_2^{-1})$.

(c) At a much higher temperature we obtain a unique shifted narrow resonance: during the characteristic time of the experiment, say T_2, all the $|k\rangle$ states are visited many times, $\tau_k \ll T_2$, thus the k dependence of $\omega(k)$ is averaged out, $\omega(k) \rightarrow \omega_0$. The line ω_0 corresponds to that obtained when using a stochastic model with a very fast transfer. We must point out that the modifications (a–c) appear even for a parallel chain and these modifications are 'missed' when using stochastic models, even refined with concepts of coherent or incoherent motion.

2. The ESR Line-Shape of a Molecular Chain

THE DIMER

Let us consider the total hamiltonian for a trap dimer [7] A–B, in a rigid lattice, depending on electronic orbital, Q, and spin, σ, variables, in the strong binding approximation:

$$H(Q, \sigma) = H_e^A(Q_A) + H_e^B(Q_B) + H_e^{AB}(Q_A, Q_B) + H_m(Q_A, \sigma_A; Q_B, \sigma_B).$$

(2)

$H_e^i(Q_i)$ notes the electronic hamiltonian for site i, with $i = A, B$. $H_e^{AB}(Q_A, Q_B)$ accounts for the electronic orbital part of the intersite interactions, $H_m^{AB} = H_m^A + H_m^A + H_m^{AB}$ accounts for the electronic spin part of intra and intersite interactions (we neglect the orbital magnetism, $\mathbf{g} = g_0 = 2.0023$).

Let us consider the most general case of coupling $H_m^A < H_e^{AB}$; then a good approximation of the electronic orbital part of the first band of exciton states may be taken to be the excited eigenfunctions of the zero-order hamiltonian $H^0 = H_e^A + H_n^B + H_e^{AB}$, with notations [2]:

$$H^0 \, \Psi^\pm(Q) = E^\pm \, \Psi^\pm(Q) \tag{3}$$

$$\Psi^\pm(Q) = 2^{-1/2}[|A\rangle \pm |B\rangle].$$

Then, we can write for the complete spin-orbit exciton states the six combinations:

$$\Psi(Q,\sigma) = \sum_{\mu=1}^{3} \{a_\mu^+ \, \Psi^+(Q) \, \tau_\mu^+(\sigma) + a_\mu^- \, \Psi^-(Q) \, \tau_\mu^-(\sigma)\}. \tag{4}$$

The τ_μ's, with $\mu = 1, 2, 3$, are three triplet spin functions, orthogonal, which constitute a complete basis for the space of $|S| = 1$ states; they are eigenfunctions of a spin hamiltonian to be defined later.

The (a_μ^+, a_μ^-) are six sets of constants to be defined from the resolution of the Schrödinger equation $(H(Q, \sigma) - W) \, \Psi(Q, \sigma) = 0$.

The matrix elements of (2) are of three types; integrating on the electronic orbital variables Q and taking into account (3), we have for these elements:

$$\langle \tau_\mu^+ \, \Psi^+ | H | \Psi^+ \, \tau_\nu^+ \rangle_{Q,\sigma} = \left\langle \tau_\mu^+ \left| E^+ + \frac{H_{ss}^A + H_{ss}^B}{2} + H_{ss}^{AB} \right| \tau_\nu^+ \right\rangle_\sigma$$

$$\langle \tau_\mu^- \, \Psi^- | H | \Psi^- \, \tau_\nu^- \rangle_{Q,\sigma} = \left\langle \tau_\mu^- \left| E^- + \frac{H_{ss}^A + H_{ss}^B}{2} - H_{ss}^{AB} \right| \tau_\nu^- \right\rangle_\sigma$$

$$\langle \tau_\mu^+ \, \Psi^+ | H | \Psi^- \, \tau_\nu^- \rangle_{Q,\sigma} = \left\langle \tau_\mu^+ \left| \frac{H_{ss}^A - H_{ss}^B}{2} \right| \tau_\nu^- \right\rangle_\sigma \tag{5}$$

with

$$H_{ss}^A(\sigma) = \langle A|H_m|A\rangle_Q; \qquad H_{ss}^B(\sigma) = \langle B|H_m|B\rangle_Q;$$

$$H_{ss}^{AB}(\sigma) = \langle A|H_m|B\rangle_Q. \tag{6}$$

In the exciton description where the sites states are taken orthogonal, $\langle A|B\rangle = \delta_{AB}$, H_{ss}^A and H_{ss}^B are the dipolar interactions averaged on the electronic distributions $|A\rangle$ and $|B\rangle$ hence, they are the spin hamiltonians of the sites, H_{ss}^{AB} is an average dipolar interactions with one of the 'unpaired' electrons in site A, the other in site B, with $H_{ss}^{AB} \ll H_{ss}^A, H_{ss}^B$.

Let us put $H_{ss}^s = 0.5 (H_{ss}^A + H_{ss}^B)$ and $H_{ss}^a = 0.5 (H_{ss}^A - H_{ss}^B)$, as symmetric and anti-symmetric parts of the spin polarizations with respect to an exchange operation $GH_{ss}^A = H_{ss}^B$.

Then, with condition $H_{ss}^{AB} \ll H_{ss}^A, H_{ss}^B$ and from investigation of Equation (5), we conclude that the spin functions which diagonalize the best the hamiltonian (2) are eigenfunctions of H_{ss}^s, with

$$H_{ss}^s \tau_a = a\tau_a \quad \text{with} \quad a = X^*, Y^*, Z^*. \tag{7}$$

Most of the crystal cells, with two non-equivalent molecules, have an axis of symmetry and one rotation C_2 which interchanges the spin polarizations of the sites, $C_2 H_{ss}^A = H_{ss}^B$.

In a naphthalene crystal, it is usual to note by Z^* the axis parallel to **b** and by Y^* and X^* the orthogonal complements, see Figure 1. Then, we can write the spin polarizations of the sites on the dimer frame $OX^*Y^*Z^*$:

$$H_{ss}^A = H_{ss}^s + H_{ss}^a; \qquad H_{ss}^B = H_{ss}^s - H_{ss}^a \tag{8}$$

with ($X \equiv X^*$, $Y \equiv Y^*$, $Z \equiv Z^*$):

$$H_{ss}^s = T_{XX}S_X^2 + T_{YY}S_Y^2 + T_{ZZ}S_Z^2 + T_{XY}(S_XS_Y + S_YS_X)$$

$$H_{ss}^a = T_{XZ}(S_XS_Z + S_ZS_X) + T_{YZ}(S_YS_Z + S_ZS_Y). \tag{9}$$

It can easily be shown that the eigenfunctions τ_a transform like unit vectors along OX, OY, OZ (or like the spin operator components S_X, S_Y, S_Z) and they are symmetric (τ_Z) or antisymmetric (τ_X, τ_Y). Then, on the basis of the aggregate (dimer) spin states, the calculation of the elements (5) leads to the matrix representation of hamiltonian (2), with $x \equiv X^*$, $y \equiv Y^*$ and $z \equiv Z^*$:

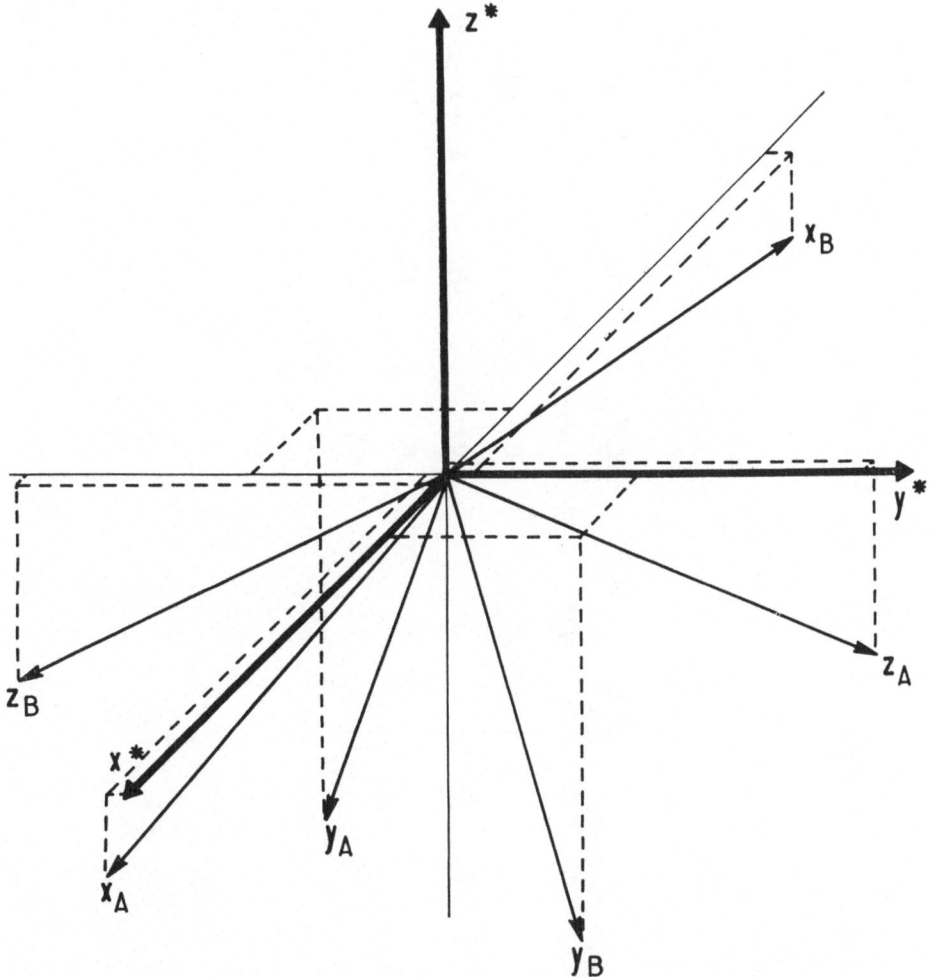

Fig. 1. Three systems of axes $(OXYZ)_{A'}$ $(OXYZ)_{B'}$ $(OX^*Y^*Z^*)$, where the spin hamiltonians H_{ss}^A, H_{ss}^B and H_{ss}^s are respectively diagonal (example a naphthalene pair A–B).

$$[\tau_x,\ \tau_y,\ \tau_z,\ \tau_x,\ \tau_y,\ \tau_z]
\begin{bmatrix}
E^+ + x + H_{ss}^{AB} & H_{ss}^{AB} & 0 & 0 & 0 & H_{ss}^{a} \\
H_{ss}^{AB} & E^+ + y + H_{ss}^{AB} & 0 & 0 & 0 & H_{ss}^{a} \\
0 & 0 & E^+ + z + H_{ss}^{AB} & H_{ss}^{a} & H_{ss}^{a} & 0 \\
0 & 0 & H_{ss}^{a} & E^- + x - H_{ss}^{AB} & -H_{ss}^{A} & 0 \\
0 & 0 & H_{ss}^{a} & -H_{ss}^{AB} & E^- + y - H_{ss}^{AB} & 0 \\
H_{ss}^{a} & H_{ss}^{a} & 0 & 0 & 0 & E^- + z - H_{ss}^{AB}
\end{bmatrix}
\begin{bmatrix} \tau_x \\ \tau_y \\ \tau_z \\ \tau_x \\ \tau_y \\ \tau_z \end{bmatrix} \tag{10}$$

From consideration of matrix (10) we can draw the following conclusions:

(1) In a general case electronic orbital and spin variables do not separate, hence we cannot define spin hamiltonians either for electronic collective excitation states $|k\rangle$, or, a fortiori, for individual sites. Transitions occur among the six spin-orbit states (4), see Figure 2.

(2) For aggregates with parallel sites, electronic orbital collective variables Q separate from spin variables and we can define a spin hamiltonian for each state $|k\rangle$, where the ESR resonances are k dependent.

(3) To describe correctly the ESR response of a linear chain with N parallel sites, we need N spin hamiltonians and $3 \times N$ Boltzmann populations, assuming Boltzmann equilibrium to exist.

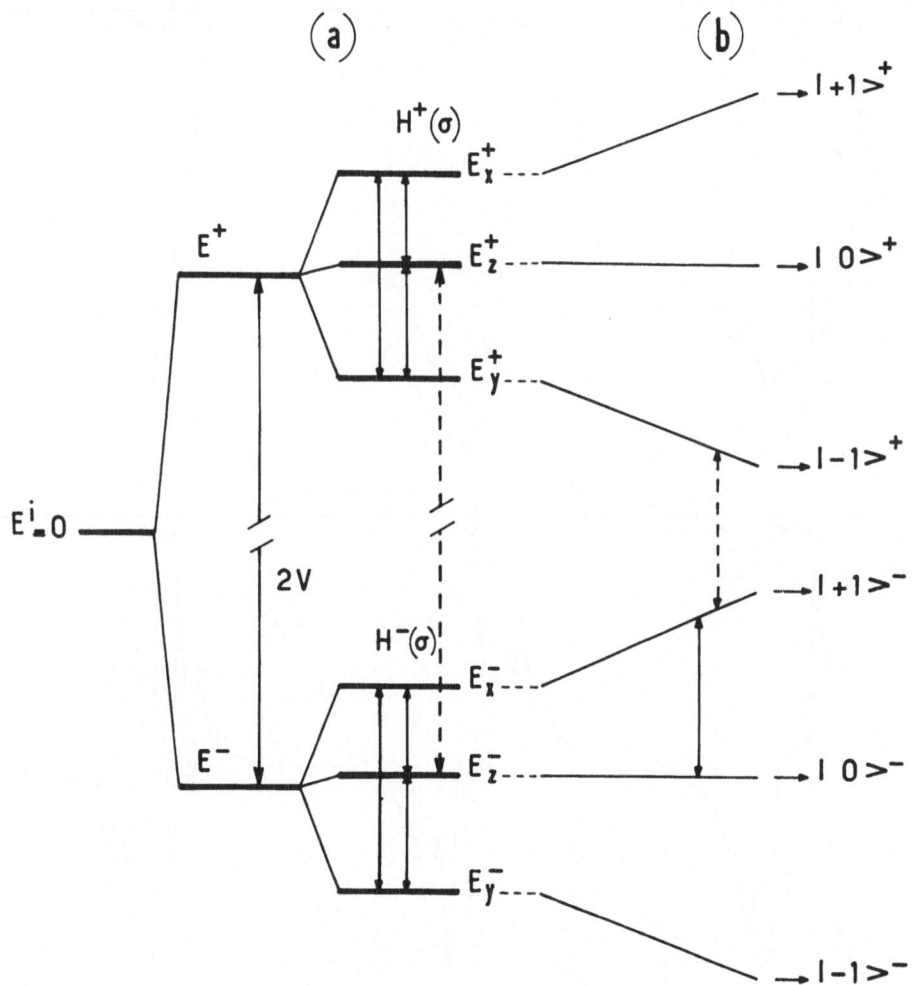

Fig. 2.   The six spin-orbit energy levels in the absence (a) and in the presence (b) of a magnetic field ($\mathbf{H} \| \mathbf{b}$). We can define spin hamiltonians, $H^+ (\sigma)$ and $H^- (\sigma)$, for the two Davydov components, only when the off-diagonal blocks of (10) are neglected.

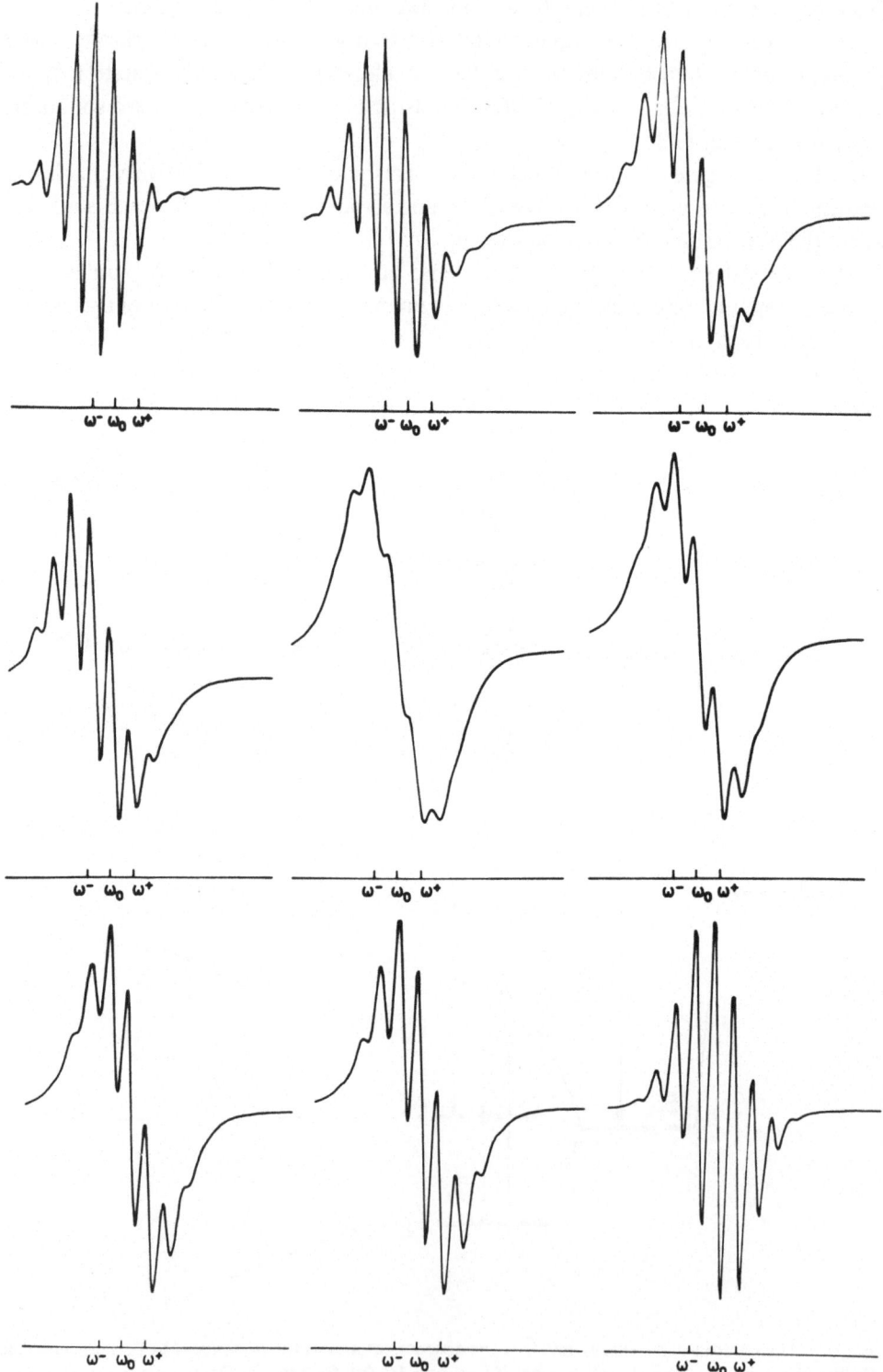

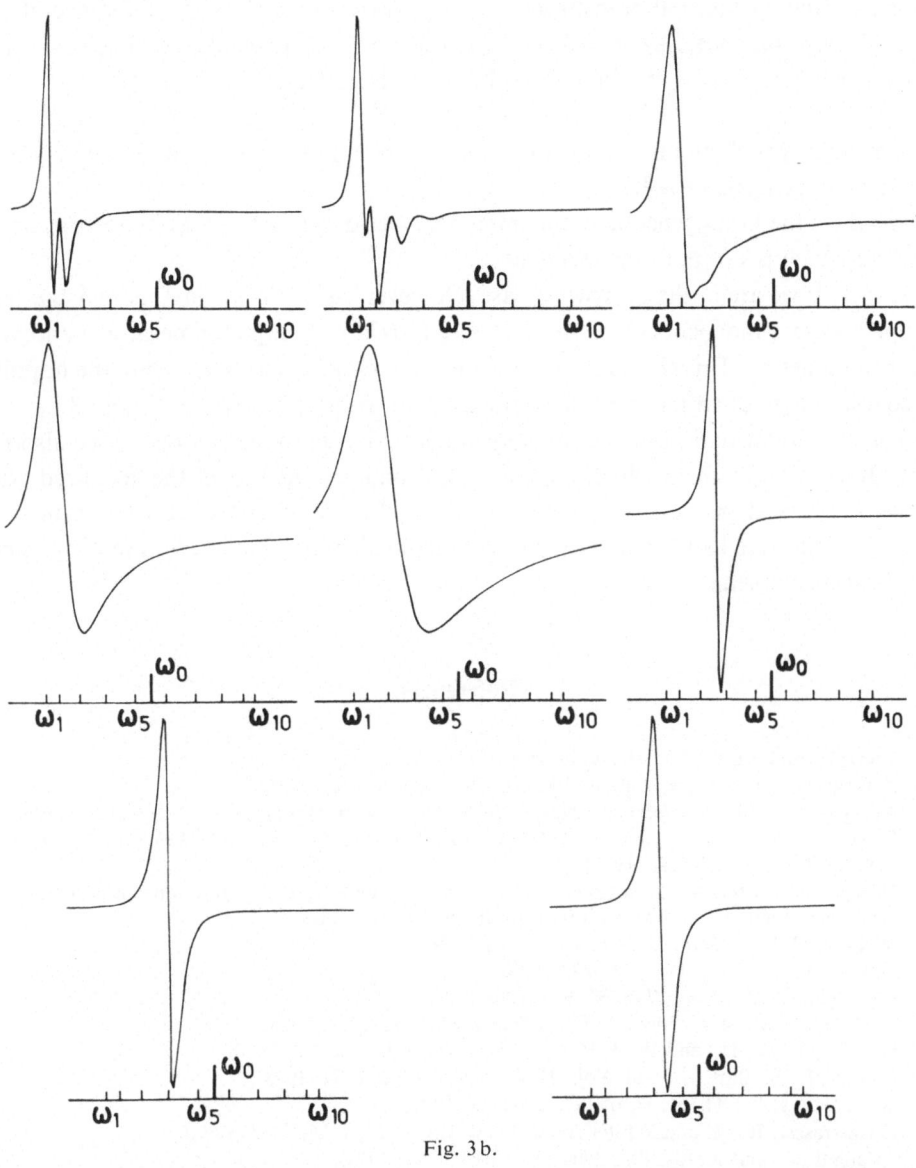

Fig. 3 b.

←

Figs. 3 a-b.   The variation of the ESR line of a dimer (a) and of a chain (b) as a function of the temperature T and average lifetimes $\tau(T)$.

We put $U_{kk'} = U_T^0 \exp\left((E_k - E_{k'})/KT\right)$ for the electronic relaxation probabilities. The different parameters used for these calculations are:

(a) $V = 1.25$ cm^{-1} $H_{ss}^{AB} = 0.0005$ cm^{-1} and $[U_T^0(10^{-4}$ cm$^{-1})$, $T(K)] = (1,2)$; $(2,4)$; $(3,6)$
        $(4,8)$; $(6,15)$; $(7,20)$; $(8,30)$; $(9,40)$; $(20,50)$

(b) $V = 7$ cm^{-1} $H_{ss}^{AB} = 0.0030$ cm^{-1} and $[U_T^0(10^{-4}$ cm$^{-1})$, $T(K)] = (1,6)$; $(1,10)$; $(1.5,15)$;
        $(2,15)$; $(3,30)$; $(6,30)$; $(6,40)$; $(6,50)$

Note: for the dimer we considered the case of a magnetic field $\mathbf{H} \| OX^*$ for which nine hyperfine lines have been observed for pairs of naphtalene in a deuterated crystal matrix [10, 11].

Let us call τ_k an average lifetime of these states $|k, m\rangle$ (m notes a spin quantum number) due to thermal transitions and $\Delta_{kk'} = (\omega(k) - \omega(k'))$ the difference of the same ESR resonance, say $m = 0 \to m = 1$, due to the k dependence. Then, we may distinguish three situations; in absence of saturation effects:

$\tau_k > \Delta^{-1}$: we observe inhomogeneous broadening; the spin of each $|k\rangle$ state absorbs independently with an intensity proportional to $\sim \exp[-(E_k - E_0)/KT]$; we may have some homogeneous broadening if $\tau_k \sim \lambda_k^{-1}$.

$\tau_k \sim \Delta^{-1}$: the homogeneous broadening becomes dominant: the ESR absorption of each $|k\rangle$ state is no more independent.

$\tau_k \ll \Delta^{-1}$: we are in the narrowing case. However, as is shown in Figure 3 the line-shape becomes independent of the k's only at relatively high temperatures and very short lifetimes τ_k. This shows from which limit and on we can start using the hopping model which gives an absorption symmetric with respect to ω_0, see Figure 3.

This is a confirmation of the theoretical discussion presented in ref. 1 (see also ref. 2, 8, 9) as to know from which temperature T and on, the use of the localized state basis (the time dependent stochastic model) or the use of collective excitation state basis (the effective hamiltonian method) are equivalent for the description of the same physical phenomena.

References

1. Voltz, R. and Kottis, Ph.: this volume, p. 187.
2. Aslangul, Cl., Lemaistre, J.-P., and Kottis, Ph.: this volume, p. 209.
3. Abragam. A.: *The Principles of Nuclear Magnetism*, Oxford (1961) Chap. X. See also: Anderson, P. W.: *J. Phys. Soc. Japan* **9**, 316 (1954). Freed, K.: in *Electron Spin Relaxation in Liquids*, Plenum Press, New York, (1972).
4. Note, however, that when isotopic traps are present, the-trap to-trap resonance interaction V is very weak, then we can observe broad resonances of trap sites.
5. McLure, D. S.: in *Solid State Physics* **8**, 1 (1959).
6. Kaplan, J. : *J. Chem. Phys.* **28**, 278 (1958).
   Alexander, S.: *J. Chem. Phys.* **37**, 967 (1962).
   Hudson, A. and McLachlan, A. D.: *J. Chem. Phys.* **43**, 1518 (1965).
7. Port, H., Vogel, D., and Wolf, H. C.: *Chem. Phys. Let.* **33**, 23 (1975).
   Mauser, K. E., Port, H., and Wolf, H. C.: *Chem. Phys.* **1**, 74 (1973).
   Whiteman, J. D.: Thesis, U. of Pennsylvania (1971).
   Hochstrasser, R. M. and Whiteman, J. D.: *J. Chem. Phys.* **56**, 5945 (1972).
8. Aslangul, Cl. and Kottis, Ph.: *Phys. Rev.* **B10**, 4364 (1974).
9. Lemaistre, J.-P. and Kottis, Ph.: in *Electron Spin Relaxation*, Plenum Press, New York (1972).
10. Hutchinson, C. A.: private communication.
11. King, J. S.: Thesis, the University of Chicago, 1973.
12. Hutchison, C. A. and King, J. S.: *J. Chem. Phys.* **58**, 392 (1973).

MOTION OF LOCALIZED EXCITATIONS IN
ORGANIC SOLIDS

V. ERN* and M. SCHOTT

*Groupe de Physique des Solides de l'École Normale Supérieure**, Université Paris VII,
Paris 75005, France*

The collective electronic excitation, the exciton, which is optically prepared in a molecular crystal, with a vector $\mathbf{k} \simeq 0$, undergoes rapid subsequent localization: Bloch functions are solutions of the wave equation, and $\mathbf{k}$ can be used in the description of the crystal stationary states, only as long as perfect translational symmetry is present. At any non-zero temperature, phonons destroy partially this symmetry, and states of the system are now described by a superposition of Bloch functions which may span the whole accessible range Δk of $\mathbf{k}$ values, depending on the strength of exciton-phonon coupling. By the uncertainty principle, this may amount to a localization, say in 1-dimension Δx such that $\Delta x \Delta k \sim 2\pi$, and a wave packet has been built. This is a process common to all electronic states in solids, and the same line of thought is followed if Wannier functions are constructed.

The degree of localization will of course depend on the extent of Δk. Different $\mathbf{k}$ states can be reached by phonon scattering with a probability proportional to a Boltzmann factor $\exp(-\varepsilon(\mathbf{k})/k_B T)$ (k_B the Boltzmann constant). For conduction electrons in usual semiconductors, or metals, Δk is in general limited to a very narrow range, either near an extremum of the band, or around the Fermi surface, and the corresponding wave packets, although 'localized', are still extending over many lattice cells. Exciton bands in molecular solids are in general much narrower, and their total width ε is often narrower than or comparable to $k_B T$. Very roughly speaking, consider as an illustrative example, Figure 1, showing a band of a 1-dimensional crystal of lattice constant a, where the accessible energy is some fraction $\Delta\varepsilon$ of the total bandwidth (shadowed region) and the corresponding uncertainty on k, Δk is $\frac{1}{4}$ the length of the Brillouin zone $2\pi/a$; the electronic excitation can then be roughly thought as localized over a region of 4 lattice spacings. Extreme localization will occur at temperatures sufficiently high such that $k_B T \gg \varepsilon$. In this case all possible $\mathbf{k}$ states in the Brillouin zone can be reached with equal probability and the excitation can be thought as momentarily localized in a volume that corresponds to one molecule in the unit cell. Indeed, as will be seen below, at high temperatures, e.g. room temperature, excitonic processes for the lowest bands, especially for triplet states, in many organic solids can be described using a 'billiard ball' model in which the excitation is visualized as a particle localized on a molecular site and which transfers carrying with it the electronic excitation energy, without an effective transport of electrical charge, from molecule to molecule.

---

* Visiting Professor, Université Paris VII.
** Laboratoire associé au CNRS.

O. Chalvet et al. (eds.), Localization and Delocalization in Quantum Chemistry, Vol. II, 249–284. All Rights Reserved
Copyright © 1976 by D. Reidel Publishing Company, Dordrecht-Holland

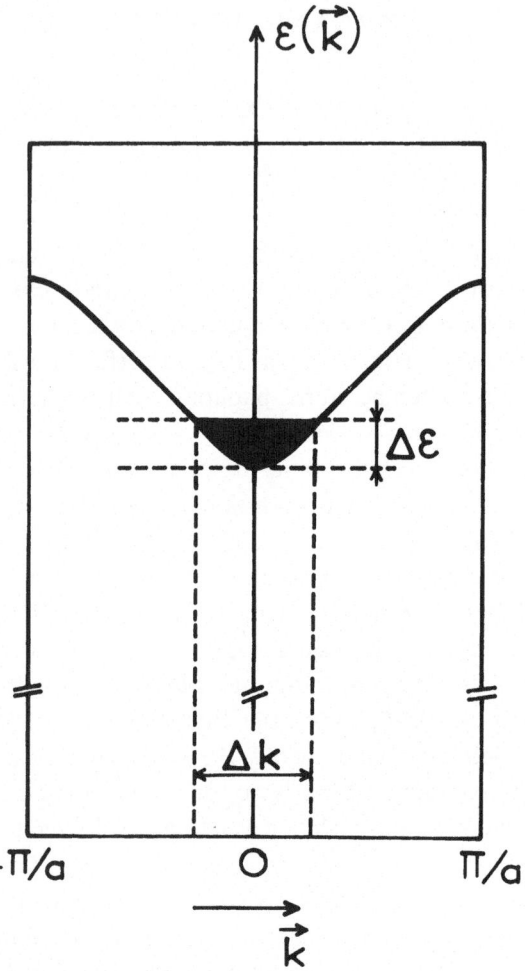

Fig. 1. Exciton band in a one-dimensional crystal of lattice constant a in which the accessible energies due to phonon scattering are schematically indicated by the width $\Delta \varepsilon$ (shadowed region). For the shape of the band shown, this leads to a spread $\Delta \mathbf{k}$ of possible $\mathbf{k}$ values of $\frac{1}{4}$ of the length of the Brillouin zone $2\pi/a$.

The existence of exciton motion in molecular solids is now a well established experimental fact. Using purposely doped crystals, many experiments have demonstrated using ESR [1] and sensitized fluorescence [2, 3] and phosphorescence [4, 5] techniques that energy transfer can take place from the host to the guest molecules or from guest to guest via the exciton band of the host crystal. That is, the energy stored at some place in the crystal can be partially retrieved at some other place, in the form of photons having an energy which is lower than that of the photons used to excite the crystal. Excellent reviews of these experiments can be found in the work by Lower and El-Sayed [6], Wolf [7] and Birks [8].

Perhaps the most clear-cut effect which gives evidence for the existence of excitation migration in organic solids is the phenomenon of mutual annihilation of excitons which can lead to either the disappearence of the excitation, the creation of charge carriers, or the striking effect of the creation of an excitation of higher energy which can then reemit a photon of energy which is above that of those used to excite initially the crystal. The existence of such a mutual annihilation for singlet excitons was conjectured as far back as 1958 [9]. In the same year, delayed fluorescence, that is, an emission in the same spectral range as the prompt normal fluorescence but which is detected after the usual fluorescence has ceased, was observed in several crystals [10, 11] and later interpreted in terms of mutual annihilation of triplet excitons [12, 13]. In 1963 Kepler *et al.* [14], using direct excitation with a laser beam into the triplet band of anthracene crystals, gave conclusive evidence for the interpretation of delayed fluorescence as due to the mutual annihilation of triplet excitons leading to the creation of the fluorescing singlets. Since these pioneering discoveries an extensive work on singlet-singlet, triplet-triplet, and singlet-triplet annihilation has appeared in the literature, the thorough review of which is outside the scope of this presentation.

If the exciton density $n(\mathbf{r}, t)$ at a point $\mathbf{r}$ in the crystal and at time t is sufficiently small as compared to the density of molecules packed in the crystal (which is the usual case in practical experiments, except possibly in scintillation), so that the probabilities of finding three and more particles at neighboring sites can be neglected [15], then the disappearance of the exciton population due to a mutual annihilation process can be described by the rate which is conventionally written as

$$\left(\frac{\partial n(\mathbf{r}, t)}{\partial t}\right) \text{ due to pair annihilation} = -\gamma_{\text{tot}}\, n^2(\mathbf{r}, t) \tag{1}$$

where γ_{tot} is a total phenomenological rate constant expressed in $\text{cm}^3\,\text{s}^{-1}$ and which includes all channels that are energetically open to the pair-annihilation. For example, for the mutual annihilation of the lowest triplet states if the sum of the energies of the two annihilating triplets $(T_1 + T_1)$ lies above the first excited singlet S_1 and of the same higher excited triplet T^*, one can visualize the annihilation channels shown schematically in Figure 2, where the possibility of forming quintet states has been neglected since they, very likely, lie above the sum of the energies of the two triplets. The total mutual annihilation rate of Equation (1) will be then given by $\gamma_{\text{tot}} = \frac{1}{2}\gamma_3 + \frac{1}{2}(1+\eta)\gamma_1 + \gamma_1'$, and the total delayed fluorescence photon flux (number of photons per second) emitted by the crystal will be given by

$$\Phi(t) = \tfrac{1}{2} f \gamma_{\text{tot}} \int_V n^2(\mathbf{r}, t)\, \mathrm{d}^3\mathbf{r} \tag{2}$$

where the integral is taken over the volume of the crystal and where $f \equiv \eta \gamma_1/\gamma_{\text{tot}}$ [16, 17] is the fraction of the total annihilation rate constant leading to these singlets observed by fluorescence, a number which can be between 0 and 1. As pointed out clearly by Suna [15], the rates of two processes determine the magnitude of γ_{tot}, namely, that of migration of two excitons toward, as well as away from, each other and that of the annihilation when they are sufficiently close to interact. This last rate is different for the different channels open to annihilation depending on the final density of states available for the process. The existence of an exciton migration is essential to bring the pairs to a sufficiently close separation (e.g. at nearest neighbor sites) so that interaction and mutual annihilation can take place, as can be recognized from the fact that, for instance in anthracene, the delayed fluorescence given by (2) can be easily observed by conventional techniques with the crystal uniformly excited with triplet exciton densities of $n \simeq 10^{12}$ cm^{-3} or lower, that is, even when the excitons are created at an average separation of $1\,\mu$ which is about 2 000 lattice spacings.

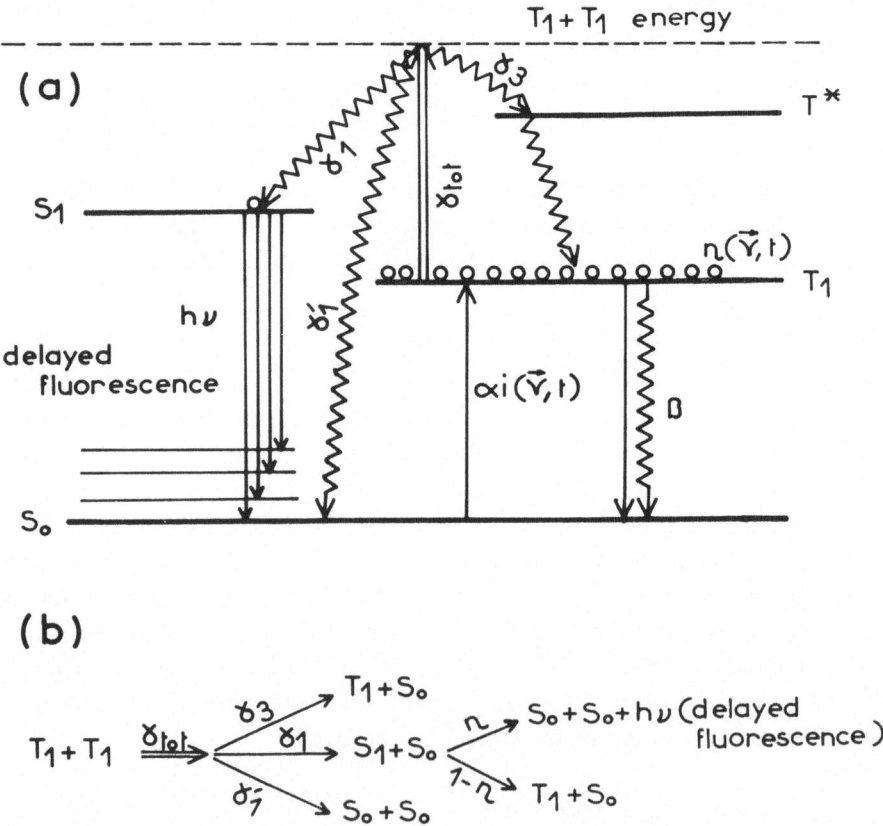

Fig. 2. Schematic representation of a simplified energy level diagram and of the possible channels leading to mutual annihilation of a pair of triplet excitons and to delayed fluorescence. The possibility of forming quintet states has been neglected. The rates γ_1' and γ_3 are for the annihilations via the excited singlet and triplet channels, respectively, γ_1 for the direct radiative or radiationless transition of the pair to the ground state. The other quantities shown are: α, the ground state (S_0) – first excited triplet (T_1) absorption coefficient; $i(r, t)$, the exciting intensity; $n(r, t)$, the concentration of the triplet exciton population; β, its monomolecular decay rate including radiative and radiationless processes; and η, the quantum efficiency of fluorescence. The virtual level of energy $T_1 + T_1$ for the pair is placed at the double energy of the lower (0,0) vibronic band of T_1.

Under spatially uniform excitation so that the exciting intensity $i(t)$ and the triplet exciton density $n(t)$ are only functions of time, and if the excited volume is sufficiently big so that the loss of excitons due to migration through the boundaries can be neglected as compared to the other loss mechanisms, one has using (1) that the total rate for the buildup or decay of triplet exciton population will be given by

$$\frac{dn(t)}{dt} = \alpha i(t) - \beta n(t) - \gamma_{tot} n^2(t) \tag{3}$$

and the delayed fluorescence photon flux per unit volume of the crystal as

$$\phi(t) = \tfrac{1}{2} f\gamma_{tot} n^2(t). \tag{4}$$

In Equation (3) α is the ground state singlet-first excited triplet absorption coefficient and β is the total monomolecular decay rate constant which includes all processes, radiative or radiationless, which produce a direct deactivation to the ground state, as shown in Figure 2, and which lead to a finite triplet exciton lifetime

$$\tau \equiv \beta^{-1}$$

in the crystal. Since the pioneering work of Kepler et al. [14] on anthracene crystals, the predictions of Equations (3) and (4) have been verified by many workers in this [16] as well as in other materials [17]. An interesting limit of Equation (3) occurs when the exciting intensity is so low that the rate of depletion by the mutual annihilation process becomes negligible with respect to the monomolecular decay term, that is, when $\gamma_{tot} n^2 \ll \beta n$, or when the triplet exciton density $n \ll (\gamma_{tot}\tau)^{-1}$. In this low intensity limit Equation (3) becomes simply

$$\frac{dn(t)}{dt} = \alpha i(t) - \beta n(t) \tag{5}$$

which for a continuous steady-state excitation of intensity i_0 predicts with (4) a steady-state fluorescence flux emitted from the excited volume V of the crystal

$$\Phi_0 = \tfrac{1}{2} f\gamma_{tot}\alpha^2 \tau^2 i_0^2 V, \tag{6}$$

that is, proportional to the square of the exciting intensity. If the crystal is excited with chopped light, that is, with a square-wave excitation such that $i(t) = i_0$ for $0 \leqslant t < T/2$ and $i(t) = 0$ for $T/2 < t \leqslant T$, and if T, the period of the excitation is long compared to triplet lifetime, $T \gg \tau$, one obtains the simple delayed fluorescence waveform shown in Figure 3 which has a temporal behavior given by

$$\Phi_B(t) = \Phi_0(1 - e^{-t/\tau})^2 \tag{7}$$

for the buildup, and

$$\Phi_D(t) = \Phi_0 e^{-2t/\tau} \tag{8}$$

for the decay portions, respectively, and which can be used to deduce the triplet exciton lifetime.

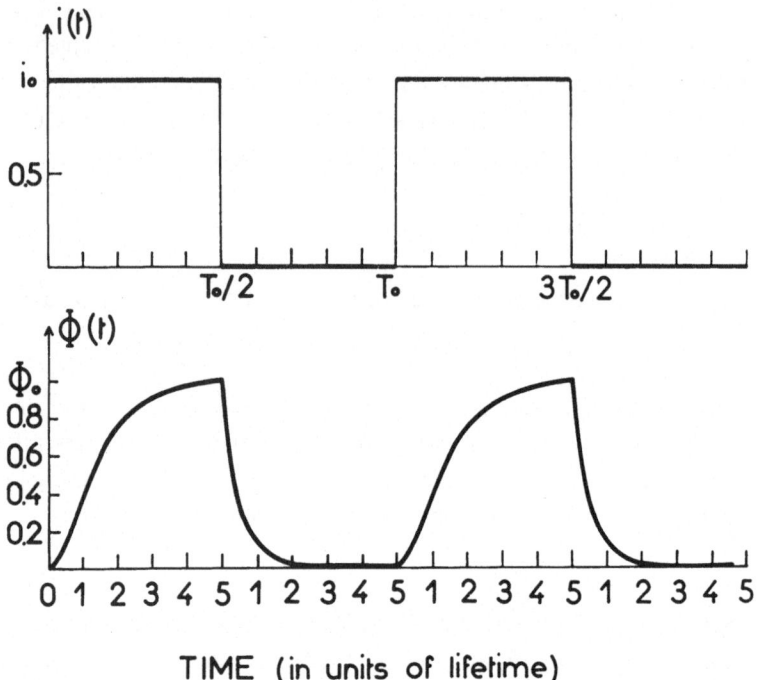

TIME (in units of lifetime)

Fig. 3. Waveforms for the exciting light intensity (top) for chopped light with a period $T_0 \gg \tau$, the triplet lifetime, and for the delayed fluorescence (bottom) observed from the crystal under a spatially uniform illumination as given by Equations (7) and (8) in text.

It should be now apparent to the reader that if the size of the excited volume becomes sufficiently small, or if the crystal is excited with a spatially inhomogeneous beam $i(\mathbf{r}, t)$, e.g. in spots through a mesh, so that spatial gradients of the triplet exciton concentration $n(\mathbf{r}, t)$ are created in the crystal, the effects of exciton migration can no longer be neglected in Equation (3) or in its more simplified form (Equation (5)) in the low intensity limit. The evolution in time of the triplet exciton concentration $n(r, t)$ will reflect the exciton motion and the total delayed fluorescence photon flux from the crystal (Equation (2)) will have a temporal dependence $\Phi(t)$ which will be sensitive to the spatial inhomogeneity of the exciting beam and to the degree of the exciton motion. In principle, delayed fluorescence can thus be used as a probe to detect triplet exciton motion in pure crystals. It should be noted that this is not the case with phosphorescence or the usual prompt fluorescence since then the total light emitted at a given instant by the crystal is proportional to

$$f(t) = \int_V n(\mathbf{r}, t) \, \mathrm{d}^3 r$$

a quantity which depends only on the instantaneous *total* number of emitting centers and which is independent of their motion and of the details of the spatial inhomogeneity of the exciting beam.

Unfortunately, the delayed fluorescence technique is not applicable for singlet excitons. The state generated by singlet-singlet annihilation, a highly excited singlet, is not radiant, and radiationlessly decays back to the lowest singlet. For singlets, therefore there is no such effect as delayed fluorescence, and the only emission, fluorescence, plays the same role as the phosphorescence for triplets. As was mentioned above (and will be seen in more detail below), this does not contain the necessary information for the measurement of motion. Therefore, in the case of singlets, one must use the much less reliable method of transfer to selective traps or accepting centers [2] which may severely distort the crystal around them. Furthermore, singlet transitions being, in general, dipole allowed the energy transfer in the singlet state may be a combination of exciton migration and resonant-Förster-transfer, from which exciton transport properties can not be easily extracted [3].

Another advantage of working with triplets is that, during an exciton lifetime, the energy is transferred much farther in the triplet state than in the singlet [13]. This is so, in spite of the well known fact that the bandwidth for the latter, or the corresponding transfer matrix elements, are two to three orders of magnitude larger than for triplets, since this is much more than compensated by the fact that triplet lifetimes (10^{-3} to 1 s) are five to nine orders of magnitude greater than the singlet lifetimes. Triplet excitons, as has been first shown experimentally by Avakian and Merrifield [18], do indeed travel during their lifetime over macroscopic distances of several thousands of lattice spacings. In this sense, they represent a clear-cut property of the organic solid and their motion can be detected using experimentally achievable spatially-inhomogeneous exciting light intensity distributions in the solid.

The delayed fluorescence technique to detect triplet exciton motion has been extensively used in the last years and a wealth of experimental information, as well as of the corresponding theoretical interpretations, has been obtained leading to a considerable improvement of our understanding of the dynamics of the localized excitations in the organic solid state. The high-lights of this work will be presented in what follows of this paper.

1. Experimental Verification of the Validity of a Macroscopic Time-dependent Exciton-Diffusion Equation

The first question that we should ask ourselves is, which is the *macroscopic* transport equation which describes the exciton motion in the crystal. Clearly, the answer to this question is a difficult one since the exciton migration should depend, besides on the crystal transfer matrix elements, on the details of the exciton-phonon interactions which must be included in the problem. In recent years the problem has been subjected to several theoretical approaches. On one hand, Haken and his group [19–23] have developed a treatment in which the details of the exciton-phonon coupling are treated phenomenologically, others [24–27] investigated special forms of the hamiltonian, and lately more formal treatments have been presented using generalized master equations which include a memory function for the exciton [28–30]. The aim of the theories is to find, for a given temperature, that is, for a given strength of the exciton-phonon coupling, a relation which gives the mean-square displacement of the exciton as a function of time $\langle R^2(t) \rangle = f(t)$.

At sufficiently high temperatures, or for strong exciton-phonon coupling, the lifetime of the exciton at a given **k**-state, the coherence time of the exciton, can become so short, namely much shorter than the exciton transit time between the molecules and than the exciton lifetime, that the motion is incoherent with the exciton suffering many scattering events before transferring from molecule to molecule. The mean-square displacement of the exciton is then a linear function of time

$$\langle R^2(t) \rangle = 2Dt \tag{9}$$

where D (in $cm^2 s^{-1}$) is the so-called diffusion coefficient and on a macroscopic scale the motion of the excitons in the crystal must be described by a *diffusion equation*. On the other hand if the exciton-phonon coupling is very weak, e.g. at very low temperatures and if the crystal is pure enough so that exciton trapping as well as exciton scattering by impurities is not important, the lifetime of an exciton **k** state can become much longer than the intermolecular transit time and comparable to or longer than the exciton lifetime and the excitons can then be thought as wave packets propagating coherently during the exciton lifetime with group velocities

$$v_g(\mathbf{k}) = \hbar^{-1} \, \mathrm{grad}_k \, \varepsilon(\mathbf{k}) \tag{10}$$

determined by the band structure for each **k**. In this case the exciton mean-square displacement is $\langle R^2(t) \rangle \propto t^2$ [31, 20–22] and the motion *cannot* be described by a macroscopic diffusion equation. At intermediate temperatures or exciton-phonon coupling strengths the lifetime of a **k** state can be still long as compared to the inter-molecular transfer time, but become short as compared to the exciton lifetime. Under these conditions the overall motion during the exciton lifetime will be incoher-ent but with a long-range coherent propagation between the scattering events. In this case the possibility of observing a macroscopic diffusion equation in a real crystal could depend on the actual physical size of the sample under observation and on the time scale of the measurements.

It should be apparent to the reader that, before interpreting experiments on exciton migration in terms of an exciton diffusion coefficient or in terms of an exciton diffusion length, the distance travelled by the exciton during its lifetime, which from (9) is given by:

$$L = (2D\tau)^{1/2} \tag{11}$$

it is essential to ascertain, by a *direct* experimental verification, that a macroscopic exciton diffusion equation is indeed a valid description of the exciton migration process in the material. Only after this has been established one can safely interpret a travel length of the exciton inferred by an indirect experiment as that given by Equation (11). A diffusion coefficient can be then derived which, in turn, can be subjected to a theoretical analysis based on the known exciton band structure [32] and on the exciton-phonon coupling strength in the material. In what follows we will present experiments which show that in a typical molecular solid, anthracene, the triplet exciton concentration $n(\mathbf{r}, t)$ does indeed obey a macroscopic time-dependent diffusion equation at room temperature [33].

Before going into the experiments it is convenient to point out some features of the problem due to the crystalline symmetry. In a crystal, because of its anisotropy, the diffusion process must be characterized by a diffusion tensor rather than by a single number D. The tensor must be of 2nd rank, and as, for instance, the case of dielectric constant, will be represented by a 3×3 symmetric matrix

$$\mathbf{D} = \{D_{ij}\} \tag{12}$$

where the 9 components are referred to an arbitrary cartesian coordinate system in the crystal. As is well known, if the coordinate system is taken so as to coïncide with the so-called principal axes of the tensor, the orientation of which with respect to the crystallographic axes is determined by the crystal symmetry, then the matrix (12) takes a diagonal form, that is, one has $D_{ij} = 0$ for $i \neq j$ and the three diagonal elements

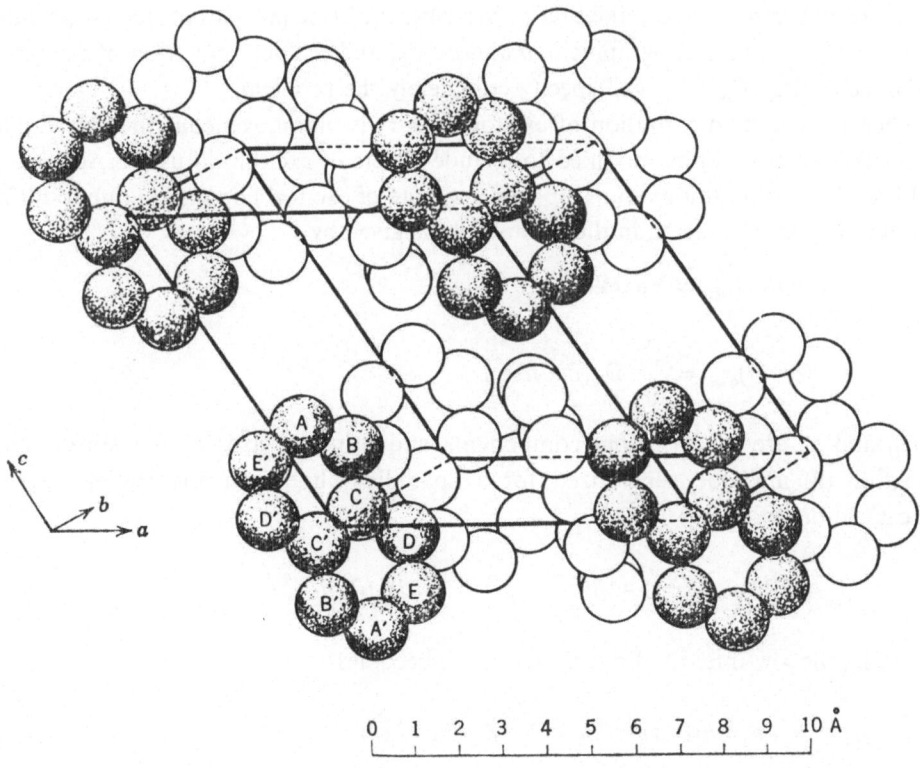

Fig. 4.   The crystal structure of naphthalene: monoclinic, space group $P2_1/a$, two molecules per unit cell (From Abrahams, S., Robertson, J. M., and White, J. G.: *Acta Cryst.* **2**, 238 (1949)).

D_{ii} are the components of the tensor along the principal axes x_i, $i = 1, 2, 3$. For instance, for crystals in the monoclinic system, like anthracene or naphthalene whose crystal structure is illustrated in Figure 4, and which is very common in molecular crystals, the crystal b axis is a two-fold rotation symmetry axis, and must be one of the principal axes of the tensor. The other two axes must be then in the ac plane and the diffusion tensor can already be written in the simplified form

$$\mathbf{D} = \begin{pmatrix} D_{xx} & 0 & D_{xz} \\ 0 & D_{bb} & 0 \\ D_{xz} & 0 & D_{zz} \end{pmatrix} \tag{13}$$

if the y axis is taken to coïncide with b, x and z lying in arbitrary directions in the ac plane. The component of the diffusion tensor along some direction in the ac plane making an angle α with the principal axis x_1 will then be given by

$$D_\alpha = D_{x_1 x_1} \cos^2 \alpha + D_{x_3 x_3} \sin^2 \alpha \tag{14}$$

and the exciton diffusion length in this direction can be then obtained with (11) as

$$L_\alpha = (2 D_\alpha \tau)^{1/2}. \tag{15}$$

In an orthorhombic crystal, like that of benzene, the principal axes of the tensor must coïncide with the crystallographic a, b and c axis, and the determination of the three components D_{aa}, D_{bb}, D_{cc} will specify completely the problem.

When the spatial distribution of excitons $n(\mathbf{r}, t)$ is inhomogeneous the rate Equations (3) or (5) should be modified to include effects of exciton diffusion. According to Fick's law the additional buildup or decay rate of exciton population due to diffusion into or out of a region in the crystal will be given by

$$(\partial n/\partial t)_{\text{diff}} = \nabla n \mathbf{D} \nabla n$$

that is,

$$(\partial n/\partial t)_{\text{diff}} = \sum_{i=1}^{3} D_{ii}(\partial n^2/\partial x_i^2)$$

if D_{ii} and x_i are the principal components and principal axes of the tensor, and Equation (3) must be generalized for a spatially inhomogeneous triplet exciton concentration to

$$\partial n/\partial t = \alpha i(\mathbf{r}, t) - \beta n - \gamma_{\text{tot}} n^2 + \sum_{i=1}^{3} D_{ii}(\partial n^2/\partial x_i^2). \tag{16}$$

Similarly, the low-intensity limit Equation (5) becomes

$$\partial n/\partial t = \alpha i(\mathbf{r}, t) - \beta n + \sum_{i=1}^{3} D_{ii}(\partial n^2/\partial x_i^2). \tag{17}$$

The term $\gamma_{\text{tot}} n^2$ in (16) shortens the effective lifetime of the exciton in the crystal and will therefore reduce the distance traveled by the exciton, and hence the effects of its diffusion. It is advantageous therefore to perform the experiments in the limit given by Equation (17), that is, at sufficiently low intensities of excitation so that the delayed fluorescence Equation (2) is used solely as a passive probe to detect the behavior of the triplets. Clearly, this will also simplify the mathematical problem for obtaining the solution of the time-dependent diffusion equation to which the experimental results are to be fitted. Facility of solving Equation (17) and experimental considerations for obtaining adequate delayed fluorescence signals and non-negligible effects due to the last term should also guide the choice of the spatial inhomogeneity of the exciting intensity $i(r, t)$. For instance, excitation with $i(\mathbf{r}, t) = i_0(t)$ for $r < R_0$, $i(\mathbf{r}, t) = 0$ for $r > R_0$, that is, with a cylindrical beam of constant intensity and radius R_0, will give sizeable effects due to the diffusion term in Equation (17) on the time dependence of delayed fluorescence only if R_0 becomes comparable to, or smaller than, the triplet exciton diffusion length L. Diffusion of excitons from the illuminated cylindrical region into the dark region (from the high to the low-exciton concentration region) will lead to a sink term which adds to the monomolecular decay and hence to a faster buildup and decay of the exciton population. However, if L is in the micron region or less, this arrangement will be impractical since it will require working with very weak delayed fluorescence signals due to the smallness of the excited volume in the crystal. Moreover, because of the cylindrical symmetry of the beam, and since the photomultiplier detects the total fluorescence flux from the sample, only an average effect of the components of the diffusion tensor in the plane of the sample will be seen, and the experiments will not provide information on the details of the anisotropy in this plane. It is clear that to avoid these shortcomings it is advantageous to use an excitation intensity $i(\mathbf{r}, t)$ which is: (a) periodic in space, that is, a repetitive pattern of equal, small, excited regions which, when viewed all together with a photomultiplier, add all equally to build up a sufficiently strong delayed fluorescence signal; and (b) spatially inhomogeneous in one-dimension only, so as to create an exciton concentration gradient only in a given direction in the sample. In short, the sample should be illuminated in a pattern of repetitive strips which are perpendicular to the direction along which one wishes to observe the exciton diffusion.

Such a spatially inhomogeneous excitation can be achieved by exciting the crystal with an expanded (e.g. diameter ~ 1 cm) laser beam through a Ronchi ruling (a grating with alternating transparent and opaque strips) and using a wavelength which is in the ground state – first excited triplet absorption region of the sample. The absorption into the first triplet is sufficiently weak (typically $\alpha = 10^{-2}$–10^{-4} cm^{-1}) so that one can assume a uniform excitation, within each strip, in the bulk of the sample. If the dimensions of the sample are sufficiently big (e.g. typically 1 cm × 1 cm × 1 mm) as compared to the diffusion length the only operative diffusion effect will be that in a direction perpendicular to the ruling lines, say x, and the anisotropy of diffusion in the plane of the sample can be investigated by rotating the ruling about the laser beam axis. Figure 5 shows the idealized exciting light intensity distribution in the crystal for a ruling of spatial period x_0 with equal widths for the open and opaque strips. In the first experiments [33] to test the validity of the time-dependent

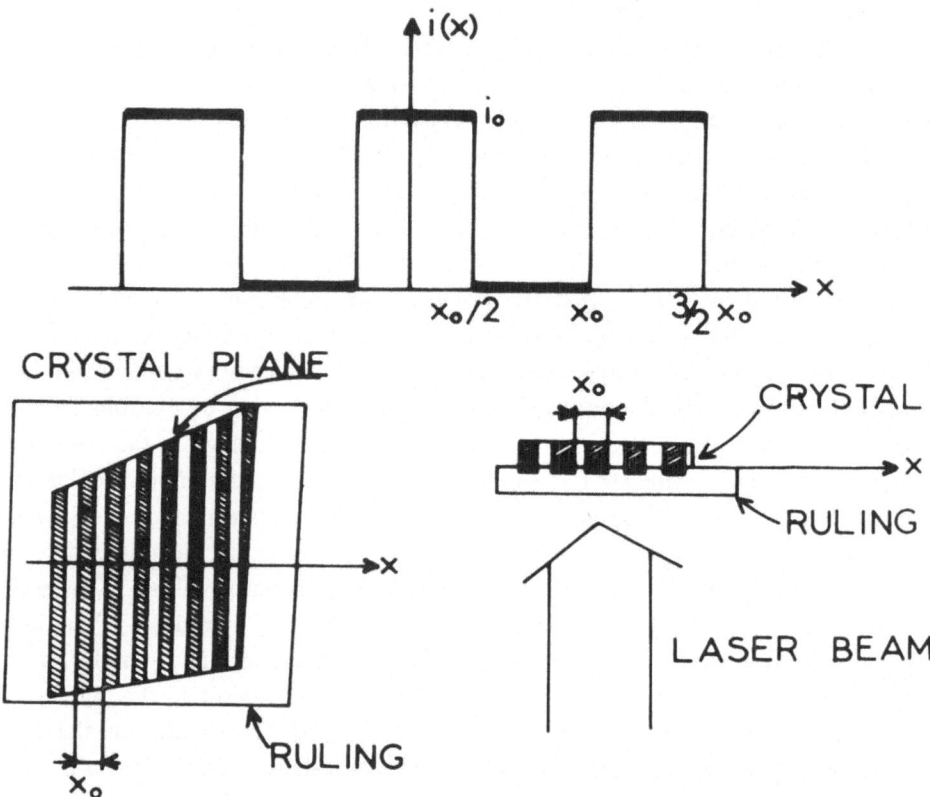

Fig. 5. Idealized representation of the spatial distribution of the intensity of an expanded laser beam exciting a crystal plate through the transparent and opaque strips of a Ronchi ruling of period x_0. Typically the crystal is held a fraction of a millimeter above the ruling. The dimensions of the crystal and of ruling period are not to scale. Typically the crystal is 1 cm × 1 cm × 1 mm and x_0 from 30 to 100 μ.

diffusion Equation (17) chopped light was used with a period τ_0 sufficiently greater that the triplet exciton lifetime τ so as to obtain, without rulings, a delayed fluorescence wave-form given by Equation (7) and (8) for the buildups and decay portions, respectively. The experimental setup is illustrated in Figure 6. Since the same chopper arrangement with a period T_0 is used for the different spatial geometries created by introducing, at stationary positions, rulings of different periodicities x_0 in the beam one can write the exciting intensity function in (17) as

$$i(x, t) = i_0 g(x) \, f(t) \tag{18}$$

where i_0 is the laser beam intensity and $g(x)$, the spatial dependence given by the ruling, is given by

$$g(x) = \tfrac{1}{2} + \sum_{l=\text{odd}} S_l \cos\left(\frac{2\pi l}{x_0} x\right) \tag{19}$$

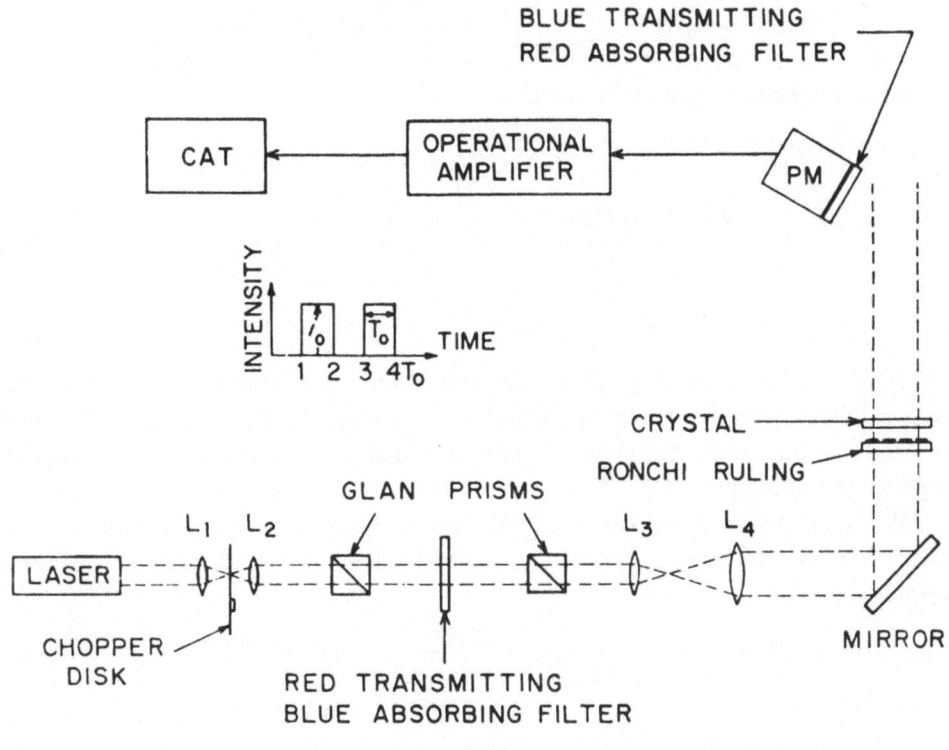

Fig. 6. Typical experimental setup to detect changes in the time dependence of delayed fluorescence due to exciton motion. The signal from the photomultiplier (PM) detecting the delayed fluorescence waveform is fed through an operational amplifier to a computer of average transients (CAT) to improve the signal-to-noise ratio (From Ern, V., Avakian, P., and Merrifield, R. E.: ref. (33)).

in which the coordinate origin is taken at the center of an opaque strip and where

$$S_l = -\frac{2}{l\pi} \sin \frac{l\pi}{2}, \quad l = 1, 3, 5 \dots . \tag{20}$$

The time-dependent part $f(t) = 1$, $f(t) = 0$, during the chopper's on and off half cycles, respectively. Since $T_0 \gg \tau$ one can assume, to a very good approximation, that $n(t) = 0$ at the beginning $(t = 0)$ of an on (buildup) half cycle, as well as at the end of the off (decay) half cycle. Introducing now (18) through (20) into the one-dimensional form of (17)

$$\frac{\partial n}{\partial t} = \alpha i(x, t) - \beta n + D_x \frac{\partial n}{\partial x^2} \tag{17'}$$

where D_x is the component of the diffusion tensor along x, one can write the exciton concentration as the sum

$$n(x,t) = n_0(t) + \sum_{l=1(\text{odd})}^{\infty} n_l(t) \cos\left(\frac{2\pi l}{x_0} x\right) \tag{21}$$

where for the buildup half-cycle one has

$$n_0(t) = \tfrac{1}{2}\alpha i_0 \, \beta^{-1}\{1 - e^{-\beta t}\} \tag{22}$$

and

$$n_l(t) = \frac{2\alpha i_0 \sin(l\pi/2)}{l\pi\beta(1 + l^2 a^2)}\{1 - e_{l \text{ odd}}^{-\beta(1 + l^2 a^2)t}\} \tag{23}$$

with

$$a \equiv 2\pi(D_x\tau)^{1/2}/x_0 = \sqrt{2}\pi L_x/x_0. \tag{24}$$

For the decay half-cycle one has similar expressions to (22) and (23) but with the temporal dependence in brackets substituted by a simple exponential decay. It should be noted that the parameter a plays an important role and scales the exciton diffusion length to the ruling period x_0. Taking the square of these expressions and integrating over the ruling period obtains from (2) the delayed fluorescence flux from the crystal, which after normalization to the steady-state value has, for the buildup, the temporal dependence

$$\Phi^b(t) \equiv \frac{\Phi(t)}{\Phi_0} = \frac{1}{N(a)}\left\{(1 - e^{-\beta t})^2 + \sum_{l=1}^{\infty} A_l(1 - e^{-\beta(1 + l^2 a^2)t})^2\right\} \tag{25}$$

where

$$A_l = 8/l^2\pi^2(1 + l^2 a^2)^2 \quad (l \text{ odd}) \tag{26}$$

and the quantity

$$N(a) \equiv 1 + \sum_{l=1}^{\infty} A_l \tag{27}$$

is proportional to the steady-state intensity of delayed fluorescence with the rulings. A similar expression to (25) is obtained for the normalized decay of delayed fluorescence but with the time-dependences in brackets substituted with $e^{-2\beta t}$ and $e^{-2\beta(1 + l^2 a^2)t}$. Comparing (25) and (7) one sees that, due to the spatial inhomogeneity of the excitation, exciton diffusion effects should be now observable in the time dependence of the delayed fluorescence and are contained in the second term of (25). Clearly, this term doesn't affect the delayed fluorescence if $a \to 0$, that is, when there is no diffusion, or when the ruling period becomes $x_0 \gg L$. As stated before, no effects of diffusion can be observed on a phosphorescence signal from the triplets since, as the reader can easily verify, integration of (21) without squaring yields a time dependence $\{1 - e^{-\beta t}\}$, independent of the exciton motion.

The above relations can be easily generalized for a ruling whose open and opaque strips are not equal. Furthermore, one can expect some smearing of the geometrical shadow pattern of the ruling due to the presence of some stray light in the shadow regions, due to diffraction and scattering by crystal imperfections. For a ruling whose window opening to period ratio is r and assuming that a certain amount of stray light

Δi is uniformly present in the shadow portion due to scattering and diffraction from the illuminated regions under conservation of the total incident flux, one has an exciting light intensity distribution shown in Figure 7, for which Equation (25) is modified to

$$\Phi^b(t) = \frac{1}{N(a)}\left\{(1-e^{-\beta t})^2 + \sigma_r \sum_{l=1}^{\infty} A_l(1-e^{-\beta(1+l^2 a^2)t})^2\right\} \tag{28}$$

where now

$$A_l = \frac{2}{r^2\pi^2}\frac{\sin^2(l\pi r)}{l^2(1+l^2 a^2)}, \quad (l \text{ odd}) \tag{29}$$

$$N(a) = 1 + \sigma_r \sum_{l=1}^{\infty} A_l \tag{30}$$

in which

$$\sigma_r = (1-\Delta i/ri_0)^2 \tag{31}$$

is a stray light factor which in (28) reduces the observable diffusion effects due to the smearing of the pattern. For a ruling with equal open and opaque strips $r=\frac{1}{2}$ and σ_r varies from 1 to 0 when Δi goes from zero (perfect geometrical pattern) to 0.5 (complete smearing of the pattern). The normalized decay of delayed fluorescence is given by

$$\Phi^d(t) = \frac{1}{N(a)}\left\{e^{-2\beta t} + \sigma_r \sum_{l=1}^{\infty} A_l e^{-2\beta(1+l^2 a^2)t}\right\}. \tag{32}$$

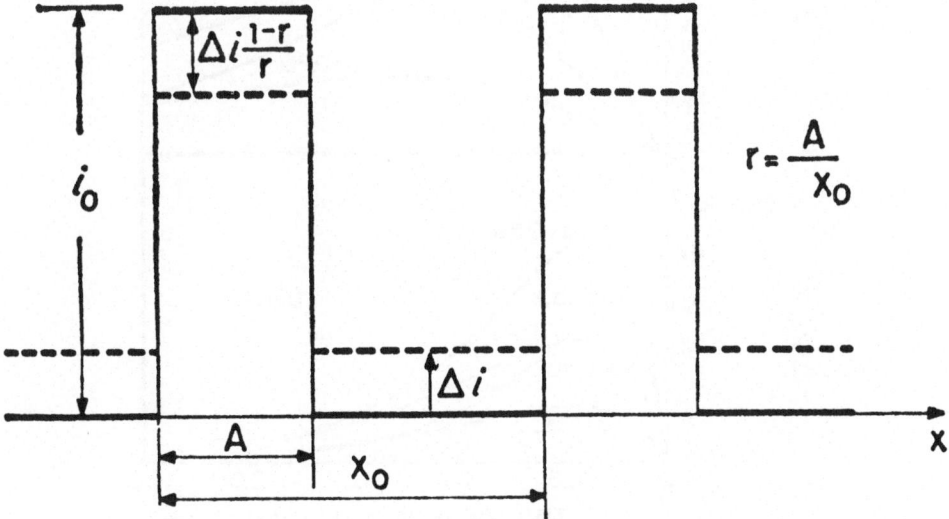

Fig. 7. Spatial distribution of the exciting light produced by a ruling of arbitrary window-to-period ratio r and in which a certain amount of stray light of constant intensity Δi is assumed to be present in the shadow regions (dashed curve) under conservation of the total photon flux. The pattern in solid lines is the ideal geometrical shadow of the ruling (From Ern, V., Avakian, P., and Merrifield, R. E.: ref. (33)).

The presence of some stray light in the shadow regions can only *reduce* the size of the observable time effects due to diffusion, that is, one can only infer apparently smaller diffusion lengths if the eventual presence of stray light is not taken properly into account. This is not the case with steady-state delayed fluorescence experiments [18]. The steady-state delayed fluorescence flux, which is proportional to the quantity $N(a)$ given by (30), varies from a normalized value of 1 in the absence of diffusion and stray light in the shadow regions to the value of $\frac{1}{2}$ in the presence of very strong diffusion ($L \gg x_0$) *or* of a complete smearing of the pattern by the presence of stray light in the shadow regions and which could be mistaken for a diffusion effect. Only time dependent experiments can properly distinguish the relative slow spreading of the excitons by diffusion into the shadow regions from their direct instantaneous generation there by stray light.

The predicted effects on the time-dependence of delayed fluorescence due to diffusion, that is, the contribution of the second term in (28) and (32) for $r = \frac{1}{2}$ and $\Delta i = 0$ are shown in Figure 8. The quantities $\Delta(t)$ plotted for different values of the

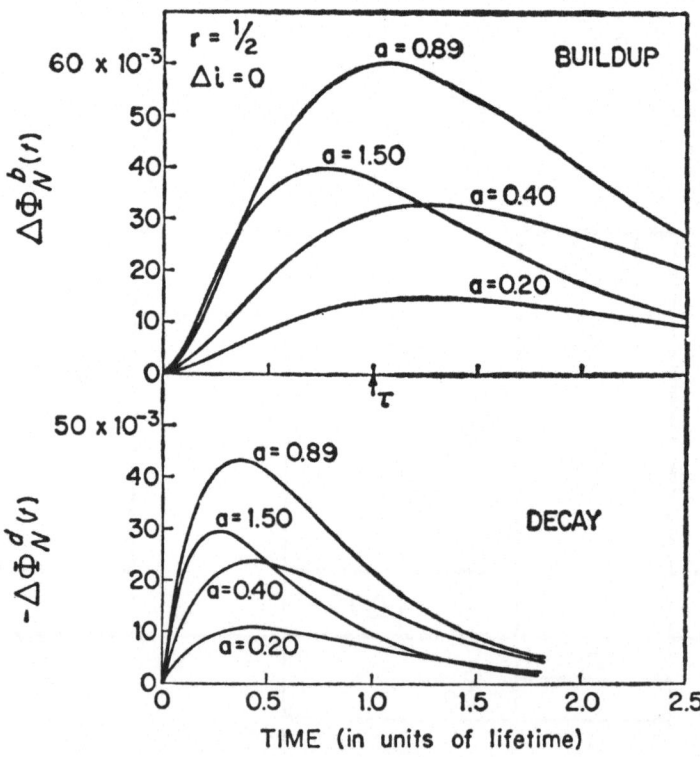

Fig. 8. Predicted differences (Equation (33) in text) in the normalized buildup and decay of delayed fluorescence due to the diffusion of triplet excitons when the crystal is excited through a Ronchi ruling of equal open and opaque strips ($r = \frac{1}{2}$) and assuming no scattering or diffraction of stray light into the shadow regions ($\Delta i = 0$). The parameter a is given by Equation (24) in text. (From Ern, V., Avakian, P., and Merrifield, R. E.: ref. (33)).

parameter a which compares the diffusion length with the ruling period are the differences of (28) and (32) and the corresponding normalized behaviors in time given by (7) and (8) under spatially uniform excitation or in absence of diffusion. That is,

$$\Delta\Phi^b(t) \equiv \Phi^b(t) - (1 - e^{-\beta t})^2$$
$$\Delta\Phi^d(t) \equiv \Phi^d(t) - e^{-2\beta t}$$

(33)

where $\Phi^b(t)$ and $\Phi^d(t)$ are given by (28) and (32), respectively. It is seen that the diffusion effects increase as the parameter a or the ratio L/x_0 increases and attain the maximum value for $a \simeq 0.9$ when the diffusion length is about $\frac{1}{4}$ or the ruling period, that is, about one half of the width of the dark region. The maximum effects occur at times in the vicinity of the triplet exciton lifetime τ (at $t \simeq 1.06\,\tau$ for the buildup and at $t \simeq 0.36\,\tau$ for the decay, respectively). As expected, when the parameter a increases further, that is, as the exciton diffusion length L becomes comparable to or larger than the ruling period, the observable effects begin to decrease, with the maximum of the effect occuring at increasingly shorter times, since now the excitons diffusing out of a given illuminated strip can reach during their lifetime the neighboring illuminated regions and the triplet exciton concentration becomes more rapidly uniformized over the whole crystal. The overall change in the uniformly excited delayed fluorescence waveform of Figure 3 due to diffusion effects when the crystal is excited with the periodic pattern of a ruling is shown in Figure 9. Both the buildup and the decay appear now to be faster since diffusion provides an additional sink term to the exciton concentration in the strongly excited regions.

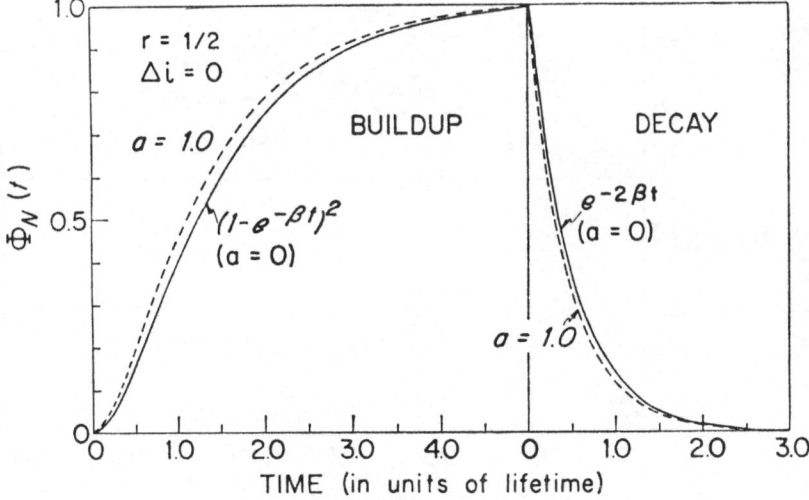

Fig. 9. Overall modification of the normalized temporal dependence of delayed fluorescence (dashed curve) predited by Equations (28) and (32) in text as compared with the case of uniform illumination when diffusion effects are negligible (solid curve). The example is for (Equation (24)) $a = 1$ and no stray light in the opaque strips. The ruling has equal open and opaque strips. (From Ern, V., Avakian, P.. and Merrifield, R. E.: ref. (33)).

In buildup the effects are bigger ($\sim 6\%$) and appear at longer times from the start of the process than in decay. It is advantageous therefore to use preferentially, whenever possible, the buildup portion of the waveform to detect the diffusion process. The buildup expression (28) can be rewritten in a more convenient form as the function [33].

$$F(t) = \frac{1}{N(a)} \left\{ e^{-\beta t} + \sigma_r \sum_{l=1}^{\infty} A_l e^{-\beta(1 + l^2 a^2)t} \right\} = \sum_{n=1}^{\infty} 2^{-n}\{1 - \Phi^b(2^n t)\} \quad (34)$$

which shows the contribution of the different exponential terms to the diffusion effect. Clearly, adequate signal-to-noise ratios and time-resolutions are necessary for experimentally extracting diffusion effects from the time dependence of the delayed fluorescence photon flux.

The agreement between the predicted effects of triplet diffusion (Equation (33)) and the experimental observations [33] for an anthracene crystal at room temperature are shown in Figures 10 and 11. The data were obtained by accumulating repetitively the delayed fluorescence waveforms in a computer of average transients (CAT) which was fed by an operational amplifier (Figure 6). The triplet lifetime was $\tau = 24$ ms. The sample plane was an ab plane and the ruling lines were parallel to a so as to

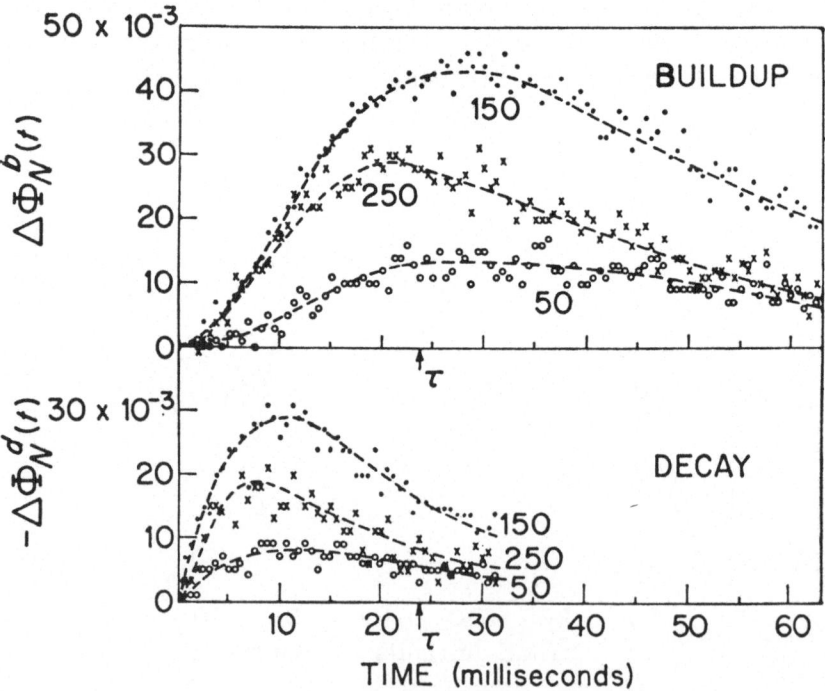

Fig. 10. Typical experimental results (points) for the changes in the time-dependence of delayed fluorescence due to exciton diffusion effects for an anthracene crystal at room temperature. The different $r = \frac{1}{2}$ rulings are labelled by their number of lines per inch. (From Ern, V., Avakian, P., and Merrifield, R. E.: ref. (33)).

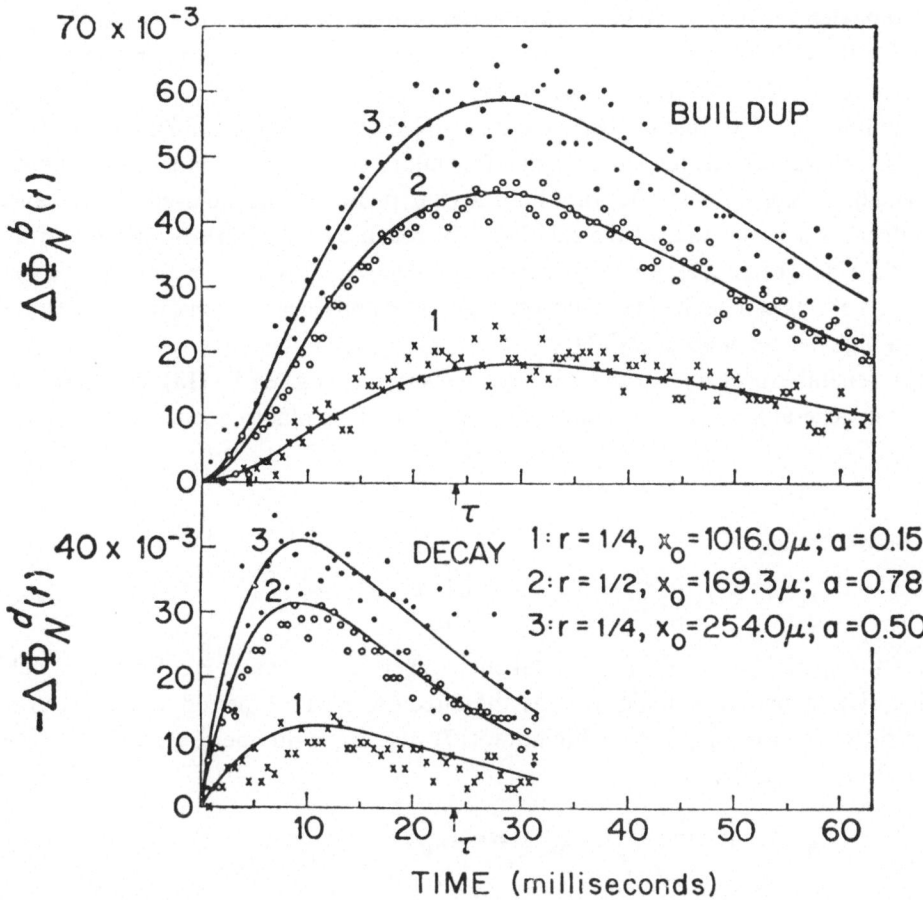

Fig. 11. Examples of the fits (curves) that are obtained, using Equation (33) in text, of the typical experimental results obtained for the differences in the time dependence of delayed fluorescence due to exciton diffusion along the b axis of an anthracene crystal at room temperature having a $\tau = 24$ ms triplet lifetime. The assumed intensity of stray light in the shadow regions is $\Delta i/i_0 = 8\%$. The inferred values of the parameter a (Equation (24) in text) are shown for each type of ruling. The deduced diffusion length is $L_b = 30 \pm 5 \, \mu$. (From Ern, V., Avakian, P., and Merrifield, R. E.: ref. (33)).

observe the exciton diffusion along the crystal b axis. All features of the expected time-dependence are observed in the results. The best fit is obtained by taking $\Delta i/i_0 = 0.08$, that is, 8% of the exciting light scattered and diffracted into the dark regions. The triplet exciton diffusion length deduced from the data is $L_b = 30 \pm 5 \mu$ which yields for the component of the exciton diffusion tensor along the b (principal) axis the value $D_{bb} = (2.0 \pm 0.5) \times 10^{-4}$ cm^2 s^{-1}. No alternative interpretation can be off-hand offered to explain the results of such direct experiments on triplet exciton motion, it being apparent that the macroscopic time-dependent diffusion equation is a correct description of triplet exciton migration in molecular solids at sufficiently high temperatures.

2. Experimental Investigation of the Triplet Diffusion Tensor

The time-dependent experiments described above and which provide a simple direct proof of the validity of the diffusion equation are cumbersome to perform and are not suitable for a complete investigation of the diffusion tensor of the crystal at different temperatures. It has further been realized [34] that analogous information on the changes of the time-dependence of delayed fluorescence can be more conveniently extracted by considering the variation of the phase of the delayed fluorescence waveform which lags that of the exciting light because of the intrinsic lifetime of the triplets. Changes in the time-dependence of delayed fluorescence due to exciton diffusion effects can be then related to changes in the phase lag which can be measured with a high degree of accuracy [34].

The general expression for the exciting intensity $i(x, t)$ given by (18), in which now $f(t)$ is in general a periodic excitation in time of fundamental frequency ω given by

$$f(t) = \sum_{l=-N}^{N} I_l \, e^{il\omega t}, \tag{35}$$

can be written as

$$i(x, t) = i_0 \left\{ \sum_{m=-\infty}^{\infty} S_m \, e^{im2\pi x/x_0} \right\} \left\{ \sum_{l=-N}^{N} I_l \, e^{il\omega t} \right\}, \tag{36}$$

where the spatial part $g(x)$ has been taken also periodic in space, e.g. for the Ronchi ruling, with a spatial periodicity along the direction x with a period x_0 and Fourier coefficients S_m. Inserting (36) in Equation (17) one can write the general form of the exciton concentration as

$$n(x, t) = \sum_{l=-N}^{N} \sum_{m=-\infty}^{\infty} n_{lm} \, e^{im2\pi x/x_0} \, e^{il\omega t} \tag{37}$$

where the coefficients n_{lm} are given by

$$n_{lm} = \alpha S_m T_l \, \tau \{1 + il\omega\tau + m^2 \pi^2 \chi^2\}^{-1} \tag{38}$$

where the parameter

$$\chi \equiv 2L_x/x_0 \tag{39}$$

scales the diffusion length in the direction x to the spatial period x_0. After squaring (37) and integrating over the spatial period one obtains from (2) that the delayed fluorescence waveform will be given by

$$\Phi(t) \propto \operatorname{Re} \sum_{n=0}^{\infty} H_n \, e^{in\omega t} \tag{40}$$

where the complex amplitudes of the different harmonics are given by the restricted sums

$$H_n = \sum_{l=-N}^{N} \sum_{l'=-N}^{N} A_{ll'} I_l I_{l'} \tag{41}$$

with $l + l' = n$. The coefficients $A_{ll'}$ depend solely on the particular spatial intensity distribution of the exciting light beam and can be in general written as the sum

$$A_{ll'} = \sum_{m = -\infty}^{\infty} S_{m,l} S_{-m,l'} \qquad (42)$$

in which

$$S_{m,l} \equiv \tfrac{1}{2} \alpha S_m \tau \left\{ \beta_l + m^2 \frac{\pi^2}{8} \chi^2 \right\}^{-1} \qquad (43)$$

where the complex, dimensionless, parameters,

$$\beta_l \equiv \tfrac{1}{2}(1 + il\omega\tau) \qquad (44)$$

scale the frequencies of the harmonics of the time-dependence to the exciton lifetime.

For a ruling with equal open and opaque strips ($r = \tfrac{1}{2}$) and an amount of stray light Δi in the shadow regions (Figure 7), and for which $S_0 = \tfrac{1}{2}$, $S_m = 0$ for m even, and $S_{\pm m} = \pm \sigma^{1/2}/m\pi$ for m odd, where $\sigma = (1 - 2\Delta i/i_0)^2$ is the stray light factor (Equation (31)), one has [34]

$$A_{ll'} \propto (\beta_l \beta_{l'})^{-1} (1 + \sigma) +$$
$$+ \chi\sigma(\beta_l - \beta_{l'})^{-1} \{\beta_l^{-3/2} \tanh(\beta_l^{1/2}/\chi) - \beta_{l'}^{-3/2} \tanh(\beta_{l'}^{1/2}/\chi)\} \qquad (45)$$

which when inserted in (42) allows the determination, for a given type of temporal excitation, of the complex amplitudes of the delayed fluorescence wave form and therefore the phase-lag for the nth harmonic

$$\phi(\chi) = \tan^{-1} (\operatorname{Im} H_n / \operatorname{Re} H_n), \qquad (46)$$

as a function of the parameter χ (39) which scales the measured diffusion length to the ruling period. To study the diffusion, the delayed fluorescence signal should be then Fourier analyzed and the phase lag of a given harmonic measured as a function of the ruling period x_0. In practice the fundamental harmonic is used since it usually provides the best signal-to-noise ratio.

Under a spatially uniform excitation or in absence of diffusion ($\chi = 0$) Equation (45) reduces to

$$A'_{ll'} = 1/\beta_l \beta_{l'} \qquad (47)$$

and the phase shift, for a given angular frequency ω, is solely a function of the lifetime of the triplet exciton in the crystal.

For the simple square wave modulation (Figure 3) provided by a chopper rotating at the angular frequency ω one has from (41) that the complex amplitude of the first harmonic of the delayed fluorescence wave form is

$$H_1 \propto A_{1,0}. \qquad (48)$$

For spatially uniform excitation one has

$$A'_{1,0} \propto (1 + i\omega\tau)^{-1} \qquad (49)$$

that is, the phase lag ϕ' is simply given by

$$\tan \phi' = -\omega\tau. \tag{50}$$

The *changes* in this phase lag

$$\Delta\phi_d \equiv \phi - \phi' \tag{51}$$

due to diffusion when rulings are present can be written then as

$$\Delta\phi_d = \arg\{A_{1,0}/(1 + i\omega\tau)\} \tag{52}$$

where $A_{1,0}$ as a function of χ is given by (45). Another convenient excitation [34] is that provided by a linear polarizer which is rotated at a mechanical frequency ω_0 and which is placed in the laser beam between two fixed polarizers set for maximum transmission [34]. This arrangement provides an intensity modulation $i = i_0 \cos^4 \omega_0 t$, that is, in (35)

$$f(t) = \sum_{l=-2}^{2} I_l e^{il\omega t} \tag{53}$$

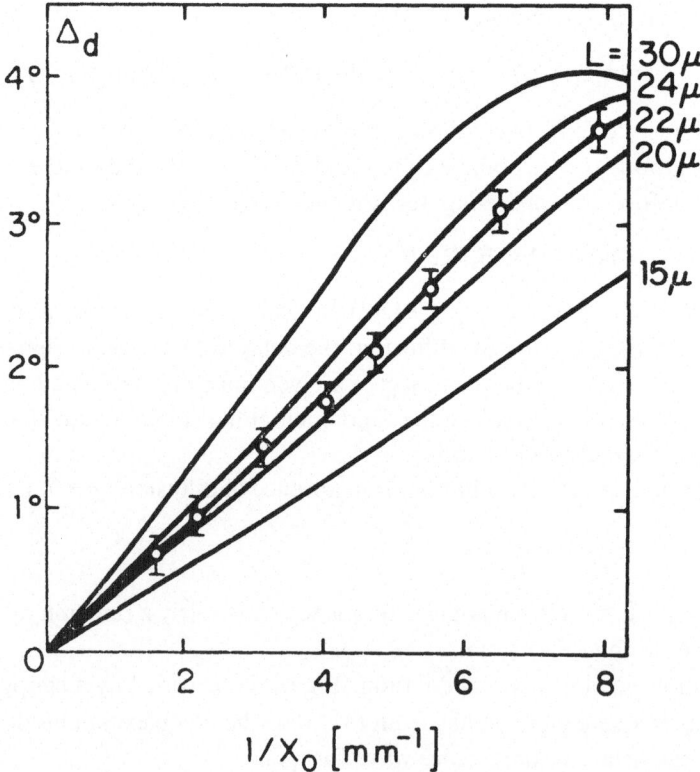

Fig. 12. Expected behavior of the differences in the phase of the first harmonic of delayed fluorescence due to exciton diffusion (Equation (55) in text), as a function of the spatial frequency x_0^{-1} of the rulings, and taking $\Delta i/i_0 = 3\%$. The experimental points are for an anthracene crystal at room temperature excited with a rotating polarizer and ruling lines perpendicular to the crystal b axis. The triplet exciton diffusion length is $L_b = 22.0 \pm 1.5\,\mu$. (From Ern, V.: ref. (34)).

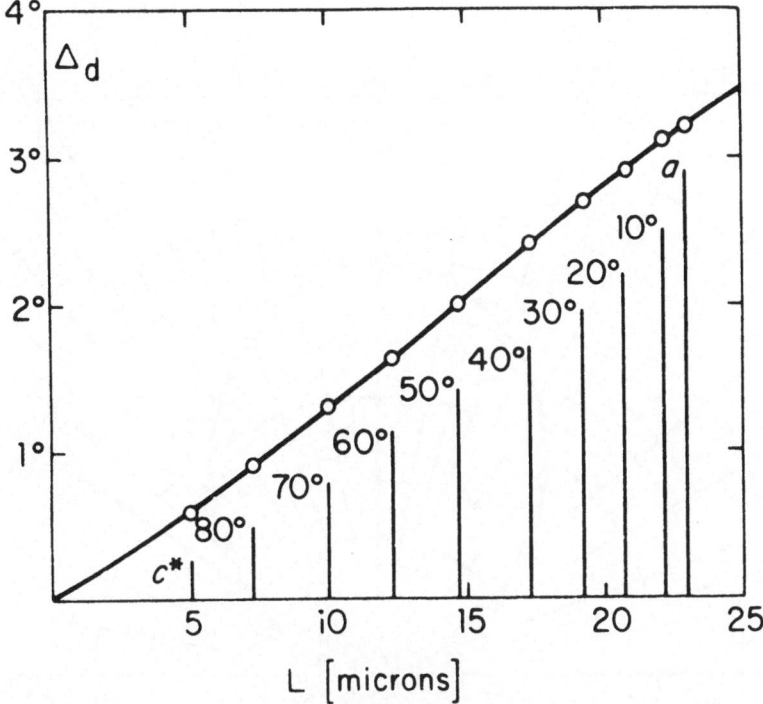

Fig. 13. Typical anisotropy in the changes of the phase of the first harmonic of delayed fluorescence (points) observed when the triplet exciton concentration gradient (e.g. the ruling lines) is rotated about the laser beam in the *ac* plane of an anthracene crystal at room temperature. The corresponding exciton diffusion lengths in this plane are shown on the abscissa, as inferred from Equations (55) and (45) in text, with $\Delta i/i_0 = 3\%$ of stray light. (From Ern, V.: ref. (34)).

with $\omega = 2\omega_0$, $I_0 = \frac{3}{8}$, $I_{\pm 1} = \frac{1}{4}$, $I_{\pm 2} = \frac{1}{16}$, which has the advantage of having only two harmonics and therefore of improving the signal-to-noise ratio in the detection of the first harmonic used in the measurements. For the excitation given by (53) one has

$$H_1 \propto 6A_{1,0} + A_{2,-1} \tag{54}$$

and instead of (52) the changes in phase due to diffusion are obtained as

$$\Delta\phi_d = \arg\{(6A_{1,0} + A_{2,-1})/(6(\beta_1\beta_0)^{-1} + (\beta_2\beta_{-1})^{-1})\}. \tag{55}$$

In all cases the diffusion effects cause a *decrease* in phase lag with respect to that observed under uniform illumination since as seen before the diffusion term in (17) leads to faster buildups and decays in the delayed fluorescence waveform.

The application of the method is illustrated for anthracene crystal in Figures 12 and 13. Figure 12 shows the expected behavior of the changes in the phase given by (55) for the fundamental harmonic of delayed fluorescence as a function of the spatial frequency of the rulings under a rotating polarizer excitation. The maximum effect

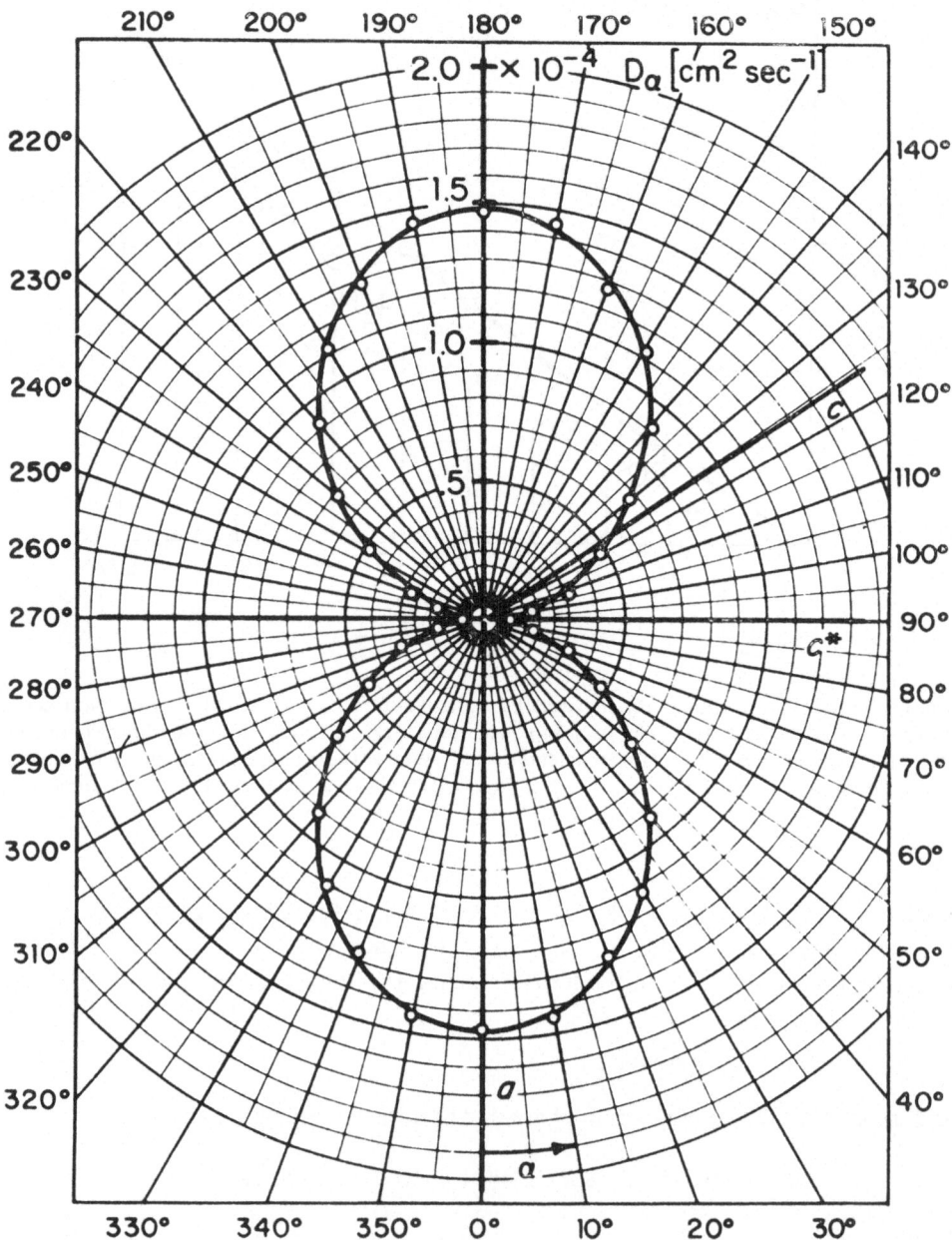

Fig. 14.   The measured anisotropy of the triplet exciton diffusion tensor in the *ac* plane of an anthra-cene crystal at room temperature. The *a* and *c** crystal axes are, within 5%, the principal axes of the tensor. (From Ern, V.: ref. (34)).

($\sim 4°$) is observed again when the diffusion length is approximatively $\frac{1}{4}$ of the ruling period. The points indicate the actually measured values [34]. The first harmonics of the delayed-fluorescence and of the excitation waveforms are extracted with a pair of phase-matched tracking filters which are tuned to the fundamental frequency and the phaselag is measured digitally with an uncertainty of $0.2°$. The phase-angle under spatially uniform excitation is $-53.1°$ corresponding to the triplet lifetime of $\tau = 16.9$ ms. The measurements shown yield a diffusion length of $L = 22.0 \pm 1.5\,\mu$. Figure 13 illustrates results of a measurement, in steps of $10°$, of the anisotropy of diffusion in the ac plane. As seen before, such measurements, in conjunction with the measurements of D_{bb}, should specify completely the principal components and the principal axes of the tensor and hence the anisotropy of triplet diffusion in a monoclinic crystal. As seen, diffusion is maximum along the a axis and minimum along the c^*, perpendicular to the ab cleavage plane, axis. These directions are thus, within the experimental uncertainty of $\pm 5°$, the principal axes of the tensor in the ac plane. The anisotropy in this plane is shown in Figure 14 where the experimental points are fitted to Equation (14),

$$D_\alpha = D_{aa} \cos^2 \alpha + D_{c*c*} \sin^2 \alpha \qquad (56)$$

in which the principal axes x_1 and x_3 of the tensor are taken to be the crystal a and c^* axes, respectively. It is seen that the exciton motion is strongly anisotropic in the ac plane and can be considered as nearly two-dimensional, restricted to the crystal ab plane. The value of $D_{Ic\,Ic} \leqslant (1.2 \times 0.5) \times 10^{-5}$ cm^2 s^{-1}, that is, at least 10 times smaller than the values in the ab plane, gives only an upper limit for triplet diffusion perpendicular to the ab plane since small misalignments and crystal imperfections can cause a projection of the other two strong tensor components on the direction taken as c^* and therefore contribute to D_{c*c*}. These measurements gave the first *direct experimental* proof that triplet exciton migration can be strongly anisotropic (e.g. two-dimensional, one-dimensional) in organic solids, which is, as will be seen later, in accord with theoretical predictions. This removed the puzzle posed by previous experiments which, using steady-state [35] and indirect techniques near the crystal surface [36, 37], led to reports of an apparently isotropic triplet exciton diffusion in anthracene crystals.

In the results described above a *stationary* spatially periodic excitation was used. It is clear that motion of the periodic pattern at a uniform velocity v, e.g. $g(x)$ replaced by $g(x-vt)$ in (18), in the direction of the exciton diffusion will affect the instantaneous exciton concentration distribution in the crystal and lead to a time-dependence, and hence to changes in the phase lag of delayed fluorescence due to diffusion, which is different from that obtained for a stationary spatially periodic excitation. The validity of the macroscopic time-dependent diffusion equation can thus be further tested by studying the changes of the time-dependence of delayed fluorescence not only as a function of the spatial period x_0 but also as a function of the velocity v of the pattern with respect to the stationary crystal. This introduces a new, externally controllable variable, in the problem. Such experiments have been recently reported [38]. The experimental setup is illustrated in Figure 15. The lines of the rulings are normal to the long dimension of the grating and of the direction of motion. The ruling is placed in a slider at 1 mm below the crystal and is driven at uniform speeds by a lead screw with switches (SW) providing a reversal of motion at each end of the travel for repetitive measurements of the phase lag of the first harmonic of delayed fluorescence at a given v. For the ruling of equal open and opaque strips of Figure 7 moving now along x with a velocity v, the formula giving the parameters $A_{ll'}$ needed in (41) and (46) is now modified from (45) to [38]

$$A_{ll'} \propto \tfrac{1}{8}(1+\sigma)\,(\beta_l \beta_{l'})^{-1} + \tfrac{1}{2}\chi\sigma\{Y_{ll'}(\mu) + Y_{l'l}(\mu) + Y_{ll'}(-\mu) + Y_{l'l}(-\mu)\}$$

$$(57)$$

where

$$Y_{ll'}(\pm\mu) = a_l^{-1}(a_l \pm \mu)^{-2}\,\frac{\tanh\{\tfrac{1}{2}\chi^{-1}(a_l \pm \mu)\}}{\beta_l - \beta_{l'} \pm \mu(a_l \pm \mu)}$$

$$(58)$$

in which

$$a_l \equiv (\mu^2 + 4\beta_l)^{1/2}$$

$$(59)$$

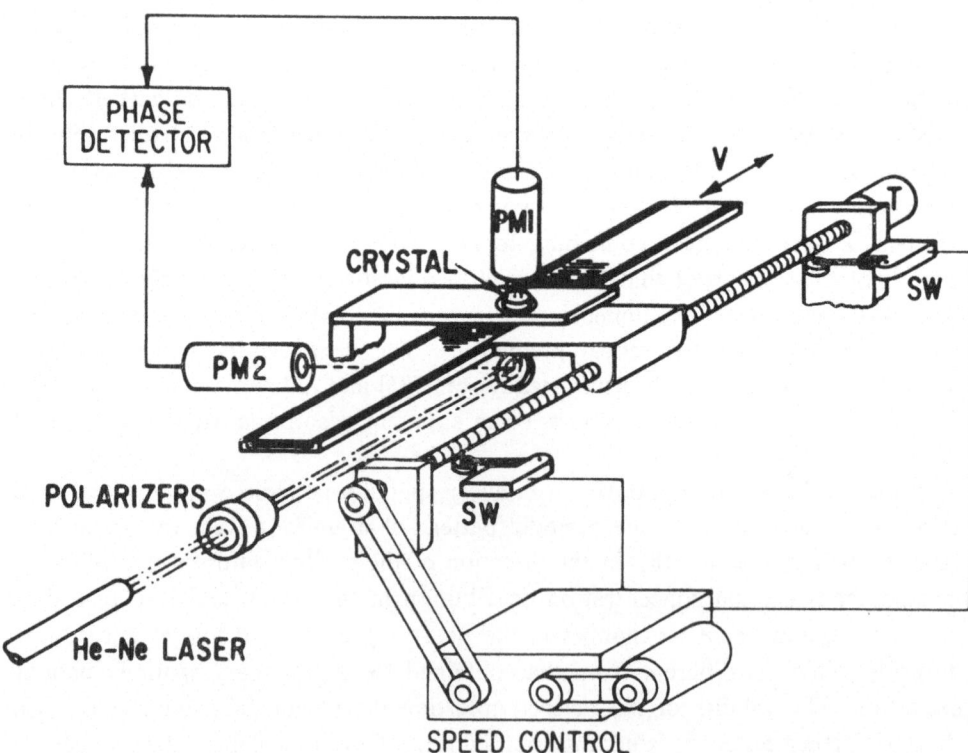

Fig. 15. Experimental setup to detect the changes in the phase of the first harmonic of delayed fluorescence as a function of the velocity of a moving pattern of light scanning the crystal. (From Ern, V., Suna, A., and Merrifield, R. E.: ref. (38)).

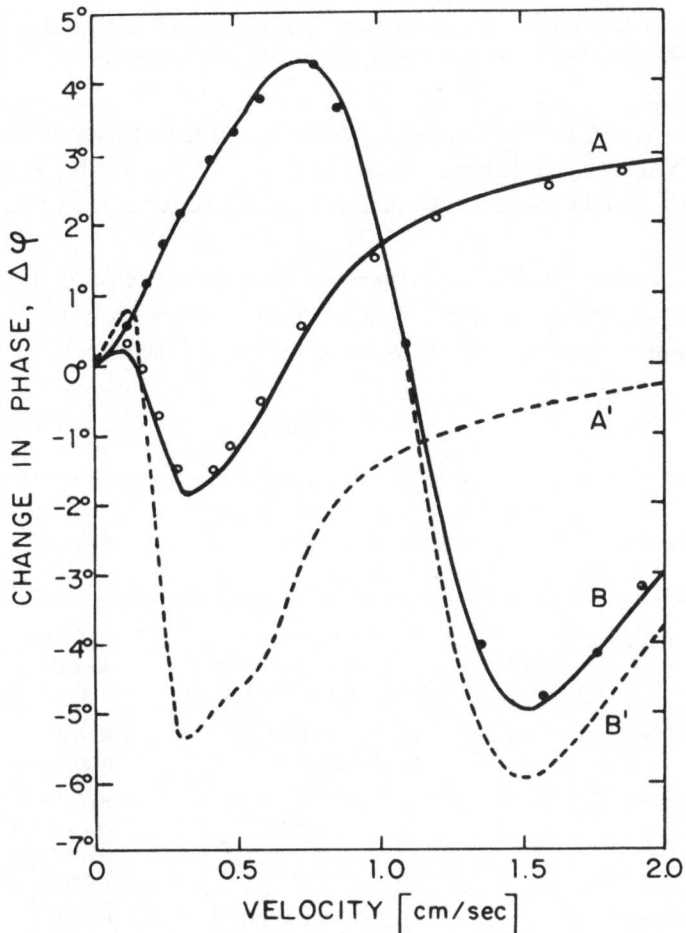

Fig. 16. Typical dependence of the changes in phase of delayed fluorescence as a function of the velocity of two periodic light intensity distributions produced by two rulings of spatial frequencies $x_0^{-1} = f_s = 59$ cm^{-1} (A) and $f_s = 20$ cm^{-1} (B). The measurements (points) are for an anthracene crystal of 20.2 ms triplet lifetime. The dashed curves are the expected effects if exciton diffusion were not operative. (From Ern, V., Suna, A., and Merrifield, R. E.: ref. (38)).

and the variable

$$\mu \equiv v\tau/L \qquad (50)$$

scales the velocity v of the pattern to the exciton average 'diffusion velocity' $v_D \equiv L/\tau$. All other quantities are defined as before. The quantity v_D is typically of the order of 10^{-2} cm s^{-1} in molecular crystals and delayed fluorescence will be insensitive to the motion of the pattern for velocities which are smaller than this typical value. Figure 16 illustrates the excellent agreement obtained between the experimental observations and the behavior predicted for the phase-lag by the exciton time-dependent diffusion equation as a function of the velocity of the moving pattern of light. It should be noted that because of the finite triplet lifetime delayed fluorescence is always sensitive to the motion of a spatially inhomogeneous pattern across the crystal, even in the absence of diffusion. Diffusion decreases the observable effects by smearing out the pattern. The highest sensitivity of the phase to velocity is obtained in the region where $v \approx f/f_s$ where f and f_s are the temporal and the spatial frequencies of the excitation, respectively. With delayed fluorescence one can thus detect directly the velocity of a moving pattern of light without resorting to separate measurements of distance and time [38].

The direct technique to obtain the triplet exciton diffusion tensor has been recently applied to several molecular solids. Table I shows the presently known results at room temperature. As will be seen below, the knowledge of the triplet exciton diffusion

TABLE I

Principal components of the triplet exciton diffusion tensor for several solids (300 K)

Material	Components of tensor (cm^2 s^{-1})	Remarks
Anthracene	$D_{aa} = 1.5 \pm 0.2 \times 10^{-4}$ $D_{bb} = 1.8 \pm 0.2 \times 10^{-4}$ $D_{c*c*} < 1.2 \pm 0.5 \times 10^{-5}$	Ref. 33 nearly two-dimensional
1,4-Dibromonaphtalene	$D_{cc} = 3.5 \pm 0.8 \times 10^{-4}$ $D_{aa}, D_{bb} < 1 \times 10^{-5}$	Ref. 39 nearly one-dimensional
Naphtalene	$D_{aa} = 3.3 \pm 0.4 \times 10^{-5}$ $D_{bb} = 2.7 \pm 0.4 \times 10^{-5}$	Ref. 39 D_{c*c*} not measured should be nearly 2-dim. as anthracene
Trans-stilbene	$D_{aa} = 0.9 \pm 0.2 \times 10^{-4}$ $D_{bb} = 0.7 \pm 0.2 \times 10^{-4}$	Ref. 39 D_{c*c*} not measured
Pyrene	$D_{aa} = 3.0 \pm 1 \times 10^{-5}$ $D_{bb} = 1.25 \pm 0.3 \times 10^{-4}$ $D_{c*c*} = 3.0 \pm 1 \times 10^{-5}$	Ref. 40 nearly isotropic, axes in *ac* plane not known

tensor can be used to provide, in conjunction with spectroscopic measurements [41, 42], a deeper insight into the exciton migration and the exciton-phonon scattering mechanisms, as well as to verify the existing first-principle band structure calculations. Knowledge of the tensor, or at least a partial knowledge of its anisotropy, is also essential for any theory of the triplet-triplet annihilation rate constant (Equation (1)) which takes the motion of the excitons into account [15]. The temperature dependence of triplet diffusion (118–371 °K) has been also recently obtained for anthracene [42]. Temperature dependent direct measurements are now in progress [43] for 1,4-dibromonapthalene, a one-dimensional system.

3. Theoretical Interpretation of the Triplet Exciton Diffusion Tensor

The validity of a macroscopic diffusion equation implies that on a macroscopic scale the exciton motion contains some degree of incoherence, that is, the exciton is scattered several times during its lifetime with loss of phase coherence. However, the motion could be still coherent on a microscopic scale if the wavepacket, not necessarily localized on a crystal site, travels many lattice spacings between scattering events; that is, if the mean free path $\lambda_s = v\tau_s$, where v is the exciton velocity and τ_s the scattering or relaxation time, is much bigger than the lattice spacing a. In general, from a measurement of the diffusion tensor alone one cannot discriminate [41] if on a *microscopic* scale the motion will be described in a band (coherent) model or in a completely incoherent or hopping model in which one can assume that the exciton passes from a site n in the crystal to a neighboring site n' by suffering many scattering events ($\lambda_s \ll a$). In the hopping model [44, 45] one can write the diffusion tensor as

$$D_{ij} = \tfrac{1}{2}\sum_n \{\psi(\mathbf{R}_n)\}\,(\mathbf{R}_n)_i\,(\mathbf{R}_n)_j \qquad (61)$$

where $\mathbf{R}_n$ is the lattice vector from a reference molecule, say 0, to the site n in the crystal. The quantities $\psi(\mathbf{R}_n)$ are the hopping rates between two molecules separated by the distance $\mathbf{R}_n$ and which are determined by the rate equation

$$\frac{\mathrm{d}P_n(t)}{\mathrm{d}t} = \sum_{n'} \psi(\mathbf{R}_{n'} - \mathbf{R}_n)\,P_{n'}(t) - P_n(t)\sum_{n'}\psi(\mathbf{R}_n - \mathbf{R}_{n'}) \qquad (62)$$

for the probability $P_n(t)$ of finding an excitation at site n at the time t.

A formalism which leads to a formula for the diffusion tensor containing both the coherent and incoherent contributions has been recently developed by Haken and his group [19–23, 42]. (See P. Reineker's contribution in the present volume.) In this theory the exciton Hamiltonian is assumed to consist of the sum of an average time independent part and a time varying part $h(t)$ which is due to the statistical fluctuations in the lattice spacings and the mutual orientations of the molecules in the lattice due to the phonons. This fluctuating term in the Hamiltonian is treated phenomenologically by assuming a specific form for its correlation functions, namely

$$\langle h_{nm}(t'+t)\,h_{n'm'}(t')\rangle = 2\gamma_{m-n}\,\delta(t)\,|\delta_{nn'}\delta_{mm'} + \delta_{nm'}\delta_{mn'}\,(1-\delta_{nm})| \qquad (63)$$

where $h_{nm}(t)$ are the fluctuating hamiltonian matrix elements, the brackets denote an average over t', $\delta(t)$ is the delta-function and γ_{m-n} are phenomenological parameters, which will measure the strength of the correlation in the fluctuations of the transfer matrix elements $\beta_{mn}\,(m \neq n)$ between molecules at sites m and n, and are functions of the distance $\mathbf{R}_m\,\mathbf{R}_n$ between the sites. The scattering of excitons by these fluctuations is usually termed non-local scattering in contrast to the local scattering which is due to the fluctuations of the exciton energy ε_0 at a given site, the correlation function of which will be specified by the parameter γ_0, that is when $m = n$ in (63). The fact that the correlation function of the fluctuations is given by a delta-function in (63) implies the assumption that all frequencies are present in the phonon spectrum (white noise).

It can be also shown [21, 46, 47, 42] that assumption (63) implies a Lorentzian optical absorption spectrum for the exciton line

$$\alpha(\omega) = \frac{\alpha_0}{\omega^2 + \Gamma^2} \tag{64}$$

with a half-width at half-maximum in units of energy given by

$$\Gamma = \sum_{n=0}^{N} \gamma_n. \tag{65}$$

For centrosymmetric crystals where the inversion is a symmetry operation of the site, the exciton diffusion tensor can now be expressed as [42, 48]

$$D_{ij} = \hbar^{-1} \sum_n \left(\frac{1}{2} \frac{\beta_n^2}{\Gamma + \gamma_n} + \gamma_n \right) (\mathbf{R}_n)_i \, (\mathbf{R}_n)_j \tag{66}$$

where the sum extends over all lattice vectors $\mathbf{R}_n$, $\gamma_n (n \neq 0)$ being the strength of the correlation in the fluctuations of the transfer matrix element β_n between the reference molecule situated at the origin ($\mathbf{R}_0 = 0$) and the nth molecule situated at the distance $\mathbf{R}_n$ from it. $(\mathbf{R}_n)_i$, $(\mathbf{R}_n)_j$ denote the projections of this vector on the directions i, j along which the component of the tensor is being calculated. In practice, the correlation, in the fluctuations of the matrix elements (which are small by themselves) are much smaller than the correlation in the fluctuations in energy at the site and one has $\gamma_{n \neq 0} \ll \gamma_0$. Equation (65) can be written now

$$\Gamma = \gamma_0 \tag{67}$$

that is, the width of the Lorentzian absorption line is a measure of the local scattering and the expression for the diffusion tensor can be written as

$$D_{ij} = \hbar^{-1} \sum_n \left(\frac{1}{2} \frac{\beta_n^2}{\Gamma} + \gamma_n \right) (\mathbf{R}_n)_i \, (\mathbf{R}_n)_j. \tag{68}$$

Now, it is clear that since Γ can be obtained from a spectroscopic measurement (if the experimental line is to a good approximation a Lorentzian) and since the β_n's are either known from a band structure calculation or in favorable cases can also be obtained from spectroscopic data (e.g. Davydov splittings), Equation (68) provides a means of obtaining the γ_n's if the components of the diffusion tensor are known.

In terms of the hopping model (Equation (61)) one can visualize (68) as predicting an 'effective' hopping rate

$$\psi_{\mathrm{eff}}(\mathbf{R}_n) = \hbar^{-1} \left(\frac{\beta_n^2}{\Gamma} + \gamma_n \right) \tag{69}$$

which measures the probability that the exciton will hop the distance $\mathbf{R}_n$ in the lattice. However, strictly speaking formula (68) cannot be considered as applicable to the hopping model of Trlifaj [45] in which the jump rate for *any* shape of the absorption and emission lines $\alpha(\omega)$ and $\alpha'(\omega)$ is given by

$$\psi(\mathbf{R}) = h^{-1}\beta_n^2\, 2\pi \int_{-\infty}^{\infty} \alpha(\omega)\, \alpha'(\omega)\, d\omega \tag{70}$$

which for Lorentzian line shapes of half-width Γ, as required by Haken's *et al.* model, will give

$$\psi(\mathbf{R}_n) = h^{-1}\frac{\beta_n^2}{\Gamma} \tag{71}$$

and hence from (61)

$$D_{ij} = \tfrac{1}{2}h^{-1} \sum_n \frac{\beta_n^2}{\Gamma}\, (\mathbf{R}_n)_i\, (\mathbf{R}_n)_j. \tag{72}$$

That is, only the first term in (68) is predicted and the non-local scattering is neglected. It should be noted that Equation (72) can also be written as [41]

$$D_{ij} = \tau \langle v_i v_j \rangle = \tau h^{-2} \sum_k \frac{\partial \varepsilon}{\partial k_i}\, \frac{\partial \varepsilon}{\partial k_j}. \tag{73}$$

Where τ is a constant relaxation time and $\langle\ \rangle$ denotes an average over the exciton band $\varepsilon(k) = \sum_n \beta_n\, e^{i\mathbf{kR}}\, R_n$ in which all $\mathbf{k}$ states are supposed to be evenly accessible ($k_B T \gg$ bandwidth) and where v_i is the group velocity along k_i given by (10). Formula (73) can be considered as a band (or plane wave description) [41, 49] of the diffusion tensor in the constant (independent of $\mathbf{k}$) relaxation time approximation. As has been recognized in the first paper [41] which used a spectroscopic approach to exciton dynamics, τ can be regarded as the lifetime of the optically prepared $\mathbf{k} = 0$ exciton state and its value should be approximately given, by the uncertainly principle, as $\tau = \hbar/2\Gamma$ where Γ is the half-width of Lorentzian absorption line expressed in units of energy. This brings (73) in agreement with Equation (72). In this sense the first term of the more general Equation (68) can be thought as 'derivable' from a band picture, that is, it bears a relation to the coherent part of the motion, while the second term $\sum_n \gamma_n(\mathbf{R}_n)_i\,(\mathbf{R}_n)_j$ is the 'completely' incoherent term, which by (61) has a jump rate $\psi(\mathbf{R}_n) = 2\gamma_n$ [50]. At high temperatures this term should dominate the first one since the optical line width Γ is great with respect to the exciton bandwidth ($\Gamma \gg \beta_n$) and the motion is highly incoherent. As the temperature lowers, Γ decreases and the degree of coherence increases in the motion. Grover and Silbey [24] have presented a theory of the temperature dependence of the diffusion constant using a exciton-phonon coupling scheme in which, however, the contribution of the local scattering, that is, of γ_0 has been neglected.

Formula (68) has been recently applied to analyze triplet exciton migration using data from spectroscopic and direct diffusion measurements in anthracene crystals. As is well known for triplet excitons, only the transfer matrix elements between the nearest-neighbors are important [48, 32]. This simplifies considerably the exciton band structure and the application of Equation (68). For anthracene, the **k**-dependent part of the exciton band structure for the first triplet can be written as [48, 32]

$$\varepsilon^{\pm}(k) = 2\beta_b \cos(\mathbf{k}\cdot\mathbf{b}) \pm 2\beta_d\{\cos\{\mathbf{k}\cdot\tfrac{1}{2}(\mathbf{a}+\mathbf{b})\} + \cos\{\mathbf{k}\cdot\tfrac{1}{2}(\mathbf{a}-\mathbf{b})\}\} \qquad (74)$$

where β_b is the transfer matrix element between the reference molecule and the molecule at **b** and β_d the one with the molecule at $\tfrac{1}{2}(\mathbf{a}\pm\mathbf{b})$. The $\pm$ signs define the two Davydov branches. The Davydov splitting at $\mathbf{k}=0$ is given by

$$\varDelta = 8\,|\beta_d| \qquad (75)$$

and is a directly measurable quantity, from the separation of the two absorption components in polarizations with the electric vector of light $\|b$ and $\perp b$, respectively [51, 32]. Using (68) the triplet exciton diffusion tensor can now be written to a good approximation as [42]

$$D_{aa} = \hbar^{-1}\{\tfrac{1}{2}\beta_d^2/\Gamma + \gamma_d\}\,a^2 \qquad (76)$$

$$D_{bb} = \hbar^{-1}\{(\beta_b + \tfrac{1}{2}\beta_b)/\Gamma + 2\gamma_b + \gamma_d\}\,b^2 \qquad (77)$$

$$D_{c*c*} = \hbar^{-1}\{2\beta_{d+c}^2/\Gamma + 4\gamma_{d+c}\}\,c^{*2} \qquad (78)$$

where a, b, c^* are the lattice constants and β_{d+c} is the transfer matrix element between the reference molecule and the molecule at $\{\tfrac{1}{2}(\mathbf{a}+\mathbf{b})+\mathbf{c}\}$, and which has been neglected in (74). From (75) one obtains for (76)

$$D_{aa} = \hbar^{-1}\left\{\frac{\varDelta^2 a^2}{128\,\Gamma} + \gamma_d a^2\right\} \qquad (79)$$

from which an unambiguous determination of the non-local fluctuation rate γ_d can be made using only experimental data. Figure 17 illustrates the observed Davydov splittings and line shapes in the first $(0, 0)$ vibronic line of the polarized ($E\|a$, full curves; $E\|b$, dashed curves) singlet-triplet absorption spectra for delayed fluorescence in anthracene, from which the Davydov splitting $\varDelta$ and an effective half-line width can be obtained [42]. At low temperatures the spectra deviates markedly from a Lorentzian line shape and an effective Γ_{eff} is computed as required by Trlifaj's formula (Equation (70))

$$\Gamma_{\text{eff}}^{-1} = 2\pi \int\limits_{0,\,0\text{ peak}} f(\omega)\,F(\omega)\,\mathrm{d}\omega \qquad (80)$$

where $f(\omega)$ is the function giving the absorption spectrum for one of the components and where $F(\omega)$ is the emission spectrum obtained as $F(\omega) = f(\omega)\,e^{-\hbar\omega/k_B T}$. Equation (80) gives the Γ required by the theory if the line shapes are Lorentzian and there is no Stokes-shift between the absorption and the phosphorescence emission, which is known to be the case at high temperatures [42]. The gross features of the results of reference (42) are summarized in Table II. In the table the Davydov splitting has been taken to be $\varDelta = 18$ cm^{-1} and constant in the temperature range studied. Γ_{eff} is the effective line width taken from the stronger a polarized spectra. The deduced value

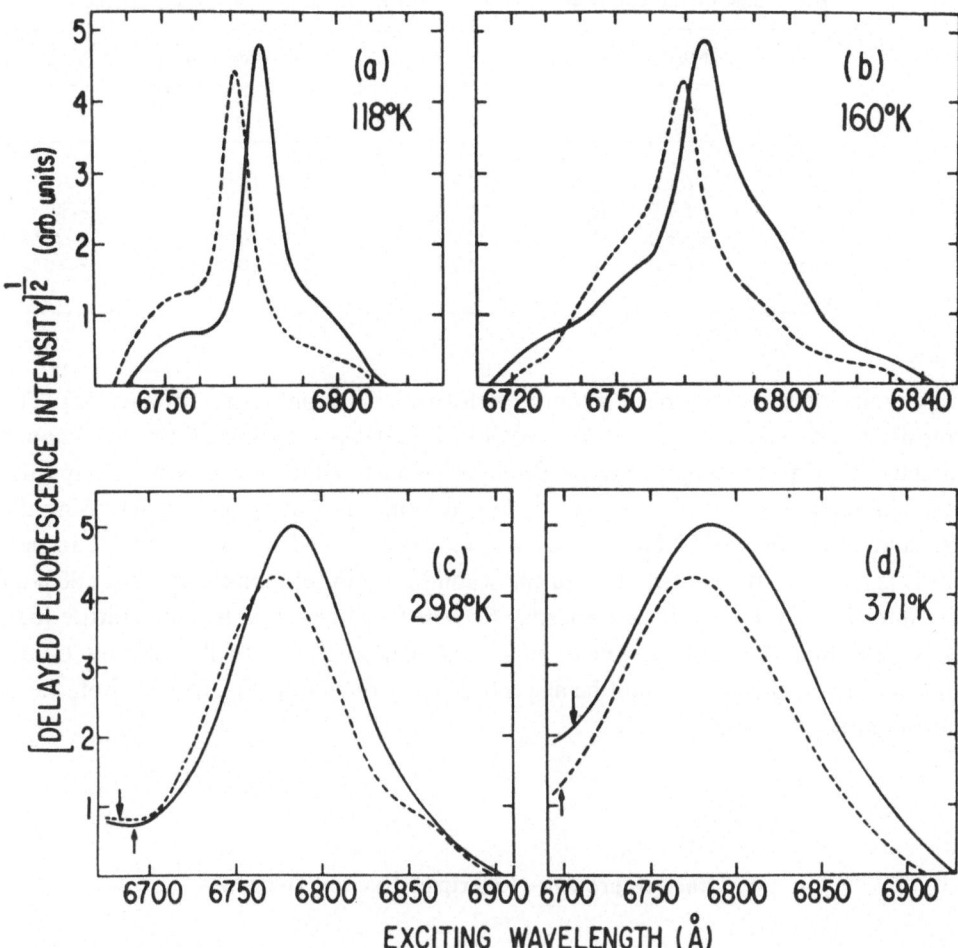

Fig. 17.   Polarized excitation spectra of delayed fluorescence obtained by absorption in the first (0,0) triplet band in anthracene crystals at different temperatures. Note the marked deviation from the Lorentzian line shape in (a) and (b). The vertical scale is arbitrary, and the wavelength scale in (a) and (b) is different from that in (c) and (d). The arrows in (c) and (d) indicate the cutoffs used in calculating Γ_{eff}^{-1} with Equation (80) in text. The full and dashed curves correspond to the a and b polarizations, respectively. (From Ern, V., Suna, A., Tomkiewicz, Y., Avakian, P., and Groff, R. P.: Ref. (42)).

for the correlation strength $\gamma_d \approx 0.1$ cm^{-1} appears, within the accuracy of the experiments, to be independent of the temperature. On the other hand, Γ, that is γ_0, is strongly temperature-dependent, decreasing nearly linearly with T as the temperature is lowered in the range studied. It is seen from Table II that at high temperatures the non-local scattering term $\hbar^{-1}\gamma_d a^2$ in (68) makes up about 60% of the diffusion tensor. The contribution diminishes to $\sim 40\%$ when the temperature is lowered to about 100 K.

TABLE II

Percentage of the contribution of first term in Equation (79) in anthracene crystals

$T(K)$	Measured values		Deduced values	
	D_{aa} cm^2 s^{-1}	Γ_{eff} cm^{-1}	$a^2 \Delta^2/128 \Gamma \hbar$ cm^2 s^{-1}	% of D_{aa}
371	$1.6 \pm 0.3 \times 10^{-4}$	65 ± 2	0.6×10^{-4}	$\sim 37\%$
298	$1.5 \pm 0.2 \times 10^{-4}$	51 ± 1	0.6×10^{-4}	$\sim 40\%$
160	$2.5 \pm 0.3 \times 10^{-4}$	30 ± 2	1.1×10^{-4}	$\sim 45\%$
118	$4.0 \pm 0.5 \times 10^{-4}$	14 ± 1	2.5×10^{-4}	$\sim 60\%$

A similar analysis has been recently performed on naphthalene crystals [52]. The transfer matrix elements β_d and β_b needed in Equations (76) and (77) were obtained directly from spectroscopic data on d_8-naphthalene crystals doped with h_8-naphthalene. At 300 K one gets $\beta_d = 1.2$ cm^{-1}, $\beta_b = 0.6$ cm^{-1}, $\Gamma = 65$ cm^{-1}, which using the D_{aa} and D_{bb} values of Table I yields $\gamma_d = 0.015$ cm^{-1} and $\gamma_b = 0.006$ cm^{-1}, respectively [52]. Again the contribution of the second term in (68) makes up about 60% of the diffusion tensor. As in anthracene, Γ decreases linearly with temperature [52]. It is clear that the contribution of non-local scattering due to fluctuations of the transfer matrix elements cannot be neglected in any theory dealing with exciton transport in molecular solids.

4. Other Approaches to Triplet Exciton Dynamics

Triplet excitons have a spin whose relaxation and magnetic dipolar interaction with the protons of the molecules forming a crystal must be sensitive to the exciton migration through the lattice. Hence, ESR and NMR techniques can, in principle, provide also a tool to investigate exciton dynamics. Studies of these type have been recently performed by Wolf and his group on anthracene, naphthalene and 1,4 dibromonaphthalene crystals [53, 54, 55, 56] and by Harris and coworkers [57] using PMDR techniques on other materials. The basic quantity which can be derived from the homogeneous line width of ESR resonance lines is the exciton correlation or residence time τ_c which, in turn, can be related to the exciton hopping rate (69) and hence to the parameters defining the exciton motion [58]. In NMR experiments of an excited crystal a shortening of the longitudinal proton relaxation time T_1 due to exciton migration is observed which in turn can be related to the dimensionality and kind of exciton motion [48]. In all the cases studied up to now a complete agreement with the results of the direct measurements has been obtained. Some aspects of these approaches to exciton dynamics are given in other papers of this Symposium.

5. Concluding Remarks

The methods presented here constitute a powerful tool for the study of the dynamics of excitons, at least of triplets, in molecular solids. Exciton diffusion tensor measurements in conjunction with optical or magnetic spectroscopy can yield sufficient information to test the validity of microscopic phenomenological models of exciton motion. It is hoped that a systematic application of these techniques in a wide temperature range on one-, two- and three-dimensional systems will allow in the near future testing in a consistent manner the theories of the exciton-phonon coupling mechanism and the exciton band structure calculations in the organic solid state.

References

1. Brandon, R. W., Gerkin, R. E., and Hutchison, C. A. Jr.: *J. Chem. Phys.* **37**, 447 (1962).
2. Northrop, D. C. and Simpson, O.: *Proc. Roy. Soc. (London)* **A234**, 124 (1956).
3. Powell, R. C.: *J. Chem. Phys.* **58**, 920 (1973) and references therein.
4. El-Sayed, M. A., Wauk, M. T., and Robinson, G. W.: *Mol. Phys.* **5**, 205 (1962).
5. Colson, S. D. and Robinson, G. W.: *J. Chem. Phys.* **48**, 2550 (1968).
6. Lower, S. K. and El-Sayed, M. A.: *Chem. Rev.* **66**, 199 (1966).
7. Wolf, H. C.: in *Advances in Atomic and Molecular Physics* (ed. by D. R. Bates and I. Estermann), vol. 3, p. 119, Academic Press Inc., New York (1967).
8. Birks, J. B.: *Photophysics of Aromatic Molecules*, Wiley-Interscience, London (1970).
9. Northrop, D. C. and Simpson, O.: *Proc. Roy. Soc. (London)* **A244**, 377 (1958).
10. Sponer, H., Kanda, Y., and Blackwell, L. A.: *J. Chem. Phys.* **29**, 721 (1958).
11. Blake, N. W. and McClure, D. S.: *J. Chem. Phys.* **29**, 722 (1958).
12. Nieman, G. C. and Robinson, G. W.: *J. Chem. Phys.* **37**, 2150 (1962).
13. Sternlicht, H., Nieman, G. C., and Robinson, G. W.: *J. Chem. Phys.* **39**, 1127 (1963).
14. Kepler, R. G., Caris, J. C., Avakian, P., and Abramson, E.: *Phys. Rev. Letters* **10**, 400 (1963). Hall, J. L., Jennings, D. A., and McClintock, B. M.: *Phys. Rev. Letters*, **11**, 364 (1963).
15. Suna, A.: *Phys. Rev.* **B1**, 1716 (1970).
16. Avakian, P. and Merrifield, R. E.: *Mol. Crystals* **5**, 37 (1968), a review.
17. See for instance, Ern, V., Bouchriha, H., Bisceglia, M., Arnold, S., and Schott, M.: *Phys. Rev.* **B8**, 6038 (1973).
18. Avakian, P., and Merrifield, R. E.: *Phys. Rev. Letters* **13**, 541 (1964).
19. Haken, H. and Strobl, G.: in *The Triplet State* (1967) pp. 311–314 and in *Excitons, Magnons and Phonons in Molecular Crystals* (1968) pp. 185–194 both edited by A. Zahlan, Cambridge U.P., Cambridge, England.
20. Haken, H. and Strobl, G.: *Z. Physik* **262**, 135 (1973).
21. Haken, H. and Reineker, P.: *Z. Physik* **249**, 253 (1972).
22. Schwarzer, E. and Haken, H.: *Phys. Letters* **42A**, 317 (1972).
23. Reineker, P.: *Phys. Letters* **42A**, 289 (1973).
24. Grover, M. K. and Silbey, R.: *J. Chem. Phys.* **54**, 4843 (1971).
25. Munn, R. W.: *J. Chem. Phys.* **58**, 3230 (1973).
26. Rackovsky, S. and Silbey, R.: *Mol. Phys.* **25**, 61 (1973); *ibid.* **26**, 857 (1973).
27. Pearlstein, R. M., Lindenberg, K., and Hemenger, R. P.: in *Excited States of Biological Molecules*, J. B. Birks editor, Wiley, N.Y., in press.
28. Kenkre, V. M. and Knox, R. S.: *Phys. Rev.* **B9**, 5279 (1974).
29. Kenkre, V. M.: *Phys. Letters* **47A**, 119 (1974).
30. Munn, R. W.: *Chem. Phys.* **6**, 469 (1974).
31. Merrifield, R. E.: *J. Chem. Phys.* **28**, 647 (1958).
32. Tiberghien, A. and Delacote, G.: *J. Phys. (Paris)* **31**, 637 (1970).
33. Ern, V., Avakian, P., and Merrifield, R. E.: *Phys. Rev.* **148**, 862 (1966).

34. Ern, V.: *Phys. Rev. Letters* **22**, 343 (1969).
35. Levine, M., Jortner, J., and Szöke, A.: *J. Chem. Phys.* **45**, 1591 (1966),
36. Kepler, R. G. and Switendick, A. C.: *Phys. Rev. Letters* **15**, 56 (1965).
37. Williams, D. F., Adolph, J., and Schneider, W. G.: *J. Chem. Phys.* **45**, 575 (1966);
    Williams, D. F. and Adolph, J.: *ibid.* **46**, 4252 (1967).
    These experiments have been later reinterpreted by Hoesterey D. C. and Robinson, G. W.:
    J. Chem. Phys. **54**, 1369 (1971), who brought them into agreement with the result for $D_{c^*c^*}$ in Ref.
    (34).
38. Ern, V., Suna, A., and Merrifield, R. E.: *J. Appl. Phys.* **42**, 2770 (1971).
39. Ern, V.: *J. Chem. Phys.* **56**, 6259 (1972).
40. Arnold, S., Fave, J. L., and Schott, M.: *Chem. Phys. Letters* **28**, 412 (1974).
41. Avakian, P., Ern, V., Merrifield, R. E., and Suna, A.: *Phys. Rev.* **165**, 974 (1968).
42. Ern, V., Suna, A., Tomkiewicz, Y., Avakian, P., and Groff, R. P.: *Phys. Rev.* **B5**, 3222 (1972).
43. Fave, J. L.: of this laboratory.
44. Dexter, D. L.: *J. Chem. Phys.* **21**, 836 (1953),
45. Trlifaj, M.: *Czech. J. Phys.* **6**, 533 (1956).
46. Schwarzer, E. and Haken, H.: *Optics Communications* **9**, 64 (1973).
47. Aslangul, C. and Kottis, Ph.: *Phys. Rev.* **B10**, 4364 (1974).
48. Haken, H. and Schwarzer, E.: *Chem. Phys. Letters* **27**, 41 (1974).
49. Jortner, J. Rice, S. A., Katz, J. L., and Choi, S. I.: *J. Chem. Phys.* **42**, 309 (1965).
50. Schwarzer, E.: Thesis Dissertation, University of Stuttgart, Germany (1974).
51. Hochstrasser, R. M.: *J. Chem. Phys.* **47**, 1015 (1967).
52. Port, H.: Thesis Dissertation, University of Stuttgart, Germany (1974).
53. Schwoerer, M. and Wolf, H. C.: *Mol. Cryst.* **3**, 177 (1967).
54. Haarer, D. and Wolf, H. C.: *Mol. Cryst.* **10**, 359 (1970).
55. Kolb, H. and Wolf, H. C.: *Z. Naturfursch.* **27a**, 51 (1972).
56. Schmidberger, R. and Wolf, H. C.: *Chem. Phys. Letters* **16**, 402 (1972); *ibid.* **25**, 185 (1974).
57. See, for instance, Harris, C. B.: *Pure Appl. Chemistry* **37**, No. 1, 2, p. 73 (1974).
58. Lemaistre, J. P. and Kottis, Ph.: in *Electron Spin Resonance in Liquids* (ed. by Muus, L. T. and
    Atkins), P. A. Plenum, N.Y., 1972, p. 455.
59. Reineker, P. and Haken, H.: *Z. Physik* **250**, 300 (1972).

THE COUPLED COHERENT AND INCOHERENT MOTION
OF FRENKEL EXCITONS IN MOLECULAR CRYSTALS

P. REINEKER

Abteilung für Theoretische Physik I, Universität Ulm, Germany

and

H. HAKEN

Institut für Theoretische Physik, Universität Stuttgart, Germany

1. Introduction

As has been well known for many years, excitons play an important role in energy transport in solids, especially in connection with fluorescence, phosphorescence and sensitized luminescence phenomena [1 to 7]. From the detailed study of excitonic energy transfer we may obtain insight into the energy transport in macromolecules. Furthermore, it seems now to be fairly sure that migration of excitons is one possible way of the energy transfer in biological systems.

For the experimental investigation of excitons in molecular crystals a variety of methods are available, such as diffusion measurements [8 to 10], optical absorption [8 to 12], electron spin resonance [13 to 17], and nuclear spin resonance [18, 19]. While it seems to be quite obvious that the results of diffusion measurements depend on the way in which the exciton migration occurs, the connection between exciton motion and optical absorption [20 to 26], ESR [20, 22, 24, 27 to 31], and NSR [20, 22, 25 to 27] is not so clear and theoretical considerations are needed.

For the description of exciton motion two different models have been proposed, namely that of coherent [32] and that of incoherent [8, 33] exciton motion. In the coherent case the exciton is assumed to move through the lattice in complete analogy to a Bloch wave which is eventually scattered by phonons. In the incoherent case it is assumed that the phase of the wave function of the exciton is very quickly destroyed by the interaction with the vibrations in the crystal so that the exciton undergoes a hopping process. In mathematical terms this process may be described by a master equation in its narrow sense or in the limit of a continuum by a diffusion equation.

Some time ago, Haken and Strobl [34, 35] developed a stochastic model which allows to treating not only both limiting cases but also the whole range in between. The coherent contribution to the energy transport is described by the transition matrix element between the molecules of the crystal containing the Coulomb and the exchange interaction integral. The phonons destroy the coherence, and their influence is taken into account by letting fluctuate the transition matrix element and the energy of the localized excitation. This model allowed deriving an exact criterion for the occurrence of the coherent or the incoherent motion depending on the parameters of the system. Meanwhile this model was developed further in order to get results comparable with the above mentioned experiments. Thus, contact to optical absorption measurements was made by calculating the influence of the coupled coherent and incoherent exciton motion on the absorption line shape [20 to 26]. Furthermore, starting from the exact equation for the density matrix of the model, it was possible to derive a diffusion equation for the motion of the exciton which is valid within a certain range of the parameters of the model [20 to 22, 26]. Using results of optical absorption and diffusion measurements, from the theoretical expressions for the optical line shape and the diffusion constant all parameters of the model could be determined [20 to 22, 26]. The application of the model of the coupled coherent and incoherent exciton motion to electron spin resonance allowed calculating the angular dependency of the line position and line width of the ESR line [29 to 31]. From the comparison with experimental data the fine structure parameters and the effective hopping rate in anthracene crystals could be determined. Finally the influence of the motion of triplet excitons on the longitudinal relaxation time and the correlation time of NSR-measurements at the protons of organic crystals has been calculated [20, 22, 25 to 27]. In this contribution, however, we shall not report on all these applications of the model and also not on two other treatments of the coupled coherent and incoherent motion of excitons by Grover and Silbey [44] and by Kenkre and Knox [45] published previously, but stress will be laid upon the assumptions [34, 35] of the Haken-Strobl model and quantities directly connected to exciton migration, such as the time behaviour of the occupation probability [26, 36] and of the mean square displacement [37 to 39] of exciton transport. Furthermore the conditions under which the exciton motion may be described by a diffusion equation [20 to 22] will be discussed, and the expressions describing optical absorption [23, 24, 26] within the Haken-Strobl model are given. Using data derived from the comparison of the theoretical expressions with experimental results, we may discuss whether the motion of triplet excitons in anthracene crystals is coherent or incoherent.

2. Model of the Coupled Coherent and Incoherent Exciton Motion

a. THE COHERENT MOTION OF THE EXCITON

Haken and Strobl [34, 35] considered an arrangement of molecules whose sites are denoted by the index n (see Figure 1). It was assumed that each molecule may be

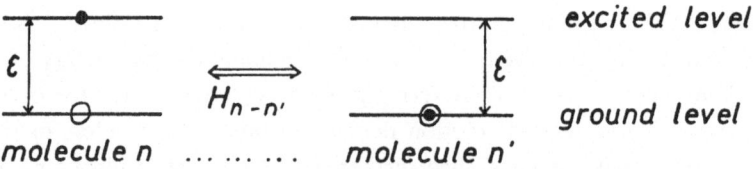

Fig. 1.   Energy levels and coherent interaction of the molecules.

divided into a leucht-electron and the rest and that each molecule has only two equally spaced electronic energy levels, a ground level and an excited level with an energy ε above the ground level. Thus the electron of each molecule may be either in the ground level or in the excited level. As one wants to consider Frenkel excitons, only such excitations were allowed, where the electron and the hole are situated at the same molecule, therefore the molecules always remain neutral. On account of the interaction $H_{n-n'}$ between the molecules, such a localized excited state is not stationary, and therefore the electron and its hole will move together to another molecule.

It is advantageous to use second quantization for the mathematical description of our system. b_n^+ and b_n are creation and annihilation operators for a localized electron-hole pair at molecule n. Then the Hamiltonian describing the coherent motion of the system is given by

$$H_0 = \sum_n \varepsilon b_n^+ b_n + \sum_{n \neq n'} H_{n-n'} b_n^+ b_{n'}. \tag{2.1}$$

The terms of the first sum represent the energy of the exciton when sitting on the molecule at site n. The terms of the double sum annihilate an exciton at site n' and generate one at site n; therefore, these terms are responsible for the coherent energy transport. Under the assumptions of the model, the matrix element $H_{n-n'}$ contains the Coulomb and the exchange interaction integral. When considering triplet excitons, the Coulomb integral is zero and the only contribution to $H_{n-n'}$ stems from the exchange interaction integral which, as usual, will be denoted by $J_{n-n'}$.

b. The Incoherent Motion of the Exciton

This coherent motion is disturbed by the vibrations within and between the molecules. Their influence is taken into account by letting fluctuate the energy of the exciton at the molecule and the transition matrix element. By doing this the molecular vibrations are treated as a heatbath pushing the excitonic system in a stochastic manner. This procedure may be justified, when the molecules are not too small and the temperature is not too low. However, it is not possible to take into account the reaction of the exciton on the molecular vibrations. The Hamiltonian of this part of the motion is

$$H_1 = \sum_n h_{nn}(t) b_n^+ b_n + \sum_{n \neq n'} h_{nn'}(t) b_n^+ b_{n'}. \tag{2.2}$$

The diagonal elements $h_{nn}(t)$ describe fluctuations of the energy of the exciton, and the non-diagonal elements $h_{nn'}$ express fluctuations of the transition matrix elements.

$h_{nn'}(t)$ is assumed to be a Gaussian Markov process with disappearing mean value. Physically this means that the fluctuations are generated by many independent vibrations with broad frequency spectrum. The correlation functions of the stochastic process $h(t)$ are given by the following expressions:

$$\langle h_{nn'}(t) h_{n''n'''}(t') \rangle = \delta(t-t') 2\gamma_{|n-n'|} \Lambda(n, n', n'', n''') \tag{2.3}$$

$$\Lambda(n, n', n'', n''') = \delta_{nn''} \delta_{n'n'''} + \delta_{nn'''} \delta_{n'n''}(1 - \delta_{nn'}). \tag{2.4}$$

The δ-function expresses the Markovian character of the stochastic process, and $\gamma_{|n-n'|}$ measures the strength of the fluctuations. The quantity Λ expresses the fact that in the Haken-Strobl model energy fluctuations at different molecules and fluctuations of the interaction matrix element between different pairs of molecules are not correlated. Two explicit examples are given here:

$$\langle h_{nn}(t) \, h_{nn}(t') \rangle = \delta(t-t') \, 2\gamma_0 \tag{2.5}$$

$$\langle h_{nn'}(t) \, h_{nn'}(t') \rangle = \delta(t-t') \, 2\gamma_{|n-n'|}. \tag{2.6}$$

γ_0 is the strength of the energy or local fluctuations, and $\gamma_{|n-n'|}$ the strength of the fluctuations of interaction matrix element, i.e. the strength of the non-local fluctuations.

c. THE EQUATION OF MOTION

From the total Hamiltonian

$$H = H_0 + H_1 \tag{2.7}$$

we obtain the equation of motion for the density operator of the system

$$i\hbar \dot{\tilde{\rho}} = [H, \tilde{\rho}] = [H_0, \tilde{\rho}] + [H_1, \tilde{\rho}]. \tag{2.8}$$

The first term of the equation of motion describes the coherent motion, the second term represents the fluctuating part.

One is not interested in the operator $\tilde{\rho}$ still containing the fluctuations of the heat bath, but in the quantity $\rho = \langle \tilde{\rho} \rangle$ averaged over the fluctuations. Taking into account that only H_1 contains fluctuations, we arrive at the equation of motion for the averaged quantity:

$$i\hbar \dot{\rho} = [H_0, \rho] + \langle [H_1, \tilde{\rho}] \rangle. \tag{2.9}$$

What one wants to have is ρ instead of $\tilde{\rho}$ also in the second term. This requires some more sophisticated calculation. It may be done either by perturbation theory to infinite order and summing up the series again [35]. Another procedure may be carried through in closed form using Feynman's time ordering operator and the characteristic functional of the stochastic process [20, 24, 40, 41]. We do not want to go into the details of the calculation but give only the result. In order to represent the density operator, we use as basis functions the states $|0\rangle$ with no exciton in the crystal and $|n\rangle = b_n^+ |0\rangle$ with the exciton sitting at molecule n. The equation of motion for the density matrix may then be written in this form

$$i\hbar \dot{\rho} = L\rho. \tag{2.10}$$

Here L is the Liouville-operator. Explicitly the equations for the diagonal and non-diagonal elements are given by

$$\hbar\dot{\rho}_{nn} = -i[H_0, \rho]_{nn} - 2\Gamma\rho_{nn} + 2\sum_{n'} \gamma_{|n-n'|}\rho_{n'n'} \tag{2.11}$$

$$\hbar\dot{\rho}_{nn'} = -i[H_0, \rho]_{nn'} - 2\Gamma\rho_{nn'} + 2\gamma_{|n-n'|}\rho_{n'n} \tag{2.12}$$

$$\Gamma = \sum_{n} \gamma_{|n|}.$$

The first term on the right side of each equation describes the coherent, the following terms stand for the incoherent part of the motion. The diagonal elements ρ_{nn} of the density operator are the probability of finding the exciton at site n. The non-diagonal elements $\rho_{nn'}$ describe phase relations between sites n and n'. Thus the first equation describes the change of the occupation probability at site n, and the second equation describes the change of phase relations. The coupling between the equations for diagonal and non-diagonal elements is given by the coherent part of the Hamiltonian.

3. Solutions of the Density Matrix Equation

a. THE TWO-MOLECULE MODEL

At this point it is illustrative to consider instead of the extended arrangement of molecules a two-molecule model [20 to 22, 42]. The equations of motion then simplify to the following expressions:

$$\hbar\dot{\rho}_{11} = -2\gamma_1\rho_{11} + 2\gamma_1\rho_{22} - iJ(\rho_{21}-\rho_{12})$$

$$\hbar\dot{\rho}_{22} = -2\gamma_1\rho_{22} + 2\gamma_1\rho_{11} - iJ(\rho_{12}-\rho_{21})$$

$$\hbar\dot{\rho}_{12} = -2(\gamma_0+\gamma_1)\rho_{12} + 2\gamma_1\rho_{21} - iJ(\rho_{22}-\rho_{11})$$

$$\hbar\dot{\rho}_{21} = -2(\gamma_0+\gamma_1)\rho_{21} + 2\gamma_1\rho_{12} - iJ(\rho_{11}-\rho_{22}). \tag{3.1}$$

The first two equations describe the change of the occupation numbers of the excitons at the two molecules. From the last equations we may get the change of the phase relation between the two molecules.

Let us first consider the situation where the exchange interaction integral J vanishes, that is we consider the purely incoherent case. Then the equations for the diagonal elements of the density matrix ρ_{11} and ρ_{22} are completely decoupled from the equations for the non-diagonal elements ρ_{12} and ρ_{21}.

The equations for the diagonal elements represent a set of rate equations for the occupation numbers of the excitons. The transition probability for an exciton between the two molecules is given by $2\gamma_1/\hbar$, which means that it is determined by non-local fluctuations. On the other hand, from the equations for the non-diagonal elements ρ_{12} and ρ_{21} of the density matrix, we see that the phase of the excitons decays exponentially with the exponent given by $2(\gamma_0+\gamma_1)$. Thus the phase of the exciton is destroyed both by local and non-local fluctuations.

Now we consider the situation of non-vanishing exchange interaction integral. Then the equations for the diagonal and non-diagonal matrix elements are coupled and we have to solve a four-dimensional eigenvalue problem. In this way we get the occupation probability for molecule number 1 with the exciton initially sitting at this molecule as

$$\rho_{11}(t) = 1/2 + \frac{1}{4}\left(1 + i\ \frac{\gamma_0}{\sqrt{4J^2 - \gamma_0^2}}\right) e^{-(1/\hbar)\, R_8 t} \tag{3.2}$$

$$+ \frac{1}{4}\left(1 - i\ \frac{\gamma_0}{\sqrt{4J^2 - \gamma_0^2}}\right) e^{-(1/\hbar)\, R_9 t}.$$

The eigenvalues, describing the behaviour in time, are written down here

$$\frac{1}{\hbar}\, R_{8/9} = \frac{1}{\hbar}\, (\gamma_0 + 4\gamma_1) \pm \frac{i}{\hbar}\, \sqrt{4J^2 - \gamma_0^2}. \tag{3.3}$$

Considering the exponents (3.3) we remark that for small values of the local fluctuations γ_0, i.e. for $\gamma_0 < 2J$, the occupation number exhibits damped oscillations. Of course, the same is true for the non-diagonal elements of the density matrix ρ_{12} describing phase relations, because their time behaviour is determined by the same eigenvalues. For values of γ_0 in this range, it is reasonable to denote the exciton motion as coherent. On the other hand, when $\gamma_0 > 2J$ the occupation number and the phase relation decrease exponentially and for such values of γ_0 the motion of the exciton is called incoherent. The transition from coherent to incoherent motion happens when $\gamma_0 = 2J$.

b. Exciton Motion on a Linear Chain with Nearest Neighbour Interaction

We now make the transition from the two-molecule model to a linear chain of molecules with nearest neighbour interaction [26, 35, 36]. The time behaviour of the exciton motion naturally again is determined by the eigenvalues of the density matrix equation. But instead of the discrete set of eigenvalues of the two-molecule model, we now have a quasi-continuum of eigenvalues R^{kl}:

$$R^{kl} = -2\gamma_0 + i\ 4J\ \sin\frac{k}{2}\ \cos l. \tag{3.4}$$

The real part of each of these eigenvalues is $-2\gamma_0$, but for each value of k the imaginary parts of the eigenvalues form a band of width $8J$. This behaviour is completely analogous to that known from band theory in solid state physics, when going from a set of two interacting molecules to a linear chain of interacting molecules. However, in addition to these band eigenvalues, for each value of k satisfying the condition

$$\left|\sin\frac{k}{2}\right| < \frac{\gamma_0}{2J} \tag{3.5}$$

there exists an additional eigenvalue R^{ks}

$$R^{ks} = -2\gamma_0 + \sqrt{(2\gamma_0)^2 - \left(4J \sin \frac{k}{2}\right)^2} \tag{3.6}$$

which is purely real.

As usual the general solution of the density matrix equation is then given by the superposition of all eigensolutions. This means that we have to sum over all values of k and l of the band solutions and over all allowed values k of the additional solutions. The summation over the band solutions gives in the limit of an infinitely long chain a double integral. When executing the sum over the additional solutions, we have to differentiate between $\gamma_0 > 2J$ and $\gamma_0 < 2J$. For $\gamma_0 > 2J$, which in the case of the two-molecule model was the condition for incoherent motion, the inequality (3.5) is satisfied for all values of k, and therefore the integration variable runs over the whole range of 2π. When $\gamma_0 < 2J$, corresponding to coherent motion, (3.5) is satisfied only for values of k between $-\bar{k}$ and $+\bar{k}$, where $\bar{k}$ is given by

$$\sin \frac{\bar{k}}{2} = \frac{\gamma_0}{2J} \tag{3.7}$$

and a corresponding range around π. Therefore in the case $\gamma_0 < 2J$ the summation over the additional solutions runs only over these two allowed ranges of k-values around π and around 0.

With the initial condition

$$\rho_{nn'}(0) = \delta_{no}\delta_{n'o}$$

the solutions in the two cases are given by (3.8) and (3.9).

$\gamma_0 > 2J$

$$\rho_{nn}(t) = \frac{1}{2\pi} \int_{-\pi}^{\pi} dk\, e^{ikn} \frac{2\gamma_0}{\sqrt{(2\gamma_0)^2 - (4J \sin k/2)^2}} e^{R^{ks}t} \tag{3.8}$$

$$+ \frac{1}{4\pi^2} \int_{-\pi}^{\pi} dk\, e^{ikn} \int_{-\pi}^{\pi} dl \frac{4J \sin \dfrac{k}{2} \sin l}{4J \sin \dfrac{k}{2} \sin l - 2\gamma_0} e^{R^{kl}t}$$

$\gamma_0 < 2J$

$$\rho_{nn}(t) = \frac{1}{2\pi} \left\{ \int_{\pi-\bar{k}}^{\pi+\bar{k}} dk + \int_{-\bar{k}}^{\bar{k}} dk \right\} e^{ikn} \frac{2\gamma_0}{\sqrt{(2\gamma_0)^2 - (4J \sin k/2)^2}} e^{R^{ks}t}$$

$$+ \frac{1}{4\pi^2} \int_{-\pi}^{\pi} dk\, e^{ikn} \int_{-\pi+i\varepsilon}^{\pi+i\varepsilon} dl \frac{4J \sin \dfrac{k}{2} \sin l}{4J \sin \dfrac{k}{2} \sin l - 2\gamma_0} e^{R^{kl}t}. \tag{3.9}$$

Considering the limiting case of $\gamma_0 = 0$, which means completely coherent motion, we see from (3.7) that $\bar{k}$ has to be zero, and therefore there is no contribution from the first part of the solution (3.9). It may be shown that in this case the double integral reduces to the square of the Bessel function $J_n^2(2Ht)$, which is in complete agreement with Merrifield's result [32].

Now we wish to discuss the behaviour of the solution for large values of the time t. The real parts of the eigenvalues describing band solutions are always $-2\gamma_0$, while the real part of the additional solutions assumes values between $R^{ks} = -2\gamma_0$ and $R^{ks} = 0$. Thus the real parts of the band solutions are always larger and therefore, in the course of time, the contributions of the band eigensolutions are damped more rapidly away than the contributions of the additional solutions. The eigenvalues of the additional solutions, however, are purely real and therefore describe no oscillations. Thus, for large enough times the exciton motion will be diffusion-like, no matter whether $\gamma_0 < 2J$ or $\gamma_0 > 2J$. This behaviour of the coupled coherent and incoherent exciton motion may be seen more explicitly when considering the time behaviour of the mean square displacement

$$\langle n^2 \rangle = \sum_n n^2 \rho_{nn},$$

a quantity which is characteristic for transport processes.

4. Mean Square Displacement of Exciton Motion
on a Linear Chain [26, 37 to 39]

Starting from the equation of motion for the density operator one may derive a differential equation for the mean square displacement. The solution of this differential equation with the exciton initially sitting at site $n = 0$ is given by the following expression [39]:

$$\langle n^2 \rangle = \left(2 \sum_m m^2 \gamma_{|m|} + \sum_m \frac{m^2 H_m^2}{\Gamma + \gamma_{|m|}} \right) t +$$

$$+ \sum_{m>0} \frac{m^2 H_m^2}{(\Gamma + \gamma_{|m|})^2} \left(e^{-2(\Gamma + \gamma_{|m|})t} - 1 \right). \tag{4.1}$$

In the purely incoherent case, i.e. $H_m = 0$, this reduces to

$$\langle n^2 \rangle = 2 \left(\sum_m m^2 \gamma_{|m|} \right) t, \tag{4.2}$$

a result which is well known from the investigation of hopping processes. $2\gamma_{|m|}$ is the hopping rate over a distance m.

On the other hand in the purely coherent case, we obtain this expression:

$$\langle n^2 \rangle = \left(\sum_m m^2 H_m^2 \right) t^2. \tag{4.3}$$

This quadratic time dependence is also well known from Merrifield's result [32].

Considering now the time dependence of the general solution, we obtain for small times by expanding to second order in t:

$$\langle n^2 \rangle \approx 2 \left(\sum_m m^2 \gamma_{|m|} \right) t + \left(\sum_m m^2 H_m^2 \right) t^2. \qquad (4.4)$$

From there we see that for small times the mean square deviation is given by a superposition of the contributions of the purely incoherent and of the purely coherent motion. For very small times the contribution linear in t determines the time behaviour, whereas for larger times the contribution of the coherent interaction predominates. On the other hand, for large times we may neglect the exponential expression and arrive at this equation

$$\langle n^2 \rangle \approx 2 \left(\sum_m m^2 \gamma_{|m|} + \sum_m \frac{1}{2} \frac{m^2 H_m^2}{\Gamma + \gamma_{|m|}} \right) t - \sum_m \frac{1}{2} \frac{m^2 H_m^2}{(\Gamma + \gamma_{|m|})^2}. \qquad (4.5)$$

Thus for large times we find also the time dependence of a diffusion process. The diffusion constant however is now modified by a contribution of the coherent interaction.

From the discussion of the mean square displacement we see that the question whether the exciton motion is coherent or incoherent depends not only on whether γ_0 is smaller or larger than the interaction matrix element. It is also a question of the time scale, which means that for very small and for very large times we observe diffusional motion with two different diffusion constants, whereas in a time interval in between the contribution of the coherent motion may be the most important.

5. Diffusion-Approximation [20–22, 26, 46, 47]

a. Derivation of a Diffusion Equation

In order to get a diffusion equation, we start from our equation of motion for the density matrix [20 to 22, 26]. Separating for diagonal and non-diagonal elements, we have these two equations

$$\hbar \dot{\rho}_{nn} = -i \sum_{n' \neq n} H_{nn'} \rho_{n'n} + i \sum_{n' \neq n} \rho_{nn'} H_{n'n} - 2\Gamma \rho_{nn} + 2 \sum_{n'} \gamma_{n-n'} \rho_{n'n'} \qquad (5.1)$$

$$\hbar \dot{\rho}_{nn'} = -i \sum_{n'' \neq n} H_{nn''} \rho_{n''n'} + i \sum_{n'' \neq n'} \rho_{nn''} H_{n''n'} - 2\Gamma \rho_{nn'_i} + 2\gamma_{n-n'} \rho_{n'n}. \qquad (5.2)$$

We wish to consider diffusive motion, and from our previous considerations of the model of coupled coherent and incoherent motion we know that in the case of diffusive motion $\gamma_0 \gg 2J$. Then phase relations are destroyed after a very short time. If we consider only time intervals longer than these decay times of the phase, in the equations for the non-diagonal elements we may neglect the time derivative $\dot{\rho}_{nn'}$ compared to $\Gamma \rho_{nn'}$, which is a kind of adiabatic approximation. After solving for $\rho_{nn'}$ and inserting into the equation for the diagonal element ρ_{nn}, we may neglect all non-diagonal elements compared to the diagonal ones and arrive at an equation, containing only diagonal elements:

$$\hbar \dot{\rho}_{nn} = \sum_{n'} \frac{1}{2} \left\{ \frac{\gamma_{n'-n}}{\gamma_{n'-n}^2 - \Gamma^2} H_{nn'}^2 + \frac{\gamma_{n-n'}}{\gamma_{n-n'}^2 - \Gamma^2} H_{n'n} \right\} (\rho_{n'n'} - \rho_{nn})$$

$$- \sum_{n'} \frac{1}{2} \left\{ \frac{\Gamma}{\gamma_{n'-n}^2 - \Gamma^2} |H_{nn'}|^2 + \frac{\Gamma}{\gamma_{n'-n}^2 - \Gamma^2} |H_{nn'}|^2 \right\} (\rho_{n'n'} - \rho_{nn})$$

$$- 2\Gamma \rho_{nn} + 2 \sum_{n'} \gamma_{n-n'} \rho_{n'n'} . \tag{5.3}$$

The first two lines have their origin in the coherent interaction, the terms in the third line follow from the stochastic part of the Hamiltonian.

Now we assume that the density of the excitons varies only slowly in space, so that we may expand $\rho_{n'n'}$ in a Taylor series in $(\mathbf{x}_{n'} - \mathbf{x}_n)$ and neglect terms higher than second order:

$$\rho(\mathbf{x}_{n'}) = \rho(\mathbf{x}_n) + (\mathbf{x}_{n'} - \mathbf{x}_n) \nabla \rho(\mathbf{x}_n) + \tfrac{1}{2} \left[(\mathbf{x}_{n'} - \mathbf{x}_n) \nabla \right]^2 \rho(\mathbf{x}_n) . \tag{5.4}$$

After inserting this into (5.3) we arrive at the following diffusion equation

$$\dot{\rho}(\mathbf{x}) = \sum_{i,j} D_{ij} \frac{\partial^2}{\partial x_i \partial x_j} \rho(\mathbf{x}) \tag{5.5}$$

with the diffusion tensor given by:

$$D_{ij} = \sum_m \frac{1}{\hbar} \left(\gamma_m + \frac{1}{2} \frac{H_m^2}{\gamma_m + \Gamma} \right) x_{mi} x_{mj} . \tag{5.6}$$

The first term in the bracket of the diffusion tensor stems from the non-local fluctuations, the second from the coherent interaction in connection with the fluctuating terms of the Hamiltonian. x_{mi} and x_{mj} are the i and j components of the lattice vector of molecule m and the sum runs over all lattice points.

b. Comparison with Directly Measured Diffusion Tensors

In order to get the parameters of the system, we use the D_{aa}-component of the diffusion tensor of triplet excitons, measured by Ern et al., in anthracene crystals at various temperatures. It is very easy to evaluate the sum in the diffusion tensor for triplet excitons. For triplet excitons the coherent interaction matrix element contains only the exchange interaction integral, which decreases exponentially with increasing distance. Thus, for triplet excitons, the sum runs only over nearest neighbours as regards the coherent interaction, and the same is assumed for the fluctuation parameters γ_m. In these crystals the interaction between translationally inequivalent molecules is much larger than between equivalent ones. Therefore, after summing over the

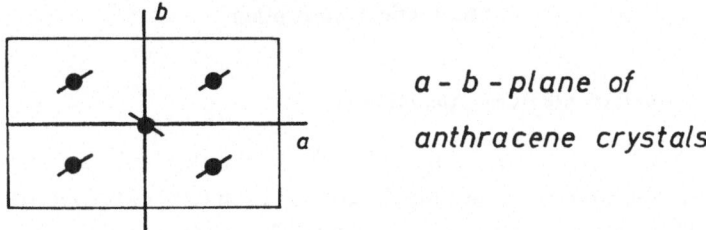

Fig. 2.   The four nearest neighbours of a molecule in anthracene crystals.

four nearest neighbours of Figure 2, we arrive at this diffusion constant

$$D_{aa} = \frac{a^2}{2}\left(2\gamma_1 + \frac{J^2}{\Gamma + \gamma_1}\right). \tag{5.7}$$

Here a is the lattice constant in a-direction, and γ_1 and J are the interactions between the above defined nearest neighbours.

Measured values of the diffusion constant in anthracene crystals are given in Table I. The first column gives the temperatures for which measurements have been done, and the second column the measured values for the diffusion constants.

TABLE I

Values for $(2\gamma_1 + J^2/(\gamma_1 + \Gamma))$ derived from diffusion measurements of (a) Avakian et al. [8] and (b) Ern et al. [10]

T/K	$D/(\text{cm}^2\ \text{s}^{-1})$	$\dfrac{2\gamma_1 + J^2/(\gamma_1 + \Gamma)}{\text{s}^{-1}}$	$\dfrac{2\gamma_1 + J^2/(\gamma_1 + \Gamma)}{\text{cm}^{-1}}$	
371	1.6×10^{-4}	4.3×10^{10}	0.23	b
298	1.5×10^{-4}	4.1×10^{10}	0.22	b
160	2.5×10^{-4}	6.8×10^{10}	0.36	b
118	4×10^{-4}	10.8×10^{10}	0.57	b
room temperature	2×10^{-4}	5.4×10^{10}	0.29	a

Using the value $a = 8.6$ Å of the extension of the unit cell of anthracene in a-direction, the combination of the model parameters

$$\left(2\gamma_1 + \frac{J^2}{\gamma_1 + \cdot}\right) \tag{5.8}$$

representing an effective hopping rate, may be determined. We see that we have an effective hopping rate of about several times $10^{10}\ \text{s}^{-1}$. The fourth column gives the same quantity but now in units of cm^{-1}, and the last column finally denotes the authors of the measurements.

6. Optical Absorption

a. THE LINE SHAPE OF OPTICAL ABSORPTION

In order to decide whether the motion of excitons in anthracene is coherent or incoherent, it is not sufficient to know only the combination (5.8) of the model parameters, but we should have the parameters separately. Therefore we should have additional information, and this information may be derived from optical absorption experiments. From these measurements we may derive J and Γ separately, and thus using the results of the diffusion measurements, γ_1 may be calculated.

As is well known, the line shape of optical absorption is given by

$$\chi''_{kl}(\omega) = \frac{1}{2h} \int\limits_{-\infty}^{+\infty} d\tau \, e^{-i\omega\tau} \, Tr_s[\langle \rho(-\infty) \, p_k(0) \, p_l(\tau)\rangle -$$

$$- \langle \rho(-\infty) \, p_l(\tau) \, p_k(0)\rangle] \tag{6.1}$$

i.e. by the imaginary part of the dielectric susceptibility tensor.

Using Kubo's formalism $\chi''_{kl}(\omega)$ has been expressed by the Fourier-transform of the two-times correlation functions of the operators of the electric dipole moment. $p_k(0)$ and $p_k(\tau)$ are the k and l components of the dipole moment operators in the Heisenberg picture. The angular brackets mean quantum mechanical and statistical averaging.

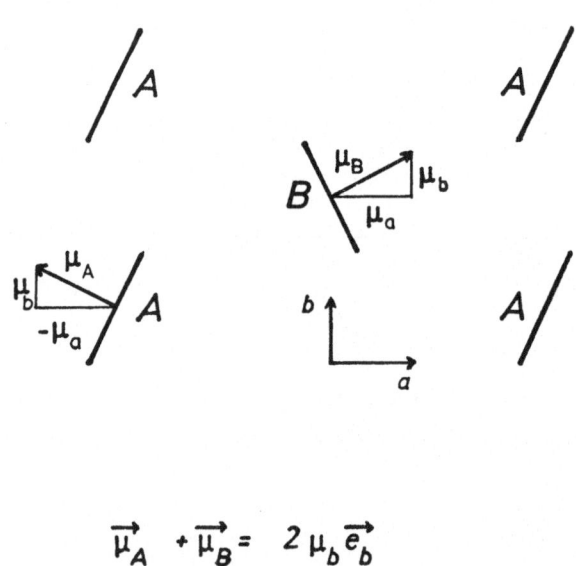

$$\vec{\mu}_A + \vec{\mu}_B = 2\,\mu_b\,\vec{e}_b$$

$$\vec{\mu}_A - \vec{\mu}_B = -2\,\mu_a\,\vec{e}_a$$

Fig. 3. Orientation of the dipole moments at the two inequivalent molecules in the unit cell.

Knowing the solution of the density matrix equation, the above correlation functions may be calculated [20, 21, 23, 26, 43]. Frequently, however, it is rather difficult to find the solution of the density matrix equation. Therefore, starting from the stochastic Hamiltonian of the Haken-Strobl model, a set of equations of motion for such two-times correlation functions has been derived. In the case of optical absorption, these equations may be solved very simply [24].

The result of this calculation is an expression for the dielectric susceptibility, which depends in a rather complicated way on the parameters of the model. However this expression may be simplified considerably, after introducing some assumptions, consistent with the experimental results and referring to the orientation of the molecular dipole moments. We shall assume that the dipole moments are oriented as shown in Figure 3, that is we assume that the b-components of the molecular dipole moments at the two inequivalent molecules are parallel and that the a-components are antiparallel. With these assumptions, the only non-disappearing components of the tensor of the dielectric susceptibility are given by

$$\chi''_{aa} = 2\mu_a \left(\frac{\Gamma}{[\omega - (\varepsilon + \alpha - \beta)]^2 + \Gamma^2} - \frac{\Gamma}{[\omega + (\varepsilon + \alpha - \beta)]^2 + \Gamma^2} \right) \tag{6.2}$$

$$\chi''_{bb} = 2\mu_b \left(\frac{\Gamma}{[\omega - (\varepsilon + \alpha + \beta)]^2 + \Gamma^2} - \frac{\Gamma}{[\omega + (\varepsilon + \alpha + \beta)]^2 + \Gamma^2} \right) \tag{6.3}$$

where

$$\alpha = \tilde{H}_{k,0} \qquad \beta = |\tilde{H}_{k,1}| \tag{6.4}$$

and $\tilde{H}_{k,0}$ and $\tilde{H}_{k,1}$ are the Fourier-transforms of the interaction matrix elements between equivalent and inequivalent molecules, respectively. (6.2) and (6.3) show that we obtain two absorption lines, one polarized parallel to the a-axis with a maximum at $\varepsilon + \alpha - \beta$, the other polarized parallel to the b-axis with a maximum at $\varepsilon + \alpha + \beta$. The distance of the two lines, the so-called Davydov or factor group splitting, is given by 2β and the line width is Γ. On account of the small wave vector of the light wave, in (6.4) we may use $k \approx 0$. Considering in addition only nearest neighbour interaction, which is a good approximation for triplet excitons we have

$$\Gamma \approx \gamma_0 + 4\gamma_1, \quad \alpha = 0 \quad \text{and} \quad \beta = \tilde{H}_{k,1} \approx 4J. \tag{6.5}$$

b. COMPARISON WITH EXPERIMENTAL VALUES

From the comparison of experimental results with our theoretical expressions we derive the numerical values of Table II. The first column shows the temperatures, for which experimental results have been found. The second column gives the values of the Davydov-splittings Δ in units of cm^{-1}. From these values the exchange interaction integral J in the third column may be calculated. The fourth column gives the line widths Γ of the optical absorption line, also in units of cm^{-1}. The next column shows once more the values of the effective hopping rate, derived from diffusion measurements. Using now the values of J and Γ, γ_1 may be calculated from the effective hopping rates. The result is represented in the sixth column. The next column gives the contribution of the coherent interaction to the effective hopping rate and the last column refers to the authors of the experimental values.

TABLE II

Values of the parameters of the model inferred from measurements of (a) Avakian *et al.* [8], (b) Ern *et al.* [10], and (c) Clarke *et al.* [11]

$\dfrac{T}{K}$	$\dfrac{\Delta}{cm^{-1}}$	$\dfrac{J}{cm^{-1}}$	$\dfrac{\Gamma}{cm^{-1}}$	$\dfrac{2\gamma_1+J^2/(\Gamma+\gamma_1)}{cm^{-1}}$	$\dfrac{\gamma_1}{cm^{-1}}$	$\dfrac{\frac{1}{2}J^2/(\Gamma+\gamma_1)}{cm^{-1}}$	
371	19	2.4	65	0.23	0.071	0.044	b
298	17	2.1	51	0.22	0.065	0.043	b
160	18	2.25	30	0.36	0.095	0.084	b
118	18	2.25	14	0.57	0.11	0.18	b
4	22	2.75	<1				c
room temperature	17	2.1	70	0.29	0.11	0.03	a

From Tabel II we may see the following:

(1) Γ is given by $\Gamma = \gamma_0 + 4\gamma_1$. On account of $\gamma_1 \ll \Gamma$, $\gamma_0 \approx \Gamma$. Above we have seen that the exciton motion may be coherent, if $\gamma_0 < 2J$, and that the motion is incoherent for $\gamma_0 > 2J$. From the table we see that at all temperatures except at $4K$ we have $\gamma_0 > 2J$. Therefore at all these temperatures the exciton motion in anthracene crystals is incoherent and may be approximated by a hopping process with an effective hopping rate which includes also the influence of the coherent interaction. At $4K$ however, $\gamma_0 < 2J$, and therefore the exciton motion should be coherent.

(2) The line width of the optical absorption is determined by Γ. On account of $\gamma_1 \ll \gamma_0$, this line width is determined mainly by local fluctuations.

(3) The diffusion constant, however, is determined by the sum of two quantities, namely by the non-local fluctuations γ_1 and by $\frac{1}{2}J^2/(\Gamma+\gamma_1)$. The table shows that at all temperatures considered, except at $118\,K$, the contribution of the non-local fluctuations γ_1 is larger than the contribution of the term containing the exchange interaction integral. Thus, whereas the non-local fluctuations may be neglected as concerns optical absorption, they play an important role in exciton migration.

Finally we only wish to mention the results of the application of the model of the coupled coherent and incoherent exciton motion to the description of ESR- and NSR-experiments [20 to 22, 24 to 31]. From the results of these measurements, which were carried through by Wolf and coworkers [14 to 19], the model parameters have been determined too. In table 3 we have gathered the data for room temperature, derived from the various experiments.

In the first two columns we have represented the data derived from optical absorption experiments. In the last three columns, the values for γ_1 are given obtained by using the data known from optical absorption and from diffusion, electron spin resonance and nuclear spin resonance, respectively.

TABLE III

Comparison of the non-local fluctuation parameter γ_1 infered from various experiments at room temperature

room temperature				
optical absorption		diffusion	ESR	NSR
J/cm^{-1}	Γ/cm^{-1}	γ^1/cm^{-1}	γ_1/cm^{-1}	γ_1/cm^{-1}
2.1	51	0.07	0.06	0.03

Taking into account the rough assumptions made in the original evaluation of NSR [27], we may say that the data for γ_1 are in satisfying agreement. Thus using our model, we may describe completely different experiments in a consistent manner.

References

1. Haken, H.: *Fortschr. Physik* **6**, 271 (1958).
2. Wolf, H. C.: in *Solid State Phys.* **9**, 1959 (eds. F. Seitz, D. Turnbull).
3. Davydov, A. S.: *Theory of Molecular Excitons*, New York-San Francisco-Toronto-London: Mc Graw-Hill Book Co. Inc. 1962.
4. Knox, R. S.: *Theory of Excitons. Solid State Phys. Suppl.* **5**, 1963.
5. *Physics and Chemistry of the Organic Solid State* (ed. by Fox, D., Labes, M. M., and Weissberger, A.), New York-London-Sydney: Interscience Publishers 1967.
6. Stepanov, D. I. and Gribkovskii, V. P.: *Theory of Luminescence*, London: ILIFFE Books Ltd. 1968.
7. Davydov, A. S.: *Theory of Molecular Excitons*, New York-London: Plenum Press 1971.
8. Avakian, P., Ern, V., Merrifield, R. E., and Suna, A.: *Phys. Rev.* **165**, 974 (1968).
9. Avakian, P. and Merrifield, R. E.: *Mol. Cryst. Liquid Cryst.* **5**, 37 (1968).
10. Ern, V., Suna, A., Tomkiewicz, Y., Avakian, P., and Groff, R. P.: *Phys. Rev.* **B5**, 3222 (1972).
11. Clarke, R. H. and Hochstrasser, R. M.: *J. Chem. Phys.* **46**, 4532 (1967).
12. Port, H.: Thesis, University of Stuttgart, 1974.
13. Hutchison, C. A. and Mangum, B. W.: *J. Chem. Phys.* **34**, 908 (1961).
14. Schwoerer, M. and Wolf, H. C.: *Mol. Cryst. Liquid Cryst.* **3**, 177 (1967).
15. Haarer, D. and Wolf, H. C.: *Mol. Cryst. Liquid Cryst.* **10**, 359 (1970).
16. Schmidberger, R. and Wolf, H. C.: *Chem. Phys. Letters* **16**, 402 (1972).
17. Schmidberger, R.: Thesis, University of Stuttgart, 1974.
18. Maier, G. and Wolf, H. C.: *Z. Naturforsch.* **23a**, 1068 (1968).
19. Kolb, H. and Wolf, H. C.: *Z. Naturforsch.* **27a**, 51 (1972).
20. Reineker, P.: Thesis, University of Stuttgart, 1971.
21. Haken, H. and Reineker, P.: *Z. Physik* **249**, 253 (1972).
22. Haken, H. and Reineker, P.: *Acta Universitatis Carolinae-Mathematica et Physica* **14**, 23 (1973).
23. Schwarzer, E. and Haken, H.: *Opt. Commun.* **9**, 64 (1973).
24. Reineker, P.: *Z. Naturforsch.* **29a**, 282 (1974).
25. Haken, H. and Schwarzer, E.: *Chem. Phys. Letters* **27**, 41 (1974).

26. Schwarzer, E.: Thesis, University of Stuttgart, 1974.
27. Reineker, P. and Haken, H.: *Z. Physik* **250**, 300 (1972).
28. Reineker, P.: *Phys. Stat. Sol. (b)* 52, 439 (1972).
29. Reineker, P.: *Z. Physik* **B21**, 409 (1975).
30. Reineker, P.: *Phys. Stat. Sol. (b)* **70**, 471 (1975).
31. Reineker, P.: *Phys. Stat. Sol. (b)* (to be published).
32. Merrifield, R. E.: *J. Chem. Phys.* **28**, 647 (1958).
33. Trlifaj, M.: *Czech. J. Phys.* **8**, 510 (1958).
34. Haken, H. and Strobl, G.: in *The Triplet State* (ed. by Zahlan, A. B.), Cambridge University Press, 1967.
35. Haken, H. and Strobl, G.: *Z. Physik* **262**, 135 (1973).
36. Reineker, P.: *Phys. Letters* **44A**, 429 (1973).
37. Schwarzer, E. and Haken, H.: *Phys. Letters* **42A**, 317 (1972).
38. Reineker, P.: *Phys. Letters* **42A**, 389 (1973).
39. Reineker, P.: *Z. Physik* **261**, 187 (1973).
40. Aslangul, C. and Kottis, Ph.: *Phys. Rev.* **B10**, 4364 (1974).
41. Kubo, R.: *J. Phys. Soc. Jap.* **17**, 1100 (1962).
42. Strobl, G.: Diplom-Thesis, University of Stuttgart, 1966.
43. Haken, H. and Weidlich, W.: *Z. Physik* **205**, 96 (1967).
44. Grover, M. and Silbey, R.: *J. Chem. Phys.* **54**, 4843 (1971).
45. Kenkre, V. M. and Knox, R. S.: *Phys. Rev.* **B9**, 5279 (1974).
46. Reineker, P. and Kühne, R.: *Z. Physik* **B22**, 193 (1975).
47. Kühne, R. and Reineker, P.: *Z. Physik* **B22**, 201 (1975).

COMMENTS ON EXCITONS AND LOCALIZATION

R. VOLTZ

Laboratoire de Physique des Rayonnements et d'Electronique Nucléaire, Centre de Recherches Nucléaires et Université Louis Pasteur, 67200 Strasbourg-Cronenbourg, France

1. In perfect crystals, an exciton is usually considered as a bound electron positive hole pair, moving together as an unique quasi-particle, with a definite value of momentum. The description necessarily distinguishes between the relative motion of the hole and the electron that make up the exciton, and the motion of the exciton as a whole. In both of these motions, localization has of course a different practical significance.

2. *For motion of the excitons in the crystal*, the meaning of localization has been repeatedly discussed in the foregoing contributions. It will be sufficient to recall here that the relevant exciton properties are identical with those of free particles in quantum mechanics. Translational invariance of the crystal implies that the stationary excited states are also eigenstates of crystal momentum, which corresponds to extreme delocalization. More or less localized excitations may be prepared as wave packets with more or less extended momentum distributions. The extreme case of an excited state located at point $\mathbf{R}_i$, in particular, is given by the superposition of all the momentum eigenstates $|\mathbf{k}\rangle$:

$$|\mathbf{R}_i\rangle = N^{-1/2} \sum_{\mathbf{k}} |\mathbf{k}\rangle \exp(-i\mathbf{k}\mathbf{R}_i).$$

Conversely, the pure delocalized exciton state is expressed in terms of the locally excited configurations as:

$$|\mathbf{k}\rangle = N^{-1/2} \sum_{i} |\mathbf{R}_i\rangle \exp(i\mathbf{k}\mathbf{R}_i). \tag{1}$$

3. *The relative motion of the electron and the positive hole* determines the internal structure and nature of the excitons. Here localization of the electron relative to the hole helps to classify the various types of excitons.

For so doing, one starts with the picture of a localized excitation in the crystal consisting of a hole at lattice site $\mathbf{R}_i$ and an electron at another site $\mathbf{R}_j = \mathbf{R}_i + \boldsymbol{\beta}$. Formally this state may be represented by

$$b_n^+(\mathbf{R}_j)\, b_0(\mathbf{R}_i)\, |\Phi_0\rangle,$$

where $|\Phi_0\rangle$ denotes the crystalline quasivacuum state with a filled valence band; $b_0(\mathbf{R}_i)$ destroys an electron (creates a hole) at site $\mathbf{R}_i$ in the upper filled orbital, while $b_n^+(\mathbf{R}_j)$ creates an electron in a higher empty orbital (index n) at site $\mathbf{R}_j$. The creation operator of such a localized excitation is hence of the form:

$$B_{n\boldsymbol{\beta}}^+(\mathbf{R}_i) = b_n^+(\mathbf{R}_i + \boldsymbol{\beta})\, b_0(\mathbf{R}_i).$$

O. Chalvet et al. (eds.), Localization and Delocalization in Quantum Chemistry, Vol. II, 301–303. All Rights Reserved

More generally, a locally excited configuration in the crystal can be defined as the expansion over the $B_{n\beta}^+(\mathbf{R}_i)$ configurations, *viz.*:

$$B_{nv}^+(\mathbf{R}_i) = \sum_{\beta} U_{nv}(\boldsymbol{\beta})\, B_{n\beta}^+(\mathbf{R}_i). \tag{2}$$

The square of the expansion coefficients $|U_{nv}(\boldsymbol{\beta})|^2$ is the probability of localizing the electron at the position $\boldsymbol{\beta}$ with respect to the hole. The different exciton types may be defined as corresponding to definite forms of the envelope functions $U_{nv}(\boldsymbol{\beta})$; in particular, it is usual to distinguish the limiting cases of 'small radius' and 'large radius' excitons.

4.   *The small radius excitons* are introduced in tight binding situations which apply to materials where the intersite electronic interactions are very much smaller than the intrasite binding interactions. This is the case for organic molecular crystals and for some rare gas or ionic lattices, in which the molecules or atoms are quite far apart with almost non overlapping electronic orbitals. Typical small radius excitons are the *Frenkel excitons*, that may be characterized by $U_{nv}(\boldsymbol{\beta}) = \delta(\boldsymbol{\beta})$ in the general expression (2). In this limiting case, the creation operator for the locally excited states, $|\mathbf{R}_i\rangle = B_n^+(\mathbf{R}_i)|\,\Phi_0\rangle$, is hence of the simple form

$$B_n^+(\mathbf{R}_i) = b_n{}^+(\mathbf{R}_i)\, b_0(\mathbf{R}_i) \tag{3'}$$

which describes an excited atomic or molecular state at lattice site $\mathbf{R}_i$. The related collective exciton state $|\mathbf{k}\rangle = B_n^+(\mathbf{k})|\,\Phi_0\rangle$ is, according to Equation (1), generated by operating on $|\Phi_0\rangle$ with:

$$B_n^+(\mathbf{k}) = N^{-1/2} \sum_{i} B_n^+(\mathbf{R}_i)\, e^{i\mathbf{k}\mathbf{R}_i} \tag{3''}$$

Well-studied examples of Frenkel excitons are the low lying singlet and triplet electronic excitations in aromatic crystals, where $B_n^+(\mathbf{R}_i) = b_n{}^+(\mathbf{R}_i)\, b_0(\mathbf{R}_i)$ is understood to describe the promotion of a π-electron from a bonding to an antibonding molecular orbital. Under the same heading of small radius excitons, one can classify the *electron-transfer excitons*, which involve the transfer of an electron from one site $\mathbf{R}_i$ to another nearby site, $\mathbf{R}_j = \mathbf{R}_i + \boldsymbol{\rho}$: they are described in terms of localized configurations given by Equation (2) with $U_{nv}(\boldsymbol{\beta}) = \delta(\boldsymbol{\beta} - \boldsymbol{\rho})$, so that one has:

$$|\mathbf{R}_i\rangle = B_{n\rho}^+(\mathbf{R}_i)\,|\Phi_0\rangle,$$

where:

$$B_{n\rho}^+(\mathbf{R}_i) = b_n^+(\mathbf{R}_i + \boldsymbol{\rho})\, b_0(\mathbf{R}_i). \tag{4'}$$

The exciton states are accordingly represented by $|\mathbf{k}\rangle = B_{n\rho}^+(\mathbf{k})|\,\Phi_0\rangle$, with:

$$B_{n\rho}^+(\mathbf{k}) = N^{-1/2} \sum_{i} B_{n\rho}^+(\mathbf{R}_i)\, e^{i\mathbf{k}\mathbf{R}_i}. \tag{4''}$$

These excitons are typical for the alkali halide crystals (e.g., NaCl, where an electron is transferred from a Cl^- to a Na^+ ion); it is interesting to note that this was already suggested in the very first studies of solid state spectroscopy by Hilsch and Pohl (1928). More recently, evidence for the occurence of such charge-transfer excitons was also found in the organic molecular materials, where an electron is removed from the highest bonding orbital of one molecule to the lowest antibonding orbital of a neighboring molecule.

5.   *The large-radius excitons* are generally described in the well-documented *Mott-Wannier* model. They are observed in semiconductors (e.g., Cu_2O, Ge, ...) characterized by small interband energy gaps, and relatively large values of the static dielectric constant ε_0 which tends to reduce the effective Coulomb interaction and to increase the mean distance in the correlated electron-hole pair. Usually, the Wannier excitons are constructed using Bloch states for the hole and the electron, rather than localized molecular orbitals: the exciton is taken to consist of an electron wavepacket made up of states from the conduction band and a hole wavepacket constructed with valence band states, these being multiplied by an envelope function depending on the relative separation β of the electron-hole pair. If, conversely, the localized representation of Section 3 is used for the electron and the hole, the excitons are expressed in terms of locally excited electron-hole pair configurations of the form (2), where the expansion coefficients $U_{n\nu}(\beta)$ now depend upon the relative coordinate β as *hydrogen-like wave functions*, with the index ν representing the characteristic set of the *nlm* quantum numbers. From this the Wannier exciton states are obtained as $|k\rangle = B_{n\nu}^+(k)|\Phi_0\rangle$, with:

$$B_{n\nu}^+(\mathbf{k}) = N^{-1/2} \sum_i B_{n\nu}^+(\mathbf{R}_i)\, e^{i\mathbf{k}\mathbf{R}_i}, \quad \nu = n, l, m. \tag{5}$$

LOCALIZED EXCITONS IN A FLUOROBENZENE CRYSTAL

M. PIERRE

Laboratoire de Spectrométrie Physique associé au C.N.R.S.,
Université Scientifique et Médicale de Grenoble,
B.P. 53, 38041 Grenoble-Cedex, France

There are two descriptions for the motion of elementary excitations in molecular crystals: the band model, where collective exciton states are assumed to be nearly stationary (delocalized states) [1–4] and the molecular mode, where localized molecular states are perturbed by a stochastic time-dependent hamiltonian [5–6].

At low temperatures we have to use the first model, initially suggested by A. S. Davydov: the fluctuations in energy and in orientation of the molecule are so small, that in the first approximation, we can neglect molecular kinetic and librational energies in the complete crystal hamiltonian (delocalized exciton).

1. Delocalized Exciton in an Infinite Crystal of Fluorobenzene

1.1. EXPERIMENTAL FOUNDATIONS

It is well known that the spectra of molecular crystals are quite similar to the vapour phase spectra: the molecules preserve their individuality [1] (Figure 1). In fluoroben-

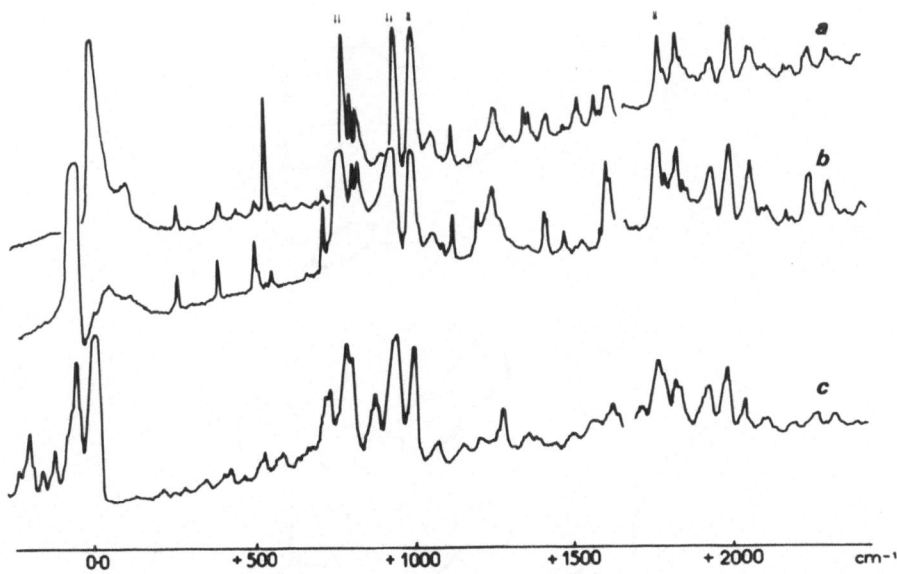

Fig. 1.   Comparison between absorption spectra of fluorobenzene: (a) crystal at 4 K; the field is perpendicular to the optical axis Z. (b) crystal at 4 K; the polarization is parallel to Z. (c) vapour phase.

zene the oscillator strength of the O—O band of the first singlet electronic transition
($f \simeq 0,01$ [7]) is large enough to justify the use of the dipolar model [1], and it is
sufficiently small to allow a perturbation treatment.

1.2. The dipolar model

The coulombic interaction operator $\hat{V}$ between fixed molecules can be expanded in a
power series of inverse distances between molecules $1/\rho$ (with $\boldsymbol{\rho} = \mathbf{a}' - \mathbf{a}$). The first
term corresponds to the dipolar approximation:

$$\hat{V} = \rho^{-5}(\hat{\mathbf{d}}_{\mathbf{a}}\hat{\mathbf{d}}_{\mathbf{a}'}\rho^2 - 3(\hat{\mathbf{d}}_{\mathbf{a}}\boldsymbol{\rho})\,(\hat{\mathbf{d}}_{\mathbf{a}}\boldsymbol{\rho})) = M(\hat{\mathbf{d}}_{\mathbf{a}}\hat{\mathbf{d}}_{\mathbf{a}'})$$

with

$$\hat{\mathbf{d}}_{\mathbf{a}} = \begin{pmatrix} \mathbf{d}_{\mathbf{a}}^0 & \mathbf{d}_{\mathbf{a}}^{01} \\ \mathbf{d}_{\mathbf{a}}^{01} & \mathbf{d}_{\mathbf{a}}^1 \end{pmatrix}$$

where $\mathbf{d}_{\mathbf{a}}^0$ and $\mathbf{d}_{\mathbf{a}}^1$ are the static dipole moments of the molecule in the fundamental
and in the first singlet electronic excited state. $\mathbf{d}_{\mathbf{a}}^{01}$ is the transition moment.

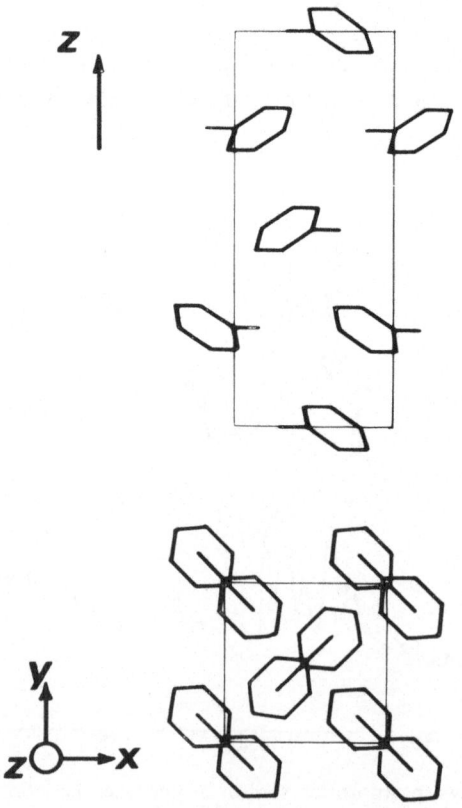

Fig. 2.   The elementary cell of fluorobenzene crystal of symmetry D_4^4 (14).

For a rigidly fixed infinite crystal, the energy and the eigenfunctions take always the form (for symmetry reasons):

$$E(\mathbf{k}) = \varepsilon^1 - \varepsilon^0 + D + \tfrac{1}{2} \sum_{\mathbf{a} \neq 0} \left(m(\mathbf{a})\, e^{-i\mathbf{k}\mathbf{a}} + m^*(-\mathbf{a})\, e^{i\mathbf{k}\mathbf{a}} \right)$$

$$= \varepsilon^1 - \varepsilon^0 + D + \mathscr{E}(\mathbf{k})$$

$$\psi_{\mathbf{k}}(\mathbf{r}) = \sum_a f_L(\mathbf{r} - \mathbf{a})\, e^{i\mathbf{k}\mathbf{a}}$$

(for brevity we supposed just one molecule in the elementary cell).

But, in the dipolar model, if

$$f_L(\mathbf{r} - \mathbf{a}) = \chi^1(\mathbf{r} - \mathbf{a}) \prod_{\mathbf{a}' \neq \mathbf{a}} \chi^0(\mathbf{r} - \mathbf{a}')$$

then

$$D \quad = 2 \sum_a M(\mathbf{d}_\mathbf{a}^1 - \mathbf{d}_\mathbf{a}^0, \mathbf{d}_{\mathbf{a}'}^0)$$

$$\mathscr{E}(\mathbf{k}) = \sum_a M(\mathbf{d}_\mathbf{a}^{01}, \mathbf{d}_{\mathbf{a}'}^{01}) \cos(\mathbf{k}\rho).$$

Such a calculation for a fluorobenzene crystal (Figure 2) quadratic with 4 molecules in the elementary cell gives 4 us excitonic bands which are solutions of the secular equation:

$$\begin{vmatrix} L_{11} - \mathscr{E} & L_{12} & L_{13} & L_{14} \\ L_{12} & L_{22} - \mathscr{E} & L_{14} & L_{24} \\ L_{13} & L_{14} & L_{11} - \mathscr{E} & L_{12} \\ L_{14} & L_{24} & L_{12} & L_{22} - \mathscr{E} \end{vmatrix} = 0.$$

The term D and the matrix elements L_{ij} were calculated by the well known Ewald method [8, 9].

1.3. THEORETICAL RESULTS AND DISCUSSION

The shift between crystal and vapour spectra is found experimentally v(crystal) − v(vapour) = 17 cm^{-1}.

But a correct comparison between theory and experimental results must take into account molecular deformation in the crystal. By the means of MO calculations in the CNDO approximation [10], we have evaluated the variation of the transition energy for this deformation (350 cm^{-1}), the sum for D yields −347 cm^{-1}, and from this we find for the net shift 3 cm^{-1} in good agreement with the experimental value.

The stationary states $\mathscr{E}_\mu(\mathbf{k})$ are given in Figure 3 for some points of the first Brillouin zone (Figure 4).

As usual [11, 9], for small $\mathbf{k}$, the energy bands are strongly dependent on the direction of $\mathbf{k}$. The bands (1, 3 and 4) have a resultant dipolar transition moment respectively parallel and perpendicular to the tetragonal axis, band 2 is forbidden.

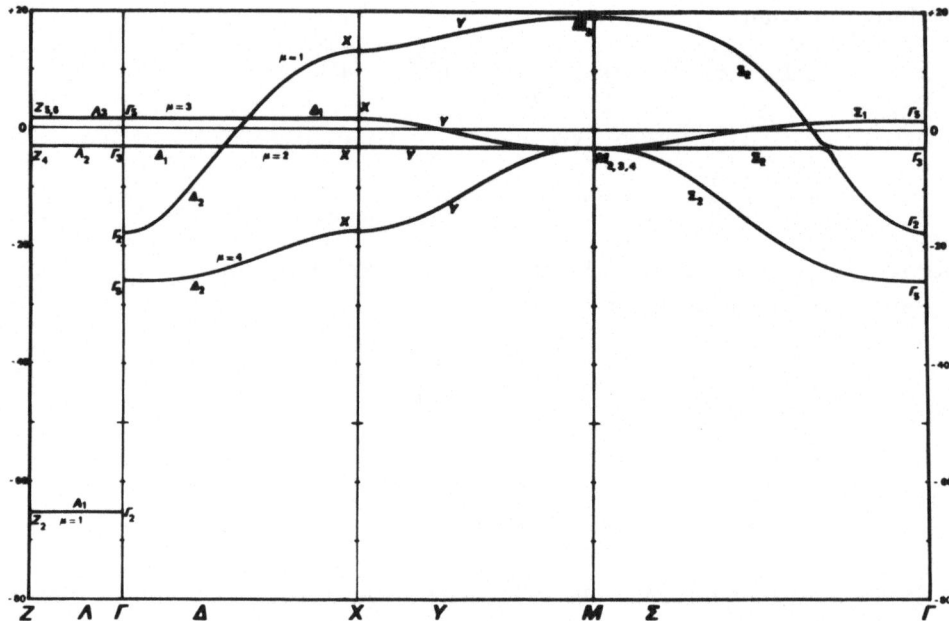

Fig. 3.   Excitonic energy $\mathscr{E}_\mu$ versus **k** for some directions of the Brillouin zone. (The value of the oscillator strength is 0.01).

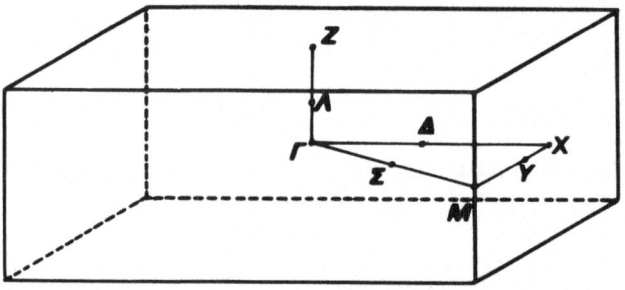

Fig. 4.   First Brillouin zone of fluorobenzene.

The O—O line in the crystal splits [12] into 2 polarized components whose profile and breadth cannot be explained by the simple delocalized exciton.

2. **Localized Exciton in the Crystal of Fluorobenzene**

2.1. INTRODUCTION

The dipolar model of the preceding section was based upon the assumption that the dimensions of the excitation are large compared to the wavelength of the exciting light.

For an idealized lattice whose molecules are rigidly fixed (that is OK if we neglect the zero-amplitudes of vibration) this approximation is always realized. At higher temperatures, the amplitudes of molecular fluctuations increase and the dimension of the exciton may thereby be smaller than the wavelength.

In this situation we suppose that the exciton can described by a wavevector $\bar{\mathbf{k}}$ averaged in the following way:

$$\Phi_{\bar{\mathbf{k}}}(\mathbf{r}) = \int_{\infty} c_{\mathbf{k}} \, \psi_{\mathbf{k}}(\mathbf{r}) \, \mathrm{d}^3 \mathbf{k} \, .$$

We can define now energy bands as functions of $\mathbf{k}$:

$$E(\bar{\mathbf{k}}) = \int_{\infty} |c_{\bar{\mathbf{k}}, \mathbf{k}}|^2 \, \mathscr{E}(\mathbf{k}) \, \mathrm{d}^3 \mathbf{k} \, .$$

The selection rules will be $\bar{\mathbf{k}} = n\mathbf{Q}_0$ where n is the refractive index of the crystal and $\mathbf{Q}_0$ the wavevector of the incident light (we take $n \approx 2$).

The physical meaning of the representation $\Phi(\mathbf{k})$ is that the actual exciton is a superposition of stationary states with different $\mathbf{k}$. The perturbation (phonon, impurities ...) introduced by this model causes the broadening of the excitonic components, and a quantitative line shape may be determined knowing the coefficients $c_{\mathbf{k}, \bar{\mathbf{k}}}$.

2.2. APPLICATION TO THE FLUOROBENZENE

It was assumed that the distribution coefficients $c_{\mathbf{k}, \bar{\mathbf{k}}}$ were isotropic in $\mathbf{k}$, with a Gaussian decay (spherical Gaussian approximation). The larger the delocalization in the k-space, the smaller the dimensions of the part of the crystal embraced by the excitation. The new energy bands corresponding to 3 different dimensions of localized excitons ($R_0 = 0.05$; $R_0 = 0.10$; $R_0 = 0.15\,\mu$, where R_0 is the width of the Gaussian distribution) are described on the higher part of Figure 5. They determine the theoretical line shape of the absorption spectrum. The best agreement with experimental results (Figure 6) is obtained taking $R_0 \simeq 0.10\,\mu$ and adjusting the oscillator strength to 0.02.

In the polarization perpendicular to Z, the Davydov component splits into 2 bands: a large band IV ($37\,817\,\mathrm{cm}^{-1}\ \delta = 30\,\mathrm{cm}^{-1}$) and a sharp line III ($37\,838\,\mathrm{cm}^{-1}$) partially superposed with the phononic structure.

In the polarization parallel to Z, the other Davydov component splits into 2 bands: a very large structured band II ($37\,769\,\mathrm{cm}^{-1}$, $\delta = 50\,\mathrm{cm}^{-1}$) preceding a strong line I ($37\,796\,\mathrm{cm}^{-1}$, $\delta = 2\,\mathrm{cm}^{-1}$). These bands correspond to the $\mu = 1$ dispersion curve (Figure 5).

In the band II, the exciton wave vector $\mathbf{k}$ is oblique with respect to $\bar{\mathbf{k}}$ (longitudinal part of the wave packet). Such a structure is better resolved for samples whose Z axis forms small angles with the incident wave vector of light $\mathbf{Q}_0$. On Figure 7, this angle is $5°$ nearly, and we can see only the sharp line I (see ref. 13).

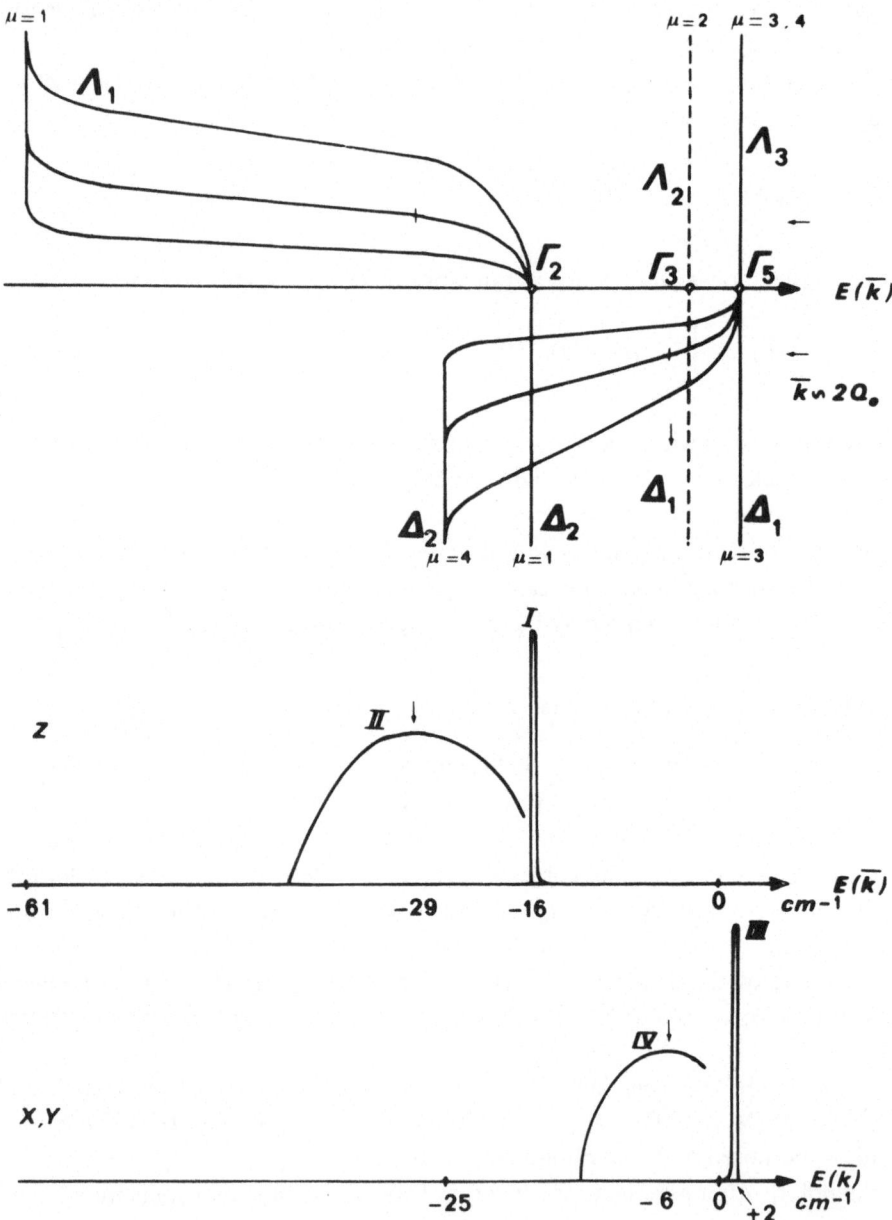

Fig. 5.   Dispersion curves for 3 dimensions of the excitation zone: (a) $R_0 = 0.05\,\mu$. (b) $R_0 = 0.10\,\mu$. (c) $R_0 = 0.15\,\mu$. The lower part shows the theoretical spectrum for $f = 0.01$.

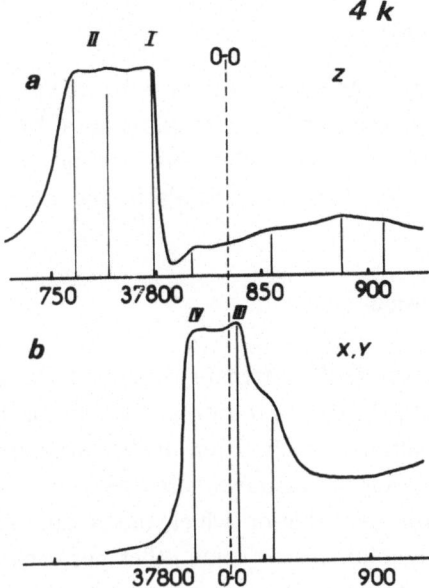

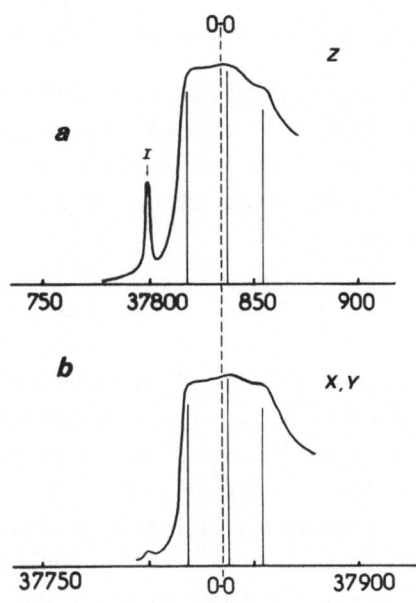

Fig. 6. Microdensitometric trace of the spectrum (obtained on photographic plates under resolution of 20.5 Å/mm at 3300 Å). (a) Polarization parallel to Z. (b) Polarization perpendicular to Z.

Fig. 7. Spectrum of sample whose Z axis forms a little angle with the direction of propagation of light.

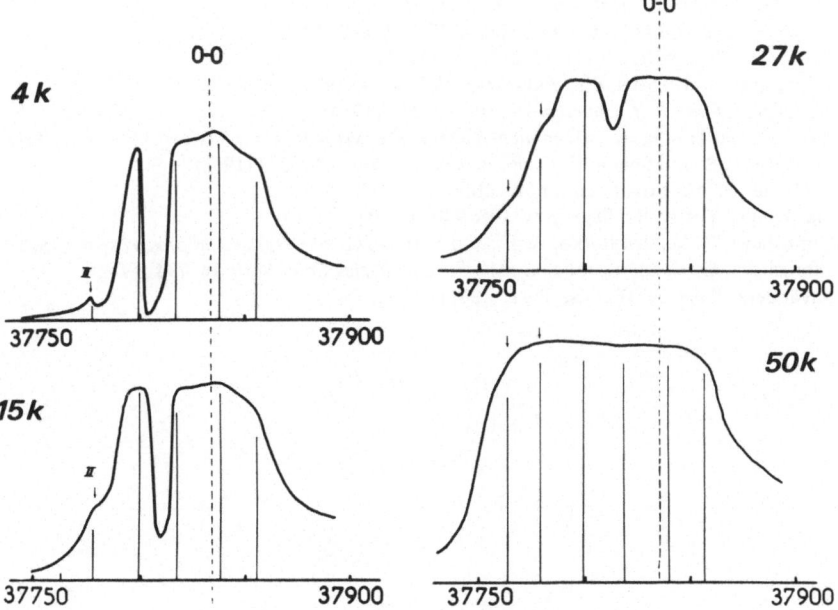

Fig. 8. Spectra obtained at different temperatures for samples whose Z axis is 15° oblique with respect of Q_0.

(1) At 4 K; (2) at 15 K; (3) at 27 K; (4) at 50 K.

2.3. Temperature variation of the spectrum

These interpretations have been confirmed by temperature studies: when the temperature increases, the bands II and IV become broader and stronger (Figure 8). This result does not leave any doubt on the phononic origin of the broadening: at 50 K the librational levels are excited and we cannot separate bands in the Davydov components.

3. Conclusion

The purpose of the present work was to interprete the internal structure of Davydov components as a phenomenon correlated with the delocalization and the localization of the exciton. This theory gives a good qualitative explanation of the experimental results. We note that the model of localized excitons assumed here does not contain any assumption concerning the origin of the perturbation which mixes different k states: the coefficients c_k reflect indifferently phonons or some other perturbation. In the case of the fluorobenzene crystal, it seems that the structure is essentially due to lattice vibrations.

References

1. Davydov, A. S.: *Theory of Molecular Excitons*, Plenum Press (1971).
2. Craig, D. P. and Walmsley, S. H.: *Physics and Chemistry of the Organic Solid State*, Vol. 1, Interscience Publishers (1963).
3. Mc Clure, D.S.: *Solid State Physics* 8, 1 (1959).
4. Davydov, A. S.: *Bulletin Academy Sci. USSR* 34, 416 (1970).
5. Haken, H.: *Zeitschrift für Physik* 262, 135 (1973).
6. Aslangul, C. and Kottis, Ph.: *Phys. Rev.* 10B, 4364 (1974).
7. Sponer, H.: *Journal of Chemical Physics* 22, 234 (1954).
8. Born, M. and Huang, K.: *Dynamical Theory of Crystal Lattices*, Oxford University Press (1968).
9. Davydov, A. S. and Sheka, E. F.: *Physica Status Solidi* 11, 877 (1965).
10. Merienne, M. F.: private communication.
11. Fox, D. and Yatsiv, S.: *Phys. Rev.* 108, 940 (1957).
12. Klimusheva, G. V., Prikhotko, A. F., and Soroka, G. M.: *Optics and Spectroscopy* 25, 197 (1968).
13. Klimusheva, G. V. and Soroka, G. M.: *Soviet Physics Solid State* 14, 794 (1972).
14. Clavaguera-Playa, N.: Thesis, Paris (1970).

PART IV

ELECTRON LOCALIZATION AND CHEMICAL REACTIVITY

'SPIN-DEPENDENCE' OF THE CHEMISTRY
OF OPEN-SHELL REACTANTS

M. STEPHENS

Centre de Mécanique Ondulatoire Appliquée, 23 Rue du Maroc, 75019 – Paris, France

Abstract. An attempt to describe the observed different chemical reactivity of atomic oxygen in its isoconfigurational lowest singlet and triplet states with paraffins and olefins through interpretation of the state charge and triplet spin density distributions is presented. Ambiguities in the description obtained and in attempts to generate qualitative model descriptions of the consequences of the electron pair structure of reacting species are described.

1. Introduction

Since the studies of methylene addition to olefins by Skell and coworkers [1], much debate has ensued over the relationship between the reaction mechanisms and spin multiplicity of open-shell reactants [2, 3].

It has been established, for instance, that $O(^1D)$ [4], $S(^1D)$ [5] and $CH_2(^1A_1)$ [6] insert into paraffinic CH bonds, while $O(^3P)$ and $CH_2(^3\Sigma_g^-)$ only abstract H atoms and $S(^3P)$ is unreactive with paraffins under the same conditions. The three singlet species add stereospecifically to olefin double bonds [8–10], whereas triplet oxygen or methylene addition results in some loss of reactant geometrical isomeric purity.

The qualitative interpretation of the methylene results originally proposed by Skell *et al.* [1] has been applied to the other cases as well [7, 11]. The singlet species, it is assumed, can insert into paraffinic CH bonds through concerted formation of CX and XH bonds ($X = {}^1O$, ${}^1S$, or ${}^1CH_2$), and add stereospecifically to olefin bonds through concerted formation of two CX bonds. Thus

The triplet species, on the other hand, cannot undergo such a 'spin-forbidden' concerted formation of two bonds in either case and instead initially forms only one bond, to an exposed paraffinic hydrogen or to an olefin carbon (in the latter case producing an open-ring diradical intermediate capable of internal rotation around the now formal 'single' carbon-carbon bond). The result is predicted to be H-atom abstraction from paraffins and non-stereospecific cycloaddition to olefins (after the diradical undergoes an intersystem crossing from the triplet to singlet surface). Thus

These model mechanisms have been strongly criticized [3] and an alternate symmetric path for the triplet cycloaddition reaction proposed on several theoretical grounds [12–14].

If one can in fact define a spin-dependence of these reactions, it must be of a subtle nature. Thus $S(^3P)$ adds to olefins with retention [9] of reactant stereochemistry .*

Furthermore, one can relate the distribution of observed reaction products to the shape of the potential surfaces linking reactants to products in the various electronic states. The Hamiltonian operator containing only kinetic and electrostatic potential terms provides a sufficiently accurate description of the contributions to the system energy to predict the potential surfaces accounting for the observed chemical behaviour. Thus the singlet-triplet energy difference cannot be attributed to an explicit spin-dependent contribution to the energies, but rather the restrictions on the allowable spatial form of the system wave function imposed by the antisymmetry requirement. This point has been repeatedly emphasized by Matsen [16]. Hence arguments based on 'spin-flips' or the like can provide only a model or facsimile of the actual mathematical workings of the Schrödinger equation [17].

The most detailed studies of the potential surfaces for the reactions of $CH_2^{[2]}$, S [18], and NH [19] all indicate clearly differentiable energetically preferred symmetric lowest singlet and asymmetric lowest triplet reaction paths (with a barrier to rotation of the terminal methylene in the sulphur case sufficient to preserve reactant stereochemistry).

Thus the question remains as to whether one can find a faithful monitor of the differences between singlet and triplet electronic structure applicable over the complete potential surfaces, whence a simple predictor of the qualitative differences in the surface topologies and hence reaction channels and products. Such continuous monitors were sought in terms of the state charge densities and triplet state spin density distribution for reasons outlined below.

---

* Explanations for this observation consistent with the Skell mechanisms have invoked [15] bonding between the sulphur and terminal methylene 'non-bonding' valence orbitals, or an enhanced intersystem crossing due to the heavy sulphur spin-orbit coupling.

2. Model Reactions and Potential Surfaces

To characterize the differing tendencies of the lowest isoconfigurational (p^4) states of atomic oxygen, ^3P and ^1D, to insert into or abstract a H atom from CH bonds, Bader and Gangi [20–22] computed the lowest triplet and lowest and first excited singlet state surfaces for the model system $H_2(^1\Sigma_g^+)+O(^3P, {}^1D)$. A similar study was made by Bader and Stephens [23, 24] of the lowest singlet and triplet surfaces for the model cycloaddition system $C_2H_4(^1A_g)+O(^3P, {}^1D)$.

The H_2+O calculations were computed in the spin-unrestricted LCAO-MO-SCF approximation using a (53/2) gaussian basis to map the general features of the surfaces, with important points recalculated in a $(10\ 5\ 2/4\ 1)\to[5\ 3\ 2/2\ 1]$ double-zeta plus polarizing function gaussian basis. Limited CI was performed where necessary to describe a separate O (^1D) atom, linear open-shell singlet H—O—H and H—H—O adducts, and low-symmetry portions of the surface where triplet contamination of an UHF singlet function may be severe.

The energetically most favorable path (Figure 2, reference 22) for the lowest singlet state was found to be an approach of the oxygen atom initially perpendicular to the bond. The optimum approach angle decreases (from 90°) to about 70° for the oxygen three to two AU from the midpoint of the H—H bond, but increases to 90° again as the adduct approaches the equilibrium geometry of ground state H_2O. The path is energetically downhill throughout. Abstraction of a H atom by oxygen in the lowest singlet state was shown to be possible for a strongly unsymmetric path. In contrast, both the excited singlet and lowest triplet states showed (Figure 3, reference 22) large energy barriers for a symmetric approach (61 kcal mole^{-1} for the singlet, 80 kcal mole^{-1} for the triplet) leading in both cases to a loosely bound linear structure unstable to unsymmetric vibration. The triplet state was found (Figure 2, reference 20) to face a minimum energy barrier of 23 kcal mole^{-1} for a linear H atom abstraction producing H plus CH radicals.

The C_2H_4+O model cycloaddition potential surfaces [23, 24] were also computed by an UHF-LCAO-MO-SCF calculation, using a $(13\ 6/4)\to[7\ 3/2]$ double-zeta equivalent basis. As the computations were much larger than for the H_2+O system, polarization functions could not be used and the geometric optimization of the paths was less complete. Again, limited CI was performed to properly describe O (^1D) and determine the low-symmetry sections of the singlet surface.

The most favorable singlet path found was an energetically smoothly downhill symmetric approach of the oxygen along the bisector of the carbon-carbon axis and perpendicular to the plane of the ethylene. In contrast, the most favorable (but still repulsive) triplet path found was strongly asymmetric. The lowest available triplet-singlet surface crossing, hence channel to the known singlet cyclic products, was found to be over a barrier of 36 kcal mole^{-1}. The symmetric triplet path was found to be highly repulsive, and the singlet on the asymmetric path (optimal for the triplet state) moderately so; thus a clear differentiation between lowest singlet and triplet state paths was found possible. No stable triplet intermediate was found.

After the essential features of the $H_2 + O$ potential surfaces were established, a description of the mechanisms of the reaction paths was sought in terms of the charge and spin density distributions [21]; the initial interpretation of the $H_2 + O$ surfaces was later extended to the case of the $C_2H_4 + O$ surfaces [24].

3. The Spin Density Distribution

The Pauli exclusion principle is a prime factor determining allowable electronic wave functions, hence properties, of the system eigenstates. This includes the electronic charge and pair distributions which together fully describe the chemically significant electronic structure (since there are only one- and two-electron operators in the Hamiltonian required to attain chemical accuracy (1–3 kcal mole^{-1}) in the calculations).

A meaningful continuous monitor of the energy difference between singlet and triplet states requires some indicator of the difference in the electronic pair distribution. One may then hope to explain the energy difference due to the different effects of the fermi correlation restrictions on the pair density for each state,

$$\lim_{r_2 \to r_1} \Gamma^{(2)} \overset{\alpha \quad \alpha}{(\mathbf{r}_1, \mathbf{r}_2)} = \lim_{r_2 \to r_1} \Gamma^{(2)} \overset{\beta \quad \beta}{(\mathbf{r}_1, \mathbf{r}_2)} = 0.$$

It was thought that the triplet state spin density might provide such a qualitative measure, and perhaps suggest a simple mechanistic terminology to explain the singlet-triplet differences in reactivity. If found possible, this would provide a useful link between the experimental results and the mass of information contained in the abstract mathematical description of the wave function.

For states of 'good' S (total electronic spin angular momentum) and M_s (its z-component) one can rigorously define contributions to the distributions of total electronic charge density, $\rho(\mathbf{r})$, and spin density, $\sigma(\mathbf{r})$, in terms of component α- and β-spin component functions, $\rho^\alpha(\mathbf{r})$ and $\rho^\beta(\mathbf{r})$,

$$\rho(\mathbf{r}) = \rho^\alpha(\mathbf{r}) + \rho^\beta(\mathbf{r})$$

$$\sigma(\mathbf{r}) = \rho^\alpha(\mathbf{r}) - \rho^\beta(\mathbf{r}).$$

Integration of $\rho(\mathbf{r})$ and $\sigma(\mathbf{r})$ over all space yields the total number of electrons, N, and excess of α- over β-spin electrons, $(N_\alpha - N_\beta)$, respectively.

$$\int_{\text{All Space}} \rho(\mathbf{r}) \, d\mathbf{r} = N$$

$$\int_{\text{All Space}} \sigma(\mathbf{r}) \, d\mathbf{r} = N_\alpha - N_\beta.$$

For spin-unrestricted single determinantal wave functions (constructed from distinct sets of singly occupied α-spin and β-spin spatial orbitals, $\{\phi_i^\alpha\}$ and $\{\phi_i^\beta\}$), $\rho(\mathbf{r})$ and $\sigma(\mathbf{r})$ can be defined in terms of the orbital densities for the state of highest $M_s(=S)$ value as follows:

$$\rho(\mathbf{r}) = \sum_i^\alpha \phi_i^{*\alpha}(\mathbf{r})\,\phi_i^\alpha(\mathbf{r}) + \sum_i^\beta \phi_i^{*\beta}(\mathbf{r})\,\phi_i^\beta(\mathbf{r}) \equiv \rho^\alpha(\mathbf{r}) + \rho^\beta(\mathbf{r})$$

$$\sigma(\mathbf{r}) = \sum_i^\alpha \phi_i^{*\alpha}(\mathbf{r})\,\phi_i^\alpha(\mathbf{r}) - \sum_i^\beta \phi_i^{*\beta}(\mathbf{r})\,\phi_i^\beta(\mathbf{r}) \equiv \rho^\alpha(\mathbf{r}) - \rho^\beta(\mathbf{r})\ .$$

For closed-shell species, the Restricted Hartree Fock (RHF) method employs $N/2$ pairs of doubly-occupied spatial orbitals, an α- or β-spin function associated with each $\phi_i(\mathbf{r})$. In this case

$$\rho^\alpha(\mathbf{r}) = \rho^\beta(\mathbf{r})$$

$$\rho(\mathbf{r}) = 2\rho^\alpha(\mathbf{r}) = 2\rho^\beta(\mathbf{r})$$

$$\sigma(\mathbf{r}) = 0\,.$$

For an open-shell system, a RHF ($M_s = S$) wave function will yield a $\sigma(\mathbf{r})$ showing only a distribution of the excess α-spin over the molecule. However, an UHF function for the same system describes regions of excess α- and β-spin throughout the molecule (with an integrated value still $N_\alpha - N_\beta$). The 'polarization' patterns thus defined have been shown for several cases [21] to agree qualitatively with the results from more detailed functions[‡].

The maps used in the interpretive work [21, 24] may be expected[‡‡] to faithfully reproduce the qualitative features of the spin polarizations of a fully correlated function.

The spin distributions for all states of an electronic term differ from that of the state of highest M_s value by only a multiplicative constant [25], M_s/S[†]. A qualitative interpretation can be generated from the properties for the (computationally convenient) state of highest M_s value ($= 1$ for the degenerate triplet term).

The spin density maps were interpreted to provide a descriptive model relating the effects of the Pauli principle on the pair distribution, whence on the (one particle) charge distribution. From the latter (plus the Hellmann-Feymann theorem) one can

[‡] There may be quantitative errors in the LCAO-UHF spin density at the nuclei of the order of 100% (in comparison with experimental results on ESR Fermi contact term hyperfine splittings) due to the limitations of the UHF method [26] and (necessarily) incomplete basis [27]. However, because gaussian functions (used in this work) are notoriously poor in describing nuclear density cusps, the densities for points removed from the nuclei should be of greater accuracy.

[‡‡] With reservations due to possible mixing, or 'contamination', of spin states in the UHF method. Contamination was found to be small in the calculations made, as $\langle S^2 \rangle$ was always found close to 2.00 (the value for a pure triplet), being in the range 2.00–2.03 for the $H_2 + O$ work, and 2.01–2.43 for the $C_2H_4 + O$ studies.

[†] The implications of this for the $M_S = 0$ component of the triplet are discussed below in Section 4.

then rationalize the difference in the forces exerted on the nuclei (hence reaction paths) between singlet and triplet states.

In this spirit two types of spin polarization were defined [21] from the variations in the computed spin density pattern for the triplet surface. In a linear approach of the triplet species to the (closed-shell) substrate, the unpaired α-density may be said to 'uncouple' the spins of a bonded α-β pair in the substate, inducing excess β-density near the closer nucleus and α-density near the more distant centre. This corresponds, then, to a description of the substrate 'bond' being 'broken' by localization of its constituent α- and β-spin density at opposite nuclei and concomitant decrease of the binding charge density between the nuclei.

In an approach of the triplet species along the bisector of the substrate internuclear axis, a 'transfer' of unpaired α-density to both substrate nuclei defines an unpairing of the spins of the substrate bonding electrons and bond weakening. The localization of charge density of identical spin on each substrate nucleus parallels the great reduction in charge in the substrate binding region.

At intermediate approach angles, a polarization combining features of both extremes may be expected.

A strong parallel was found [24] between the interpretations possible for the singlet insertion and cyclo-addition mechanisms. The energetically downhill symmetric path in each can be likened to smooth concerted bond formation. Reaction with a 'single bond' leads to separation of the substrate nuclei and insertion of the oxygen atom. The reaction with an olefin 'double bond' results instead in cyclic product formation due to the resistance to separation of the substrate nuclei because of the larger residual binding density between them.

The triplet spin distributions for symmetric insertion and addition can be interpreted as describing a different mechanism. Excess α-spin is 'transferred' from the oxygen to both nuclei of the substrate, at the same time as the substrate binding density is localized on the nuclei rather than between them. In the insertion case, sufficient binding density is removed to permit separation of the nuclei. Reaccumulations of density in the two new binding regions between the approaching atom and separating nuclei lead to the two new accumulations of binding density, the 'bonds' of the insertion product.

Thus can be explained the formation of a linear H—O—H intermediate unstable to dissociation into OH plus H radicals predicted by the topology of the potential surface. In the cycloaddition case, the charge decrease is insufficient to completely disrupt the stability of the more strongly bonded substrate nuclei. The symmetric path is then simply highly repulsive and unproductive.

Strong similarities were found between the linear triplet abstraction and asymmetric cycloaddition mechanisms as well. In the former case, abstraction of a hydrogen atom occurs since there is sufficient removal of the density binding it to the substrate molecule to force scission of the original bond. This is accompanied by a spin 'uncoupling' polarization. In the latter case, an apparent transfer of charge density from the carbon-carbon binding region to a carbon-oxygen binding region was found to be paralleled by a spin pattern suggesting α-spin 'transfer' from the oxygen to the terminal methylene, 'uncoupling' of density in the carbon-carbon binding region, and 'recoupling' of density in the carbon-oxygen binding region. The charge decrease is insufficient to lead to complete rupture of the carbon-carbon bonding. Thus can be rationalized the computed energetically preferred open-ring diradical transition state rather than separated radicals.

In both cases, an unsymmetric singlet attack defines a slightly repulsive surface. This leads to an unfavourable (but experimentally observed) abstraction 'by chance' from a paraffinic bond.

Comparing the polarization description to Skell's intuitive argument, one can see a strong analogy between his rationale and that of spin polarizations defining preferred regions of coupled binding density. However, both descriptions contain limitations and problems of consistent interpretation as described below.

4. Limitations

With use, model descriptions tend to acquire the attribute of causality. Thus, for instance, 'concerted bond formation' may be 'forbidden' for a triplet state without a preceeding 'spin-flip' to allow 'pairing' and hence formation of a new 'bond'. As long as the model predictions consistently correlate with the experimentally observed results claimed to depend on the mechanism defined, then the model may be accepted as a faithful predictor. As long as the model does not violate any of the postulates or derived predictions of a more detailed mathematics known to accurately describe the observed results, then the model provides a valid (but certainly not necessarily unique) representation of that more detailed theory.

Several problems were encountered in finding consistent connections from the model description presented above to both experiment and theory.

To assert that $\sigma(\mathbf{r})$ contains the necessary information to differentiate singlet from triplet reactions implies that the chemistry of the $M_S = 0$ component of the three degenerate (in the approximation used) triplet states differs from that of the $M_S = \pm 1$ components, since $\sigma(\mathbf{r}) = 0$ for the $M_S = 0$ component. One can ascribe a rotation of the spin quantization axes to consistently work with an $M_S = +1$ component. However, if the experimental system were subjected to a magnetic field, the small energy differences between the now non-degenerate triplet components would be insufficient to lead to radically different chemical behaviour.

For states of identical 'frozen-orbital' configuration (i.e. open-shell singlet built from triplet orbitals), to first order one can relate the singlet/triplet energy difference strictly to the '$2K$' exchange contribution. This leads to two problems of interpreting energy difference through $\sigma(\mathbf{r})$ (assuming that $\sigma(\mathbf{r})$ correlates in some manner with the energy contributions due to the exchange pair density).

The first lies in the consistency of the electronic configuration on the reaction path. Thus, for the $H_2 + O$ symmetric (C_{2v}) path, the following correlations were computed:

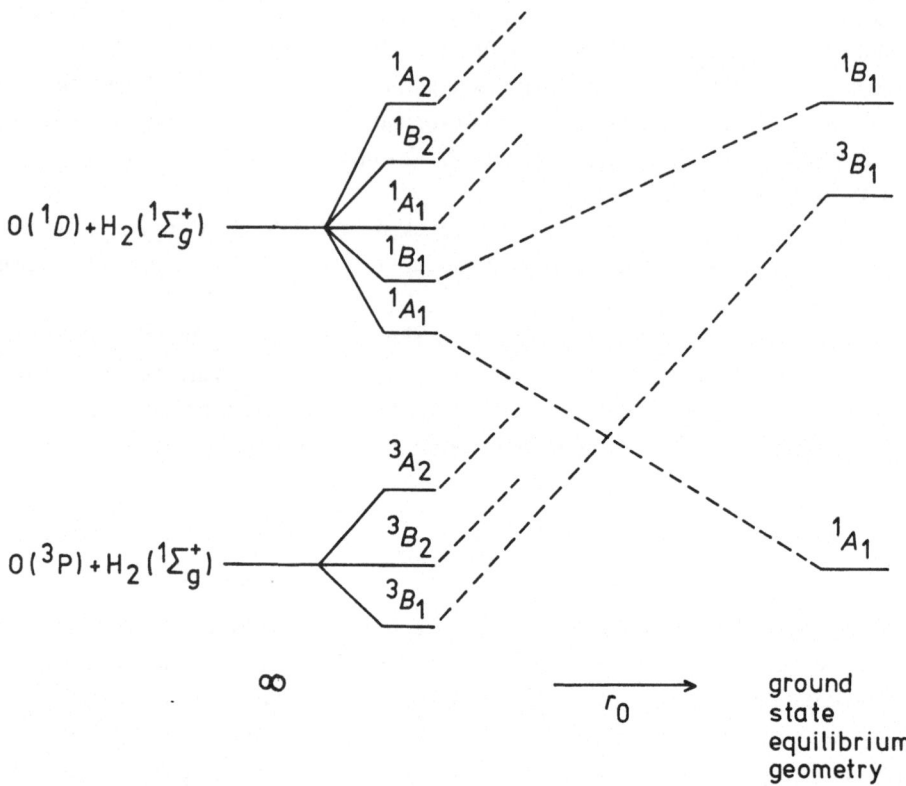

As r_0 (the distance of the oxygen atom to the midpoint of the H_2 bond) decreases from ∞ to a value typical of 'normal' bonded distances, the configuration of the lowest system state changes from p^4 for the oxygen in a spherical field to A_1 in the C_{2v} field provided by the hydrogen molecule. Whereas the lowest triplet state, also p^4 in a spherical field at $r_0 = \infty$, retains a $^3B_1$ configuration along its path the $^1B_1$ configuration becomes the first excited singlet state, diverging from its long-range degeneracy with the lowest $^1A_1$ singlet. Attributing differences in adduct energy to $\sigma(\mathbf{r})$ would be more appropriate in distinguishing $^3B_1$ and $^1B_1$ surfaces rather than those for the $^3B_1$ and $^1A_1$ states. The $^1B_1$ symmetric path was, in fact, also calculated [22] and found to have the same qualitative features as the $^3B_1$ path. The slightly more compact $^1B_1$ geometry and charge density were related to the spin distribution of the triplet, but the differences are not large enough to be qualitatively significant on the chemical scale of energy.

The second problem in the spin polarization mechanisms proposed is an implicit use of the 'electrons of like spin keep further a part than those of opposite spin' description of fermi correlation. A connection to the observed singlet/triplet energy difference is tempting via the corollary, '... thus the electrostatic repulsions between electrons of like spin are lower than for electrons of opposite spin'. This is the basis for a common interpretation of Hund's rule, for instance.

However, it has recently been shown [28] that the open-shell singlet/triplet energy difference can be ascribed to a $2K$ exchange contribution to the energy only to first order. In fact, when energy components correct to second order are computed [29], it is found that $\langle 1/r_{ij} \rangle$, hence the inter-electronic repulsion energy, may be larger for the triplet than singlet for low Z values (including the classic $He(^3P)$ case!). (The lower triplet total energy is due to a greater electron-nuclear potential energy sufficient to outweigh the higher repulsion energy.)

Thus building models on the idea of fermi correlation keeping electrons of like spin spatially localized (explicitly or implicitly) must be considered with considerable caution.

Finally, this has been an attempt at a 'two-level' description. The effects of a spin polarization are interpreted to explain the observed charge density differences between singlet and triplet. In turn the different charge distributions are interpreted (via the Hellmann-Feynman theorem) to obtain the quasi-classical electrostatic description of the forces defining the preferred nuclear rearrangements of the reaction paths. For the reasons given above, the correlations observed between the polarizations defined and reaction paths computed cannot be considered to be unambiguous.

5. Conclusions

The potential surfaces computed for the $H_2 + O$ and $C_2H_4 + O$ model systems correlate with the different reactivities of oxygen in its lowest isoconfigurational singlet and triplet states. Spin 'polarization' as a monitor was found to provide a rationale for the singlet/triplet reaction differences. The interpretation proposed was found not unambiguous and underlines the difficulty in finding appropriate descriptions of the effects of the electron pair density on chemical reactivity.

Acknowledgment

The author wishes to express his sincere thanks to Dr R. Constanciel for initiating several most helpful discussions.

References

1. Woodworth, R. C. and Skell, P. S.: *J. Am. Chem. Soc.* **81**, 3383 (1953) and references therein.
2. Hoffmann, R.: *J. Am. Chem. Soc.*. **90**, 1475 (1968) and references therein.
3. Gaspar, P. G. and Hammond, G. S.: *Carbene Chemistry*, Chapter 12, W. Kirmse, ed., Academic Press, New York, N.Y. (1964).
4. Paraskevopoulos, G. and Cvetanović, R. J.: *J. Chem. Phys.* **50**, 590 (1969).
5. Knight, A. R., Strausz, O. P., and Gunning, H. E.: *J. Am. Chem. Soc.* **85**, 2349 (1963).
6. Bader, R. F. W. and Generosa, J. I.: *Can. J. Chem.* **43**, 1631 (1965).
7. Yamasaki, H. and Cvetanović, R. J.: *J. Chem. Phys.* **41**, 3703 (1964).

8. Cvetanović, R. J.: *Can. J. Chem.* **36**, 623 (1958).
9. Sidhu, K. S., Lown, E. M., Strausz, O. P., and Gunning, H. E.: *J. Am. Chem. Soc.* **88**, 254 (1966).
10. Anet, F. A. L., Bader, R. F. W., and Van der Awera, A. M.: *J. Am. Chem. Soc.* **82**, 3217 (1960).
11. Cvetanović, R. J.: *Adv. Photochem.* **1**, 115 (1963).
12. Demore, W. B. and Benson, S. W.: *Adv. Photochem.* **2**, 219 (1964).
13. Hoffmann, R., Wan, C. C., and Neagu, V.: *Mol. Phys.* **19**, 113 (1970).
14. Leppin, E. and Gollnik K.: *Tet. Lett.* **43**, 3819 (1969).
15. Gunning, H. E. and Strausz, O. P.: *Adv. Photochem.* **4**, 143 (1966).
16. Matsen, F. A. and Cantu, A. A.: *J. Phys. Chem.* **72**, 21 (1968).
17. Ruedenberg, K.: *Rev. Mod. Phys.* **34**, 326 (1962).
18. Strausz, O. P., Gunning, H. E., and Csizmadia, I. G.: *J. Am. Chem. Soc.* **94**, 8317 (1972).
19. Haines, W. J. and Csizmadia, I. G.: *Theoret. Chim. Acta (Berlin)* **31**, 283 (1973).
20. Bader, R. F. W. and Gangi, R. A.: *Chem. Phys. Lett.* **6**, 312 (1970).
21. Bader, R. F. W. and Gangi, R. A.: *J. Am. Chem. Soc.* **93**, 1831 (1971).
22. Gangi, R. A. and Bader, R. F. W.: *J. Chem. Phys.* **55**, 5369 (1971).
23. Bader, R. F. W., Stephens, M. E., and Gangi, R. A.: in preparation.
24. Stephens, M. E.: Ph. D. Thesis, Mc Master University, Hamilton, Ont. Canada, March, 1975.
25. Mc Weeny, R. and Sutcliffe B. T.: *Methods of Molecular Quantum Mechanics*, Vol. II, Academic Press, New York, N. Y., 1969.
26. Schaeffer, H. F., III: *The Electronic Structure of Atoms and Molecules*, Addison-Wesley, Don Mills, Ont., 1972, p. 95.
27. Meher, W.: *J. Chem. Phys.* **51**, 5149 (1969).
28. Colpa, J. D. and Islip, M. F. J.: *Mol. Phys.* **25**, 701 (1973).
29. Snow, R. L. and Bills, U. L.: *J. Chem. Ed.* **51**, 585 (1974).

LOCALIZED ANALYSIS OF THE STEREOSPECIFICITY
OF CONCERTED REACTIONS

J. LANGLET

Laboratoire de Biochimie Théorique associé au CNRS,
Institut de Biologie Physico-Chimique, 75005 Paris, France

and

J. P. MALRIEU

Laboratoire de Physique Quantique, Université Paul Sabatier, 31077 Toulouse, France

Abstract. It is pointed out that in the delocalized Woodward-Hoffmann demonstration of the concerted reactions stereospecificity (i) some energetical hypotheses concerning the transition state region are implicitly required, (ii) the MO's correlations are ambiguous if no spatial criterion is taken into account, and (iii) the demonstration itself and its application to real unsymmetric systems require some relocalization of the canonical MO's. Using a perturbative configuration interaction in a basis of localized orbitals (PCILO), one tries to analyze the energy of the transition state region. The stereospecificity for a n-bond polyenic chain results from a nth order correction in the perturbation expansion representing circular delocalization effects around the reacting circle. The stereospecificity is linked directly to the parity of n without any symmetry consideration, but the energy difference between the two modes may vanish when n increases. PCILO-CNDO/2 and *ab-initio* numerical calculations for $n = 2$ support the theoretical analysis.

The *ab-initio* approach allows a distinction between the σ, π and $\sigma\pi$ effects and confirms that the π electronic factors are responsible for the stereospecificity However a stereospecific short range repulsion effect also appears, due to the overlap between the π bonds. The localized demonstration of the Woodward-Hoffmann rules is generalized to cycloadditions and sigmatropic reactions.

0. Introduction

Elucidation of the mechanisms of chemical reactions is one of the goals of Quantum Chemistry. But the calculation of reaction paths, and reaction barriers do not seem sufficient to understand the reaction mechanism. As was enunciated by Goddard [1], 'it is not enough to perform energy calculation, but one has to extract from the wave function the key parts that determine the stability and structure of the molecule and that determine whether and how the molecule can react'.

In these last years, the interpretation of the stereoselectivity of concerted reactions by Woodward and Hoffmann has been one of the great successes of quantum chemistry. Woodward and Hoffmann in a series of preliminary communications [2–4], have laid down some fundamental bases for the theoretical treatment of all concerted reactions; the basic principle they have enunciated was that: 'reactions occur readily when there is congruence between orbital symmetry characteristics of reactants and products, and only with difficulty when that congruence does not obtain'. This principle may be enunciated more succinctly by: Orbital symmetry is conserved in concerted reactions. This principle has been found to be an exceptionally powerful predictive and interpretative tool for the analysis of concerted reactions.

This success has incited a great number of theoretical works [5–17], some of them are based on symmetry arguments [5–7] and some of them, not [8–17].

O. Chalvet et al. (eds.), Localization and Delocalization in Quantum Chemistry, Vol. II, 325–363. All Rights Reserved
Copyright © 1976 by D. Reidel Publishing Company, Dordrecht-Holland

Our contribution takes into account that:

(a) The absence or presence of molecular symmetry in an absolute sense cannot be the ultimate source of the allowedness or forbiddenness of a reaction: the symmetry may be lacking either as a result of a substitution (by a methyl group for instance) or from the lack of symmetry of the component: most experimental tests involve systems without useful symmetry.

Furthermore Bryan *et al.* [18] have shown that even in reactions which are easily handled by symmetry arguments, the geometries of useful symmetry may never be physically achieved.

(b) Most theoretical investigations are performed using delocalized molecular orbitals, and Trindle [12–14] has shown that the use of localized molecular orbitals leads to considerable simplifications in the analysis of concerted reactions.

(c) Most theoretical arguments are in fact validated by their final success. Actually many properties are linked to the number of bonds involved.

The symmetry of the highest or lowest molecular orbital, and the nodal properties of molecular orbitals directly result from the parity of the number of bonds involved in the reacting system. But these correlations do not imply that one has touched the physical phenomenon. The Woodward-Hoffmann rules finally may be enunciated from parity considerations, and one may try to derive an energetic demonstration of the role of the parity.

Speaking of symmetry as the leading phenomenon suggests that one must use *delocalized* MO's. If symmetry is simply connected to the parity, a direct localized demonstration should be possible, without any symmetry consideration. This was the challenge which led us to this problem.* So starting from a zero order description of the molecule which uses fully localized molecular orbitals, we have shown that it is possible to analyse the origin of the stereospecificity in terms of interaction processes.

The *first section* is a general discussion about the use of (de)localized molecular orbitals in the study of the stereospecificity. The *second section* concerns the electrocyclic reactions; a demonstration is given for the butadiene⇌cyclobutene reaction ($N = 2$), numerically verified through semi empirical and ab-initio calculations, then extended to all values of N. *Section* 3 concerns cycloadditions and sigmatropic reactions.

In order to make the reading easier, let us recall briefly a few definitions. Among the different stereospecific concerted reactions one may distinguish: *electrocyclic reactions* (or intramolecular cyclizations), *cycloaddition reactions* (or intermolecular cyclizations) and *sigmatropic transpositions* (or σ bond migrations).

(a) *Electrocyclic reactions* are defined as the formation of a σ bond between terminal carbons of a linear system containing k π-electrons and the inverse process (Figure 1 a), for instance the reaction butadiene⇌cyclobutene (see Figure 1 b). This isomerization is the result of a concerted rotation of the hydrogens of the terminal CH groups with

---

* One of us (J.P. M) is indebted to his students of the Cuban Summer School in La Havana (1970) who presented the Woodward–Hoffmann rules as an objection to his apology of localized descriptions.

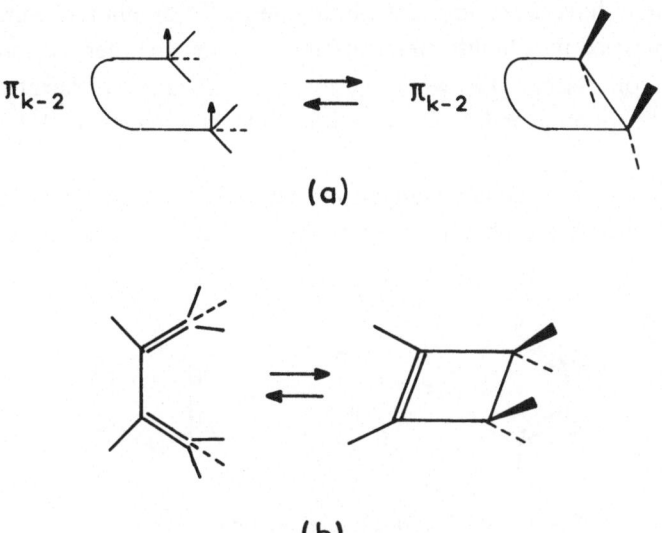

Fig. 1. Electrocyclic reaction exemplified.

an exchange of the π interaction between the two terminal carbon atoms into a σ bond (or reverse). Depending whether these concerted rotations are performed in the same direction or in opposite direction, one may speak of a conrotatory or disrotatory evolution (Figure 2). In the disrotatory mechanism, the system preserves a symmetry plane, and in the conrotatory mechanism, a twofold axis of symmetry is maintained (see Figure 2).

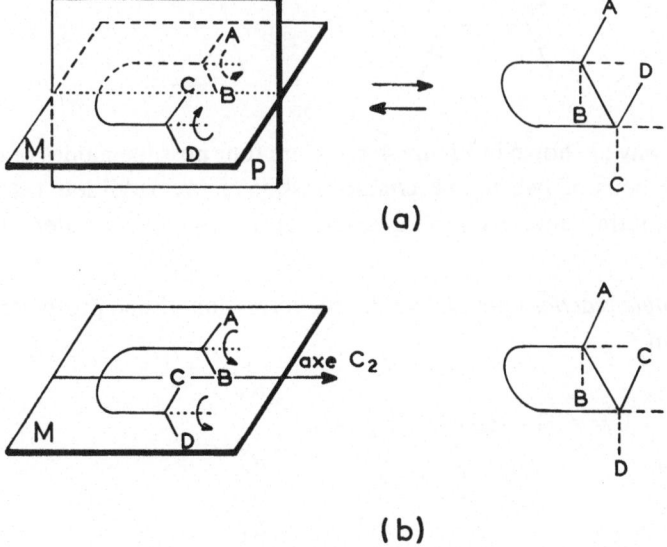

Fig. 2. Electrocyclic reaction. (a) Disrotatory mode – The symmetry plane P is conserved; (b) Conrotatory mode – The symmetry axis is conserved.

These reactions have been brought about thermally or photochemically, but all known cases proceed in a highly stereospecific manner; the thermal iomerization of cyclobutene is conrotatory [19], while the thermal cyclization of hexatriene is disrotatory [20]. The photochemical iomerizations of cyclobutene and hexatriene are respectively disrotatory [21] and conrotatory [22].

(b) *Intermolecular cycloadditions* of an *m*-electron system to an *n*-electron system (*m* + *n* cycloadditions) are characterized by concomitant formation of two σ bonds (Figure 3).

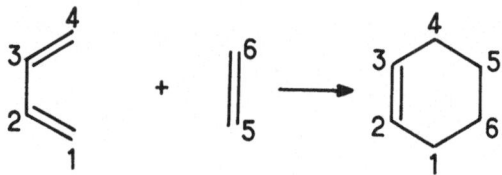

Fig. 3.   (2 + 1) cycloaddition reaction exemplified.

These reactions are stereospecific. One may distinguish:

(i) *Suprafacial processes* in which the newly formed or broken bonds lie on *the same face* of the reacting process.

(ii) *Antarafacial processes* in which the newly formed or broken bonds lie *on opposite faces* of the reacting process.

In the same way as shown in Figure 4, there are four possible modes of combination of terminal carbons of two unlike components in a cycloaddition reaction: supra-supra, supra-antara, antara-supra, antara-antara. The Diels-Alder reaction is a $(4_s + 2_s)$ process.

(c) *Sigmatropic reactions* are defined by the migration of one proton from carbon *j* towards carbon *i*

$$\begin{array}{c} R_1 \diagdown \\ {} \diagup \\ R_2 \end{array} C_i {=} CH{-}(CH{=}CH)_k{-}C_jH \begin{array}{c} \diagup R_3 \\ {} \\ \diagdown R_4 \end{array}$$

$$\downarrow$$

$$\begin{array}{c} R_1 \diagdown \\ {} \diagup \\ R_2 \end{array} C_i {-}(CH{=}CH)_k{-}CH{=}C_jH \begin{array}{c} \diagup R_3 \\ {} \\ \diagdown R_4 \end{array}$$

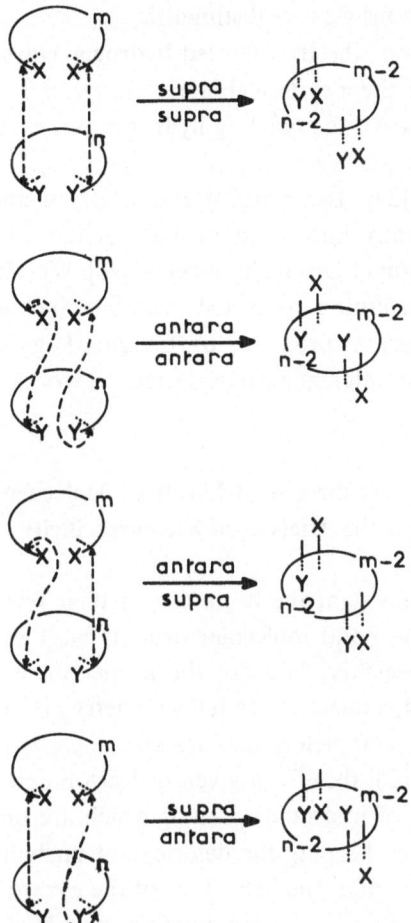

Fig. 4.   Stereochemical consequences of cycloadditions. These figures are schemes and do not give the real geometry of transition states.

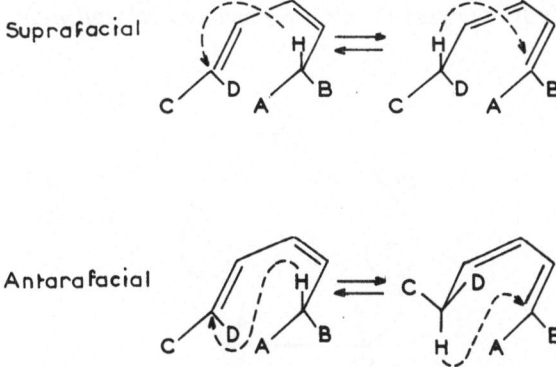

Fig. 5.   [1–5] Sigmatropic reactions exemplified.

In the sigmatropic reactions we may distinguish:

(i) *The suprafacial process*: the transferered hydrogen stays during all the reaction course in the same face of the π system (Figure 5a).

(ii) *The antarafacial process*: the reacting hydrogen passes from the top face of one terminal carbon to the bottom face of the other (Figure 5b).

The papers by Syrkin [23], Balaben [24] and Mathieu and Valls [25] give many references for experimentally known concerted reactions. In 1968, in their review paper entitled 'Conservation of Orbital Symmetry' [26], Woodward and Hoffman have collected their results concerning concerted reactions, including in this publication the use of correlation diagrams proposed by Longuet-Higgins and Abrahamson [5]. This paper also gives many references to experimental works.

1. Discussion of the Use of Localized Molecular Orbitals for the Analysis of Stereospecificity

Woodward and Hoffmann say at the beginning of their review [27] that 'The Semi-Localized Orbitals' i.e. the bond molecular orbitals built from two atomic orbitals properly hybridized if necessary, 'are not the proper molecular orbitals, the latter are completely delocalized, subject to the full symmetry of the molecule'.

In fact, one must leave completely the idea that there is an unique set of proper orbitals: it is well known [28] that for a given n-electron determinant, there exists an infinity of equivalent sets of molecular orbitals which are obtained from each other by unitary transformations, keeping the determinant unchanged.

Quantum Chemistry describes the behaviour of the electrons of a molecules, using a wave function ψ, which, in a first approximation, may be described as a product of monoparticulate functions (the orbitals). The orbitals are only mathematical objects and have no physical reality at all. So the delocalized or localized molecular orbitals are both convenient to describe a given physical phenomenon as long as they are properly used.

Furthermore we will show that most applications of the Woodward-Hoffman rules are relevant if, and only if, one carries out a preliminary relocalization of the delocalized molecular orbitals.

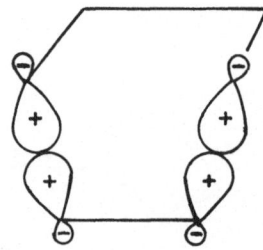

Fig. 6.   Semi-localized σ orbitals used by Woodward-Hoffmann to describe the cyclohexene molecule.

(a) *A partial localization is necessary even in the simplest cases considered in the theoretical demonstration of Woodward-Hoffmann.* Their σ molecular orbitals used in the correlation diagrams are never completely delocalized, they are semilocalized, i.e., delocalized on a small number of chemical bonds. For instance, in the Diels-Alder $4+2$ cycloaddition of ethylene and butadiene molecules (Figure 6), the σ orbitals of the cyclohexane are not really delocalized. The diagonalization of any hamiltonian of this molecule would lead to symmetrical and antisymmetrical molecular orbitals equally spread over the whole skeleton. With such molecular orbitals, even the $(\sigma - \pi)$ separation in cyclohexene would require some relocalization.

(b) *A more important localization is necessary when one considers unsymmetrical substituted compounds.* As noticed by Hoffmann, a substitution by a methyl group may destroy the total symmetry, but cannot be expected to change dramatically the mechanism of a reaction. Effectively we may show that to work with the butadiene molecular orbitals for the study of the reaction of a substitued butadiene is not so odd as it would appear when looking at the fully delocalized molecular orbitals of the substituted molecule; *in fact the use of symmetrical orbitals for unsymmetrical compounds is correct if one performs a localization of the molecular orbitals of this molecule on the basic symmetrical system.* Let us consider for instance 2-methyl-butadiene. In this molecule, the three delocalized π molecular orbitals obtained by diagonalization of the one electron hamiltonian look completely different from the butadiene molecular orbitals, each of them having an important weight on the CH_3 system (86, 9, 5%), but the same determinant may be written with CH bond molecular orbitals and two π delocalized molecular orbitals which are very close to butadiene π molecular orbitals, having only 1% weight on the CH bond.

(c) *A relocalization of the delocalized molecular orbitals inside and outside the reaction cycle itself allows applying directly the Woodward-Hoffmann rules, demonstrated on very small systems, to very large compounds.* On the butadiene molecule for instance, one may find two ethylene-like localized π molecular orbitals, and if the butadiene molecule enters the reaction by only one double bond, as in the concerted $_\pi 4_s + _\pi 2_s$ Diels-Alder cycloaddition of butadiene to itself, it is not necessary to make another demonstration than for butadiene+ethylene reactions using two systems of molecular orbitals.

So one may apply directly the Woodward-Hoffmann rules, demonstrated on very small systems, to very large conjugated systems, if one performs a preliminary relocalization of the molecular orbitals; for instance, it would be tedious to build a correlation diagram of the transformation of cis- and trans-benzocyclobutene (see Figure 7), using delocalized molecular orbitals. But the localization gives a system of molecular orbitals of cyclobutene type, which are the only molecular orbitals undergoing a transformation during the reaction, and the demonstration performed for the reaction butadiene⇔cyclobutene remains valid.

(d) *When using localized molecular orbitals for analysing the concerted reaction, the ambiguities and contradictions which may appear in correlation diagrams (when topological criteria are not introduced) will disappear.* Actually, the correlation is in principle simply based on the symmetry and energetic characters of the molecular

Fig. 7. Transformation of cis benzocyclobutene: electrocyclic reaction + a Diels-Alder cycloaddition.

orbitals; such a correlation sometimes introduces a correspondance between mole-cular orbitals of the initial and final states which are not defined on common atomic orbitals, i.e. which are topologically completely exclusive. One may find two examples of such ambiguities:

(i) In the butadiene-ethylene $4+2$ Diels-Alder reaction, as already noticed by Millie (29), in the Woodward-Hoffmann demonstration, the π molecular orbital of ethylene defined on the $\chi_5\chi_6$ atomic orbitals corresponds to the ethylenic orbital of the final product defined on the $\chi_2\chi_3$ atomic orbitals (see Figure 3); furthermore the π_1 orbital of butadiene corresponds to the σ orbital of cyclobutene. Millie has shown that these correspondances are not satisfying. This author, considering that one may not invoke the non crossing rules in the correlation diagrams, has proposed another correlation diagram where the $\pi_1(s)$ molecular orbital of butadiene corresponds to the bonding $\pi(s)$ molecular orbital. However, although these authors have built correlation diagrams between different molecular orbitals, they have drawn the same conclusion, and their results are in agreement with experiment. This is not always the case, as shown in the following example.

(ii) The transformation of the cyclooctatetraene into cubane is a priori conceivable. This reaction would be the sum of two cyclizations between two ethylenic systems almost orthogonal to each other. Experimentally this reaction cannot take place.

However, the correlation rules applied without consideration of any spatial problem describe the reaction as thermally possible; this is due to the fact that the correlation occurs between MO's of the two systems.

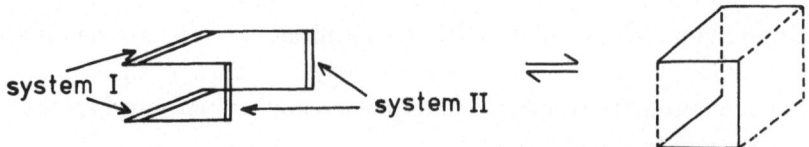

Woodward and Hoffmann are therefore compelled to introduce a supplementary rule: 'each basic process must be isolated and analyzed separately'. They explain that 'in fact whether a reaction is symmetry allowed or symmetry forbidden is determined by the height of the electronic hill that a reactant or product orbitals must climb in reaching the transition state. And the presence or absence of a hill is a function of the intended correlation, or the initial slope of the levels'. By such sentences the authors introduce topological criteria, as they do when building the correct correlation diagram, for this reaction [26].

A question then arises. Is it necessary to take systematically into account this spatial criterion in the correlation diagram? Why do Woodward and Hoffmann neglect this 'intended correlation' in the Diels-Alder $4+2$ reaction? When, and to what extent, is it necessary to introduce the spatial character of the MO's?

Such a problem will not appear in our localized demonstration, in which the spatial localization of the MO's is well defined and plays the leading role.

Several types of localized MO's may be considered, among which one may distinguish between localized SCF-MO's obtained by unitary transform from the canonical SCF-MO's [30] and fully localized MO's. The localized SCF-MO's are mainly defined on one bond but they have small 'tails' on the other bonds. These tails may be significant, and, for instance, the tails may be of different orders of magnitude in the two modes of a concerted reaction. Our method will be different; we shall start from fully localized MO's each of them being completely localized on a given chemical bond. These MO's are no longer self-consistent, and they do not satisfy Brillouin's theorem. A lot of calculations have shown now [31] that the determinant built from these MO's is a good enough approximation to the exact wave function to be considered as the zero-order approximation of a perturbative process. The delocalization of the fully localized MO's toward SCF-MO's is obtained by a perturbation procedure which gives small tails on the neighboring bonds.

Therefore we want to start from a zero-order description of the molecule which uses typical MO's independent of the considered mode.

2. Electrocyclic Reactions

A. LOCALIZED DEMONSTRATION OF THE STEREOSPECIFICITY
OF THE BUTADIENE⇌CYCLOBUTENE REACTION

One tries to evaluate directly the energy of intermediate states between final and initial states, giving a likely representation of the transition state. Our qualitative demonstration of stereospecificity is based on a direct approach of the energy of the intermediate state. In this theoretical approach we have not attempted to optimize the geometry of the transition state, but we have attributed to this state an intermediate geometry between the final and initial state: for instance we may suppose that each bond length and each bond angle have been subjected to a fraction of the total change when going from the initial to the final state.

The calculation method we have employed is the PCILO method [32] which is a perturbative configuration interaction in a basis of bond localized orbitals. This method is a semi-empirical method using the CNDO approximations [33]. The calculation process can be summarized as follows.

For every nuclear configuration (i) one builds a set of likely bond molecular orbitals both bonding (φ_i) and antibonding (φ_i^*), each of them defined on one or two properly hybridized atomic orbitals; (ii) The bonding orbitals are used to build a fully localized determinant. This determinant represents the zero-order wave function Φ_0 for the given geometry.

(iii) With the antibonding orbitals, one may build singly or doubly excited determinants (or configurations). These configurations interact with the Φ_0 wave function, and the interaction between these configurations represents the configuration interaction matrix.

(iv) The ground state energy of the system will be obtained as the lowest eigenvalue of this matrix found by a Rayleigh-Schrödinger perturbation development. This mathematical process physically means that one considers the interaction between the zero-order state and the other configurations Φ_c as a perturbation. The unperturbed hamiltonian has been chosen according to the Epstein–Nesbet definition [34].

(a) *In a first step*, one must choose a zero-order wave-function giving a good representation of the transition state.

One can represent the electronic zero-order wave-function with only one determinant. In the course of the reaction, some molecular orbitals of the initial state undergo only quantitative transformations (for instance the π bonds in the cyclizations of linear polyenes). One will call these bonds 'invariant bonds'.*

One may build these bonds by a criterion such as the maximum overlap [35] between two atomic hybrid orbitals. In the minimal basis set at least, this procedure defines some atomic orbitals, one on each carbon atom of the cycle, which are orthogonal to the hybrids of the same center entering the invariant bonds. The reacting (or π) bonds will be defined on these AO's. For instance, in the butadiene-cyclobutene reaction, one gets four atomic orbitals orthogonal to the three σ hybrids of the same center; two of them (on the central carbon atoms) are pure π atomic orbitals, while the two others (on the external carbon) have an intermediate hydridization between pure $\pi(p_z)$ and sp^3 (Figure 8).

For the electrocyclic reaction involving the n double bonds of a polyene, one gets thus $2n$ atomic orbitals χ_i orthogonal to the hybrids of the invariant bonds. With these $2n$ atomic orbitals one has to build $2n$ bonding molecular spin orbitals in order to obtain the zero-order determinant. If one calls φ' the invariant molecular orbitals and φ the reacting molecular orbitals constructed on the $2n$ atomic orbitals χ_i, the approximate ground-state determinant Φ_0 has the following form.

$$\Phi_0 = |\varphi'_1 \bar{\varphi}'_1 \dots \varphi'_p \bar{\varphi}'_p \dots \varphi_1 \dots \varphi_{2n}|. \tag{1}$$

---

* These invariant bonds may be some π bonds which are not involved in the cyclization process.

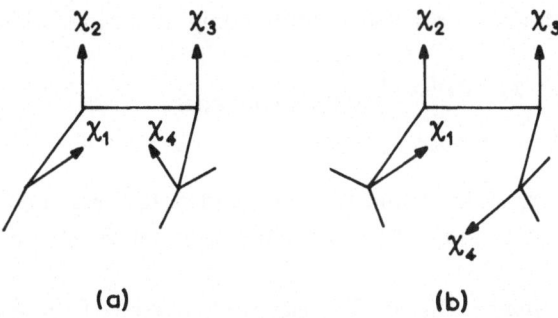

Fig. 8.   Definition of the atomic orbitals for the intermediate states in the butadiene⇔cyclobutene reaction. The arrows represent the orientation of the π atomic orbitals, orthogonal to the hybrids of invariant bonds. Their direction defines the chosen positive lobe. (a) denotes disrotatory mode; (b) denotes conrotatory mode.

We can consider two set of localized MO's φ, representing two extreme descriptions of the transition state.*

(i) *Distorted Initial State.* The molecular orbitals φ are defined on the AO's χ_i in the same way as in the initial state. So in the butadiene-cyclobutene reaction, the molecular orbitals will be in the disrotatory mode

$$\varphi_1 = (\chi_1 + \chi_2)/\sqrt{2} \quad \text{and} \quad \bar{\varphi}_1,$$
$$\varphi_2 = (\chi_3 + \chi_4)/\sqrt{2} \quad \text{and} \quad \bar{\varphi}_2, \tag{2}$$

if the orientation of the AO's χ_i is chosen according to Figure 8. In the conrotatory mode, φ_2 is changed into

$$\varphi_2 = (\chi_3 - \chi_4)/\sqrt{2}. \tag{3}$$

The corresponding antibonding molecular orbitals are built on the same pair of atomic orbitals

$$\varphi_1^* = (-\chi_1 + \chi_2)/\sqrt{2}, \tag{4}$$
$$\varphi_2^* = -(\chi_3 - \chi_4)/\sqrt{2} \quad \text{in the disrotatory mode}, \tag{5}$$
$$\varphi_2^* = -(\chi_3 + \chi_4)/\sqrt{2} \quad \text{in the conrotatory mode}. \tag{6}$$

These molecular orbitals are orthogonal in the CNDO approximations [33].

(ii) *Distorted Final State.* The bonding molecular orbitals are a π bond between χ_2 and χ_3 and a partially broken σ bond between χ_1 and χ_4 (in the two modes),

$$\left.\begin{aligned}\varphi_1 &= (\chi_2 + \chi_3)/\sqrt{2}\\\varphi_2 &= (\chi_1 + \chi_4)/\sqrt{2}\end{aligned}\right\} \text{ in the two modes}. \tag{7}$$

* One may also consider the intermediate state as biradicalar, with two unpaired electrons on the terminal carbons. A demonstration is still possible from that description which is omitted here for sake of simplicity.

There are two corresponding π and σ antibonding molecular orbitals

$$\left.\begin{aligned}\varphi_1^* &= (-\chi_2 + \chi_3)/\sqrt{2} \\ \varphi_2^* &= (-\chi_1 + \chi_4)/\sqrt{2}\end{aligned}\right\} \text{ in the two modes}. \qquad (8)$$

(b) *In a second step*, we have studied the ground-state energy and its different energetic contributions in an attempt to clarify the origin of the stereospecificity.

Ground-State Determinant Energy. The electronic energy of the ground-state determinant Φ_0 is given by

$$E_{0e} = \langle \Phi_0 | H | \Phi_0 \rangle = \sum_i \langle \varphi_i | -\nabla^2/2 + T | \varphi_i \rangle + 1/2 \sum_i \sum_j (J_{ij} - K_{ij}), \quad (9)$$

where T represents the nuclear attraction operator and where the summations on i and j are carried out over the molecular orbitals occupied in Φ_0. J_{ij} and K_{kj} are the well-known coulombic and exchange integrals.

One adds the nuclear repulsion energy E_n to the electronic energy, E_{0e}, and one gets the total energy

$$E_0 = E_{0e} + E_n. \qquad (10)$$

It will be arbitrarily supposed here that the E_0 energy is practically the same for the two modes at the same stage of the geometrical transformation.

The E_0 energy physically represents the electrostatic interaction energy and the short distance repulsion. As the hydrocarbon bonds are not very polar, our hypothesis is equivalent to assuming that the steric repulsions are almost equal in analogous intermediate states of the two modes.

Second order energetic corrections. We have shown that:

(i) The bond polarization corrections arising from the interaction between Φ_0 and $\Phi_{i \to i^*}$ may be neglected in weakly polar molecules. For polar molecules, the interaction term $\langle i | F | i^* \rangle$, where F is the Hartree Fock operator relative to Φ_0, mainly depends on the bond i itself, and cannot depend significantly on the considered mode.

(ii) The second order corrections due to the interaction between Φ_0 and the doubly excited determinants, giving the so-called 'correlation' correction are not stereospecific in the PCILO-CNDO hypotheses.

(iii) The second order delocalization correction coming from the interaction between Φ_0 and the $\Phi\begin{pmatrix} j^* \\ i \end{pmatrix}$ delocalization singly excited configurations, arising from the electron jumps from the bonding orbitals on bond i to the antibonding orbitals of bond j are responsible for the stereospecificity.

The delocalization correction is given by:

$$E_{\text{delo}}^2 = \sum_i \sum_{j \neq i} 2\langle \Phi_0 | H | \Phi_{i \to j^*} \rangle^2 / (E_0 - E(_i^{j^*})), \qquad (11)$$

where

$$\Phi_{i \to j*} = |\ldots \bar{\varphi}(i-1) \; \varphi_{j*} \bar{\varphi}_i \ldots \bar{\varphi}_n|.$$

In the CNDO hypotheses, due to the total localization of the molecular orbitals,

$$\langle \Phi_0 | H | \Phi_{i \to j*} \rangle = \langle \varphi_i | h | \varphi_{j*} \rangle, \tag{12}$$

where h is the (kinetic energy plus nuclear attraction) monoelectronic operator.

If χ_{i1} and χ_{i2} are the two hybrid atomic orbitals entering in the fully localized bond φ_1, i.e., if

$$\varphi_i = C_{i1} \chi_{i1} + C_{i2} \chi_{i2}$$

$$\varphi_{j*} = -C_{j2} \chi_{j1} + C_{j1} \chi_{j2}, \tag{13}$$

$$\langle \varphi_i | h | \varphi_{j*} \rangle = -C_{i1} C_{j2} \langle \chi_{i1} | h | \chi_{j1} \rangle + C_{i1} C_{j1} \langle \chi_{i1} | h | \chi_{j2} \rangle -$$
$$- C_{i2} C_{j2} \langle \chi_{i2} | h | \chi_{j1} \rangle + C_{i2} C_{j1} \langle \chi_{i2} | h | \chi_{j2} \rangle. \tag{14}$$

The σ bonds are invariant, so we may suppose that the corrections due to σ delocalization are not stereospecific, but the two cyclization modes present a qualitative difference in the charge transfer $\pi \to \pi^*$ correction.

If we start from the distorted initial state representation, for the butadiene disrotatory mode

$$\langle \varphi_1 | h | \varphi_2^* \rangle = \tfrac{1}{2}(- \langle \chi_1 | h | \chi_3 \rangle - \langle \chi_2 | h | \chi_3 \rangle + \langle \chi_1 | h | \chi_4 \rangle + \langle \chi_2 | h | \chi_4 \rangle),$$

and in the conrotatory mode

$$= \tfrac{1}{2}(\langle \chi_1 | h | \chi_3 \rangle + \langle \chi_2 | h | \chi_3 \rangle + \langle \chi_1 | h | \chi_4 \rangle + \langle \chi_2 | h | \chi_4 \rangle). \tag{15}$$

From Figure 8 it immediately appears that $\langle \chi_1 | h | \chi_3 \rangle = \pm \langle \chi_2 | h | \chi_4 \rangle$ according to the disrotatory or conrotatory character of the cyclization; it follows that in the disrotatory mode

$$\langle \varphi_1 | h | \varphi_2^* \rangle = (- \langle \chi_2 | h | \chi_3 \rangle + \langle \chi_1 | h | \chi_4 \rangle)/2,$$

and in the conrotatory mode

$$= (\langle \chi_2 | h | \chi_3 \rangle + \langle \chi_1 | h | \chi_4 \rangle)/2. \tag{16}$$

The main question concerns the signs of $\langle \chi_2 | h | \chi_3 \rangle$ and $\langle \chi_1 | h | \chi_4 \rangle$ in the two modes. With a Wolfsberg-Helmoltz-like approximation (as in the CNDO hypothesis) [33]

$$\langle \chi_i | h | \chi_j \rangle = \langle \chi_i | \chi_j \rangle \, (\beta_i^0 + \beta_j^0)/2, \tag{17}$$

where β_i^0 and β_j^0 are two negative parameters characteristic of the atomic orbitals χ_i and χ_j. It appears immediately that the overlap $\langle \chi_2 | \chi_3 \rangle$ is always positive in two modes, while $\langle \chi_1 | \chi_4 \rangle$ is positive as soon as the reaction has progressed sufficiently in the conrotatory mode, as is apparent from Figure 9, and always positive in the disrotatory mode.

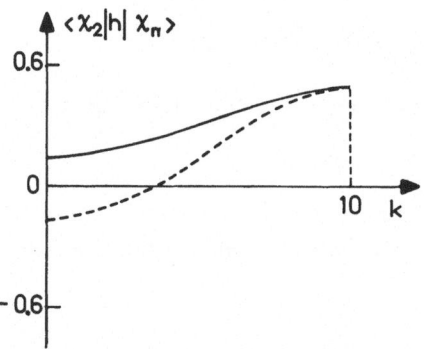

Fig. 9.   Evolution of the overlap $\langle \chi_1 |h| \chi_n \rangle$ between the terminal π atomic orbitals along the reaction path for the two modes: solid line denotes the disrotatory mode, dotted lines the conrotatory mode.

If we start from the distorted final state representation one may consider the same geometry and the same hybridization of the atomic orbitals as in the preceeding case. The only change concerns the definition of the molecular orbitals on these atomic orbitals; according to Equation (7) one may notice that in this case the bond molecular orbitals are also symmetry MO's. Let us consider the delocalization correction arising from the electron jump $\varphi_1 \rightarrow \varphi_2^*$. φ_1 is symmetrical with respect to the symmetry plane σ_v kept along the disrotatory mode, while in this mode φ_2^* is antisymmetrical. *Thus $\langle \varphi_1 |h| \varphi_2^* \rangle_d$ is zero in the disrotatory mode.* On the contrary φ_1 and φ_2 are both antisymmetrical with respect to the axis of symmetry C_2 kept along the conrotatory mode, *and one gets a non zero delocalization in the mode conrotatory mode,* since $\langle \varphi_1 |h| \varphi_2^* \rangle_c$ has a negative value.

$$\frac{\langle \varphi_1 |h| \varphi_2^* \rangle_c^2}{\Delta E_{\varphi_1 \rightarrow \varphi_2^*}} < 0 \quad \frac{\langle \varphi_1 |h| \varphi_2^* \rangle_d^2}{\Delta E_{\varphi_1 \rightarrow \varphi_2^*}} = 0. \tag{18}$$

The same conclusion holds for the inverse delocalization $\varphi_1 \rightarrow \varphi_2^*$. Again the delocalization between the reacting molecular orbitals favors the conrotatory mode.

If one does not use the CNDO approximations for the bielectronic integrals, some double excitations with a $\pi \rightarrow \pi^*$ charge transfer, i.e. the excited configurations $\Phi \begin{pmatrix} \varphi_k^* & \varphi_2^* \\ \varphi_k & \varphi_1 \end{pmatrix}$ would lead to stereospecific effects. Actually these doubly excited configurations interact with the ground-state Φ_0 by an interaction matrix element of the form $\langle \varphi_1 \varphi_k | \varphi_2^* \varphi_k^* \rangle$, and when developed by the Mulliken approximation, this integral is still proportional to the overlap $\langle \varphi_1 | \varphi_2^* \rangle$ and therefore larger in the conrotatory mode than in the disrotatory mode.

B. Numerical Study of the Butadiene⇌Cyclobutene Reaction Using Localized Molecular Orbitals

1. *Methods and Hypotheses*

(a) Method. Four types of calculations have been performed.

The first step [36a] simply confirms numerically the validity of our simple PCILO-

CNDO preliminary demonstrations attributing the origin of the stereospecificity to the π system through the second order delocalization contribution. The energies have been calculated up to the third order for a 'linear' progression of the reaction.

The second step is the ab-initio transcription of the first approach. We have performed ab-initio calculations [36b] for a few representative points on the reaction path, corresponding to equilibrium geometries of the two isomers, and to planar intermediate geometries in the transition state region.

These ab-initio calculations have been made from a fully localized single determinant Φ_0 as zeroth order wave-function which is perturbed to the second order.

The calculation was performed in a minimal basis set of Slater atomic orbitals with optimized exponents ($\zeta_c\, 2s = 1.759$, $\zeta_c\, 2p = 1.670$, $\zeta_c\, 1s = 5.679$, $\zeta_H = 1.227$). The atomic integrals are calculated using the POCYCAL program of Stevens [37]. The non-orthogonal $1s$ MO's are simply the $1s$ AO's on the carbon atom. The non-orthogonal bond MO's are constructed from two hybrids in the valence shell satisfying the Del Re criterion of maximum overlap [35]; these MO's are the same as the semi empirical PCILO MO's. These localized MO's are orthogonalized through a five step procedure: the $1s$ MO's are orthogonalized between themselves through an $S^{-1/2}$ procedure, the bonding valence-shell MO's are orthogonalized to the $1s$ MO's by projecting them in the complementary subspace, then $S^{-1/2}$ orthogonalized among themselves. The virtual antibonding MO's are first projected in the complementary subspace of the occupied MO's, then $S^{-1/2}$ orthogonalized. One may find a justification of this procedure in a previous study [38]. It represents the starting point of the ab-initio PCILO method.

In a third step, since previous calculations show that the perturbation is too strong for the transition state region and tends to diverge, a multiconfigurational zeroth order wave-function was used, including the excited determinants which have the most important coefficients in the basic first order wave function. This zeroth order multiconfigurational wave-function has been perturbed to the second order in energy.

These calculations have been made according to the CIPSI method [30] in an excitonic scheme [40] (i.e. a configuration interaction in a localized basis). We have first used the CNDO parametrization [33] and then we have performed a few ab-initio multiconfigurational calculations.

(b) Geometries. The initial and final geometries of butadiene and cyclobutene are taken from literature [41, 42], except for the fact that we used the same CH bond lengths for the two molecules.

Assuming a planar structure of carbon skeleton, including H_2 and H_3 in this plane, neglecting the changes of the CH bond lengths, assuming the same valence angle for the hydrogen atoms of the CH_2 groups and keeping the symmetry of the considered mode along the reaction path, one may consider seven fundamental parameters (5 angles and two bond lengths) defined in Figure 10. In this figure, θ is the dihedral angle of the planes $C_1C_2C_3$ and $H_1C_1C_2$. These seven parameters are sufficient to define a symmetrical configuration if the carbon skeleton is kept planar and if the hydrogen atoms H_2 and H_3 are kept respectively in the planes $C_1C_2C_3$ and $C_2C_3C_4$.

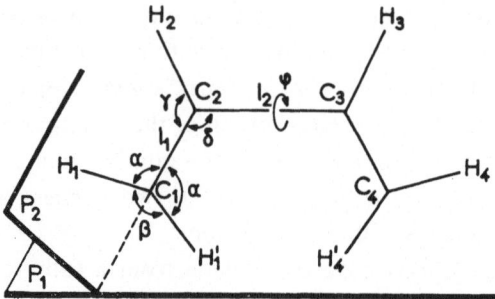

Fig. 10.   Definition of the geometrical parameters for intermediate conformations. P_1 is the plane $C_1 C_2 C_3$; P_2 is the plane $H_1 C_1 C_2$; θ is the angle (P_1, P_2).

If one removes the planarity of this skeleton, one must add an angle φ of torsion of the group $C_3H\ C_4H_5H_6$ around the C_2C_3 bond.

The seven parameters $\alpha, \beta, \gamma, \delta, 1_1, 1_2, \theta$ have not been optimized independently.

In the first semiempirical calculation, the reaction path is determined by nine intermediate geometries. The seven parameters $\alpha, \beta, \gamma, \delta, 1_1, 1_2$ and θ have been changed simultaneously in a linear way according to a parameter k varying from 0 to 1.0; if α_0 represents the value of α for the butadiene equilibrium geometry, and $\Delta\alpha$ its change when going from butadiene to cyclobutene, for a given value of k, α takes the value:

$$\alpha_k = \alpha_0 + k\ \Delta\alpha \tag{19}$$

and similarly

$$\beta_k = \beta_0 + k\ \Delta\beta. \tag{20}$$

This hypothesis is not compulsory but it allows the construction of a likely 'linear' reaction path with a small number of calculations. The ab-initio calculations have been performed for $k = 0.4, 0.5, 0.6$, which are likely to be representative of the transition state region. Hsu *et al.* have shown that the actual reaction path is not linear, the methylene groups beginning first to rotate before the breaking of the C—C bond. However, the transition state geometry of this stepwise process corresponds to a partial rotation and a partial lengthening and is rather close to our selected geometries. In their second calculation, Hsu *et al.* [43] have shown that a supplementary rotation occurs around the central C_2—C_3 bond (angle φ, Figure 1) without introducing qualitative differences.

In the semi-empirical multiconfigurational calculations, the seven geometrical parameters have been frozen into two sets. The first one, v, concerns α, β and θ. The second one, t, concerns $\gamma, \delta, 1_1$ and 1_2. The parameters v or t have been varied from 0 to 1, resulting in a two dimensional map $E = f(t, v)$ for planar configurations of the carbon skeleton. We have considered also a torsion around the central C_2—C_3 bond (angle φ) in the conrotatory mode*: for each value of t and v, φ has been optimized.

---

* After a torsion around the C_2–C_3 bond, the plane of symmetry of the disrotatory mode is destroyed and the notion of disrotatory mode loses its meaning.

We know that the CNDO parametrization fails to reproduce the geometries of strained cycles and their energies with respect to the open chain isomers [44]; after the second order correction, cyclobutene actually is more stable than butadiene, in contradiction with experiment. These series of calculations however enable us to see *qualitatively* whether the geometry takes a significant part in the stereospecificity.

Anyway, one must remember that our purpose is mainly interpretative and that the leading factors of the stereoselectivity are certainly present in the whole transition state region; the precise knowledge of the exact saddle point should not be necessary.

(c) Fully localized zero order representation

The butadiene slope has been explored from the butadiene Φ_0 representation for k (or t and v) varying from 0 to 0.6. For the cyclobutene slope, we have started from the cyclobutene Φ_0 representation for k (or t and v) varying from 0.4 to 1.0.

One can expect that the butadiene representation becomes worse when one comes close to the cyclobutene conformation, i.e., for large k. Reciprocally, the cyclobutene representation should be bad for small k. One may therefore expect that the energy curves of the two representations intersect somewhere in the transition state region. One can choose for the corresponding values of k the representation which gives the lower energy.

(d) Choice of the multiconfigurational zeroth order wave-function.

The divergence in the simple perturbation procedure comes from the fact that the diagonal Fock energies of the 'π' bonding MO's increase, while the corresponding Fock energies for the antibonding 'π' MO's decrease when the 'π' bonds are destroyed along the reaction. Excitation energies in the π system tend therefore toward zero and near degeneracies begin to occur between the ground state determinant Φ_0 and 'π' excited determinants. The new zeroth-order wave function will include, besides Φ_0, the charge transfer singly excited determinants

$$\begin{pmatrix} \pi_2^* \\ \pi_1 \end{pmatrix} = a_{\pi_2^*}^+ a_{\pi_1} \Phi_0$$

and

$$\begin{pmatrix} \pi_1^* \\ \pi_2 \end{pmatrix} = a_{\pi_1^*}^+ a_{\pi_2} \Phi_0$$

(and the β spin analogs).

On the other hand a degeneracy is progressively introduced, during this torsion between the ground-state determinant and the $\Phi \begin{pmatrix} \pi^* & \pi^* \\ \pi & \pi \end{pmatrix}$ doubly excited determinants; we have therefore introduced in the zeroth order wave function the two types of doubly excited determinants which interact with the ground state determinant in the CNDO hypotheses, namely

(i) the intra-bond doubly excited determinants

$$\begin{pmatrix} \pi_i^* & \bar{\pi}_i^* \\ \pi_i & \bar{\pi}_i \end{pmatrix} = a_{\pi_i^*}^+ a_{\pi_i} a_{\bar{\pi}_i^*}^+ a_{\bar{\pi}_i} \Phi_0 \quad i = 1, 2$$

(ii) the inter-bond doubly excited determinants

$$\begin{pmatrix} \pi_i^* \pi_j^* \\ \pi_i \pi_j \end{pmatrix} = a_{\pi_i^*}^+ a_{\pi_i} a_{\pi_j^*}^+ a_{\pi_j} \Phi_0 \quad i \neq j = 1, 2$$

with all possible spin combinations.

Let us call S this subspace of 13 determinants.

This wave function will be used for the 'butadiene' slope of the reaction surface. For the cyclobutene slope, we have included, besides Φ_0, the same type of excitations as for the butadiene slope, changing π_1 into π and π_2 into σ. If one had performed a total CI on the four bonding and antibonding π MO's (including up to the quadruply excited determinants), the butadiene and cyclobutene wave functions would give the same energy for any geometry. As we do not perform the full CI of the π system to build our zeroth-order multiconfigurational wave function, one may simply hope that the two slopes will join together smoothly in the transition state region.

The zeroth-order wave function ψ_0 of the ground state for a given geometry is obtained as an eigenvector of the CI matrix restricted to the subspace S. If P_s is the projector on the subspace S,

$$P_s H P_s \Psi_0 = E_1 \Psi_0, \tag{21}$$

$$E_1 = \langle \Psi_0 | H | \Psi_0 \rangle. \tag{22}$$

E_1 will be called the first order energy. It includes, therefore, the total (σ and π) short range repulsion, the π delocalization, and purely π correlation energies. E_0 and E_1 are upper bounds to the energy.

The ψ_0 wave function is perturbed under the influence of the other determinants of the CI, up to the second order in energy. Due to the diagonalization of S, the other components of S are orthogonal to $\psi_0 (\langle \psi_i | \psi_0 \rangle = 0)$. Therefore, they do not interact with ψ_0 and are not involved in the second order energy

$$\Delta E_2 = \sum_{IS} \frac{\langle \Psi_0 | H | \Phi_I \rangle}{E_0' - E_1}. \tag{23}$$

The choice of E_0' and of the unperturbed hamiltonian H_0 are discussed elsewhere [40]. This second-order correction includes some π correlation effects due to the purely π triply and quadruply excited configurations, which interact with Ψ_0 through its singly and doubly excited components. But most of the second-order correction represents the purely σ and $\sigma\pi$ delocalization and correlation corrections.

All the theoretical explanations of stereoselectivity attribute its origin to 'π' electronic factors, 'π' meaning the electrons of the bonds which are destroyed in the course of the reaction. The correlation diagrams only consider the corresponding MO's, and the same feature was explicitly supposed in our localized demonstration. The analysis of E_0 allows us to see the role of the short range repulsion. When going from E_0 to E_1 one gets the role of the other π electronic factors and when going from E_1 to E_2 one gets the role of the σ and $\sigma\pi$ electronic factors.

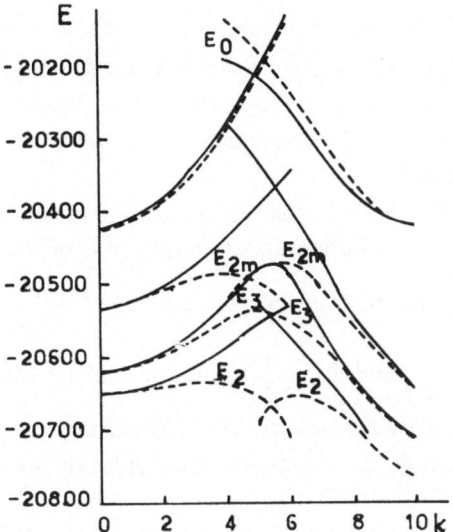

Fig. 11. *PCILO–CNDO calculations*: ground state energies in kcal/mole as a function of k. For this figure and for Figures 12, 13, the dotted lines represent the conrotatory mode and the solid lines represent the disrotatory mode. E_0 represents the zeroth-order energy. E_{2m} represents the energy corrected to the second order by the monoexcited states. E_2 represents the energy including the full second-order correction. E_3 represents the energy corrected up to the third order.

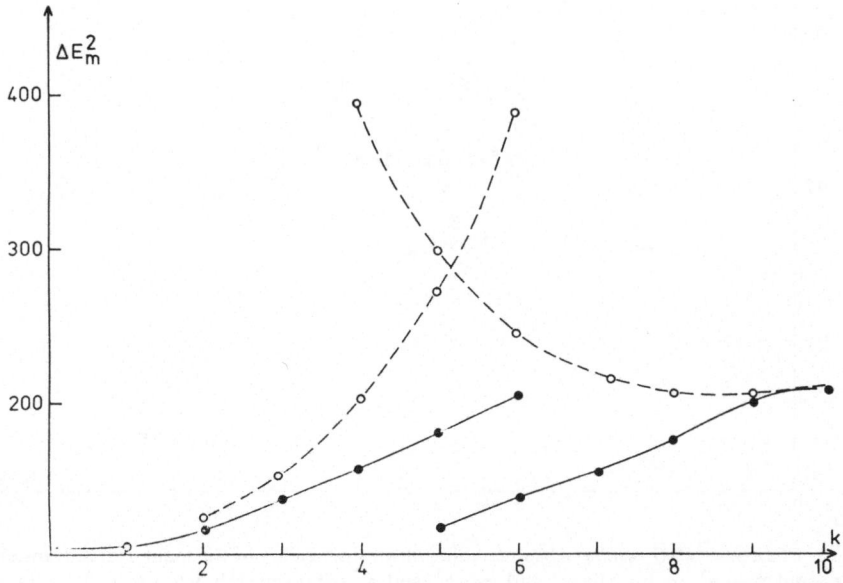

Fig. 12. *PCILO–CNDO results*. Second order delocalization energy correction (in kcal/mole^{-1}) as a function of k.

2. *Results*

The results obtained with the semi-empirical PCILO method are summarized in

Figure 11 which gives, as a function of the reaction parameter k, the evolution of (i) the zero-order energy E_0 ; (ii) the energy corrected to the second order by the singly excited configurations, E_{2m} ; (iii) the second order energy E_2 ; and (iv) the energy corrected up to the third order.

Figure 12 which gives the second order delocalization correction as a function of k.

Figure 13 which gives the evolution as a function of k of the absolute value of the coefficient of the main charge transfer configurations $\Phi\begin{pmatrix}\pi_2^*\\\pi_1\end{pmatrix}$ and $\Phi\begin{pmatrix}\pi_1^*\\\pi_2\end{pmatrix}$ in the butadiene slope and $\Phi\begin{pmatrix}\pi^*\\\sigma\end{pmatrix}$ and $\Phi\begin{pmatrix}\sigma^*\\\pi\end{pmatrix}$ in the cyclobutene slope

The results obtained with the perturbative ab-initio calculations performed from a fully localized single determinant Φ_0 as zeroth order wave function are summarized in Table I which gives

(i) the zeroth order energy E_0, and the second order corrected energy E_2,

(ii) the second order correction representing polarization and delocalization effects ΔE_m^2,

(iii) the correlation corrections due to the doubly excited determinants ΔE_d^2,

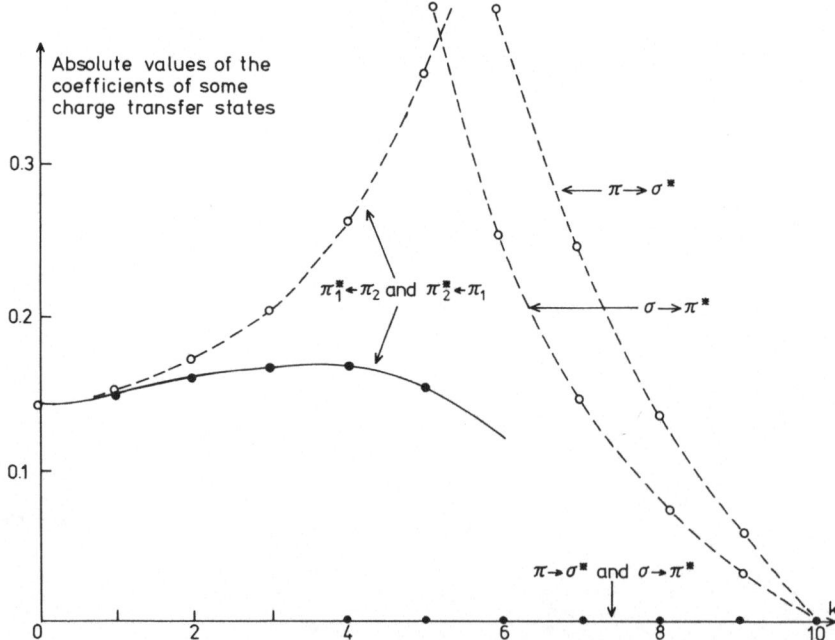

Fig. 13.   *PCILO–CNDO results.* Absolute value of the cœfficients of the main charge transfer states ($\pi_1 \to \pi_2^*$ and $\pi_2 \to \pi_1^*$ in butadiene, and $\sigma \to \pi^*$ and $\pi \to \sigma^*$ in cyclobutene) as a function of k:π_1 and π_2 are the two π bonds of butadiene, π is the π bond of cyclobutene, and σ is the cyclobutene C–C σ bond which is broken in the reaction course.

TABLE I

Ab Initio zeroth and second-order energies (all energies in AU) from single determinant descriptions of the ground state [a]

	Butadiene					Cyclobutene				
	0%	40%		50%		50%		60%		100%
		con	dis	con	dis	con	dis	con	dis	
E_0	−154.4785	−154.3120	−154.2885	−154.2485	−154.2077	−154.2105	−154.2016	−154.2553	−154.2609	−154.4493
E_2	−154.9355	−154.9058	−154.8679	−154.8991	−154.8245	−154.9512	−154.8393	−154.8668	−154.8263	−154.9003
ΔE_m^{-2}	−0.0971	−0.1586	−0.1371	−0.1884	−0.1363	−0.1957	−0.1105	−0.1559	−0.1216	−0.1178
ΔE_d^{-2}	−0.3600	−0.4352	−0.4423	−0.4622	−0.4805	−0.5451	−0.5273	−0.4556	−0.443	−0.331
$|C_{\pi_1\pi_2^*}|$ or $|\varepsilon_{\pi_1\pi_2^*}^2|$	0.137	0.263	0.180	0.356	0.176	0.449	0	0.262	0	0
$|C_{\pi\sigma^*}|$ or $|\varepsilon_{\pi\sigma^*}^2|$	−0.0079	−0.0215	−0.0098	−0.0355	−0.0080	−0.0433	0	−0.0178	0	0
$|C_{\pi_2\pi_1^*}|$ or $|\varepsilon_{\pi_2\pi_1^*}^2|$	0.137	0.263	0.180	0.356	0.176	0.407	0	0.227	0	0
$|C_{\sigma\pi^*}|$ or $|\varepsilon_{\sigma\pi^*}^2|$	−0.0079	−0.0215	−0.0098	−0.0355	−0.0080	−0.0430	0	−0.0171	0	0

[a] All quantities are defined in the text.

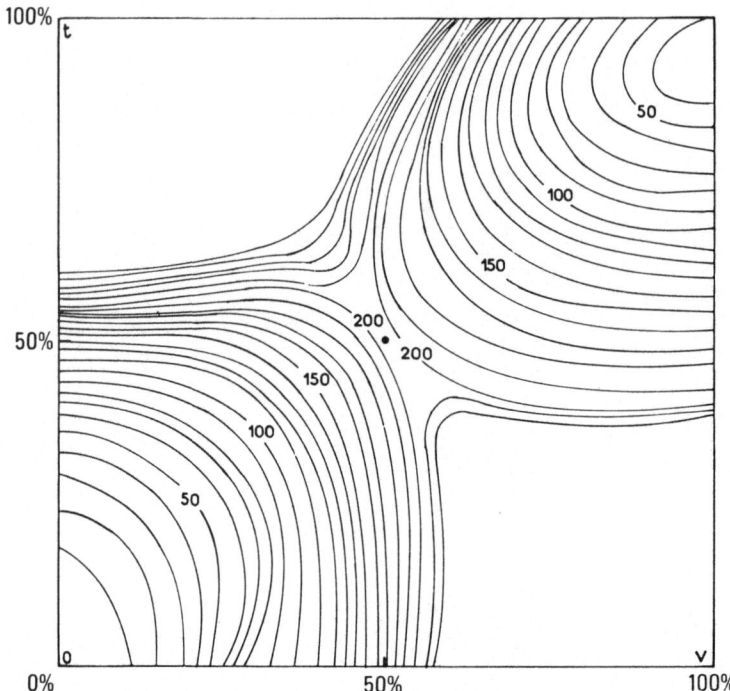

Fig. 14.   *CIPSI–CNDO results.* $E_1 = f(v, t)$. E_1 is the first order energy obtained by diagonalization of the π C.I. matrix (results in kcal/mole^{-1}). (a) disrotatory mode; (b) conrotatory mode. These results are obtained with φ ($C_1 C_2 C_3$, $C_2 C_3 C_4$) = 0.

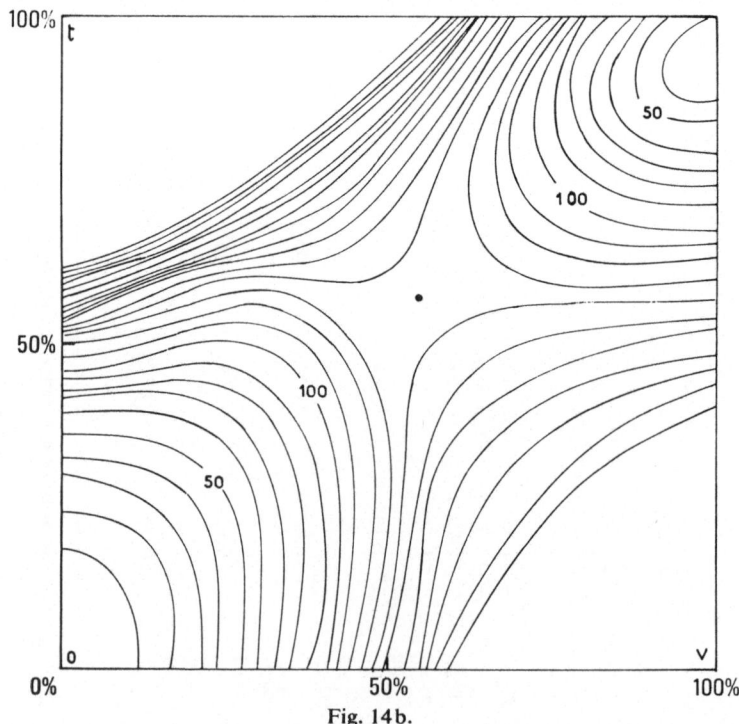

Fig. 14 b.

(iv) the coefficients of the π charge transfer configurations: $C_{\pi_1\pi_2^*}$, $C_{\pi_2\pi_1^*}$, $C_{\pi\sigma}$, $C_{\sigma\pi^*}$ in the first order wave functions, and the corresponding second order energy corrections: $\varepsilon^2_{\pi_1\pi_2^*}$, $\varepsilon^2_{\pi_2\pi_1^*}$ and $\varepsilon^2_{\sigma\pi^*}$, $\sigma^2_{\pi\sigma^*}$.

The results obtained with perturbative semi-empirical calculations performed with a multiconfigurational zero-order wave function are given in

(i) *Figures 14a and 14b* which give the first order energy $E_1 = f(v, t)$ for the conrotatory and disrotatory mode respectively,

(ii) *Figures 15a and 15b* which give the energy corrected to the second order $E_2 = f(v, t)$ for the conrotatory and disrotatory mode respectively.

Figure 15c which gives the second order energy E_2 obtained after minimization of the angle $(C_1C_2C_3, C_2C_3C_4)$ for each point (v, t) of the map.

(iii) *Table II* which gives the weight (in %) of singly excited delocalization configurations and of the doubly excited configurations in the multiconfigurational zero-order wave function.

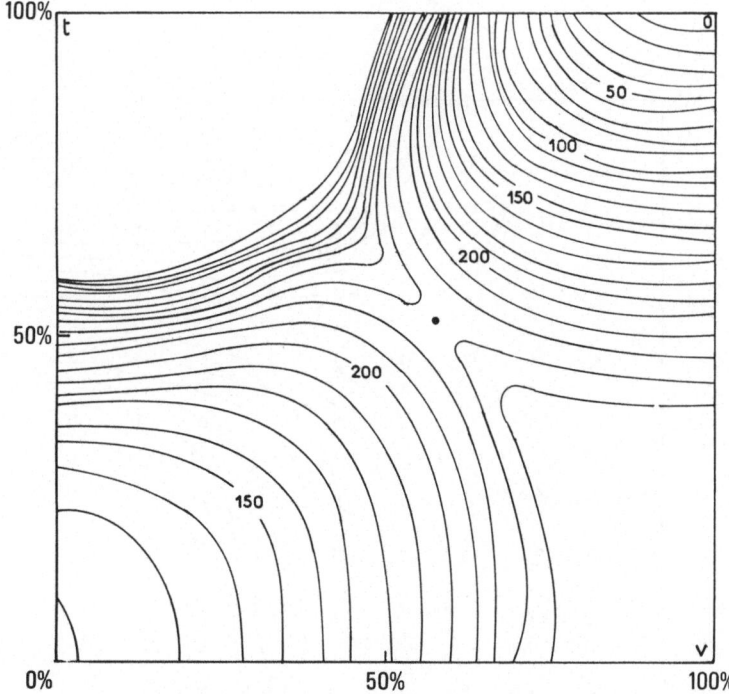

Fig. 15. *CIPSI–CNDO results. $E_2 = f(v, t)$. E_2 is the second order energy (in kcal/mole^{-1}). (a) disrotatory mode. (b) conrotatory mode. Results a and b are obtained with $\varphi = 0$. (c) $E_2 = f(v, t)$ in the conrotatory mode, with $\varphi(C_1C_2C_3, C_2C_3C_4) \neq 0$. The angle φ is indicated for each calculated point (unit = 10°)*

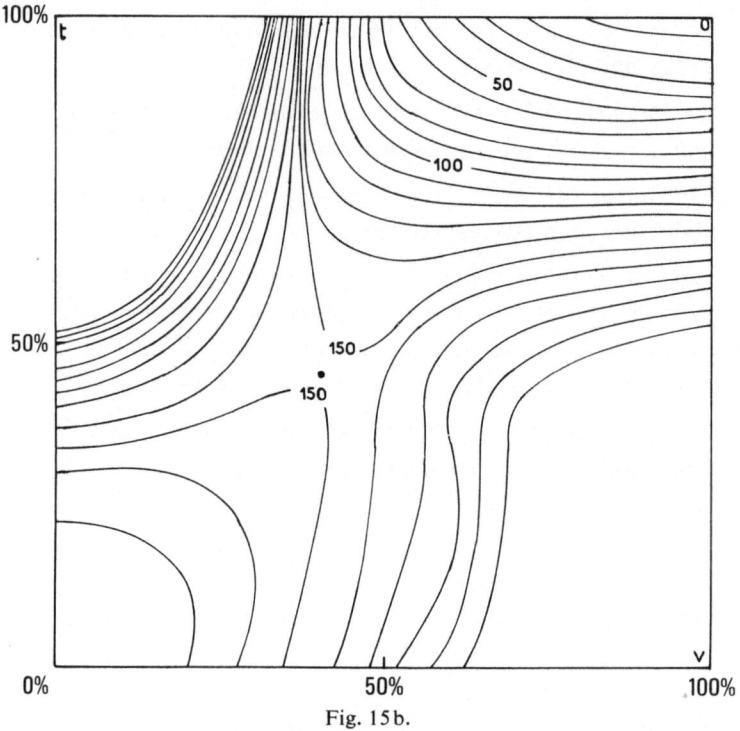

Fig. 15b.

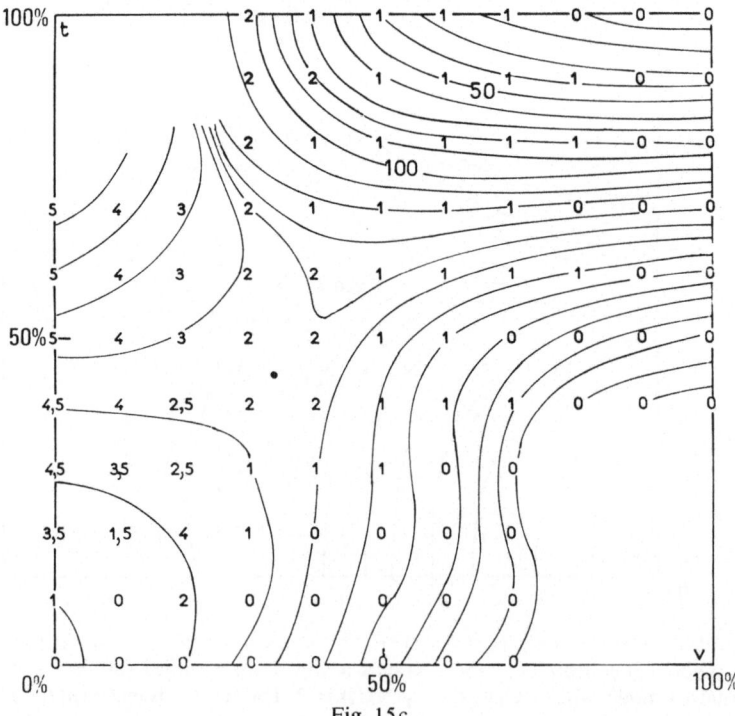

Fig. 15c.

TABLE II

Semiempirical calculations. Weight (in percent) of the singly (M) and doubly (D) excited configurations in ψ_0

Butadiene

t \ v		0%	30%	50%
$t = 0\%$	M con	6.87	8.88	12.96
	M dis		7.28	8.52
	D con	3.10	3.84	5.82
	D dis		3.72	5.52
$t = 30\%$	M con	8.28	11.68	19.00
	M dis		8.16	8.16
	D con	4.22	4.98	7.78
	D dis		4.82	6.86
$t = 50\%$	M con	8.88	14.12	23.40
	M dis		7.84	6.32
	D con	4.68	5.86	9.52
	D dis		5.58	7.98

Cyclobutene

t \ v		100%	70%	50%
$t = 100\%$	M con	0	5.82	17.54
	M dis		0	0
	D con	2.39	3.08	6.97
	D dis		2.98	3.38
$t = 70\%$	M con	0	9.94	22.46
	M dis		0	0
	D con	4.94	6.24	10.94
	D dis		6.11	8.42
$t = 50\%$	M con	0	13.32	25.90
	M dis		0	0
	D con	9.37	10.26	14.38
	D dis		11.27	15.55

TABLE III

Ab initio zeroth, first- and second-order energies (in AU) for multiconfigurational descriptions of the ground state)[a]

	Butadiene			Cyclobutene		
	0%	50%		50%		100%
		con	dis	con	dis	
E_0	−154.4785	−154.2485	−154.2077	−154.2100	−154.2010	−154.4490
ΔIC	−0.0882	−0.1604	−0.1292	−0.1893	−0.144	
$E_1 = E_0 + \Delta$IC	−154.5667	−154.4089	−154.3369	−154.3993	−154.3455	−154.5053
ΔE_2	−0.3356	−0.4225	−0.4325	−0.4239	−0.4148	−0.3735
$E_2 = E_1 + \Delta E_2$	−154.9023	−154.8314	−154.7694	−154.8232	−154.7603	−154.878
$M(\%)$	2.04	12.64	2.11	11.59	0	0
$D(\%)$	9.19	17.92	17.92	21.24	22.18	6.28

[a] All quantities are defined in the text except M and D, which represent (in percent) the weight of the singly (M) and doubly (D) excited configurations.

The results obtained with perturbative ab-initio calculations performed with a multiconfigurational zero-order wave function are reported in Table III which gives

(i) the correction due to the π CI (ΔIC) and the first order energy $E_1 = E_0 + \Delta IC$,

(ii) the second order correction ΔE_2 and the second order corrected energy $E_2 = E_1 + \Delta E_2$,

(iii) the weight of the singly excited configurations ($M\%$) and doubly excited configurations ($D\%$).

An analysis of our results has shown that:

(a) *When considering the zeroth order energy of the fully localized determinant the stereospecificity does not appear in the semi empirical calculations (Figure 1) and appears at this level in the ab-initio calculations (Table I); the conrotatory mode is then significantly favored.*

The zeroth order energy only includes electrostatic and short range repulsion effects linked to the overlap [45]. The electrostatic effects are small for these non polar molecules. The repulsive effect might come from (i) steric hindrance between the hydrogen atoms of the terminal CH_2 groups (ii) from π electronic factors.

In the two modes, the relative interatomic distances (especially between the terminal H atoms) are completely different for a given k. Actually in the semi-empirical calculation a compensation occurs between the difference of the nuclear repulsion energies and the difference of the electronic energies; it appears that the nuclear repulsion energy would favor the conrotatory mode by about 50 kcal mole^{-1}.

In fact to analyse the zeroth order energy in terms of bond interaction it is necessary to define bond nuclear fields (valence bonds receive one unitary charge on both atoms defining the bond, while inner shells and lone pairs receive two unitary charges on the bearing atoms [46, 47]) and to expand the orthogonalization effects in powers of the overlap matrix S. When applied to this problem, the decomposition shows the leading role of the $S_{\pi_1 \pi_2}$ overlap between the two π MO's of the butadiene molecule, which is larger in the disrotatory mode. *The stereospecific zeroth-order contribution in the butadiene representation is a repulsive effect coming from the large overlap between the localized π MO's occurring in the disrotatory mode. This effect is actually a π electronic effect.* In the cyclobutene representation, this overlap effect is less pronounced, the overlap between the π and σ MO's being zero in the conrotatory mode (due to symmetry considerations) and small in the disrotatory mode. This stereospecific effect did not appear in our previous CNDO calculation, since the CNDO parametrization only introduces the interbond overlap effects through delocalization [32 b].

(b) *The main stereospecific factor appears essentially through π delocalization effects,* as revealed by the perturbation development (Figures 11, 12, 13 and Table I), and the weight of the singly excited determinants in the eigenvector of the π configuration interaction matrix (Tables II and III).

(c) In agreement with the results obtained by Trindle [12–14], *the σ system does not play any important role in the stereospecificity.* The role of the σ system appears through the difference between E_2 and E_1 in Table III. One sees that the energy difference between the barriers calculated for the two modes, i.e. the stereospecificity, is 39.8 kcal mole^{-1} when considering E_1 and 42.5 kcal mole^{-1} when considering E_2. *The σ system simply has a quantitative role, lowering the activation energy in the two modes*, from 99.0 and 138.9 kcal mole^{-1} respectively in the conrotatory and disrotatory mode at the E_1 level to 43.0 and 86.0 kcal mole at the E_2 level (see Table III). The comparison between the maps of Figures 14 and 15 also confirms this conclusion.

(d) *The weight of the doubly excited determinants* in the zero order multiconfigurational wave function presents a slight difference between the two modes (Tables II and III), but the order of magnitude of these coefficients and their stereospecificity is much smaller than those of the $\pi \rightarrow \pi^*$ delocalization configurations. In fact, by analyzing the configuration interaction matrix elements, one may see that the larger weight of intrabond closed shell doubly excited configurations in the disrotatory mode is not due to larger interactions of these configurations with the ground state determinant, but to the smallness of the singly excited configurations weight, through the normalization. On the contrary the interbond doubly excited determinants $\begin{pmatrix} \pi_1^* \ \pi_2^* \\ \pi_1 \ \pi_2 \end{pmatrix}$ have larger weights in the conrotatory mode, because they interact with the charge-transfer singly excited states $\begin{pmatrix} \pi_2^* \\ \pi_1 \end{pmatrix}$ and $\begin{pmatrix} \pi_1^* \\ \pi_2 \end{pmatrix}$.

So the correlation energy does not play an important role in the stereospecificity.

(e) A detailed analysis of the results obtained for the two dimensional maps (Figures 14 and 15) has shown *that the geometry does not play any role in the stereospecificity.*

(f) *The calculations performed with the PCILO-CNDO method has shown that the third order energy correction decreases the energy difference between the two modes:* after the third order correction the conrotatory mode is still favored but by 65 kcal mole^{-1} instead of 85 kcal mole^{-1} after the second order correction.

(g) As concerns the intermediate region, the two descriptions (butadiene or cyclobutene), which were very different for the single determinant wave function energy E_0 (see Table I), have almost the same energy when using a multiconfigurational zeroth-order wave function (From Table II, one may see that $(|E_2^{but} - E_2^{cyclo}| < 0.009\ \text{AU})$. The divergence which appeared in the ordinary second order calculations has been removed and was mainly due to the strong 'π' interactions. So, when using the multiconfigurational zeroth order wave function, the energies obtained for the intermediate region become reliable.

(h) The ab-initio second order results are in rather good agreement with experiment; this fact shows the likeliness of our analysis, despite the crudeness of our transition region.

The calculated enthalpy of the reaction is 15 kcal mole^{-1}, while it was estimated to be 20 kcal mole^{-1} by Dauben [48], 15 kcal mole^{-1} by Srinivisan [49] and 8.9 kcal mole^{-1} by McIver et al. [50], using experimental measurement by Wiberg [51], in good agreement with the 7.8 kcal mole^{-1} calculated by Hsu et al. [43].

As concerns the activation energy, the calculated altitude of the point $k = 0.5$ is

42–48 kcal mole^{-1} in the conrotatory mode in reasonable agreement with the experimental evaluations: 30 kcal mole^{-1} [52] or 40 kcal mole^{-1} [53].

C. Generalization of the Demonstration to n Double Bond Polyenes

The demonstration given in Section 2A for two double bonds may be generalized to any number of double bonds, but the stereospecificity appears at higher and higher orders of the perturbation series, so that a simple technique is necessary to handle the perturbation series. The many body theory diagrammatic techniques provide such an instrument.

If we assume the definition of the atomic orbitals given in Figure 16, the bonding molecular orbitals are defined in the initial state representation by:

$$\varphi_1 = (\chi_2 + \chi_1)/\sqrt{2}$$
$$\varphi_2 = (\chi_3 \pm \chi_4)/\sqrt{2} \qquad (24)$$
$$\varphi_3 = (\chi_5 + \chi_6)/\sqrt{2}$$
$$\varphi_n = (\chi_{2n-1} + \chi_{2n})/\sqrt{2}$$

according to the disrotatory or conrotatory mode. The related antibonding molecular orbitals are obtained by changing the sign of the first atomic orbital in these expressions,

In this demonstration, we make a further hypothesis; the monoelectronic integrals between the atomic orbitals, which are not bonded, either in the initial, or in the final state, are neglected. The overlap between these atomic orbitals is actually about five times smaller than the overlap between the atomic orbitals which are linked or will be linked in the course of the reaction. Therefore the integrals $\langle \chi_1 |h| \chi_5 \rangle$, $\langle \chi_1 |h| \chi_6 \rangle$, $\langle \chi_1 |h| \chi_3 \rangle$, $\langle \chi_2 |h| \chi_6 \rangle \ldots$ will be neglected.

It results from Figure 16 and Equation (24) that the integrals $\langle \varphi_1 |h| \varphi_2 \rangle$, $\langle \varphi_1 |h| \varphi_2^* \rangle$, $\langle \varphi_2 |h| \varphi_1^* \rangle$, $\langle \varphi_2^* |h| \varphi_1^* \rangle$ are all equal to $\pm \langle \chi_1 |h| \chi_4 \rangle$ according to the con- or disrota-

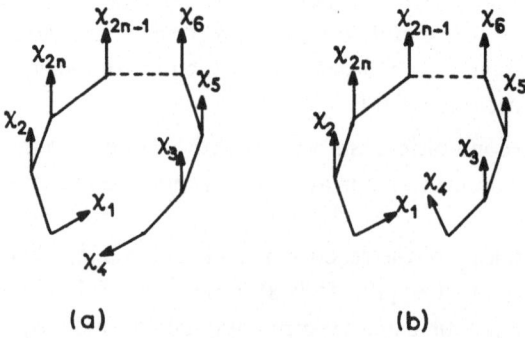

(a)   (b)

Fig. 16.   Definition of the atomic orbitals for the intermediate states of an n double bond electrocyclic reaction. Same definition of a and b as in Figure 8.

tory mode of the reaction and therefore positive in the conrotatory mode and negative in the disrotatory one. The other integrals are not stereospecific. The second order delocalization corrections are not stereospecific since the integrals are squared. *One may analyze the diagrams of order $3 < k < n$ which only imply monoelectronic interactions on their interaction lines.* We may notice that:

(i) the diagrams which do not imply φ_1 (or φ_1^*) *and* φ_2 (or φ_2^*) are not stereospecific.

(ii) the diagrams are negligible if they imply a long range interaction between the non neighbor molecular orbitals.

Therefore the non zero diagrams of order $k < n$ imply a one-step progression along the chain from one bond a to the bond $1 + a$, and then a regression to the point of departure by a one-step process, as occurs in the following diagram.

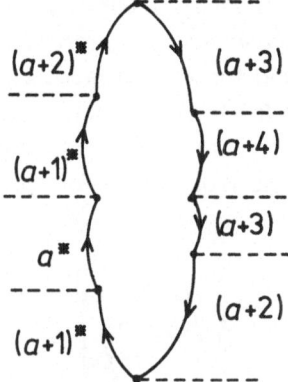

The nonzero diagrams implying only monoelectronic interactions are then necessarily of even orders. If they imply an interaction between the bonds 1 and 2 in the progression, another interaction between the bonds 2 and 1 appears in the return process: as the corresponding matrix elements are of the same sign for a given mode; these diagrams, even when they imply an interaction between the bonds 1 and 2, give the same correction for the two modes.

The stereospecific diagrams must imply an odd number of times the interaction between the bonds 1 and 2 and they necessarily are of order $k \geqslant n$. If one considers the lowest order diagrams $(k = n)$, which are supposed to be the dominant ones, one will get stereospecific contributions from the diagrams representing n interactions between the successive bonds of the chain in a cyclic process. For instance, for the hexatriene molecule, the diagrams

are stereospecific, and for the octatetraene molecule, the diagrams

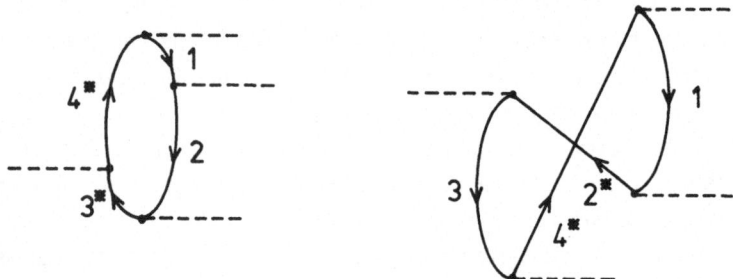

are stereospecific, since the integrals $\langle \varphi_1|h|\varphi_2 \rangle$ or $\langle \varphi_1|h|\varphi_2^* \rangle$ which appear in these diagrams have opposite signs for the two modes.

To demonstrate that the favored mode is conrotatory for an even number n of double bonds and disrotatory for an odd number n, one proceeds in the following manner.

(i) One first demonstrates that for a given mode and an n double bond polyene, all the nth-order diagrams which only imply monoelectronic interactions around the cycle have the same sign, whatever their topology and their indexing are.

In full generality these diagrams look like the following diagram.

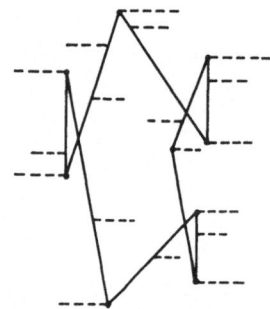

The $2v$ angular points correspond to the hole-particle interactions. There are $v-1$ crossing points of lines which are of the same nature in the diagram. When $v = 1$ one has a 'sausage' diagram implying only singly excited intermediate states, like the diagram

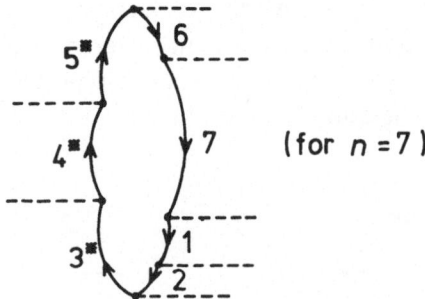

$(\text{for } n = 7)$

Moreover there are p interactions on the upward lines (between particles) and q interactions on the downward lines (between holes) so that $p + q + 2v = n$. If one arranges the molecular orbitals in an increasing order along the cyclizing polyene, the interactions will take place between the MO's of successive bonds r and $r+1$.

The signs of the molecular orbital coefficients may be chosen according to the following convention.

$$\frac{+(r)+\ \ +(r+1)+}{a\ \ a+1}$$

$$\frac{-\ \ +\ \ -\ \ +}{(r^*)\ \ (r+1)^*}$$

The interactions between holes $\langle r|h|r+1\rangle$ will be equal to $+\beta$ if $\beta = \langle \chi_a|h|\chi_{a+1}\rangle$. The interactions between particles will be equal to $-\beta$, and the interactions between holes and particles are given by

$$\langle r|h|(r+1)^*\rangle = -\beta$$

$$\langle r^*|h|(r+1)\rangle = +\beta$$

(25)

One can verify that if the diagram involves $2v$ angular points, there are v interactions $\langle r|h|(r+1)^*\rangle$ and v interactions $\langle r^*|h|(r+1)\rangle$.

Consequently, the product of the matrix elements occuring in the diagram has the sign $(-1)^{v+q}\beta^n$. The sign rules of the diagrams imply a sign change for each hole interaction and for each crossing point of the upward lines (or downward lines) between themselves. So the diagram gives a contribution

$$(-1)^{v+q+p+v+1}\beta^n/\Delta E^{n-1}$$

$$(-1)^{n-1}\beta^n/\Delta E^{n-1}$$

(26)

the sign of which depends on n only (for a given mode).

(ii) The sign of all these diagrams is obtained immediately from consideration of a particular diagram, for instance, from the diagram representing the interaction between the various excited state charge transfer states from bond 1

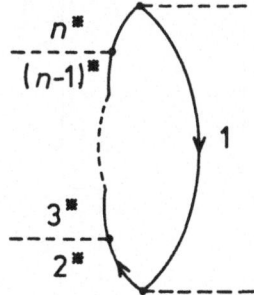

The contribution of the nth order diagram is given by

$$(-1)^{n-1}\frac{\beta^{n-1}}{\Delta E^{n-1}}\langle \varphi_1|h|\varphi_2^*\rangle$$

(27)

$(\beta/\Delta E)^{n-1}$ is positive and the sign of $\langle\varphi_1|h|\varphi_2^*\rangle$ depends on the mode so one may notice that *the preferred mode is conrotatory for even n and disrotatory for odd n, in agreement with the Woodward-Hoffmann rules.*

3. Other Stereospecific Reactions

A. CYCLOADDITION REACTIONS

It is possible, using localized molecular orbitals, to perform an analysis of the cyclo-addition reaction similar to the one performed for electrocyclic reactions and to show the importance of a circular interaction process in the determination of the approach mode of two molecules.

1. *Cycloaddition of Two Ethylene Molecules*

According to the Woodward-Hoffmann rules, the thermal cycloaddition $(2+2)$ is of supra-antara mode $(_\pi 2_s + _\pi 2_a)$. Such an example has been found experimentally by Kraft and Koltzenburg [54]. The supra-supra $(_\pi 2_s + _\pi 2_s)$ mode is forbidden in thermal reactions.

(a) *Supra-supra process*

This process is characterized by an approach of the two molecules according to Figure 17a: the two ethylene molecules are in two parallel planes A and B, and the two molecules approach parallel to each other. In such a case the localized MO's are symmetry MO's with respect to the mirror plane M, perpendicular to the molecular planes and bisecting the two molecules, which is maintained as a symmetry element during the reaction. Therefore our energetic approach is almost identical to the classical symmetry demonstration.

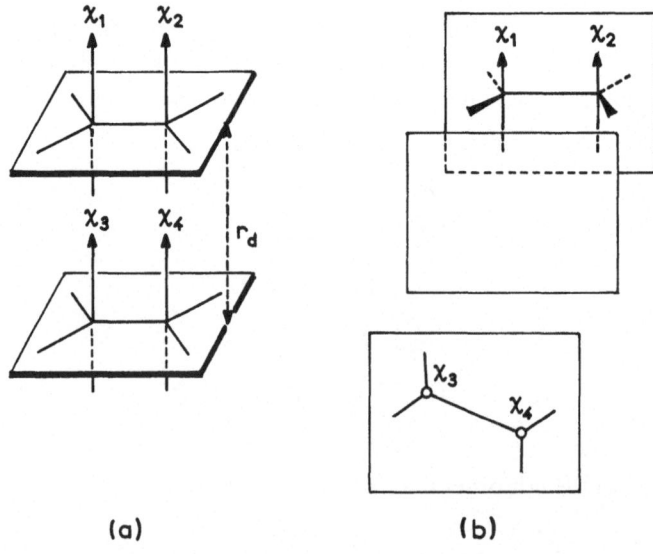

(a)   (b)

Fig. 17.   Cycloaddition of two ethylene processes (a) Supra-supra process; (b) Supra-antara process.

One may consider two bonding molecular orbitals φ_1 and φ_2 and the corresponding antibonding molecular orbitals:

$$\varphi_1 = 1/\sqrt{2}\,(\chi_1 + \chi_2) \quad \text{and} \quad \varphi_1^* = 1/\sqrt{2}\,(-\chi_1 + \chi_2)$$
$$\varphi_2 = 1/\sqrt{2}\,(\chi_3 + \chi_4) \quad \text{and} \quad \varphi_2^* = 1/\sqrt{2}\,(-\chi_3 + \chi_4). \tag{29}$$

The bonding molecular orbitals φ_1 and φ_2 are symmetrical while the antibonding molecular orbitals φ_1^* and φ_2^* are antisymmetrical.

So the second order delocalization correction

$$E_{delo}^2 = \left(\frac{\langle \varphi_1 | h | \varphi_2^* \rangle^2}{E_0 - E_{\varphi_1 \to \varphi_2^*}} + \frac{\langle \varphi_2 | h | \varphi_1^* \rangle^2}{E_0 - E_{\varphi_2 \to \varphi_1^*}} \right) \times 2 = 0. \tag{30}$$

(b) *Suppra-antara process* $(_\pi 2_s + _\pi 2_a)$

The supra-antara cycloaddition of two ethylene molecules requires a complicated twisting of one of the ethylene molecules (see Figure 17 b), the two ethylene molecules are no more in parallel planes. As one may see from Figure 17 b, the torsion of the π atomic orbital χ_4 allows a bonding interaction between the π atomic orbitals χ_2 and χ_4, and an opposite torsion of the π atomic orbital χ_3 allows a bonding interaction of the π atomic orbitals χ_3 and χ_1. According to the definition of the molecular orbitals φ_1 and φ_2 (Equation (28)), the matrix element $\langle \varphi_1 | h | \varphi_2^* \rangle$ is given by:

$$\langle \varphi_1 | h | \varphi_2^* \rangle = - \langle \chi_1 | h | \chi_3 \rangle + \langle \chi_2 | h | \chi_4 \rangle.$$

Figure 17 b shows that, in the transition state region, the integral $\langle \chi_1 | h | \chi_3 \rangle$ is positive while the integral $\langle \chi_2 | h | \chi_4 \rangle$ is negative, so the integral $\langle \varphi_1 | h | \varphi_2^* \rangle$, and the second order delocalization correction are not zero in the supra-antara process.

The second order delocalization correction tends to stabilize the supra-antara process with regards to the supra-supra process.

2. Diels-Alder Reaction

According to the Woodward-Hoffman rules, the cycloaddition of one butadiene molecule with one ethylene molecule is performed according to the supra-supra process in agreement with the experimental result [55].

(a) *Supra-supra process*

The approach of the two molecules is given by Figure 18 a.

The bonding molecular orbitals are given by:

$$\varphi_1 = (\chi_1 + \chi_2)/\sqrt{2}$$
$$\varphi_2 = (\chi_3 + \chi_4)/\sqrt{2} \tag{31}$$
$$\varphi_3 = (\chi_5 + \chi_6)/\sqrt{2}$$

Figure 18 a shows that the interaction between the atomic orbitals χ_1 and χ_5, and χ_4 and χ_6 are of $- + $ type.

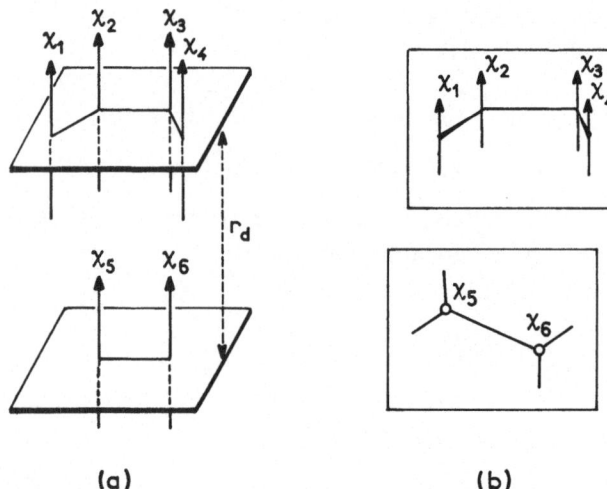

Fig. 18.   Cycloaddition of ethylene and butadiene. (a) Supra-supra process; (b) Supra-antara process.

So when considering the transition state of cycloaddition 'butadiene+ethylene' one finds an organization of atomic orbitals isomorphous to the organization of the atomic orbitals visualized by the scheme **B**: effectively one obtains scheme **B** by replacing φ_3 by $-\varphi_3$.

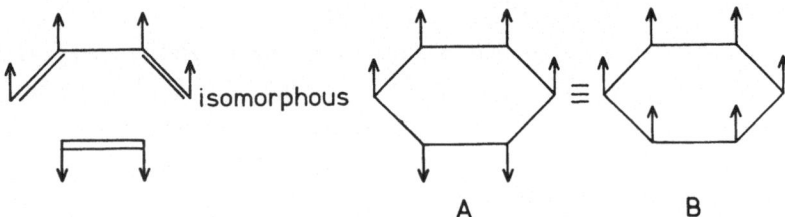

The organization of the atomic orbitals, visualized by the scheme **B** showing all the positive lobes of atomic orbitals in a parallel disposition, is isomorphous to the organization of atomic orbitals in the electrocyclic cyclization of hexatriene according to the disrotatory mode (Figure 19a): so the third order delocalization correction gives a negative correction which stabilizes the energy of the transition state obtained by the supra-supra approach process.

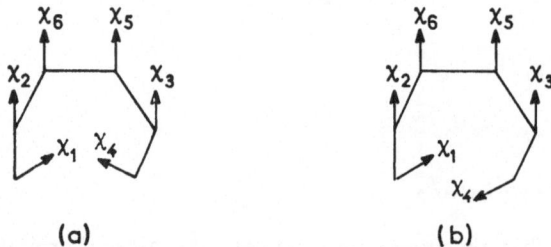

Fig. 19.   Electrocyclic cyclization of hexatriene. (a) Disrotatory mode; (b) Conrotatory mode.

(b) Supra-antara process

Figure 18b gives the approach process of the two molecules. The χ_5 and χ_6 atomic orbitals are twisted in an opposite way. So in the transition state region, one gets an organization of the atomic orbitals isomorphous to the one visualized by the scheme C

C

i.e. is isomorphous to the organization obtained in the conrotatory transition state of the electrocyclic reaction hexatriene⇔cyclohexadiene (Figure 19b).

The third order delocalization correction is positive.

So, in the Diels-Alder reaction $(_\pi 4 + _\pi 2)$, the third order circular delocalization correction favors the supra-supra process with respect to the supra-antara process.

3. Generalization to the $m+n$ Cycloadditions

It is always possible, according to the supra-supra or supra-antara process, to get a transition state with a circular organization of the atomic orbitals, organization isomorphous to the one obtained in the electrocyclic reaction of a $(m+n)/2$ polyene. So there is a correlation between the supra-supra process and the disrotatory mode and between the supra-antara process and the conrotatory mode.

So when a cycloaddition reaction involves an even number of double bonds, the supra-antara process is allowed and the supra-supra process is allowed when the cycloaddition reaction involves an odd number of double bonds. This result is in agreement with the Woodward-Hoffman rules [26].

B. SIGMATROPIC REACTIONS

One may study the simplest example, the reaction:

φ_1 and φ_2 are the two molecular orbitals implied in the reaction

$$\varphi_1 = 1/\sqrt{2}\,(\chi_1 + \chi_2)$$
$$\varphi_2 = 1/\sqrt{2}\,(\chi_3 + \chi_4)$$

where χ_1 and χ_2 are the two atomic orbitals of the $\pi(C_1=C_2)$ bond; χ_3 is a hybrid of carbon C_3 and χ_4 is the atomic orbital of the hydrogen H_4.

If the hydrogen H_4, transferred from the C_3 carbon to the C_1 carbon, stays during all the reaction course above the plane $C_1C_2C_3$ (suprafacial process), we get interactions between atomic orbitals of $+\beta$ type, and the reaction will be allowed if the number of bonds implied in the reaction is odd, when the number of double bonds is even.

In the antarafacial process, the atomic orbital χ_1 is twisted so that one obtains an interaction between χ_1 and χ_4 of type $-\beta$, leading to the reverse result.

Our results are in agreement with those of Woodward and Hoffmann [26].

4. Conclusions

(i) The stereospecificity of concerted reactions which appeared as a private property of delocalized descriptions may be demonstrated from purely local models. This success is not astonishing if one refers to the basic equivalence between the various sets of MO's, which simply are mathematical intermediates without physical significance.

(ii) The delocalized approaches to stereospecificity must choose between an analytic oversimplified model, and an almost purely numerical study. The analytic demonstrations lie on very drastic energetic hypothesis; one must implicitly assume that the MO's are regularly ordered and that their energetic evolution along the reaction path is regular enough for the correlation diagrams to be meaningful. One must also assume that the CI effects between states of the same symmetry is not too large. All these hypotheses, together with the strong $\sigma\pi$ separation postulated at the very beginning of the demonstration look somewhat dangerous to the rigourous ab-initialists. But the semi-empirical all valence electron and ab-initio calculations no longer give any interpretation of the phenomenon; even the respective roles of π and σ levels is not discussed.

On the contrary the localized analytical demonstration and localized numerical calculations progress along the same way and the second one allows a verification of the hypotheses of the former, step by step.

(iii) Symmetry does not play the key role, a purely energetic approach is possible.

(iv) The localized demonstration and calculations confirm the implicit hypothesis of the Woodward-Hoffmann demonstration of a leading role of 'π factors'. The σ part only decreases the barriers without changing the stereospecificity.

(v) The π overlap repulsive effects may play a role, but the main stereospecific correction is a circular delocalization phenomenon around the reacting circle. From that conclusion one may derive three important remarks:

—the stereospecific correction is linked to the 'aromatic diagrams' already discussed in Volume I of this book [56], and noticed a long time ago by Zimmerman [8].

—the difference between the conrotatory and disrotatory modes should decrease and tend toward zero when the number n of bonds involved in the reacting circle increases. Since it appears as a nth order perturbation correction, this prediction cannot be verified now, the experimental cases only concerning small molecules.

—the localized nth order stereospecific correction involves a circular delocalization of electrons around the reacting circle. One may notice the striking isomorphism between the stereospecific process involved (for $n = 3$ for instance) in a demonstration and the traditional dot and arrow picture of the phenomenon.

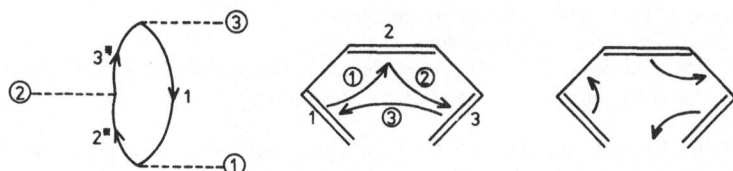

Once more the localized description appears as a possible bridge between the quantum mechanical and chemical languages.

References

1. Goddard, W. A., III: *J. Amer. Chem. Soc.* **94**, 793 (1972).
2. Woodward, R. B. and Hoffmann, R.: *J. Amer. Chem. Soc.* **87**, 395 (1969).
3. Hoffmann, R. and Woodward, R. B.: *J. Amer. Chem. Soc.* **87**, 2046 (1965).
4. Woodward, R. B. and Hoffmann, R.: *J. Amer. Chem. Soc.* **87**, 2511 (1965)
5. Longuet-Higgins, A. C. and Abrahamson, E. W.: *J. Amer. Chem. Soc.* **87**, 2046 (1965).
6. Kimichi-Fukui, K.: *Tetrahedron Letters* **24**, 2009 (1965).
7. Feler, G.: *Theoret. Chim. Acta (Berl.)* **12**, 412 (1968).
8. Zimmerman, H. E.: *J. Amer. Chem. Soc.* **90**, 543 (1968).
10. Van Der Lugt, W. Th. A. M. and Oosterhoff, L. J.: *J. Amer. Chem. Soc.* **91**, 6042 (1969).
11. Mulder, J. J. and Oosterhoff, L. J.: *Chem. Commun.* 305 (1970).
12. Trindle, C.: *J. Amer. Chem. Soc.* **92**, 3251 (1969).
13. Trindle, C.: *J. Amer. Chem. Soc.* **92**, 3255 (1969).
14. Trindle, C.: *Theoret. Chim. Acta* **18**, 261 (1970).
15. Goddard, W. A., III: *J. Amer. Chem. Soc.* **94**, 793 (1972).
16. Epiotis, N. D.: *J. Amer. Chem. Soc.* **95**, 1200 (1973).
17. Epiotis, N. D.: *J. Amer. Chem. Soc.* **95**, 1191 (1973).
18. Bryan, R. F., Doorakian, G. A., Freedman, H. H., and Weger, H. P.: *J. Amer. Chem. Soc.* **92** 399 (1970).
19. Vogel, E.: *Liebigs Ann. Chem.* **615**, 14 (1958).
   Criegee, R. Seebach, D., Winter, R. E., Börretzen, B., and Brune, H. A.: *Chem. Ber.* **98**, 2339 (1965).
20. Havinga, E., de Kock, R. J., and Rappold, M. P.: *Tetrahedron* **11**, 278 (1960).
   Havinga, E. and Schattmann, J. L. M. A.: *Tetrahedron* **16**, 146 (1961).

21. Corey, E. J. and Streith, J.: *J. Amer. Chem. Soc.* **86**, 950 (1964).
    Paquette, L. A., Barrett, J. H., Spitz, R. P., and Pitcher, R.: *J. Amer. Chem. Soc.* **87**, 3417 (1965).
22. Sanders, G. M. and Havinga, E.: *Rec. Trav. Chim.* **83**, 665 (1964).
23. Syrkin, Ya. K.: *Izv. Akad. Nauk SSSR* **238**, 389, 401, 600 (1959).
24. Balaban, A. T.: *Rev. Roum. Chimie* **11**, 1097 (1966); **12**, 875 (1967).
25. Mathieu, J. and Valls, J.: *Bull. Soc. Chim. France* **1509** (1957).
26. Hoffmann, R. and Woodward, R. B.: *Accounts of Chem. Research* **1**, 17 (1968).
27. Woodward, R. B. and Hoffmann, R.: *The Conservation of Orbital Symmetry*, Verlag Chemie, Academie Press, (1970).
28. Lennard-Jones, J. E.: *Proc. Roy. Soc., Ser. A* **198**, 114 (1949); Hall, G. G. and Lennard-Jones J. E.: *ibid.* **202**, 155 (1950); **205**, 267 (1951); Hall, G. G., *ibid.* **202**, 166 (1950); **213**, 102 (1952); (b) Edmiston C. and Ruedenberg K.: *J. Chem. Phys.* **43**, 597 (1963); *Rev. Mod. Phys.* **35**, 457 (1963); (c) Gilbert, T. L.: *Molecular Orbitals in Chemistry: Physics, and Biology*, P. O. Löwdin and B. Pullman (eds.), Academic Press, New York N.Y., 1964, p. 405. This fundamental point does not appear in recently published text books in quantum chemistry. For a pedagogical review on this question see Berthier, G.: *Aspects de la Chimie Quantique Contemporaine* (Colloques Internationaux du CNRS, 195), R. Daudel and A. Pullman (eds.), Centre National de la Recherche Scientifique, Paris, (1971), and reference (30b)).
29. Millé, P.: *Bull. Soc. Chim. Fr.* **4031** (1966).
30a. Forster, J. M. and Boys, S. F.: *Rev. Mod. Phys.* **32**, 300, (1960); Boys, S. F.: *Quantum Theory of Atoms, Molecules and the Solid State*, P. O. Löwdin (ed.), Academic Press, New York, N. Y., 1966.
30b. Millié, P. H., Levy, B., and Berthier, G.: *Localization and Delocalization in Quantum Chemistry* (ed. by O. Chalvet, R. Daudel, S. Diner and J. P. Malrieu), D. Reidel Publishing Company, Dordrecht-Holland and Boston-U.S.A., p. 59 (1975).
31. McWeeny, R. and Ohno, K.: *Proc. Roy. Soc., Ser. A* **255**, 367 (1960); Tsuchida, A. and Ohno, K.: *J. Chem. Phys.* **39**, 600 (1963); McWeeny, R. and Del Re, G.: *Theor. Chim. Acta* **10**, 13 (1968); Petke, J. D. and Whitten, J. L.: *J. Chem. Phys.* **51**, 3166 (1969); Hoyland, J. R.: *J. Amer. Chem. Soc.* **90**, 2227 (1968); Masson, A., Levy, B., and Malrieu, J. P.: *Theor. Chim. Acta* **18**, 193 (1970).
32a. Diner, S., Malrieu, J. P., and Claverie, P.: *Theor. Chim. Acta* **13**, 1 (1969);
32b. Diner, S., Malrieu, J. P., Jordan, F., and Gilbert, M.: *ibid.* **15**, 100 (1969);
32c. Jordan, F., Gilbert, M., Malrieu, J. P., and Pincelli, U.: (b) *ibid.* **15**, 211 (1969).
33. Pople, J. A. and Segal, G. A.: *J. Chem. Phys.* **44**, 3289 (1966).
34. Claverie, P., Diner, S., and Malrieu, J. P.: *Int. J. Quantum Chem.* **1**, 151 (1967).
35. Del Re, G.: *Theor. Chim. Acta* **1**, 188 (1963).
36a. Langlet, J. and Malrieu, J. P.: *J. Amer. Chem. Soc.* **94**, 7254 (1972).
36b. Daudey, J. P., Langlet, J., and Malrieu, J. P.: *J. Amer. Chem. Soc.* **96**, 3393 (1974).
37. Stevens, R. M.: *J. Chem. Phys.* **52**, 1397 (1970).
38. Masson, A., Levy, B. and Malrieu, J. P.: *Theor. Chim. Acta* **18**, 193 (1970).
39. Huron, B., Malrieu, J. P., and Rancurel, P.: *J. Chem. Phys.* **58**, 5745 (1973).
40. Simpson, W. T.: *J. Amer. Chem. Soc.* **73**, 5363 (1951); **77**, 6164 (1959); Pople, J. A. and Walmsley, S. M.: *Trans. Farad. Soc.* **58**, 441 (1962).
41. Almenningen, A., Bastiansen, and Traetteberg, M.: *Acta Chem. Scand.* **12**, 1221 (1967); Cole, A. R. H., Moroy, G. M., and Osborne G. A.: *Spectrochim. Acta* **23**, 909 (1967).
42. Elihu Goldish, E. Hedberg, K., and Schomaker, V.: *J. Amer. Chem. Soc.* **78**, 2714 (1956).
43. (a) Hsu, D. Buenker R. J., and Peyerhimhoff, S. D.: *J. Amer. Chem. Soc.* **93**, 2117, 5005 (1971); (b) *ibid.* **94**, 5639 (1972).
44. Jordan, F., Gilbert, M., Malrieu, J. P., and Pincelli, U.: *Theor. Chim. Acta* **15**, 211 (1969).
45. Murrell, J. N., Randic M., and Williams, D. R.: *Proc. Roy. Soc., Ser. A* **284**, 566 (1965).
46. Daudey, J. P., Malrieu, J. P., and Rojas, O.: *Localization and Delocalization in Quantum Chemistry* (ed. by O. Chalvet, R. Daudel, S. Diner and J. P. Malrieu), D. Reidel Publishing Company, Dordrecht-Holland and Boston-U.S.A., p. 155 (1975).
47. Musso, G. F. and Magnasco, V.: *Chem. Phys. Letters* **23**, 79 (1973).
48. Dauben, W. G.: *Reactivity of the Photoexcited Organic Molecules*, Interscience, New York, N.Y., 1967, p. 171.
49. Srinivisan, R.: *J. Amer. Chem. Soc.* **90**, 3395 (1968).
50. McIver, J. W. and Komornicki A.: *J. Amer. Chem. Soc.* **94**, 2625 (1972).

51. Wiberg, K. B. and Fenoglio, R. A.: *J. Amer. Chem. Soc.* **90**, 3395 (1968).
52. Frey, H. M., Pople, B. M., and Skinner, R. F.: *Trans. Faraday Soc.* **63**, 1166 (1967).
53. Hammond, G. S.: *J. Amer. Chem. Soc.* **77**, 334 (1955).
54. Kraft, K. and Koltzenberg, G.: *Tetrahedron Letters* **4357**, 4723 (1976).
55. Huisgen, R., Grashey, R., and Saver, J.: in S. Patai, *The Chemistry of Alkenes*, Interscience, N.Y. (1964), p. 739; A. Wasserman: *The Diels-Alder Reaction*, Elsevier, Amsterdam (1965).
56. Malrieu, J. P.: in *Localization and Delocalization in Quantum Chemistry*, Vol. I, p. 335 (1975).

MOLECULAR VALENCE STATES AND BINDING ENERGIES

P. VERMEULIN, B. LEVY, and G. BERTHIER

Laboratoire de Chimie, École Normale Supérieure de Jeunes Filles,
92120 Montrouge, France
Laboratoire de Biochimie Théorique, Institut de Biologie physico-chimique,
75005 Paris, France

Abstract. An extension of the concept of atomic valence states to molecular fragments is presented, with the object of treating chemical functional groups. The total energy of a molecular fragment is computed, using appropriate modifications of the SCF and CI theories. An example of covalent and semi-polar bonds, the fragments CF and NO of fluorinated alkanes and saturated amine-oxides have been studied; the following equilibrium distances and force constants: CF (1.365 Å and 5.7 mdyne. $Å^{-1}$), NO (1.48 Å and 2.4 mdyne. $Å^{-1}$) are found, in good agreement with the values usually assigned to these bonds. Results for CH bonds are also given.

1. Bond-Energy Calculations and Chemical Reactivity

In order to interpret chemical transformation processes, a number of quantum chemists have recently directed their efforts to the determination of potential surfaces and molecular trajectories. This direct approach to chemical reactivity, however, requires a huge amount of numerical computation especially at the ab-initio level, and up to now its practical applications concern the results of molecular beam experiments [1] rather than standard chemical problems. So, an analysis of chemical reactivity in terms of conventional theoretical concepts as bond energies, transition states, inductive and mesomeric charge transfers, solvatation effects etc... retains all its usefulness. As a matter of fact, the most popular quantity associated to the concept of chemical bond is probably its bond energy, because a knowledge of bond energies allows the chemist to predict the energy balance of a given reaction. Unfortunately, as is well known, bond energies are not strictly physical observables in polyatomic molecules: To evaluate them, it is necessary to make a partition of the formation energy of a molecule in various components [2], and this is usually performed by identifying the energy of a certain bond (for instance the C—H bond in normal alkanes) to the fractional part of the energy of a reference compound (*e.g.* the fourth part of the formation energy of methane). This procedure neglecting the interaction between the different bonds of the reference molecule is arbitrary, and it is the starting point of many disputes about the definition and transferability of a bond energy.

From the point of view of Molecular Quantum Mechanics, the situation seems to be very similar: at first appearance, there is no unambigous way of distributing the binding energy of a polyatomic molecule (*i.e.* the difference between the total electronic and nuclear energy of a molecule and the energies of its atomic components) for the very reason that the chemical bonds themselves are not defined quantum-mechanically. In the circumstances, some people are content with purely formal quantities derived from population analyses [3], even if they might be criticized on the ground they use quantities belonging to the 'plague of non-observables' [4]. In our opinion, such an attitude towards chemical phenomena is a little bit too extreme: We think that it is possible to escape from these conceptual difficulties by reversing the problem and considering that the existence of chemical bonds guaranteed by two centuries of experimental practice is enough to justify an intuitive use of the concept of chemical bond.

First of all, it has been necessary to adapt the Quantum-Chemistry methods to the calculations of bond properties by translating the concept of chemical bond itself into a wave-mechanical language. This was achieved by transposing the concept of *valence states*, originally developed by Van Vleck [5], Mulliken [6], Moffitt [7], Hinze and Jaffe [8] and others for atoms, to the case of a chemical bond or, more generally, a polyatomic *molecular fragment* [9]. Consequently, remarks or reservations about the fact that a valence state is not a spectroscopy stationary state, but may be assimilated to some kind of superposition of spectroscopic states, could be mentioned.

As first examples, we have calculated the electronic structure of the fragments N→O and C—F in amine-oxides and fluorinated alkanes as examples of a semi-polar bond and a covalent bond respectively. Our model is able to reproduce not only the very different equilibrium characteristics of these two types of bonds, as *internuclear distances* and *bond energies*, but also those properties which depend on the whole behaviour of the energy curve, as *force constants*, because possible correlation effects can be incorporated in it through standard configuration interaction methods. This is quite important for some reactivity phenomena where the vibrations of the attacked bonds play a great role [10]. Finally, we have studied the C—H bond for various 'hybridization' states of the carbon atom in order to determine to what extent a simple bond-model can account for environment effects.

2. Molecular Valence-State Theory

Let us consider a molecule, formed by a group of atoms M which are linked to atoms $A_1, A_2, A_3, \ldots A_v$. The molecular fragment is the entity M obtained when the atoms A are carried away to infinity. The molecular fragment may be composed of an arbitrary number of atoms (one, two or more ...). The preceding dissociation process involves the presence in the fragment of one or more unpaired electrons, the spin state of which is unknown (we call them random-spin electrons). Even if the molecular fragment cannot be represented by a single antisymmetrized product of spin-orbital functions, its energy may be deduced from the wave function of the original molecule we suppose, for simplicity, in a closed-shell state.

Within the *LCAO-MO* approximation the molecular wave function can be written in the following form:

$$\Phi = ((2n+2v)! \, 2^v))^{-1/2} \det|\varphi_1(1) \, \bar{\varphi}_1(2) \ldots \varphi_n(2n-1) \, \bar{\varphi}_n(2n) \times$$

$$\times \; (\xi_{n+1}(2n+1)\, \bar{\omega}_{n+1}(2n+2) + \omega_{n+1}(2n+1)\, \tilde{\xi}_{n+1}(2n+2)) \; \dots \times$$

$$\times \; (\xi_{n+v}(2n+2v-1)\, \bar{\omega}_{n+v}(2n+2v) + \omega_{n+v}(2n+2v-1) \times$$

$$\times \; \tilde{\xi}_{n+v}(2n+2v))|$$

where the φ's and ξ's are the orbitals of the fragment M occupied by two paired electrons and random-spin electrons respectively and the ω's are the orbitals of the atoms A associated with the functions ξ of M in covalent bonds.

The fragment energy is then the limit of the expectation value of the molecular Hamiltonian operator $\langle \Phi | \mathscr{H} | \Phi \rangle$, for atoms A going to infinity. Using orthogonalized spin-orbitals φ and ξ for the fragment M, we finally obtain

$$E = \sum_i n_i I_i + \sum_i \sum_j n_{ij}(2J_{ij} - K_{ij})$$

where n_{ij} is defined in terms of the occupation numbers n_i and n_j by

$$n_{ij} = n_i n_j / 4 \quad \text{if} \quad i \neq j$$

$$n_{ii} = n_i - 1$$

with $n_i = 1$ for an orbital occupied by a random-spin single electron and $n_i = 2$ for an electron pair (for any virtual orbital figuring in subsequent calculations, $n_i = n_{ii} = 0$). The preceding expression for E is identical with the Dirac formula used by Moffit in his valence-state theory.

We now consider the problem of minimizing E. The standard Lagrange multiplier technique cannot easily be applied because it is not possible to find a unitary transformation which diagonalizes the Fock matrix and at the same time keeps the Fock operator invariant. Consequently, we have followed the method of partial derivatives used by Levy in the case of MC-SCF methods [11]. Starting with a set of trial molecular orbitals φ_i, we choose a matrix X the elements of which satisfy the equation

$$X_{ij} = - X_{ji}$$

and give new orbitals φ_i by the transformation

$$\varphi_i'(X) = e^X \varphi_i$$

that is to say at the first order

$$\varphi_i' = \varphi_i + \sum_j X_{ij} \varphi_j \,.$$

Minimizing the total energy with respect to the matrix elements X_{ij} leads to the following system of equations:

$$\partial E(X) / \partial X_{ij} = 2 \langle \varphi_i | (F_j - F_i) | \varphi_j \rangle = 0$$

where F is a Fock-type operator defined by

$$F_i = n_i H_i + 2 \sum_k n_{ik}(2J_k - K_k) \,.$$

The matrix X is computed by the relation

$$X = - [\partial E(X)/\partial X]_{X=0}/[\partial^2 E(X)/\partial X^2]_{X=0}$$

and the matrix of second derivatives is determined from the first derivatives formula by the finite-difference method. In this way, a new set of molecular orbitals φ'_i can be computed and the process repeated iteratively up to the point of minimum energy.

In accordance with the definition of fragments, the following conditions for an orbital containing one random-spin electron are to be introduced:

(a) It should be kept well-localized on one of the atoms of the fragment and be little perturbed by a subsequent dissociation of the fragment. Among other things, this means it must be unchanged, except through reorthogonalization, during the energy minimization process.

(b) Such a molecular orbital should point in the direction of the bond from which it is issued. Appropriate constrains can be introduced by putting some of the elements of the matrix X equal to zero, namely those X_{ij} where i refers to an orbital occupied by a single random-spin electron and j to all other orbitals, except for the virtual ones directed in the same way as i. The orthogonalization process modifies the atomic components of random-spin orbitals slightly, but their direction after orthogonalization can be readjusted to the appropriate value by calculating their actual gravity center.

3. Illustrative Calculations for Semi-Polar and Covalent Bonds

The preceding method has been applied to the bonds NO and CF of amine-oxides and fluorinated alkanes, two isoelectronic fragments which can be taken as examples for a semi-polar bond and an ordinary covalent bond, and to various types of CH bonds. The computations have been carried out with a set of $9s\,5p$ Gaussian orbitals on each atom [12], reduced to a double-zeta form by the usual contraction technique and augmented by one d orbital [13]. For the hydrogen atoms of CH bonds, a basis of 4s Gaussians contracted in a double-zeta form was also used, but no polarization orbitals were introduced in the calculation.

A preliminary study of the NO bond without any constraint preserving the direction of random-spin orbitals gave us the abnormal potential curve of Figure 1. This is due to the change of the angles of orbitals occupied by unpaired electrons when the NO distance varies.

As a matter of fact, the dissociation of the semi-polar bond N→O described in Figure 1 leads to nitrogen and oxygen atoms in the valence states $V_3(s^2 p^3)$ and $V_0(tr^2\, tr^2\, tr^2)$ respectively. The corresponding orbital angles for the nitrogen atom are not consistent with the actual bond angles in those saturated amines which would be obtained from amine-oxides by such a dissociation process. In the present case, the constraint effects are particularly important because the total energy of the NO fragment is very sensitive to the directions assumed by the random-spin orbitals. In addition, their angles have to be fixed at the right value, in order to obtain correct equilibrium distances and force constants, as shown in Figure 2.

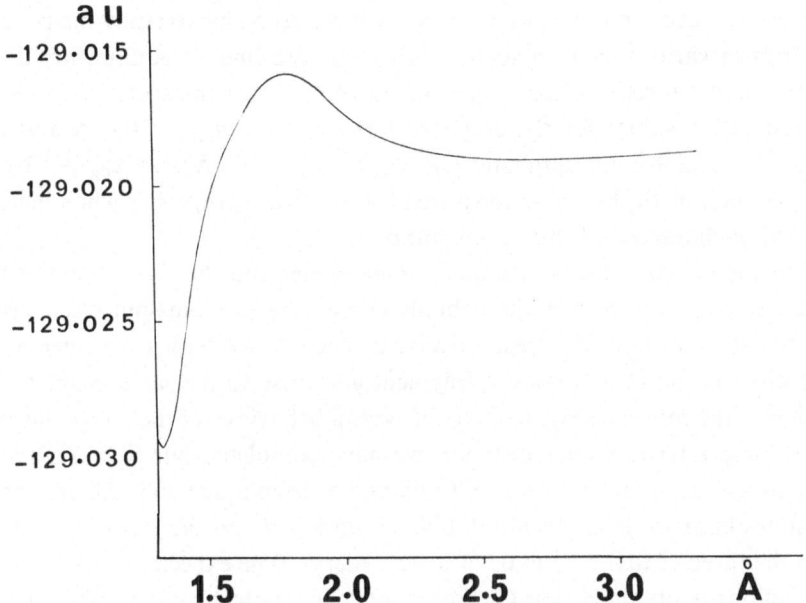

Fig. 1.   Potential curves of the NO bond without orientation constraints.

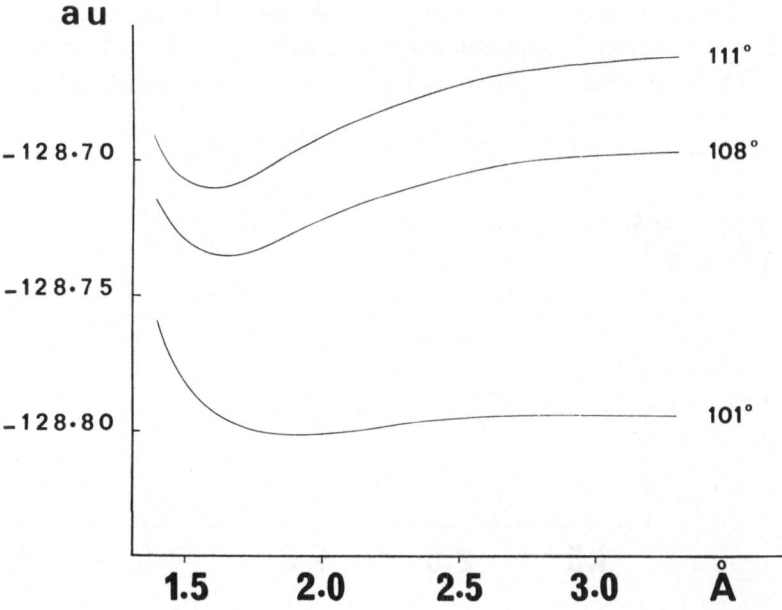

Fig. 2.   Potential curves of the NO bond with various random-spin directions.

Taking advantage of the dependence of our molecular valence states with respect to the angles between random-spin orbitals, we have been able to study the properties of CH bonds in various environments. In this way, we find an equilibrium distance $d_{CH} = 1.05$ Å and a force constant $k_{CH} = 6.3$ mdyne Å^{-1} for the acetylenic CH bond. The corresponding values for the ethylenic CH bond are $d_{CH} = 1.055$ Å and $k_{CH} = 5.1$ mdyne Å^{-1} and for the aliphatic CH bond $d_{CH} = 1.06$ Å and $k_{CH} = 4.1$ mdyne Å^{-1}, in agreement with the well-known trend of the bond properties when increasing the so-called 'p-character' of the carbon atom.

To define the N→O bond in saturated amine-oxides and the C—F bond in fluorinated alkanes, we assume that the orbitals containing random-spin electrons have angles equal to 111° and 109° respectively, in accordance with experimental data. Since the dissociation of a two-atom fragment gives rise to the same problem as the dissociation of the molecule H_2, namely the wrong behaviour of the molecular orbital method at large intermolecular distance, we have completed our SCF-MO calculations by configuration interaction. A CI mixing has been made with the ground state and the states built from single and double excitations from the σ orbital of highest energy to the three virtual σ orbitals of lowest energy. The expected correction to the SCF predictions is obtained: the CI correction appears to be essential for the covalent bond CF, but not for the semi-polar bond NO, a difference worthwhile to be noted. Energy and equilibrium distance are appreciably decreased by completing the basis with d orbitals. For the fragment NO, the equilibrium distance 1.48 Å is close to the experimental value, 1.44 Å for trimethylamine-oxide [14]. For the fragment CF, the equilibrium distance 1.365 Å is very close to the experimental value 1.385 for fluoromethane [15].

From the potential curves of Figures 3 and 4, the following force constants are obtained in the harmonic approximation: 5.72 mdyne Å^{-1} for CF and 2.43 mdyne Å^{-1} for NO. These results are in accordance with the value usually attributed to the

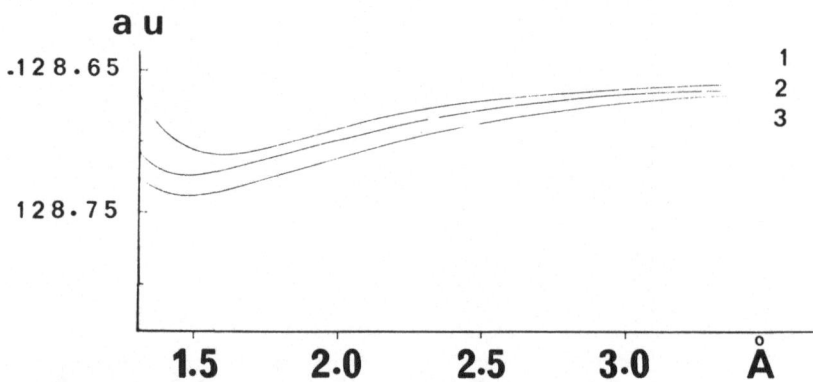

Fig. 3.   Potential curve of the N→O fragment in saturated amine-oxides (1 : SCF – 2 : SCF with d orbitals – 3 : CI with d orbitals).

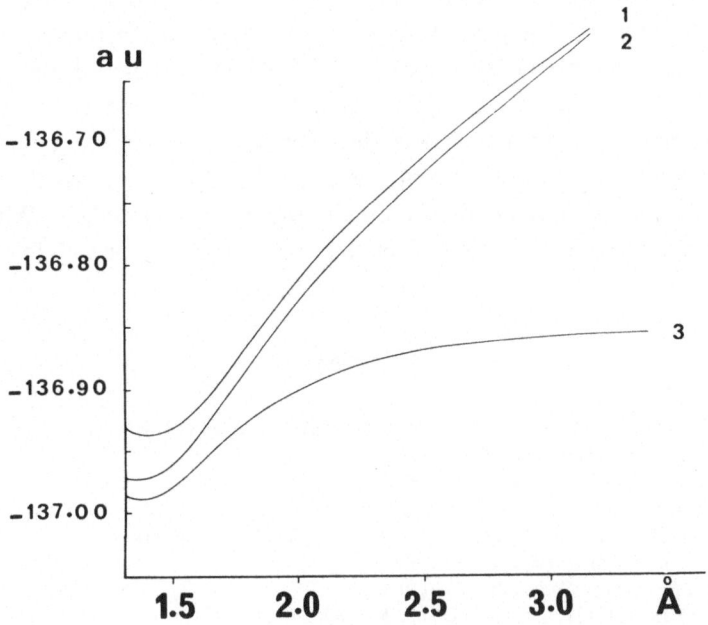

Fig. 4. Potential curve of the C–F fragment in fluorinated alkanes (1 : SCF – 2 : SCF with d orbitals – 3 : CI with d orbitals).

CF bond: 5.7 mdyne $Å^{-1}$ [16], and in fairly good agreement with the value extrapolated for the NO bond: 3.5 mdyne $Å^{-1}$, using a relationship [17] between the v_{NO} stretching frequency (in saturated amine-oxides, $950 \, cm^{-1}$) and the experimental

TABLE I

Total energies of the fragments NO and CF (in AU)

$d(Å)$	1.30	1.40	1.50	1.60	3.00
		Fragment NO			
(1)		− 128.6963	− 128.7076	− 128.7109	− 128.6635
(2)	− 128.7066	− 128.7232	− 128.7250	− 128.7219	− 128.6676
(3)		− 0.0144	− 0.0148	− 0.0145	− 0.0054
		Fragment CF			
(1)	− 136.9314	− 136.9379	− 136.9297	− 136.9126	− 136.6333
(2)	− 136.9712	− 136.9696	− 136.9553	− 136.9332	− 136.6382
(3)	− 0.0146	− 0.0180	− 0.0219	− 0.0272	− 0.2218

(1) Total energy of the molecular fragment without d orbitals (the sums of the energies of the dissociation products of NO and CF are − 128.6590 AU and − 136.8442 AU respectively.
(2) Total energy of the molecular fragment with d orbitals.
(3) Configuration interaction contribution to total energies.

bond length (1.44 Å). Actually, our force constant f_{NO} falls into line with the longer NO bond predicted by our calculations for the NO fragment. Finally, the binding energies of the various fragments $M = B - C$ considered in the present report were calculated as the difference between the energy of M and the sum of the energies of the sub-fragments B and C in corresponding valence states. The binding energy of NO is found to be smaller in absolute value than for the fragment CF (50 kcal mole^{-1} instead of 90 kcal mole^{-1}) in conformity with the bond energies usually assigned to the two bonds. For the CH bonds, we obtain about 130 kcal mole^{-1}. We hope that these values could be the beginning of the purely theoretical approach for bond energies.

References

1. Schötter, J. and Toennies, J. P.: *Chem. Phys.* **2**, 137 (1973); **4**, 24 (1974).
2. Benson, S.: *J. Chem. Educ.* **42**, 502 (1965).
3. Clementi, E.: *J. Chem. Phys.* **46**, 3842 (1967).
4. Platt, J. R.: *Handbuck Physik* **37/2**, 173 (1961).
5. Van Vleck, J. H.: *J. Chem. Phys.* **1**, 177, 219 (1933); **2**, 20, 297, (1934).
6. Mulliken, R. S.: *J. Chem. Phys.* **2**, 782 (1934).
7. Moffitt, W.: *Proc. Roy. Soc.* **A202**, 534 (1950).
8. Hinze, J. and Jaffe, H. H.: *J. Amer. Chem. Soc.* **84**, 540 (1962).
9. Vermeulin, P., Levy, B., and Berthier, G.: *Comptes Rendus* **281**, Série C, 515 (1975).
10. Coulson, C. A., Duchesne, J., and Manneback, C.: Vol. Comm. V. Henri, p. 15, Desoer, Liège (1948).
11. Levy, B.: *Chem. Phys. Lett.* **4**, 17 (1969); **18**, 59 (1972).
12. Huzinaga, S.: *J. Chem. Phys.* **42**, 1293 (1965).
13. Roos, B. and Siegbahn, P.: *Theoret. Chim. Acta* **17**, 199 (1970).
14. Rundle, R. E.: quoted in *Tables of Interatomic Distances*, The Chemical Society, London (1958).
15. Caron, A., Palenik, G. J., Goldish, E., and Donohue, J.: *Acta Cryst.* **17**, 102 (1964).
16. Mills, I. A.: in *Infrared Spectroscopy and Molecular Structure* (ed. by M. Davies), Elsevier Amsterdam, p. 192, (1965).
17. Jonathan, N. B. H.: *J. Mol. Spectrosc.* **4**, 75 (1960).

APPLICATION OF THE METHOD OF GROUP DENSITY ANALYSIS TO A STUDY OF EXCITED STATES ACID-BASE EQUILIBRIA IN CONJUGATED MOLECULES

R. CONSTANCIEL and O. CHALVET

*Centre de Mécanique Ondulatoire Appliquée du Centre National de la Recherche Scientifique,
75019-Paris, France*

1. Introduction

It was shown that perturbation theory, as applied within the framework of the π-electron approximation, is a simple and powerful tool to qualitatively account for the change in acid-base equilibria on photoexcitation [1]. However, even if this method has clearly shown that the difference between the π electronic distributions of the ground and excited states of a proton donor molecule is strongly correlated with the pK change, it remains to elaborate a physically acceptable model which serves as a support for a better understanding of the origin of such a correlation.

An initial satisfactory and simple explanation seems to be immediately provided by using classical rules governing the displacement of the electronic charges [2]. Using phenol as a particular example, we come to the following conclusion: 'The mesomeric effect leads to a drop in the electronic charge near the oxygen atom, i.e. to the existence of a positive net charge on this atom. This effect is enhanced in excited state [1], so that the electron pair of the O—H bond is attracted more efficiently in the excited state than in the ground state. It follows that the tendency to release the proton is greater in the excited state leading to an increase of the acidity.'

Such a process can be visualized in a pattern where the arrows indicate the modifications of the electronic density induced by the excitation. (Figure 1).

Two sorts of objections can be made concerning this particular example.

(1) (*Theoretical Objections*) A CNDO/S calculation shows that the change in the polarity of the O—H bond is quite negligible (the variations of the density matrix elements between the ground state and the excited one are about 10^{-4}). The inertness of the electronic density distributions of the valence bonds with respect to any change in their surroundings has been equally demonstrated through some ab initio calculations on the borazane and amino-borane molecules [3]. Both these molecular systems are formed by respectively covalent or dative bonds located between a couple of identical atoms. However, the core charges that one can define in the vicinity of each atom are very different in the two cases. More precisely, the core charges characteristic of the dative bond result from the core charges characteristic of the covalent bond by transfer of one unit of charge. Despite this strong difference between these two core fields the calculated displacement of the barycenter of the two electron bond density distribution was about one tenth of the bond length. We conclude that the modifications induced in a two electron bond by the reorgani-

zation of a π system which follows excitation is quite negligible because the corresponding changes of π atomic charges rarely exceed one tenth of an electron.

On the other hand, it is not easy to understand how merely considering variations of the electronic density alone could justify the departure of the proton. This is a process in which the nuclear interactions are equally modified and consequently must also be considered in order to provide a full explanation of the phenomenon.

(2) *Physical Objection*: In the crude model used above, the role of the solvent was not taken into account although its presence is of fundamental importance as a proton acceptor.

These above considerations led us to undertake a study with particular emphasis on the behavior of the valence bonds and especially on the bond which is weakened under the influence of excitation.

This study is essentially an analysis of a CNDO/S wave function; we have chosen this approximation in order to preserve the consistency with the previous calculations performed in the framework of this method. In a first step, where we consider an isolated acidic molecule like phenol, our purpose is to show how the weakening of the O—H bond on photoexcitation can be explained despite the fact that its own electronic density is not affected. The explicit consideration of a solvent molecule is undertaken in a second step.

2. Method of Analysis

The group density analysis method that we have reported elsewhere [4] is a tool particularly well adapted to the treatment of this type of problem. In effect, this technique allows one to theoretically partition the various interacting electronic subsystems which form as an ensemble the molecular system of interest in any electronic state. It then becomes possible to study what is the effect of the excitation on the subsystem structure; are the subsystems modified? Are some new ones formed? What are the subsystems which disappear?

We have previously detailed how to extract [4] from the study of the p-particle reduced density matrix an index capable of characterizing the separability of a given subsystem with respect to the rest of the molecule. Let us recall simply that this property derives from the separability of the total wave function itself and that it corresponds, from a physical point of view, to the possibility of describing the interactions between the various subsystems through a simple electrostatic scheme. Moreover let us point out that we can obtain simultaneously by means of the p-particle group density a complete description of the properties of the subsystem (and in particular the one-particle group density) in the field of the other constitutive parts of the molecule whose influence can be accounted for through an effective external charge field. This external charge field can be obtained by removing the 1-particle group density from the total charge distribution. This procedure proposed by Veillard *et al.* [3] has been used by Aslangul *et al.* [3] in order to emphasize the fundamental difference between covalent and dative bonds* as we have pointed out in the introduction.

Such an approach which consists of focusing attention on a particular subsystem, while the rest of the molecule (what is usually called the core) is treated with more crude approximations, has been widely used for the study of conjugated systems. Looking at a two electron subsystem is not more than a kind of π-approximation with the difference that we concentrate on the two electron group associated to a given valence bond rather than on the electrons which form the π system.

The analysis of the wave function can be completed by an energetic one. The total energy of the molecular system can be written as a sum of bond, core, and bond-core interaction contributions

$$E = E_B + E_C + E_{BC}$$

when the wave function is correctly approximated by an antisymmetrized product of bond and core functions, i.e. when the missing information $I(\mathscr{B})$ characterizing the bond is small.

The bond contribution is simply

$$E_B = \mathrm{Tr}_2 \left([\rho_n^{(2)}]_\mathscr{B} \, [H^{(2)}]_\mathscr{B} \right)$$

where $[\rho_n^{(2)}]_\mathscr{B}$ is the two-particle reduced density operator restricted to the best representative subspace $\mathscr{H}_\mathscr{B}^{(1)} \otimes \mathscr{H}_\mathscr{B}^{(1)}$ of the bond, and $[H^{(2)}]_\mathscr{B}$ is the restriction to that subspace of the two-particle hamiltonian operator.

The bond-core interaction contribution is the sum of coulombic interactions between the electrons of the bond and the electrons and nuclei of the core.

The sum of these two first terms can be considered as the effective energy of the two-electron system associated to the bond in the field of the core

$$E_B^{\mathrm{eff}}(C) = E_B + E_{BC}.$$

Then we obtain immediately the core contribution by

$$E_C = E - E_B^{\mathrm{eff}}(C).$$

However, it is interesting to divide the core contribution in two terms, having in mind a kind of diatomic model for the bond. This further partition is possible when the core charge distributions are well separated so that intercore I_C and intracore ε_C interactions can be recognized. The former can also be reasonably approximated by a sum of coulombic terms and can be added to $E_B^{\mathrm{eff}}(C)$ to form what we will call the bond effective energy ε_B.

If follows that the total energy can be considered as a sum of a diatomic molecule-like bond energy ε_B and of the self energy of each core piece ε_C:

$$E = \varepsilon_B + \varepsilon_C.$$

* The group density that they considered was derived from a molecular orbital localization procedure.

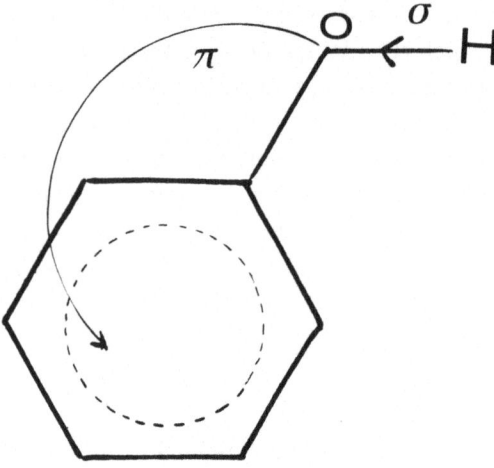

Fig. 1.

It is anticipated that the variation of the total energy E deriving from a change of the nuclear configuration, which would consist of a removal of one part of the core from another, is closely reflected in the corresponding variation of ε_B. Under these conditions, it is possible to account for the global changes of the molecular system in terms of changes of a properly chosen diatomic-like system. This provides a general way of interpreting the results of theoretical calculations on the whole system in the traditional analytical language of chemists.

In the limiting case where the two parts of the core are at infinite distance we obtain an expression for the dissociation energy

$$\Delta E = \Delta \varepsilon_B + \Delta \varepsilon_C$$

which is mainly a bond dissociation energy, corrected by the core polarization effects. The experimental evidence of a good constancy of the dissociation energy of the same given bond in certain classes of molecules implies necessarily that the core polarization energy is negligible. However, this fact must not be accepted as a general rule as we shall see later.

3. The Nuclear Interaction Problem

The introduction of nuclear interaction terms into the theoretical expression of the diatomic-like bond dissociation energy is a source of difficulties in the framework of the CNDO/S approximation of the wave function calculation. It is well known, in effect, that this method gives poor results in the search for equilibrium distances, a deficiency which can principally be imputed to the choice of electronic integrals to fit the nuclear repulsions. For the sake of consistency, it seemed to us necessary to use, in studying the implications of the pK variations at the σ bond level, the same method as the one which we worked out for the study of spectral properties at the π system level. Then we decided to introduce in the CNDO/S scheme the formula first proposed by Dewar for fitting the nuclear interactions [5].

This formula is a compromise between a point charge contribution $1/r_{ij}$ and a distribution-like contribution γ_{ij}

$$D_{ij} = (T_{ij}/r_{ij} + (1 - T_{ij})\,\gamma_{ij}).$$

The importance of each contribution varies with the internuclear distance r_{ij} according to

$$T_{ij} = \exp(-\alpha_{ij}r_{ij}).$$

All the calculations we are presenting here were carried out with the following set of parameters α_{ij} properly chosen in order that the computed equilibrium distances roughly agree with experience (see Table I).

TABLE I

Parameters used in determining the nuclear repulsion integrals

i	j	α_{ij}
H	H	0.66
H	C	1.63
H	O	1.80
C	C	2.14
C	O	2.30
O	O	1.97

4. Preliminary Study of the Total Energy

The total energy has been computed against the O—H bond distance variation for the following molecules OH^-, OH_2, OH_3^+ and ϕ—OH in the ground and first excited electronic states.

A satisfactory agreement is found between the calculated equilibrium distances and the experimental ones (see Table II and Figure 2).

In order to test the quality of the wave function for the chemical problem of interest, we have to derive from the computed dissociation energies ΔE the estimated variations of pK. To do this we considered the theoretical quantity

$$\gamma_{\text{theor}} = \frac{\Delta E(\phi OH_S) - \Delta E(\phi OH_G)}{\Delta E(\phi OH_T) - \Delta E(\phi OH_G)} = 1.627$$

and we have compared with the experimental quantity:

$$\gamma_{exp} = \frac{pK_S - pK_G}{pK_T - pK_G} = 4.267$$

both would be equal if we admit the validity of the Forster Cycle. The comparison is rather disappointing.

The same work was performed after introduction of a water solvent molecule. The nuclear configuration of the phenol-water supersystem has been fixed after the theoretical calculation of Pullman and Coll [6] (Figure 3). We supposed that the proton transfer from the acid molecule to the solvent occured through an hydrogen bridge, the other nuclei remaining fixed. The total energy curves exhibit two potential wells and again some reasonable equilibrium distances are found for the proton. Moreover, we have now an excellent agreement between the quantities γ_{exp} and γ_{theor}; the last one

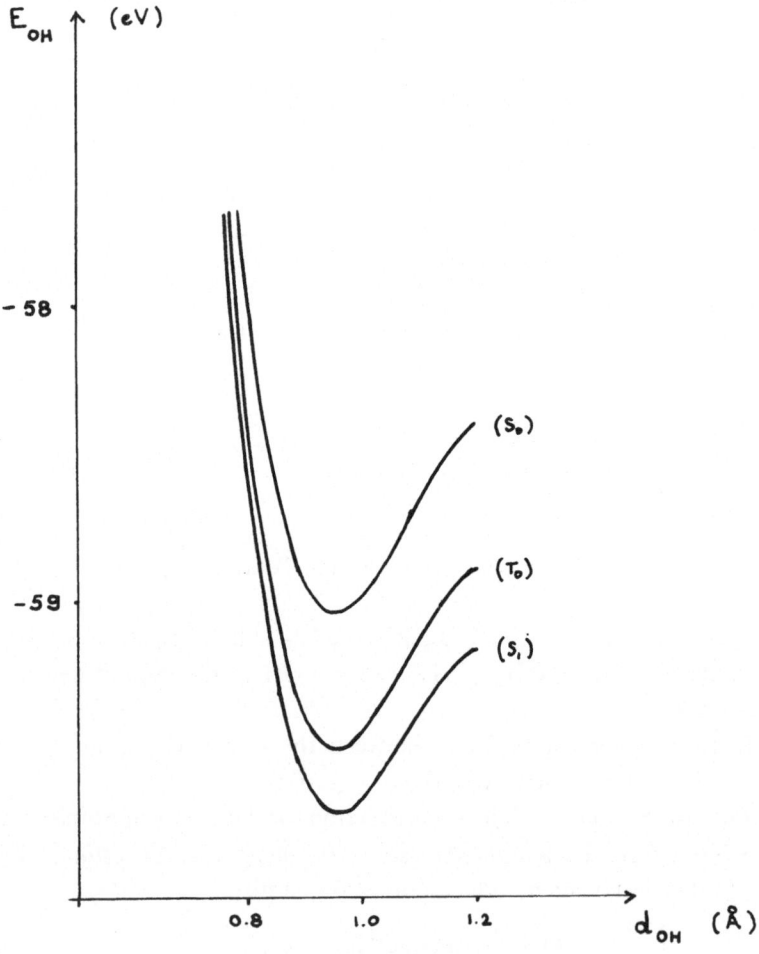

Fig. 2.   Isolated phenol.

TABLE II

(a) Minima of the total energy, the effective O—H bond energy and the O—H bond missing information function

(b) Minima of the effective O: lone pair energy and maxima of the O: lone pair population. The number in bracket indicates the corresponding value of the O—H distance

		E_{min}	ε_{min}	I_{min}
(a) OH bond	OH$^-$	−404.773 (0.9)	−56.229 (0.8)	0.005 (0.9)
	OH$_2^+$	−446.566 (0.95)	−66.681 (0.9)	0.000 (1.08)
	OH$_2$	−431.828 (0.9)	−55.154 (0.9)	0.000 (1.00)
	ϕOH(S_0)	−1401.683 (0.9)	−59.044 (0.95)	0.021 (0.8)
	ϕOH(T_0)	−1398.621 (0.9)	−59.493 (0.95)	0.021 (0.8)
	ϕOH(S_1)	−1396.958 (0.9)	−59.716 (0.95)	0.021 (0.8)

			ε_{min}	Q_{max}
(b) O: lone pair	HO$^-$	−404.773 (0.9)	−30.659 (1.10)	1.000 (0.9)
	H$_3$O$^+$	−446.566 (0.95)	−76.127 (0.85)	1.000 (1.08)
	H$_2$O	−431.828 (0.9	−54.747 (0.96)	1.000 (1.00)
	ϕO$^-(S_0)$	−1375.826	−39.25	0.975
	ϕO$^-(T_0)$	−1373.249	−46.49	0.968
	ϕO$^-(S_1)$	−1371.890	−46.25	0.975

being evaluated with the energy difference ΔE between the two equilibrium positions of the proton is found equal to 5.589. This clearly shows the fundamental role of the solvent.

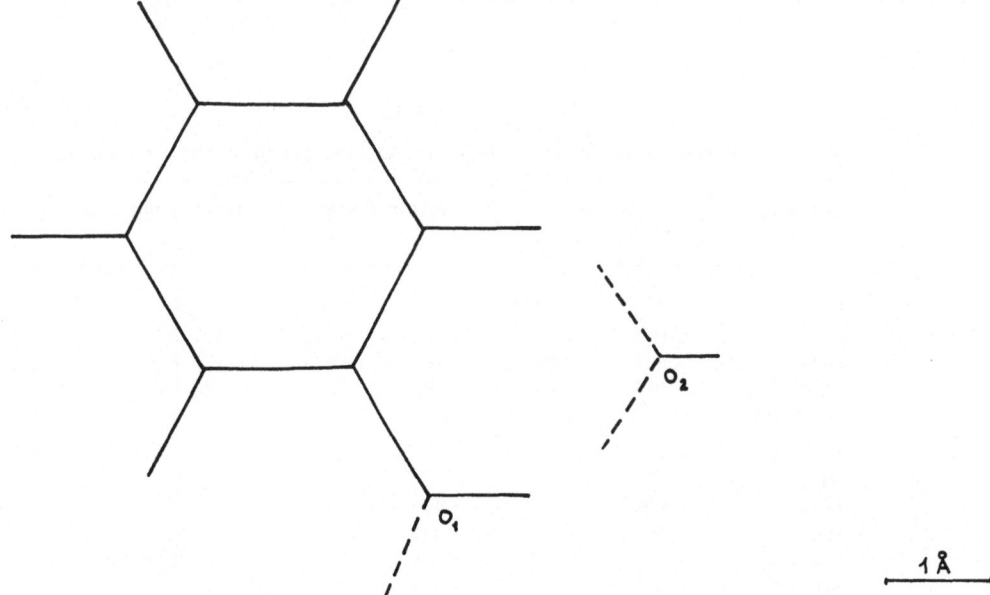

Fig. 3. Conformation used for the study of the system phenol + water.

5. General Results Concerning the Analytical Study

Most of all one finds excellent $\sigma - \pi$ separability in both phenol and phenolate using wave functions derived from the CNDO/S-CI method involving 80 monoexcitations.

As we said above a two electron group in the field of two core charges evokes immediately a diatomic bond model. In Table III we have given the values obtained for the core charges of the O—H bonds involved in various molecules. We note a very accurate constancy of the core charge values corresponding to a given type of bond, which suggests immediately a relatively good transferability of bond properties from one molecule to another (let us point out here that these core charge values have been obtained without recourse to an arbitrary partition of the nuclear charges into bond contributions). The deviations from the integer values can arise from a modification of the core charge distributions which leave the total core charges invariant but, this modification, which is strongly dependent on the core structure, leads generally to small deviations of the atomic core charges of the bond.

The subspaces able to represent the various two-electron groups have been constructed from a convenient set of hybrid atomic orbitals defined by geometrical considerations only. The further attempts to determine optimized sets did not give sufficiently significant improvements with respect to the other approximations made, so that the missing information functions have been computed in the frame of the former procedure for all the bonds considered in this work. As we have pointed out in our preceding paper a general correspondance is found between the minimal total energy distance and the minimal missing information one (see Table II and Figure 5).

As a consequence it is possible to evaluate with good accuracy the diatomic-like bond effective energy in the neighbourhood of the equilibrium distance and we observe

TABLE III

Charge distributions for O—H bond and O: lone pair in different molecules

Molecule	core charges		electron charges		polarity parameter	
	C_O	C_H	q_O	q_H	λ	
OH^-						
OH_2^+	2.006	1.003	1.308	0.701	0.302	
OH_2	1.003	1.002	1.180	0.825	0.177	
$\phi OH(S_0)$	1.013	1.009	1.199	0.823	0.186	O—H bond
$\phi OH(T_0)$	–	–	–	–	–	
$\phi OH(S_1)$	–	–	–	–	–	
HO^-						
H_3O^+	2.990		1.990			
H_2O	1.997		1.997			
$\phi O^-(S_0)$	0.975		1.975			O: lone pair
$\phi O^-(T^0)$	0.975		1.975			
$\phi O^-(S_1)$	0.968		1.968			

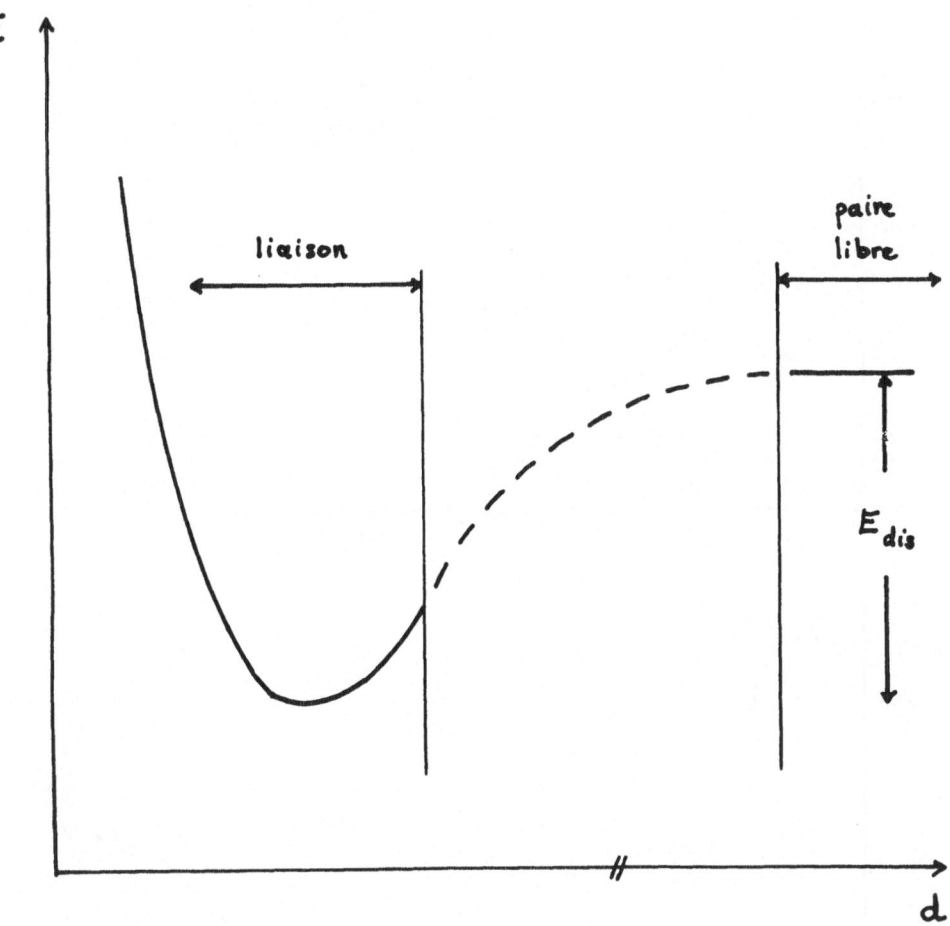

Fig. 4.   Morse curve showing effective bond energy vs. bond length.

that the minima of these curves pretty well corresponds with that of the total energy*
(see Table II).

This agreement means that the equilibrium nuclear configuration of the whole
molecular system results from the realization of the equilibrium of the considered bond
subsystem in the field of the others.

6. Analytical Detailed Study of the Isolated Phenol Molecule
and Phenolate Ion

The values of the missing information characterizing the valence bonds show that the
subsystem of two electrons associated to an O—H bond is separable in the first

---

* Care must be taken that the calculation of this effective bond energy is meaningless when the missing
information characterizing the bond is too high; in this case delocalization corrections to the energy
would be necessary.

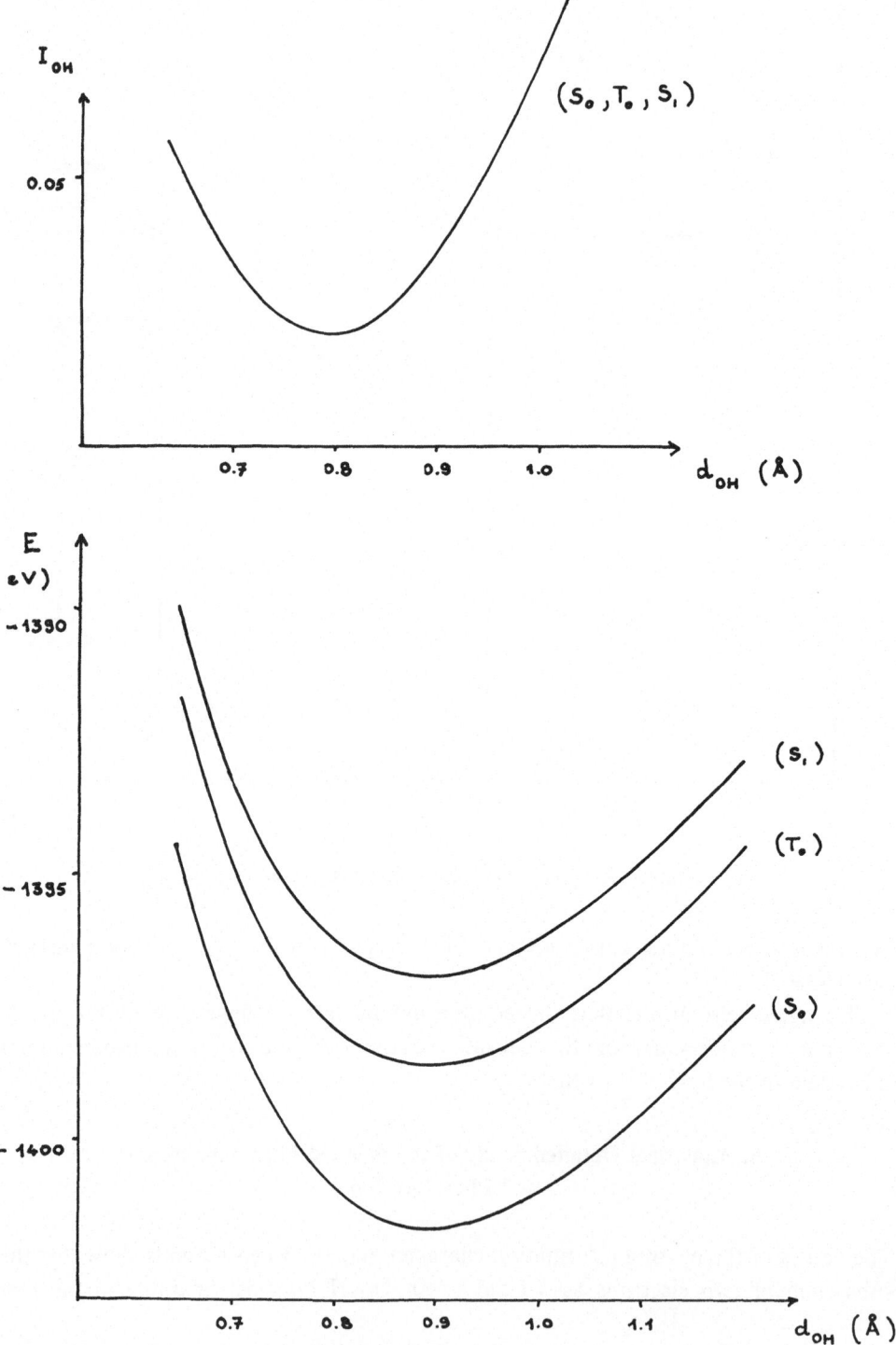

Fig. 5.   Missing information and total energies: isolated phenol.

excited singlet and triplet states with the same accuracy as in the ground state, i.e. the two electron bond does not disappear on excitation. One concludes, moreover, in accordance with the previous observations concerning the weak polarizability of the σ density distribution, that the bielectronic function which describes this sub-system is unaffected. (Table II and III and Figure 5).

The core charges computed for the three states are the same and slightly differ from the unit charge, which is a characteristic of a covalent bond. However, the space distribution of these core charges is different in each state; these differences result from the π system reorganization induced by the excitation which consists essentially, as we have seen above, of an electronic charge transfer from the vicinity of the oxygen atom towards the benzenic kernel. These differences in the core charge distributions are the essential cause of the stabilization of the two electron subsystem associated with the O—H bond under excitation as is shown by the computation of the effective energies; in effect, the decrease of the π electron density on the oxygen atom produces a more positive core field. As a consequence, a concomitant stabilization of the dia-tomic model energy is observed.

One verifies also that there exists in the phenolate ion a well separated subsystem of two electrons*, which can be identified with the lone pair on the oxygen atom, in the field of a unit core charge. The computation of the effective energy shows that this subsystem is stabilized in excited states; however, the observed stabilization is more important here than in the phenol molecule because the π charge transfer which causes this stabilization is also more important in the ion than in the molecule. As for the molecule, an important lowering of the diatomic model energy follows the two electrons effective energy stabilization.

In both cases, the lowering of the diatomic model energies are qualitatively well correlated with the variation of the atomic π electronic charge on the oxygen atom which gives a good indicator of the main feature of the π system reorganization.

TABLE IV

Energies in eV

		ϕ—OH	ϕO:		
S_0	E	− 1401.483	− 1375.826	$\Delta E = 25.657$	$\Delta \varepsilon_C = 5.87$
	ε_B	− 59.04	− 39.25	$\Delta \varepsilon_B = 19.79$	
S_1	E	− 1396.759	− 1371.890	$\Delta E = 24.87$	$\Delta \varepsilon_C = 11.42$
	ε_B	− 59.72	− 46.27	$\Delta \varepsilon_B = 13.45$	
T_0	E	− 1398.421	− 1373.249	$\Delta E = 25.17$	$\Delta \varepsilon_C = 12.17$
	ε_B	− 59.49	− 46.49	$\Delta \varepsilon_B = 13.$	

* In a minimal basis calculation one cannot attribute a definite missing information function to a lone pair subsystem, because the matrix representing the operator $[\rho_n^{(2)}]_{\mathscr{B}}$ is one dimensional. We have used as separability index the group pair density which is bounded by 1.

However, if the use of this atomic charge variation as an index of the pK difference between the singlet ground and excited states is well justified through the diatomic model of the O—H bond, it is not the case for the triplet state. In effect we found a triplet π reorganization analogous to that of the singlet whereas pK_T is comparable to pK_G and not at all to pK_S. We will return to this question later.

Using the results given in Table II it is possible to evaluate the quantities $\Delta\varepsilon_C$. These quantities are given in Table IV. As we have already pointed out, these quantities are not negligible, but good qualitative results can be obtained from the $\Delta\varepsilon_B$ as well as from the ΔE quantities.

7. Dissociation Energies

From the preceding results it is concluded that the dissociation energies of the bonds are weaker in the excited state than the ground state, this being particularly so in the excited singlet. A comparison of the various contributions to effective energy of dissociation leads one to conclude that the change of this term on photoexcitation is mainly due to the variation of the effective energy of that subsystem of electrons constituting the non-bonding pair. Otherwise stated this means that purely local electronic considerations such as we have already argued above are sufficient to justify the weakening of the O—H bond on condition that this argument is applied to the ion rather than the molecule.

8. Force Constants

An examination of the Morse curve of Figure 4, which acts as an effective bond energy for small OH displacements and as an effective lone pair energy at large, permits an understanding of how a lowering of this latter quantity is accompanied by a decrease in the O—H force constant.

9. Interpretation

If it is admitted that the increase in acidity on excitation is a direct consequence of the weakening of the OH bond, it can be seen that the results we have obtained, which are concerned with effective energies or force constants, are sufficient to explain the observed pK_a sequence.

The values we have calculated are characteristic of a molecule-ion pair (phenol-phenolate). Our calculations show that one is not able to get enough information on the variation in pK_a on excitation by looking at phenol alone: the most important changes occur between the ground and excited states of the ion.

The results obtained suggest that, in the schematic diagram of Figure 1 showing electronic reorganization on excitation, it must be understood that the σ arrow symbolizes the capacity of retaining a free electron pair (which is characteristic of the ion) rather that the polarization of the OH bond in phenol.

10. Introduction of the Solvent

We have reasonably assumed that the transfer of a proton in phenol to the solvent occurs through a hydrogen bonded intermediate. The geometrical configuration of such an intermediate in water was assigned in the same manner as reported by Pullman [6]. The analysis of the CNDO/S wave functions was accomplished for various positions of the proton while keeping other nuclei fixed. Two principle positions were shown for the proton, one near oxygen O_1 of phenol and the other near the oxygen O_2 on water. In both cases the missing information index characteristic of one of the OH bonds approaches a minimum value of zero, which indicates a good separation of the subsystem studied, whereas at the same time the same index for the other OH bond approaches the opposite limiting value of 1. The distances for O_1—H and O_2—H at which the missing information indices have minima are comparable to those which represent minima in the potential energy surfaces, i.e. the equilibrium O—H positions (Figure 6).

These characteristics are conserved in the lowest excited singlet and triplet states. Thus we have established the existence of two electron bonds which in turn justifies the calculation of energetic terms associated with these bonds. It is also interesting to note that these results cannot be obtained using just any arbitrary molecular configuration such as placing the water molecule in the molecular plane of phenol which is energetically unfavorable.

It is noted that the photoexcitation influences most of all the effective energies of the O_1—H bond and the lone pair located on O_1. This is understandable because the reorganization in the π system occurs in the region close to O_1 but rather far from O_2. As a consequence conclusions drawn from the study of the phenol-phenolate pair without solvent remain valid. We have collected together in Table V the various values of effective energies relative to the preceding study together with the corresponding water-hydroxonium and water-hydroxyl pairs.

11. Discussion of the Results (See Table V)

First of all it is seen that the OH bond of phenol is more easily broken than the OH bond of water in agreement with experimental observation.

$$\varepsilon_B(\phi - OH) < \varepsilon_B(HO - H).$$

If one considers a proton acceptor such as water the affinity is given by

$$\varepsilon_B(H_2O - H^+) > O$$

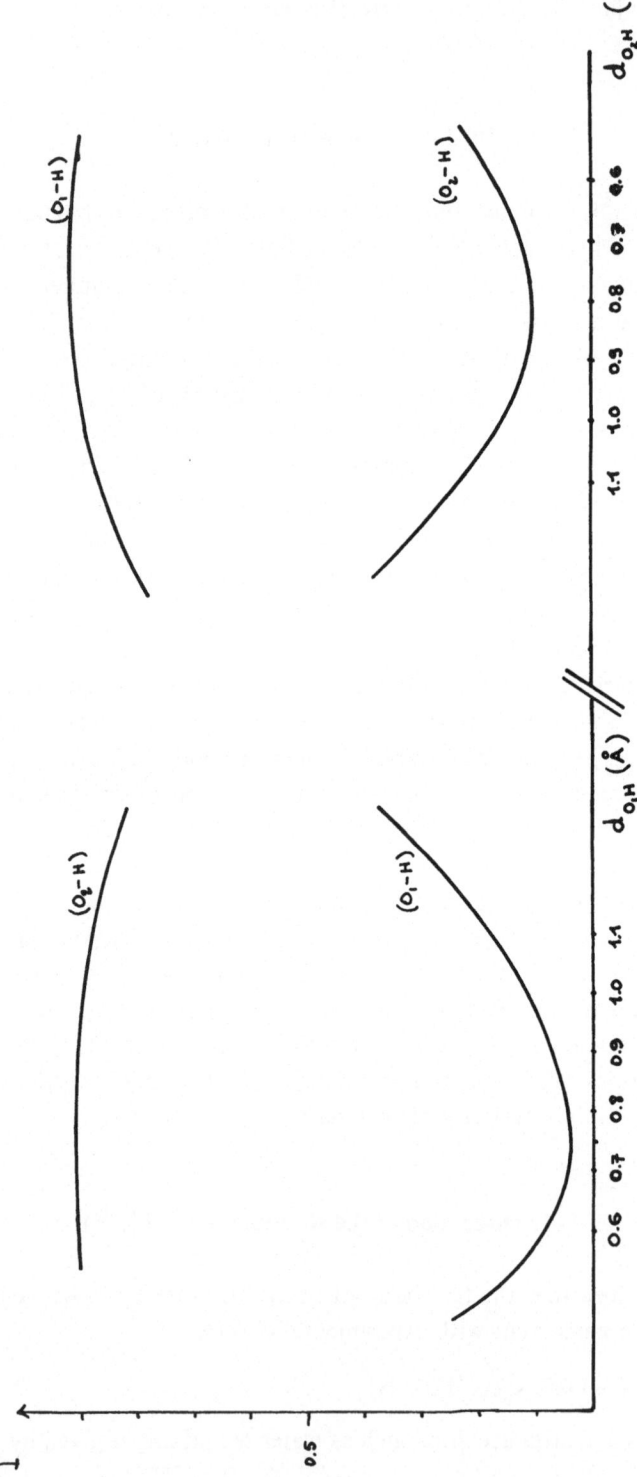

Fig. 6.   Missing information for O_1—H and O_2—H bonds vs. distances from the proton to the nuclei O_1 and O_2.

TABLE V

Numerical values relative to the couple of isolated molecules (a) water-hydroxyl; (b) water-hydroxonium; (c) phenol-phenolate

(a)
E (eV)

$\underline{(HO:)^- \quad -30.24}$

$\underline{HO—H \quad -55.15}$

$\varepsilon_B(HO—H) = 24.91$

(b)

$\underline{H_2O: \quad -54.74}$

$\underline{(H_2O—H)^+ \quad -66.68}$

$\varepsilon_B(H_2O—H^+) = 11.94$

(c)

$\varphi O:$
$\underline{\quad -39.25 \ (S_0)}$
$-46.25 \ (S_1)$
$-46.49 \ (T_0)$

$\varphi O—H$
$\underline{-59.04 \ (S_0)}$
$-59.70 \ (S_1)$
$-59.49 \ (T_0)$

$\varepsilon_B(\varphi O—H) = 19.79 \ (S_0)$
$13.45 \ (S_1)$
$13.00 \ (T_0)$

Numerical value relative to the couple phenol-water interacting through an hydrogen bridge (a) ground state; (b) excited triplet; (c) excited singlet

(a)
E (eV)

$\underline{(\varphi O:)^- \quad -42.71}$

$H_2O: \quad -54.64$
$\underline{\varphi O—H \quad -58.82}$

$\underline{(H_2O—H)^+ \quad -65.24}$

(b)

$\underline{(\varphi O:)^- \quad -50.05}$

$H_2O: \quad -54.73$
$\underline{\varphi O—H \quad -59.30}$

$\underline{(H_2O—H)^+ \quad -65.46}$

(c)

$\underline{(\varphi O:)^- \quad -49.17}$

$H_2O: \quad -54.76$
$\underline{\varphi O—H \quad -59.51}$

$\underline{(H_2O—H)^+ \quad -65.45}$

we see that

$$\varepsilon_B(\phi - \text{OH}) > \varepsilon_B(\text{H}_2\text{O}-\text{H}^+)$$

indicating that, at the thermodynamic equilibrium, the O—H bond will yield a preference for phenol, explaining the weak acidity of this molecule in water.

On the photoexcitation, the OH bond of phenol will get weaker

$$\varepsilon_B(\phi\text{O}-\text{H})^* < \varepsilon_B(\phi\text{O}-\text{H})$$

which yields an increase of acidity. In the presence of a proton acceptor such as water in which phenol is hydrogen bonded the same qualitative picture is obtained as with isolated molecules.

The minimal values of the effective energies of dissociation in the excited states do not account for the differences in behavior between the excited singlet and triplet. The analytical procedure that we have used can not make up for any deficiencies in the wave functions. These were obtained using the CNDO/S technique and we have been unable to determine any significant difference in the π electronic reorganization between these two states.

Since the calculations obtained in extended monoexcited bases set including all of $\pi \to \pi^*$ type failed to show a relationship between the differences in the atomic charge at atom O_1, Δq and the variation in transition energies $\Delta\varepsilon$, it can be tentatively concluded that it may be necessary to introduce doubly excited configurations; this conclusion is also supported by our previous arguments.

TABLE VI

	Δq	$\Delta\varepsilon$
$S_0 \to S_1$	-0.16	0.531 eV
$S_0 \to T_0$	-0.17	0.095 eV

References

1. Bertran, J., Chalvet, O., and Daudel, R.: *Theor. Chim. Acta* **14**, 1 (1969).
   Constanciel, R.: *Theor. Chim. Acta* **26** (1972) 249.
   Chalvet, O., Constanciel, R., and Rayez, J. C.: *The Jerusalem Symposia on Quantum Chemistry and Biochemistry*, VI, 1971.
   Constanciel, R., Chalvet, O., and Rayez, J. C.: *Theor. Chim. Acta* (in press).
2. Julia, M.: *Mécanismes électroniques en chimie organique*, Gauthier Villars, 1961.
3. Aslangul, C., Veillard, A., Daudel, R., and Gallais, F.: *Theoret. Chim. Acta.* **23**, 211 (1971).
   Veillard, A. and Daudel, R.: Colloque International du C.N.R.S., No. 191, Editions du C.N.R.S. (1970).
4. Constanciel, R.: Vol. 1, p. 43.
   Constanciel, R. and Esnault, L.: Vol. 2, p. 3.
   Constanciel, R.: *Phys. Rev.* **A11**, 395 (1975).
5. Bodor, N., Dewar, M. J. S., Harget, A., and Haselbach, E.: *J. Am. Chem. Soc.* **92**, 3854 (1970).
6. Port, G. N. J. and Pullman, A.: *FEBS Letters* **31**, 70. (1973).

LOGE THEORY AND CHEMICAL REACTIVITY

R. DAUDEL

Sorbonne et Centre de Mécanique Ondulatoire Appliquée du CNRS, Paris, France

1. Introduction

For a long time the analysis of relations between molecular electronic structure and chemical reactivity was based on the use of static indices (like atomic charges, bond orders, free valence numbers). A large number of interesting results have been obtained by following that way [1].

However usual static indices show severe limitations.

(a) *They are not associated with well defined regions of molecules.* It is well known (for example) that a classical electronic atomic charge is not the result of an integration of the electronic density in a given volume.

(b) *They are not usually derivatives* of the energy with respect to a reaction coordinate.

This point has been recently clearly analyzed by Chalvet and Constanciel [2].

It is easy to understand why it is difficult to generate static indices directly related to the energy.

Static indices are usually related to the first order density matrix. The energy depends on both first order and second order density matrices. This is why it is difficult to understand the difference in chemical behaviour of singlet and triplet states of a given molecule.

The difference in behaviour between $O(^1D)$ and $O(^3P)$ is an interesting example. $O(^1D)$ inserts into paraffinic CH bonds, $O(^3P)$ abstracts an H atom. The interpretation of such experimental results has been made by Gangi and Bader [3] in terms of spin density (which is a first order density matrix). But in fact it has been seen during the discussion this morning that such a procedure amounts to introducing indirectly the electron pair density.

(c) The classical static indices are not defined for any kind of wave function. They are related to certain coefficients which appear when LCAO MO wave functions are built from certain basis sets. Therefore the indices depend on the basis and are not always invariant with respect to a unitary transform which does not change the wave function.

(d) It is difficult to use classical static indices to take account in a systematical way of both reagents in a bimolecular process.

The loge theory provides a tool to go farther.

O. Chalvet et al. (eds.), Localization and Delocalization in Quantum Chemistry, Vol. II, 389–391. All Rights Reserved
Copyright © 1976 by D. Reidel Publishing Company, Dordrecht-Holland

2. Static Indices Related to the Loge Theory

Daudel *et al.* [4] have shown how it is possible to generate new static indices by using that theory.

The starting point of this analysis is the formula already established in this book* showing that it is possible to divide any expectation value $\langle \Psi | \hat{\Omega} | \Psi \rangle$ into loge contributions $\bar{\Omega}_l$ and loge pair contributions $\bar{\Omega}_{ll'}$. This formula can be used to express the total electronic energy E of a molecule as:

$$E = \sum_l \bar{H}_l + \sum_{l<l'} \bar{H}_{l,l'} .$$

If we consider now two molecules participating in a bimolecular process they can be treated as one supermolecule. Therefore the foregoing formula can be used as an expression of the total energy of the system including the interaction energy between the two molecules.

Let us concentrate the discussion on the term $H_{A,B}$ representing the interaction between loge A of the first molecule and loge B of the second.

That term can be written as:

$$\bar{H}_{A,B} = \sum_{\lambda, \lambda'} p_\lambda p_{\lambda'} e_{\lambda\lambda'}$$

if p_λ represents the probability of occurence of the electronic event λ in the loge A and $p_{\lambda'}$, the probability of occurence of the electronic event λ' in the loge B.

If furthermore the two molecules are rather far away the term $H_{A,B}$ can be expanded as:

$$\bar{H}_{A,B} = \sum_{\lambda, \lambda'} p_\lambda^0 p_{\lambda'}^0 e_{\lambda\lambda'} + \sum_{\lambda, \lambda'} \mathscr{E} \frac{\partial p_\lambda}{\partial \mathscr{E}} p_{\lambda'}^0 e_{\lambda\lambda'} +$$

$$+ \sum_{\lambda, \lambda'} \mathscr{E} p_\lambda^0 \frac{\partial p_{\lambda'}}{\partial \mathscr{E}} e_{\lambda\lambda'} + \dots$$

if $\mathscr{E}$ denotes a given reaction coordinate.

The various probabilities p_λ^0, $p_{\lambda'}^0$, and their derivatives can be considered to be the new static indices.

(a) They are associated with well-defined regions of molecules.

(b) They are related to the interaction energy between the two molecules.

(c) They are defined for any kind of wave function.

(d) They depend on both partners of the reaction.

The effective calculation of such indices is not easy because it is difficult to calculate the frontiers of good loges. But the PCILO programm could be certainly used successfully to produce approximate quantitative information if we admit that the PCILO localized functions are approximations of loge functions.

---

* See R. Daudel, p. 71.

References

1. For a review see for example Daudel, R.: *Quantum Theory of Chemical Reactivity*, Reidel Publ. Co., Dordrecht-Holland, 1973.
2. Chalvet, O. and Constanciel, R.: *Theor. Chim. Acta* **26**, 249 (1972).
3. Gangi, R. A. and Bader, R. F. W.: *J. Chem. Phys.* **55**, 5369 (1971).
4. Daudel, R., Chalvet, O., Constanciel, R., and Esnault, L.: *The Jerusalem Symposia on Quantum Chemistry and Biochemistry*, Vol. VI, p. 63 (1974).

FINAL DISCUSSION

STATISTICAL AND STOCHASTIC ASPECTS OF THE DELOCALIZATION PROBLEM IN QUANTUM MECHANICS

PIERRE CLAVERIE and SIMON DINER

Institut de Biologie Physico-Chimique, Laboratoire de Chimie Quantique,
75005 Paris, France

Abstract. The purpose of this paper is to review what it is possible nowadays to say about space-time behaviour of electrons in atoms and molecules.

In the first chapter the wave conception of the electron is criticized and the poverty of the non reductionist attitude is underlined.

In the second chapter the two main interpretations of quantum mechanics are recalled: the Copenhagen and the Statistical Interpretations. The meaning and the successes of the Statistical Interpretation are explained, and it is shown that it does not solve all problems, because quantum mechanics is irreducible to a classical statistical theory. In the third chapter, we study, in the spirit of the Statistical Interpretation, the fluctuation of the particle number and its relationship to loge theory, delocalization and correlation.

The fourth chapter reviews different stochastic models for microphysics. The markovian Fenyes-Nelson process allows us to give an interpretation of the original heuristic considerations of Schrödinger. Non markov processes with Schrödinger time evolution are shown to be equivalent to the base state analysis of Feynmann, but they are unsatisfactory from a probabilistic point of view. Stochastic electrodynamics is presented as the most satisfactory conception nowadays.

0. Prologue

As far as quantum mechanics is a calculus for random variables, the ordinary practice of quantum chemistry remains at the 'mean value level' characterization of the observables. The purpose of this paper is to examine the meaning of observable dispersion (fluctuation) in quantum chemistry and to show on specific examples the useful employ of such a concept. The problem of 'electronic localization or delocalization' in quantum mechanics seems to be more significantly expressed in term of dispersion than in the mean value formulation. This study arised from the fundamental question: 'What actually electrons do (if they do anything) in a molecule?' The answer appeared to us mainly as a problem of interpretation of quantum mechanics and quantum chemistry. It lies at the heart of the interpretation problem of the quantum formalism, generally disregarded by the majority of quantum chemists because of its philosophical flavour. The above question is reputed having no answer in quantum mechanics, but the quantum chemist, if he is really a chemist, cannot prevent himself from wondering about electrons in the molecule. Our epistemological reflection has led us to formulate some non classical conceptions. We hope that, although there are no definitive conclusions, this paper will stimulate a critical attitude among quantum chemists, delocalized between the two following points of view:

The general theory of quantum mechanics is now almost complete, the imperfections that still remain being in connection with the exact fitting in of the theory with relativity ideas.... The underlying physical laws necessary for the mathematical theory of a large part of physics and the whole of

O. Chalvet et al. (eds.), Localization and Delocalization in Quantum Chemistry, Vol. II, 395–448. All Rights Reserved

chemistry are thus completely known, and the difficulty is only that the exact application of these laws leads to equations much too complicated to be soluble.   Dirac [1]

The trouble with quantum mechanics is not only in solving the equations but in understanding what the solutions mean.   Feynman [2]

1. Introduction

The relative inability to 'explain' or 'understand' the brilliant formal results obtained is not one of the least paradoxes of modern theoretical physics. The reason is that the microphysical world, not directly perceptible, is studied with the help of mathematical models whose heuristic foundations prevent simple physical interpretation. Two attitudes coexist: calculation and explanation, a dichotomy which goes back 'to the end of the XVIIth century, when the controversy raged between the upholders of the Descartes and Newton physics. Descartes, with his whirls, his hooked atoms and so on, explained everything and calculated nothing; Newton with his gravitational law $1/r^2$, calculated everything and explained nothing' [3]. Such a situation is found nowadays in quantum theory and especially in quantum chemistry where the overflow of numerical results keeps the chemist unsatisfied, being, as Descartes, favourable to a realistic picture of the physical universe. Whereas teaching and applied research can give the feeling that physical interpretation problems of quantum mechanics are solved, the discussions on this subject are as strong as ever [4]. They have nevertheless a new character, being more technical and less philosophical, which will certainly contribute to focus the interest of the users of quantum mechanics. In this paper we shall try to see how these problems of interpretation of quantum mechanics arise in quantum chemistry. The concepts of localization and delocalization are the heart of this discussion.

1.1. THE STATUS OF MATHEMATICAL MODELS. ABUSIVE REIFICATIONS

Models play in cognitive processes – especially in those of theoretical physics – two distinct and complementary roles. Understanding a natural phenomenon always depends on a space-time model where distinct objects are defined along with their motions and their mutual relationships (interactions). Without such a model it is impossible to say that there is a complete understanding and it is then very difficult to let people understand such a phenomenon, a fact which appears clearly during teaching. 'Magic or geometry, that is the dilemma set up by every trial of scientific explanation' [3]. The absence of *pictures* gives a magical aspect to some of our physical theories from gravitation to quantum theory including electromagnetism.

Models have another role as elements of an approximation process for the phenomena under study. In that case we have essentially mathematical models and their main function is a symbolic simulation of the phenomena. Whether the mathematical model bears upon an explicative space-time model or not, the autonomy of the mathematical speech gives to it some independency of reality, which needs interpretation. The mathematical model has such an *operational* character, that it is very dangerous to try to interpret all its constituents, that is to try to make connections between the constituents of the model and the different parts of the reality under study. The mathematical model is a complex object which from some input data gives us as output some information on the possible observations. It is not a detailed model of behaviour of the system, but a way to simulate some of its properties, and one must not make *abusive reifications*.

The description of the motion of a solid by the motion of its center of gravity endowed with the total mass does not dupe anybody. But the normal vibrations of a complex vibrating system are a more subtle simulation which can be wrongly thought of acs an actual dynamical description. The same happens for quasi particle models of N interacting body systems.

The Huyghens principle allows a simulation of the behaviour of the electromagnetic field when obstacles are present, corresponding to the solution of the electromagnetic wave equation with appropriate boundary conditions. The secondary wavelets introduced by the Huyghens-Fresnel theory are fictitious, but give a qualitatively correct solution of the problem, and can even appear as a kind of explanation.

That white light can be mathematically represented as a superposition of monochromatic waves, and that these monochromatic waves seem to be revealed by a dispersive system, do not allow to say that these waves physically coexist in the white light.

All the preceding mathematical models were built up with elements with the same physical meaning as the object to be described, but it is possible to work out mathematical models with a pure simulatory function. In such models no physical identification seems possible; there are only some mathematical characteristics – e.g. values of some functions – which correspond to the physical characteristics of the system represented. This is for example the meaning of the probabilistic simulation of deterministic processes (Monte Carlo methods). In a more abstract way this is the meaning of cybernetic models for complex systems.

The relationships between theoretical physics, mathematical physics and the microphysical reality are in the heart of the great debates of contemporary epistemology. Progress of linguistics and of the theory of complex systems, influencing the whole field of philosophical thinking, have progressively displaced the topics of discussion.

Between the two world wars and up to 1960, the conception of the world was reintroducing man into nature as observator and actor; from marxism to existentialism, in some branches of catholic philosophy (Mounier's personalism) and in all the manifestations of surrealism, man, independant and willing is facing nature [5]. One is struck by the role devoted to man, through observation and measurement, in the epistemology of quantum mechanics due to the 'Copenhagen School'. The philosophical background of Bohr is known (cf. Jammer [71], p. 166 and Holton [72], p. 115).

About the sixties the simultaneous development of some human sciences (linguistics, ethnology, sociology) and of the practice of studying complex systems, characteristic of the age of computers (cybernetics), added to the diffusion of a new mathematical knowledge (modern mathematics, N. Bourbaki), completely renewed ideological positions. *The structuralist point of view* is slowly spreading, trying to find in reality autonomous levels of description, not necessarily excluding the presence of man [6]. Such an autonomy of structures appears in all contemporary conceptions where the autonomy of speech against reality is underlined. Our discussion on the cognitive function of models outlines a figurative function and an operative function. This is a fundamental conception of modern scientific psychology [7], reminding us of all the oppositions found in the most varied fields: the signifier and the signification in linguistics and semiotics [8]; the objectivist conceptions of models [9] (model must be interpreted in a literal and not in a metaphorical way [4c]); the new relations between axiomatic theories and reality which appear in quantum field theory and the theory of elementary particles [10] [11] [12].

The different attitudes toward quantum mechanics are influenced directly or indirectly by this ideological background.

The Copenhagen Interpretation as we have already said stresses the role of observation and measurement. The non-reductionist attitude (cf. 2.3) is in fact a structuralist point of view. This is the dominant point of view at the present time among theoretical physicists. Between these two extremes, the search for 'hidden variables' stems from a realistic attitude never given up by successive generations of physicists. The Statistical Interpretation (cf. Section 2) is a minimal common conception to many supporters of 'hidden variable theories'.

1.2. CRITICISM OF THE WAVE MODEL IN QUANTUM MECHANICS

The wave model in quantum mechanics has been introduced in 1923 by de Broglie [13]. It postulates a dual aspect of microphysical particles, analogous to the dual aspect of light. But the wave aspect, apparently strongly verified by electron diffraction on crystals (Germer and Davisson, 1927) does not play a true role of model in the elaboration of quantum theory. The matrix mechanics of Heisenberg *et al.* [14] is independant of wave conceptions, whereas Schrödinger does not use any wave model to get the time-independent Schrödinger equation for the hydrogen atom in his first paper of 1926 [15]. The formalisation and axiomatics due to Dirac [16] and Von Neumann [17] do not use the wave concept. As to the probabilistic interpretation of the wave function (Born 1926), it contributes to eliminating the conception of a physical wave, although the quantum speech keeps the expression: 'wave of probability density amplitude'.

In modern quantum mechanics, especially within the so called 'orthodox interpretation' of the Copenhagen School, the De Broglie wave is called upon whenever the mathematical results of the theory need some intuitive explanation: Heisenberg uncertainty relations, tunnel effect or interpretation of the role of Fourier transform.

The wave conceptions of microphysics have been often criticized. Recently, Landé has revived strongly this criticism [18] [19] and his attitude is shared by the anti-reductionists [4c] [20] (cf. 1.3).

The most elementary argument points out that time dependent Schrödinger equation is first order in time and is not a wave equation. It is not relativistic invariant. It is equivalent to a second order in time equation found initially by Schrödinger [21], but this equation is of the fourth order in coordinates and is analogous to equations found in elasticity theory:

$$-\frac{16\pi^2}{h^2}\frac{\partial^2\psi}{\partial t^2} = \left(\Delta - \frac{8\pi^2}{h^2}U\right)^2\psi.\tag{1.1}$$

Starting from this equation Schrödinger found his final equation, but by the way a complex factor i was introduced, so that any meaning of physical wave was lost. In the remarks he has written for the french translation of his fundamental papers [21], Schrödinger frankly says (p. 166):

'Les mots 'essentiellement complexe' cherchent à dissimuler ici une grande difficulté. Dans son désir d'envisager à tout prix le phénomène de la propagation des ondes comme quelque chose de réel dans le sens classique de ce terme, l'auteur s'était refusé de reconnaître franchement que tout le développement de la théorie mettait de plus en plus clairement en évidence le caractère essentiellement complexe de la fonction d'onde. Et cependant cette fonction est déterminée par une équation dont les coefficients sont essentiellement complexes.

Mais comment le $\sqrt{-1}$ a-t-il pu s'introduire dans cette équation? Une réponse, dont je n'ose indiquer ici que le sens général, a été donnée à cette question par un physicien, qui a autrefois quitté l'Autriche, mais qui, malgré de longues années passées à l'étranger, n'a pas complètement perdu son humeur mordante de Viennois et qui, en outre, est connu pour sa faculté de trouver toujours le mot juste, d'autant plus juste qu'il est plus cru. Voici cette réponse: Le $\sqrt{-1}$ s'est glissé dans l'équation comme quelque chose que nous laissons échapper par hasard, en éprouvant toutefois un soulagement inappréciable après lui avoir donné naissance involontairement.

There is a true wave equation in relativistic quantum theory: the Klein-Gordon equation for the free particle:

$$\frac{\hbar^2}{c^2}\frac{\partial^2\psi}{\partial t^2} = (\hbar^2\Delta - M^2c^2)\psi.\tag{1.2}$$

But one must remark that for a free particle neither the Klein-Gordon equation nor the Schrödinger equation, contains more information than the classical energy expression for a particle of mass M. They can be trivially derived for the mathematical object:

$$\psi = e^{i/\hbar S}\tag{1.3}$$

where S is the Hamilton Jacobi action

$$S = \mathbf{p}\mathbf{r} - Et\tag{1.4}$$

from the energy expressions

$$\frac{E^2}{c^2} = p^2 + M^2c^2 \quad \text{and} \quad E = \frac{p^2}{2M}.\tag{1.5}$$

There is no physically identified wave in such a theory. Moreover, it is dubious that the Klein-Gordon equation can be interpreted as concerning one particle, the relativistic theory introducing states with an undefined number of particles. This is due to the creation and annihilation of particles which is impossible in the non relativistic theory. That equation has another physical meaning (for the field) and it is not reasonable to look at it as a relativistic refinement of the Schrödinger equation [10] [22].

At this stage one must bear in mind the fundamental difference between a phenomenological theory as the non relativistic quantum mechanics and a more physical theory as quantum field theory (quantum electrodynamics) where there is an explicit account of the field. Ordinary quantum mechanics appears as a simulation model where the physical objects are not easily identified. One must not be surprised that its formalism does not allow any identification of physical waves.

Does physical reality really oblige us to speak of waves, because particles give rise to phenomena known for light or acoustic waves?

If ordinary quantum mechanics gives correctly the presence probability of particles in space, it must reproduce the diffraction patterns experimentally obtained. Such a calculation would be a typical scattering problem, but the difficulty would be the correct determination of interaction potentials between particles and the screen. It seems more natural to look at the 'inverse scattering problem', that is finding the potential from the results of the scattering [23]. But one must realize with Bunge [4c] that: 'Unfortunately the diffraction by a single slit has been calculated only for an infinitely long slit and a monochromatic de Broglie 'wave'. Moreover the available computation is approximate and its result conflicts head on with the Heisenberg inequalities (Beck and Nussenzweig 1958 [24]), which of course cannot be held against the latter. A fortiori the much discussed double slit experiment has never been computed exactly in quantum mechanics, let alone in quantum electrodynamics. So much so, that the most complete, accurate and recent 681 pages long work on scattering (Newton 1966 [25]) does not treat the standard diffraction gedanken experiment found in discussions on the quantum theory.' Nevertheless the same situation prevails in optics where diffraction phenomena are scarcely exactly calculated from Maxwell equations taking in account the obstacles with their shapes and specific optical properties. One generally uses an approximate method, based on the Huyghens-Fresnel postulate and which does not take account of the true boundary conditions. This *kinematical* method represents the vibration at a point beyond the obstacle as the resultant of an infinity of vibrations emitted by secondary sources spread over the non hidden part of the wave; an interference calculation then follows. This can be approximately justified if one considers waves of the form

$$f(x, y, z, t) = u(x, y, z) \, e^{i\omega t} \tag{1.6}$$

which are solutions of the Helmholtz equation

$$\Delta u + \frac{\omega^2}{c^2} u = 0. \tag{1.7}$$

The solution of this equation beyond the screen is obtained by integration from the solution on the screen and on a surface surrounding the point according to Green's theorem. The Kirchoff diffraction theory sets reasonable solutions on these surfaces* and therefore justifies the Huyghens-Fresnel postulate as an approximate procedure [65]. But all that is purely mathematical. Secondary waves are fictitious. The best proof is that the solution of the Helmholtz equation can be found by a Monte Carlo method, without the least pretention that we have necessarily diffusion markov processes involved [26].

So, without any kind of affirmation about the physical mechanisms implied, all the particle diffraction phenomena are going to be obtained in quantum mechanics if one remarks that the time independent Schrödinger equation for the free particle

$$\frac{\hbar^2}{2M} \Delta\psi + E\psi = 0 \tag{1.8}$$

is just the Helmholtz equation. With the same boundary conditions, the same intensity (probability) beyond the screen is obtained. As far as diffraction patterns are described by the stationary solutions, Maxwell and Schrödinger equations, having the same equation for these solutions, give rise to the same effects.

In that way quantum mechanics shows once more its phenomenological character. The formal use of the kinematical wave model to explain diffraction patterns, so successful in optics, will work for particles with a formal wave length $\lambda = h/p$. This wave length is defined by identifying Equations (1.7) and (1.8):

$$\frac{\omega^2}{c^2} = \frac{2M}{h^2} E = \frac{2M}{h} \frac{p^2}{2M} = \frac{p^2}{h^2} \tag{1.9}$$

where we have put $E = E_{kinetic} = p^2/2M$, the velocity being defined with respect to the screen considered at rest. Then $\lambda = 2\pi c\omega^{-1}$. This fundamental De Broglie relation plays then an analogical role and does not prove the existence of physical waves. Lande [19] and Levy-Leblond [20] have shown that invariance analysis of this relation does not allow giving a physical meaning to it. (The identification of (1.7) and (1.8) had to be made in a special reference frame, namely that in which the screen is at rest.)

Let us now comment on some experimental aspects of 'particle optics'. If different diffraction experiments are possible, and if interference experiments of the kind: 'division of a wave front' (electrical biprism [27]) have been realized, one has never observed an actual experimental interference effect of the kind 'division of amplitude' like that of the Michelson interferometer. The observation of such true interference effects has not been realized for particles although some experiments claimed it [28] [29]. As for interference effects with superconductivity currents and the Josephson effect, they need a careful examination [2].

The Young's experiment for particles, which is discussed in every book on quantum

---

* Let us recall that the usual assumptions for these boundary conditions, namely $u = 0$ behind the screen and $u = (u$ incident) at the aperture of the screen, do *not* give the *exact* solution, but an approximate one (which is good, for example, when the size of the aperture is large with respect to the wave-length).

mechanics, was not really performed before 1957 by Faget at Toulouse [30] and Jönsson at Tübingen [31]. They found it necessary to have holes or slits with a distance *of the same order of magnitude as their width:* two holes of 1 μ diameter and 1,3 μ distance between centers, two slits of 0,3 μ width at 1 μ from each other. Nevertheless after these works, as before, the Young's experiment is always presented with small holes at a large distance from each other, so that one is led to think of these holes as independent. The gedanken experiment is then used to 'prove' that probability laws of quantum mechanics are not those of classical probability calculus, especially that the law of probability addition for independent events is not satisfied and that appears a 'probability interference' [2]. This has even encouraged mathematicians to develop a probability calculus adapted to quantum theory, using as basis for axiomatization an orthocomplemented lattice of events instead of a boolean σ algebra [32]. One must in fact say that the motivations for such a new probability theory are deeper, and come essentially from the trial to develop a probability theory on the closed subspaces of a Hilbert space, which do not constitute a boolean algebra [33] [34] [35] [36]. But, is such a theory necessary from a physical point of view? Our discussion shows that there is no physically identified wave in quantum theory and that diffraction phenomena can be phenomenologically found from the time independent Schrödinger equation for the free particle viewed as an Helmholtz equation. As for the time dependent Schrödinger equation, Lax and Phillips have found the following remarkable result [37]: the scattering matrix $\mathscr{S}^S(z)$ of the Schrödinger equation with a potential $U(\mathbf{r})$, is connected to the scattering matrix of the wave equation (with the same potential) $\mathscr{S}(z)$ by the relation

$$\mathscr{S}^S(z) = \mathscr{S}(\sqrt{z}).$$ \hfill (1.10)

This result seems not to have been incorporated in quantum theory. Again this implies some structural formal analogy between wave and quantum mechanical scattering theory, and hence between resulting phenomena.

All these remarks criticizing the wave conception in quantum theory are intended to protest against the attitude of thinking that quantum mechanics introduces a non local character of the particle, as far as it accounts for phenomena which are analogous to those familiar for waves. We think that *particles are perfectly localized objects – with their proper size.* Quantum mechanics describes in a probabilistic way a complex behaviour of particles, due to complex interaction between particles and field. Particles are better to be considered as excitations or singularities of the field. That motion of particles could be accompanied in the field by wave type phenomena is neither excluded nor proved. The double solution theory of De Broglie is somewhat in that spirit [38] [39]. But ordinary quantum mechanics involves a phenomenological description of particles, and of particles only. Quantum electrodynamics introduces the field but only as a perturbation on the preceding phenomenology. As De Broglie recently strongly recalled: 'Après ma thèse, on a souvent interprété faussement mes idées en disant que, d'après moi, l'électron était une onde, ce qui escamotait la particule.' [40].

1.3. ON THE NON-REDUCTIONIST ATTITUDE

Non reductionism is an attitude based on the fundamental assumption that the description of the microphysical world cannot be performed by using the same conceptions as those which are successful for the macroscopic world. Non reductionists criticize wave pictures as well as wave corpuscule duality in quantum theory. For them, mathematical formalism is more important than pictures, their only requirement being the reproduction of experimental facts. In some way this is nowadays the dominant attitude among physicists, an attitude strengthened by the considerable successes of mathematical phenomenology and the impressive weight of mathematical physics (distinguished from theoretical physics [41]). From an historical point of view let us point out that non reductionism is spreading in physics at a time where it is practically completely thrown out from biology.

Let us give some characteristic quotations:

'Bodies and fields are now behind and for good: the things with which the quantum theories are concerned cannot be pictured as classical entities. They are *sui generis* entities deserving a special name – say *quantons* – this being why the quantum theories themselves are *sui generis*'.

[4a, p. 235]. M. Bunge

'Un autre point sur lequel nous devons beaucoup insister auprès des élèves, c'est le fait que les grandeurs et objets physiques (électrons, atomes, molécules) ne peuvent être bien représentés par des modèles macroscopiques, car leurs propriétés sont différentes.

Un électron n'est pas une bille d'acier. Dans l'expérience des trous d'Young, le même électron passe à la fois dans deux trous. Deux électrons sont rigoureusement indiscernables, etc. Là, le seul recours possible, c'est évidemment l'abstraction, seul moyen d'exprimer qu'un électron soit, à la fois, onde et corpuscule. Seul le langage mathématique est adapté au monde atomique; le langage usuel n'est nullement utilisable à l'échelle microscopique.   A. Kastler [42]

'En vérité, la terminologie ('incertitudes') et la problématique qu'elle sous-entend ('limite à la précision de la connaissance, 'perturbation incontrôlable du système par la mesure' etc.), résultent d'un refus de prendre la théorie quantique au sérieux. Ce que nous disent $\delta E\,\delta t \geqslant \hbar$ et $\delta p\,\delta x \geqslant \hbar$ en effet est très simple: *il n'est plus possible de penser les concepts de temps et d'énergie, de position et de quantité de mouvement en termes classiques*.

Reconnaissons donc la spécificité des objets quantiques et, pour y insister, appelons les ici 'partiqules'.   J. M. Levy-Leblond [43]

'In other works, we want literal interpretations – even if they are assigned no familiar visualizations – because we want objectivity. It is only in mathematics that we are interested in mirroring one conceptuel structure on to another'.   M. Bunge [4c, p. 120]

Following these exhortations to give up space-time pictures, the whole modern theoretical physics provides us with mathematical buildings in which dynamical pictures are lacking. The theory of fields and the theory of elementary particles show such an abstract description of physical phenomena. Elements of reality are dissolved in a mathematic formalism, difficult to handle and quite unable to produce its own physical interpretation. The successes of these theories are not so important as to justify these huge abstract buildings. On the other side, the experimentalist, a physicist or a chemist, is not convinced by theories very difficult to handle intuitively to guide his steps. There is a permanent revolt (often in the underground) of the customers against the prestidigitation character of the theories of microphysics. The non reductionist attitude, which claims to be a progressist attitude, appears as a deceiving agnosticism. The complexity of reality must not prevent us from trying to build clear interpretations. In his practice the physical chemist is often rejecting the non reductionist interpretations of quantum mechanics. He is an heretic, but sins in good company. Some illustration of this situation will be presented at the end of 2.2 on the specific examples of physico-chemical problems involving symmetrical double-well potentials (optical isomers, NH_3 type molecules).

2. The Statistical Interpretation of Quantum Mechanics

2.1. On the different interpretations of the quantum formalism

Beyond the different historical circumstances which have led to the universally accepted non relativistic quantum theory, there was a unique trend to find a good mathematical model for the fundamental experimental facts: stability and line spectra of atoms. The discrete spectra of a self-adjoint operator in an Hilbert space has been found as the simplest model*, giving rise to the following formalism and *mathematical interpretation*:

a microphysical observable is a random variable whose distribution is characterized with the help of a self-adjoint operator L on a Hilbert space and an element ψ of this space, corresponding to the *state* of the system; the different moments of the distribution are:

$$m_1 = \langle \psi | L | \psi \rangle \quad \text{mean value}$$

$$m_2 = \langle \psi | (L - m_1)^2 | \psi \rangle \quad \text{variance or dispersion}$$

$$m_n = \langle \psi | (L - m_1)^n | \psi \rangle$$

the distribution can on the other side be directly given by the probability law, which stems from the spectral decomposition of the operator L; for example in the discrete case:

$$L = \sum_\lambda \lambda \, |\lambda\rangle \, \langle\lambda|$$

$$m_1 = \sum_\lambda \lambda \, |\langle \psi | \lambda \rangle|^2$$

$$m_2 = \sum_\lambda \lambda^2 \, |\langle \psi | \lambda \rangle|^2 - m_1^2.$$

This can be expressed by the two rules:

(a) the observable corresponding to L can only take values from the spectrum of L.

---

* More refined models appear in the modern axiomatic formulations of quantum theory. The two major trends are: proposition calculus [35] [36] [44] [45] and C^* algebra theory [46] [11].

(b) The probability of such a value is

$$|\langle\psi|\lambda\rangle|^2 = \langle\psi|\mathbf{P}_\lambda\psi\rangle$$

where $\mathbf{P}_\lambda$ is the projection operator associated to $|\lambda\rangle$.

The different *physical interpretations* of this formalism arise from the different meanings given to the probability statements and to the concepts of state and observable. Probabilities can be interpreted as representing one of the following situations:

(a) Mean values on repeated experiments.

(b) Simplification for the study of systems with phase space of a very high dimension.

(c) A way of taking account of the influence of a 'big system' on a 'small' one coupled to it.

(d) A measure of our partial knowledge of the actual state of the system, due for example to difficulties of observation.

(e) A realistic description of nature (fundamental indeterminism).

This enumeration is not limitary. All these interpretations, which are not completely exclusive, can be found in quantum mechanics during the last fifty years. Grosso modo there are two main opposing points of view:

(A) The traditionnal interpretation of Bohr, Heisenberg and Dirac [47] [48] [16], called *the Copenhagen interpretation*, and which is the interpretation adopted in most classical textbooks on quantum mechanics [49]:

(I) ψ represents the complete description of the state of a single system (or the best possible description). It follows that it is meaningless to ascribe a definite position and momentum to a particle.

(II) the laws of microphysics (or of microphysical observation) are probabilist, non determinist.

(III) Heisenberg's uncertainty principle refers to a single system.

(B) *The Statistical Interpretation* of Einstein [50], Popper [51], Blokhintsev [52], and Ballentine [53].

(I) ψ represents a statistical ensemble of identically 'prepared' systems. It allows the calculation of the statistical distribution of values for each observable on such an ensemble. A particle has definite positions and momenta. Quantum mechanics provides us only with the statistical distribution of observed position and momentum; it is an incomplete description of reality.

(II) The laws of microphysics are determinist, but a description of a 'sub quantum level' is to be discovered in term of 'hidden variables'.

(III) Heisenberg's uncertainty principle refers to the spread of values (dispersion) in the statistical ensemble.

We are not going here to argue on interpretations. Let us just recall some great topics of discussion with fundamental references: hidden variables [54] [55] [56] [4p], theory of measurement [57], the Everett interpretation [58]. Our purpose in this paper is just to see how quantum mechanics (quantum chemistry) relates to the

Statistical Interpretation, that is to try to answer the question of the molecular physicist (chemist): 'What actually do electrons do in one molecule?' using a theory which fails the single system. Our work has been motivated by the strong support to the Statistical Interpretation recently given by Ballentine [53]. To people who would say that the Statistical Interpretation is very often used by physicists in practice we shall answer that:

(a) religious practice is often heretic. As more and more people use quantum mechanics, very different points of view appear, which do not coincide with what is taught in authoritative books. The chemist and the quantum chemist are very often in such an heretic situation, under the pressure of experimental facts, for which they try to produce a clear picture (cf. the end of 2.2).

(b) There are no textbooks completely written on the basis of the Statistical Interpretation. Even such a textbook as Blokhintsev's [52] has an ambiguous position on the meaning of the incertainty principle. Ballentine's review is to our knowledge the first systematic exposition of an unambiguous statistical interpretation of the quantum formalism. See also [62] [64].

(c) The practice of quantum chemistry gives the feeling that the individual molecule is the central object of the theory, notwithstanding the fact that no experiment concerns a unique molecule. The brute force reduction of statistical data into a *one* molecule picture seems to be at the origin of a lot of misunderstanding and dubious concepts in quantum chemistry. The purpose of this paper is to try to show how some of these unclear conceptions can be rightly stated.

2.2. THE MEANING OF THE STATISTICAL INTERPRETATION AND
THE ERGODIC HYPOTHESIS

In quantum chemistry, as in the other fields of quantum physics, it has been impossible to discover a fact which would favour in an unambiguous way one of the physical interpretations of quantum mechanics. This is due to the experimental difficulties which prevent actually measuring the observables which are described by the theory. One does not know how to isolate an atom and observe events located in a specified region (to look at an electron for example!).

Nevertheless, it seems that in quantum chemistry as in quantum physics in general, the Statistical Interpretation solves paradoxical situations and allows some understanding of facts without too many ad hoc interpretations.

This appears with full light for the traditional problem of the 'reduction of the wave packet' in the quantum theory of measurement. The 'orthodox' interpretation considers that the state vector describes the state of a single system; according to the axioms of quantum mechanics:

(1) a measurement of an observable L may give only a value l, which is necessarily an eigenvalue of L.

(2) when a value l is obtained, the single system is left after measurement in the eigenstate $|l\rangle$.

It must be emphasized that, if we accept that the measurement process is described by the Schrödinger equation, then the axiom (2) cannot have an absolute validity: as shown by Landau and Lifshitz [49, Chapter I, Section 7], Araki and Yanase [57] (see also [53, Section 4.1, footnote 10]), if a system is initially described by a state function $|l\rangle$ (eigenfunction of $\mathbf{L}$) its state function after the measurement of L will in general be a *different* function $|\varphi_l\rangle$. Several textbooks (see e.g. [49 b], [49 c]) nevertheless wrongly present axiom (2) on the same footing with the other axioms of quantum theory (such a situation, in our opinion, illustrates the falsity of the widespread affirmation according to which 'quantum theory is verified up to the slightest details', an affirmation often opposed to anyone trying to depart from the presently established formalism). This problem (function $|\varphi_l\rangle$ instead of $|l\rangle$ after the measurement of L with an initial function $|l\rangle$) must be carefully distinguished from the so-called 'problem of the reduction of the state vector', that we shall now examine: this last problem appears when the initial function ψ (before measurement of L) is not an eigenfunction $|l\rangle$ of $\mathbf{L}$: $\psi = \sum_l c_l |1\rangle$ with $c_l = \langle l|\psi\rangle$. Then, according to the orthodox theory of measurement, if the value l is obtained (its probability being $|c_l|^2$, we must admit that the interaction with the measuring device has changed ψ to the state $|\varphi_l\rangle$ (not $|l\rangle$ in general, as mentioned above), while the assumption of the linearity of the evolution of the state function (which is satisfied by the usual evolution operator) would give $\sum_l c_l |\varphi_l\rangle$. Therefore, this evolution into a *final* function $|\varphi_l\rangle$ which *differs* according to the result l of the measurement, while the *initial* function $|\psi\rangle$ is *the same* for all systems, cannot be obtained from any linear evolution operator acting directly on the state functions, and this evolution must be introduced in the theory as a special postulate: *the reduction of the state vector postulate* [53, Section 4.2].

In the Statistical Interpretation such a postulate is not necessary. Finding a value l for L means that we have selected for measurement a single system of the Statistical ensemble where the observable L has the value l.

Contrary to the Copenhagen Interpretation, the Statistical Interpretation could stop at that point. Indeed, it does not claim that Quantum Theory is a complete description of physical reality, and is consequently *not* obliged to give a detailed account of the measurement process itself. Let us consider, as an example of this conceptual situation, the theory of Brownian motion: this theory obviously cannot (and actually does not try) to give an account of the position measurements of the Brownian particle performed with a microscope: this would indeed involve nothing less than the theory of light (electromagnetic field) and its interaction with matter. This does not hinder the Brownian motion theory to describe perfectly the statistical properties of the results of these position measurements*. The Statistical Interpretation

---

* Let us remark that this situation is not at all specific to the statistical character of the theory: classical mechanics itself perfectly describes deterministic motions without any concern about the fact that, almost always, position measurements will be obtained by optical methods, the physical theory of which is completely outside its scope.

could have exactly the same modest attitude, namely, reproducing the statistics of results of measurements, without pretending to give a complete theory of the measurement process itself. But the Copenhagen Interpretation, which claims to be a complete theory of each individual system, must then try to describe this process, and this attempt leads to the well-known difficulties of the 'orthodox' theory of measurement mentioned above. As explained very clearly by Ballentine ([53] Section 4.2), in order to obtain for the 'pointer' of the measuring apparatus a single position at each measurement (instead of a mysterious 'superposition' of all possible positions!) this orthodox theory must assume that an initial *pure state* evolves into a *mixed state* in the course of the measurement process: but this is impossible if the evolution of the state obeys the Schrödinger equation (because the character 'pure' or 'mixed' is kept invariant by this equation), and therefore this abnormal evolution is assumed under the name of the 'reduction of the state vector postulate'. Strictly speaking, the orthodox theory of measurement thus avows the incomplete character of the Schrödinger evolution operator, since it introduces (in a qualitative way only) another type of time evolution in the case of measurement processes.*

As mentioned above, the Statistical Interpretation may avoid such problems by giving up any attempt to describe the measurement process itself. But if, following the example of the orthodox theory of measurement, it tries to present its own description of this process, then (contrary to the account given by Ballentine ([53], Section 4)) it runs into somewhat similar difficulties: namely, accepting that the supersystem (system + measuring apparatus) remains in a pure state after the measurement would imply, for composite observables (involving system and apparatus) rather strange probabilities involving 'interference terms' between functions corresponding to *different outcomes* of the measurement [53, Section 4.3], a situation which would contradict the essential spirit of the Statistical Interpretation. Accepting the mixed state after the measurement eliminates such funny effects, and is therefore much more appealing physically. Now, exactly as we saw above, this amounts to rejecting the universal validity of the Schrödinger evolution operator, but this is not very troublesome in the framework of the Statistical Interpretation, which does not a priori consider Quantum Mechanics as necessarily complete (in terms of the stochastic extensions of the Statistical interpretation to be discussed later, a very natural possibility is that the measurement splits one initially ergodic stochastic process into *several sub-processes*, each ergodic on its own, but with different ergodic means, so that the *whole*-process (reunion of these subprocesses) is no more ergodic after this splitting).

The individual system interpretation of quantum mechanics is the source of many conceptual difficulties, leading to never closed discussions. The measurement problem is one of the most famous examples, the resonance problem is another one. And for both, the Statistical Interpretation provides us with a satisfactory picture. The discussion about resonance is a discussion on the interpretation of stationary states in quantum mechanics. From the orthodox point of view, nothing happens in a system when in a stationary state.

* See the note added in proof at the end of the paper.

Coulson writes:

'When we speak of a structure, we mean a certain way of pairing electronic orbitals together, using for this purpose certain component atomic orbitals .. Now the first thing to say about these structures is that they do not exist in themselves: they have no objective reality, and it is quite inaccurate to speak of any kind of resonance between two or more structures as implying that each structure exists for a fraction of the time given by its weight in the compound wave function. There is no such thing as an independant ionic or covalent structure, for the simple reason that their wave functions are not eigenfunctions of the allowed stationary states'. ([59a], p. 124).

'So, too, when we introduce the Kékulé structures for benzene we do not mean that the bonds jump around, spending a certain fraction of the time in one or the other sets of position: we are using this language primarily to show why there is no bond alternation, but all C—C bonds are equivalent. In this respect, Kékulé's original concept of 'dynamic oscillation' is no longer acceptable, for the molecule is in a stationary state' ([59b], p. 372).

The same point of view prevailed in the Soviet Union where the theory of resonance was strongly criticized in the fifties [60]. Volkenstein [61] writes:

'Resonance structures do not exist only because the quantum theory of the molecule looks at stationary states with constant electron distributions'.

Here again the epistemological difficulties are due to the application of the notion of 'state' to a single physical system. From the Statistical point of view a stationary state describes a Statistical ensemble where the frequency characteristics of all the observables do not vary with time. What happens in a single system belonging to such an ensemble? There are no answers in qantum mechanics according to the Statistical Interpretation.

In fact it can appear very natural to suppose that in a single system the observables vary with time, and that time average (average on a single system for very long time) becomes equal to the ensemble average calculated by quantum mechanics. This is an 'ergodic hypothesis' which is not only yet unproved, but which cannot be proved if one doesn't know explicitly the behaviour of the single system. Nevertheless such an implicit 'ergodic hypothesis' is often made by the quantum chemist in trying to get intuitive pictures from his numerical results.

A good example of this situation is displayed by the 'double-well potential' problems: the two optical isomers of a given molecule, or the two symmetrical forms of molecules like NH_3: see for example Herzberg ([66], Chapter I, II, Section 5d, pp. 224–225). Let us denote ψ_a and ψ_b the state functions associated with the nuclear configurations (a) and (b) respectively. The expectation values with respect to the total hamiltonian $\mathbf{H}$ are equal, and since the non-diagonal matrix element $\langle \psi_a | \mathbf{H} | \psi_b \rangle$ is not strictly zero, quantum mechanics will give as true eigenstates the symmetrical and antisymmetrical linear combinations $\psi_0 = 1/\sqrt{2}(\psi_a + \psi_b)$ and $\psi_1 = 1/\sqrt{2}(\psi_a - \psi_b)$. According to these functions, the nuclear configurations would be symmetrically 'delocalized' between the two forms (a) and (b) (a symmetrical 'nuclear cloud', so to speak!). Now, if the mean time for interconversion between the two forms is one million or one billion years (for an optical isomer) or simply about one year (for e.g. H_3) no physical chemist will accept that the symmetrized functions ψ_0 (or ψ_1) have an instantaneous meaning for any given *single* molecule: for him, over any short interval of time, only the non-symmetrical functions ψ_a and ψ_b have a meaning for a single molecule; the symmetrized function ψ_0 will appear as describing a single molecule seen *over a time long* with respect to the mean interconversion time, or as giving an instantaneous description of a *racemic ensemble of molecules*. But such a physical chemist's view amounts to rejecting the Copenhagen Interpretation and adopting a Statistical Interpretation supplemented by an Ergodic Hypothesis!

In fact the Statistical Interpretation would need a Stochastic counterpart to be really satisfactory. In Section 4 we shall see how one can imagine such a stochastic picture in agreement with quantum mechanics. We shall produce 'hidden variables' pictures. But it must be born in mind that strictly speaking there is an incompatibility between quantum mechanics and any 'hidden variable picture' of the micro-world. This is what we are going to recall briefly in the next paragraph.

2.3. QUANTUM MECHANICS IS NOT A CLASSICAL STATISTICAL THEORY

From 1966 it is definitely clear that quantum mechanics cannot be formulated exactly as a classical probabilistic theory, like classical statistical mechanics. This is due to the fact that the different random variables (observables) of quantum mechanics cannot be defined on a common classical probability space. Each observable in quantum mechanics can be considered as a classical random variable, but two incompatible observables (corresponding to non commuting operators) cannot be viewed simultaneously as classical random variables defined on the same boolean algebra of events. This fundamental result has been demonstrated in lot of different ways; the work of Kochen and Specker [54 g] is an achievement in that field, Cohen [63] has shown this situation in a simpler way. He takes 'as basic variables' the position q and the momentum p. If quantum mechanics can be formulated as a classical probability theory, then there should exist a probability distribution $F(q, p)$ which satisfies the following requirements:

(a) $F(q, p) \geqslant 0$ for all q and p

(b) $\int F(q, p) \, \mathrm{d}p = |\psi(q)|^2$ where $\psi(q)$ is the coordinate state function.

(c) $\int F(q, p) \, \mathrm{d}q = |\Phi(p)|^2$ where $\Phi(p)$ is the momentum wave function, which equals:

$$\frac{1}{2\pi\hbar} \int \psi(q) \exp\left(-\frac{i}{\hbar} qp\right) \mathrm{d}q \, .$$

(d) the expectations of observables calculated in the classical manner should be equal to those obtained using the operator formalism of quantum mechanics. That is, if the quantum mechanical mean value of the operator $\mathbf{G}$ is $\langle \mathbf{G} \rangle$, then there should exist a function $g(q, p)$ such that

(d_1)

$$\langle \mathbf{G} \rangle = \iint g(q, p) \, F(p, q) \, \mathrm{d}p \, \mathrm{d}q$$

and

(d_2)

$$\langle K(\mathbf{G}) \rangle = \iint K(g(q, p)) \, F(q, p) \, dp \, dq$$

where K is any function.

The most general F which satisfies b and c, cannot also be made to satisfy d.

A simple illustration of this situation may be given by considering the position q and impulsion p of a particle along with its energy $\mathbf{E}$, choosing for representating E the classical function $g(q, p) = p^2/2m + V(q)$ (q, p and E are scalar quantities and not operators, since we ask for the possibility of a classical theory).

Let us first consider the harmonic oscillator: $V(q) = \frac{1}{2}m\omega_0^2 q^2$, and let us denote by q_0 positive the value of q such that $V(q) = E_0 = \frac{1}{2}\hbar\omega_0$ (the so-called ground state energy). Then, for $|q| > q_0$, $V(q) > V(q_0) = E_0$, and therefore $E > E_0$ since $p^2/2m \geqslant 0$ whatever p. But if we accept the ground state probability density

$$|\psi_0(q)|^2 = \left(\frac{m\omega_0}{\pi\hbar}\right)^{1/2} e^{-(m\omega_0/\hbar)q^2}$$

we have a strictly positive probability for the event $|q| > q_0$ namely:

$$\Pr(|q| > q_0) = \int_{-\infty}^{-q_0} |\psi_0(x)|^2 \, dx + \int_{q_0}^{+\infty} |\psi_0(x)|^2 \, dx$$

and consequently $\Pr(E > E_0) \geqslant \Pr(|q| > q_0) > 0$, a property which is obviously incompatible with the quantum mechanical prediction for the energy in the state ψ_0, namely that E is equal to E_0 with probability 1. The same argument can be developed in the case of the hydrogen atom: since $V(r) = -1/r$ and $E_0 = -\frac{1}{2}$ (in atomic units it is clear that $V(r)$ and therefore $E = p^2/2m + V(r)$ will be larger than E_0 for $r > 2$). Therefore, the probability laws given by quantum mechanics for the position and the energy are clearly incompatible from the point of view of a classical statistical theory as soon as the relationship $E = p^2/2m + V(q)$ is assumed (see Moyal ([67], p. 100) and Urbanik ([68], p. 124).

With Cohen's notations, we see on the specific example of the energy ($\mathbf{G} = \mathbf{H}$) that the choice $g(q, p) = p^2/2m + V(q)$ gives agreement for the mean value: $\langle \mathbf{H} \rangle = \overline{g(q, p)}$ ((d_1) is satisfied), whereas the agreement disappears already *for the second moment*: $\langle \mathbf{H} - \langle \mathbf{H} \rangle^2 \rangle = 0$, while $\overline{(g(q, p) - \overline{g(q, p)})^2} \neq 0$ ((d_2) is not satisfied). It could be argued of course, that other relationships between q, p and $\mathbf{E}$ would remove this incompatibility, and it is just the essence of the works mentioned above (Kochen and Specker, Cohen) to prove that, whatever relationships between scalar quantities treated classically are supposed, some incompatibility will appear with the quantum-mechanical results obtained in the framework of the operator formalism (but, contrary to widespread opinion (see e.g. Bub ([69], pp. 39–43)), this does not rule out classical models for microphysics: in the example that we have considered, it could be argued that the probability density $|\psi_0(q)|^2$ is essentially correct, but that the δ-peaked probability for the energy is *not*, and that E_0 is only the *mean value* of the energy of the system, which would undergo fluctuations around this value as a function of time: see Section 4 below for further detail).

The formalism of quantum mechanics is strictly speaking incompatible with any phase space picture of microphysics. So that the Statistical Interpretation, as suggestive as it may be, cannot be transformed completely into a Statistical theory of quantum mechanics. This must always be born in mind when using such an interpretation.

We emphasize strongly this fact, because it seems not to have been clearly appreciated by the supporters of the Statistical Interpretation. Thus Einstein ([70], *p.* 10)* thought that the relationship between quantum theory and a more complete underlying theory would be analogous to the relationship between the statistical theory of brownian motion and the molecular kinetic theory; this would suggest considering quantum theory as really a statistical theory, which is not the case. More recently, Ballentine ([53], p. 380), while being aware of this situation, does not stress sufficiently its meaning. Establishing a true statistical theory for microphysics remains a highly non-trivial task. Thus we are left with the following dilemma:

(a) take quantum mechanics as 'ultima verba' and try to elaborate its structure with some hope in the appearance of a 'meaning'.

(b) pursue the quest for an 'hidden variable' picture with the conviction that some part of quantum mechanics must be given up. This part of quantum mechanics would be a part which has up to now *not* been checked by experiment, and that would *not* be verified if such experiment was actually conceived and could be practically performed. Some steps on this way will be described in Section 4.

3. Delocalization and Fluctuation in Quantum Chemistry**

(in collaboration with J. P. Malrieu)

3.1. CONCEPTS OF DELOCALIZATION IN QUANTUM CHEMISTRY

The unity of the molecule – as a system of dependent parts – implies a delocalization of the physico-chemical information, by reciprocal influence of the different constituent parts. In fact experiment reveals that effects are transmitted on more or less long distances in molecular systems. Such a transmission can take place on distances

---

* 'Mir erscheint es, dass die statistische Quantentheorie ebensowenig einen brauchbaren Ausgangspunkt für die Aufstellung einer vollständigeren Theorie bildet, wie etwa die auf klassische Mechanik und das Gesetz des osmotischen Druckes gegründete Theorie der Brown'schen Bewegung einen brauchbaren Ausgangspunkt für die Aufstellung der kinetischen Molekular-Mechanik hätte bilden können, wenn die Theorie der Brown'schen Bewegung der letzteren Theorie zeitlich vorangegangen wäre.'

** This section is a work done with J. P. Malrieu. Historically it reflects the origin of our thoughts about the problems of this paper: a search for the meaning of localized orbitals, the use of which had led us to the successes of the calculation method PCILO. This work was done in the winter 1971–1972.

far beyond the normal radius of action of the implied forces (more or less screened electrostatic forces). In order to formalize the concept of delocalization one should study the way in which a molecular system responds to a local perturbation, and show the relationship between the response and the structural characteristics of the unperturbed system. Destructive analysis of a system (by breaking of physical linking) defines constitutive elements. The constructive theory of such a system tries to express the behaviour of the system in terms of the elements found by analysis. It is not guaranteed a priori that these analytical elements will be those elements which reflect in the best way the global behaviour of the system. The classical theory of chemical structure is from that point of view a successful constructive theory, founded on building blocks like atoms, bond and functional groups.

Quantum theory starts from a different analysis in terms of electrons and nuclei. But as it renounces from the very beginning describing electron trajectories, it adopts a global point of view where the relationship with the classical structural theory becomes difficult. In opposition to the analytical attitude of the chemist, the quantum physicist works with a global model of the molecular system.

The problem of delocalization is at the heart of this dialectic between the whole and the parts in which the quantum chemist is engaged.

The chemical concept of delocalization is based on the analysis of global phenomena in terms of local units as bonds or atoms.

The quantum mechanical concept of delocalization has for origin the consideration of the domain of definition of wave functions. Its meaning depends deeply on the interpretation which is given to quantum mechanics. The wave picture of the electron implies an always present delocalization, whereas the corpuscular and statistical interpretation looks at delocalization as the result of motion.

More difficulty arises from the methods used to solve the mathematical n-body problem. Mathematical physics describes complex systems with interactions by replacing dependent physical characteristics by 'independent' mathematical objects which have lost direct physical interpretation. Normal vibrations of a complex vibrating system, Fourier analysis of any motion are good examples of such a practice. The peculiar features of quantum mechanics are added to the mathematical abstractions of the n-body problem, to make difficult the interpretation of the quantum mechanical n-body problem.

So a new aspect of delocalization appears with the use of delocalized orbitals as mathematical intermediates for the building of wave functions taking into account the interaction between electrons. Orbitals are meant to represent some quasi-particles. The delocalization of the quasi-particles seems natural in quantum mechanics if they are looked upon as particles. Doing this, one forgets the conventional features of the quasiparticles which are shown by the arbitrariness in the choice of localized or delocalized orbitals.

In conclusion we can say that three concepts of delocalization appear in quantum chemistry:

(a) the chemical delocalization

(b) the quantum mechanical delocalization

(c) the many-body problem delocalization.

The complex relationships between these different aspects are the very subject of the whole seminar. Our concern in this paper is mainly with the quantum mechanical concept of delocalization.

3.2. The Fluctuation of an Observable in a Quantum Stationary State

Les us consider a system in a quantum stationary state, eigenfunction of the hamiltonian operator $\mathbf{H}$. The energy, and all the observables represented by operators commuting with the hamiltonian, have zero dispersion.

$$\langle\psi|\mathbf{H}^2 - \langle\psi|\mathbf{H}|\psi\rangle^2|\psi\rangle = 0. \tag{3.1}$$

According to the postulates of quantum mechanics, when a measure of such an observable is performed on the different elements of an ensemble of systems prepared in the state ψ, the same value should always be obtained. The ergodic hypothesis leads naturally to supposing that this observable is invariant with time in each individual system.

For an observable represented by an operator which does not commute with the hamiltonian, as position or impulsion, the situation is different. The statistical dispersion of the results of measurement on an ensemble of systems associated to the state ψ is no more zero. The ergodic hypothesis attributes the same dispersion to the values of the observable in an individual system of the ensemble at different times, supposing that in the stationary state the observable changes in time in every individual system, a fact which is not described by quantum mechanics. We shall extensively comment on that in Section 4.

As an exemple let us consider the distance of the electron to the nucleus in the ground state of the hydrogen atom: $\psi = (\pi a_0^3)^{-1/2} \exp(-r/a_0)$ with $a_0 = 0{,}529$ Å. We have

$$\langle\psi|r|\psi\rangle = 1.5\,a_0$$

$$\langle\psi|r^2|\psi\rangle = 3\,a_0^2$$

$$\langle\psi|r^2|\psi\rangle - \langle\psi|r|\psi\rangle^2 = 0.75\,a_0^2$$

so that the standard deviation, square root of the dispersion, is equal to $0.866\,a_0$.

When it is said that in its ground state the hydrogen atom has an electron described by a $1\,s$ spherically symmetric wave function, from the Statistical Interpretation point of view this means that in the corresponding statistical ensemble of hydrogen atoms we shall find the electron present in all the possible points of space, with a frequency distribution giving for the distance r a mean value $1.5\,a_0$ and a standard deviation $0.866\,a_0$. We have no right to say that the electron is at the same time spread in the whole space (charge cloud), because in quantum mechanics the self-interaction between the different parts of such a cloud never appears. We have no right to say that we know something about the trajectories of the electron in space; quantum mechanics does not give any dynamical picture of the behaviour of the electron. Nevertheless one can read in a representative book: 'An atom in an S state, such as hydrogen or helium has a spherical charge distribution, to be sure. But in view of the uncertainty principle the charge cannot be at rest. Although its motion is not organized – there is no mean angular or linear momentum – its random excursions engender instantaneous multipoles which are able to produce rapidly fluctuating fields... '([73], p. 23). Pictures are arbitrarily created to solve the apparent contradiction between the existence of a spherical charge distribution in *one* hydrogen atom and the attractive interactions between two such atoms known as Van der Waals forces. It is funny to look at the uncertainty principle as responsible for the dynamics of the electron (charge). In fact the ergodic hypothesis, added to the Statistical Interpretation, would also speak of the motion of the electron, creating an instantaneous dipole moment (and the uncertainty principle is a consequence and not a cause!). But as suggestive as such a picture may be, let us stress again that quantum mechanics does not describe such an individual motion. Its statistical prediction could (and not completely we have seen!) agree with a motion, but this motion is unknown. Nevertheless a majority of quantum chemists feel quantum results with such a motion in mind. This chapter is an illustration of that situation.

3.3. THE FLUCTUATION OF THE PARTICLE NUMBER IN A VOLUME

Consider a system of N identical particles, the electrons of an atom or a molecule for example. The observable N_V: number of particles in a volume V located in some region of space, can be defined by the operator

$$\mathbf{N}_V = \sum_i Y_V(\mathbf{r}_i) \tag{3.2}$$

where the index i numbers the different particles, and Y_V is the characteristic function of the volume V

$$Y_V(\mathbf{r}_i) = \begin{cases} 0 \text{ if the point } \mathbf{r}_i \text{ lies outside } V. \\ 1 \text{ if the point } \mathbf{r}_i \text{ lies in } V. \end{cases} \tag{3.3}$$

This definition, which is currently used in quantum mechanics, needs in fact a careful justification. The mean value of the particle number in V is

$$\overline{N}_V = \langle \psi | \mathbf{N}_V | \psi \rangle = \sum_i \int_V \mathrm{d}\mathbf{r}_i \int_{\text{space} \times \text{spin}} \psi^* \psi \, \mathrm{d}\mathbf{r}_1 \dots \mathrm{d}\mathbf{r}_{i-1} \, \mathrm{d}\mathbf{r}_{i+1} \, \mathrm{d}\mathbf{r}_N \times$$
$$\times \mathrm{d}\sigma_1 \dots \mathrm{d}\sigma_N \tag{3.4}$$

where $\mathrm{d}\sigma_i$ stands for the integration on spin coordinates. With

$$\rho(\mathbf{r}) = N \int \psi^* \psi \, \mathrm{d}\mathbf{r}_2 \dots \mathrm{d}\mathbf{r}_N \, \mathrm{d}\sigma_1 \dots \mathrm{d}\sigma_N \quad (\mathbf{r}_1 = \mathbf{r})$$

$$\overline{N}_V = \int_V \rho(\mathbf{r}) \, \mathrm{d}\mathbf{r} \tag{3.5}$$

$\rho(\mathbf{r})$ is the diagonal part of the one particle reduced density matrix; it appears as a density of mean value of particle number, or charge density; it is not a probability density as is often written:

'$\rho(\mathbf{r})$ is the probability density of finding any of the N particles at the point $\mathbf{r}$ [74].
'The total probability of finding an electron at atom r is equal to the electronic charge density q_r at that atom' ([75], p. 83).

Such a misnomer is due to the fact that the mean number of particles in V is equal to N times the probability of the particle 1 (for example) to be in V; apparent correspondance between two quantities of different kind: a mean value and a probability!
The fluctuation (dispersion) of the particle number in V is

$$\bar{N}_V = \langle\psi|N_V^2|\psi\rangle - \langle\psi|N_V|\psi\rangle^2 . \tag{3.6}$$

Due to the fact that $Y_V^2(\mathbf{r}) = Y_V(\mathbf{r})$

$$\langle\psi|N_V^2|\psi\rangle = \sum_i \langle\psi|Y_V(\mathbf{r}_i)|\psi\rangle + \sum_i\sum_{i\neq j} \langle\psi|Y_V(\mathbf{r}_i)\,Y_V(\mathbf{r}_j)|\psi\rangle$$

$$= \langle\psi|N_V|\psi\rangle + 2\int_V d\mathbf{r}_1 \int_V d\mathbf{r}_2\, \rho_2(\mathbf{r}_1,\mathbf{r}_2)$$

where

$$\rho_2(\mathbf{r}_1,\mathbf{r}_2) = \frac{N(N-1)}{2}\int \psi^*\psi\, d\mathbf{r}_3 \ldots d\mathbf{r}_N\, d\sigma_1 \ldots d\sigma_N$$

is the diagonal element of the two-particle reduced density matrix.
Then

$$\bar{\bar{N}}_V = \bar{N}_V - \bar{N}_V^2 + 2\int_V d\mathbf{r}_1 \int_V d\mathbf{r}_2\, \rho_2(\mathbf{r}_1,\mathbf{r}_2). \tag{3.7}$$

Let us add some comments:

A. *The Fluctuation of the Particle number and the orbital representation*

$$\rho(\mathbf{r}) = \sum_\alpha n_\alpha\, x_\alpha^*(\mathbf{r})\, x_\alpha(\mathbf{r}) \tag{3.8}$$

$$\bar{N}_V = \sum_\alpha n_\alpha \int_V x_\alpha(\mathbf{r})\, x_\alpha(\mathbf{r})\, d\mathbf{r}. \tag{3.9}$$

The mean value of the particle number in the volume V can be expressed as a sum of natural orbital contributions. This does not occur for the dispersion. We shall see later (3.5.c) that it is possible to define 'fluctuation orbitals' which give to the dispersion an additive form.

B. *Special Case of a Slater Determinant (Hartree-Fock)*

If $\{\varphi_\alpha(r_i,\sigma_i)\}$ is a complete basis of spin orbitals, the particle number operator $\mathbf{N}_V$ can be written in the second quantization representation [77]:

$$\mathbf{N}_V = \sum_\alpha\sum_\beta S_{\alpha\beta}^V a_\alpha^+ a_\beta \tag{3.10}$$

where a_α^+ and a_β are the creation and annihilation operators for spin-orbitals, and

$$S_{\alpha\beta}^V = \int d\sigma \int_V \varphi_\alpha(\mathbf{r}\sigma)\, \varphi_\beta(\mathbf{r}\sigma)\, d\mathbf{r}. \tag{3.11}$$

If ψ is a Slater determinant built with N spin orbitals taken from the set $\{\varphi_\alpha(\mathbf{r}_i, \sigma_i)\}$, using Wick's theorem we find for the dispersion:

$$\tilde{\tilde{N}}_V = \sum_\alpha S_{\alpha\alpha}^V - \sum_\alpha (S_{\alpha\alpha}^V)^2 - \sum_{\alpha \neq \beta}\sum (S_{\alpha\beta}^V)^2 \tag{3.12}$$

all the summations being restricted to the occupied spin-orbitals.

This formula shows clearly the term responsible for the deviation to orbital contribution additivity. This term depends on the local overlap between orbitals and can be minimized by using localized orbitals.

C. The Number of Pairs in a Volume V

Following the same way as for N_V, it is possible to associate to the observable N_{2V}: number of particle pairs in the volume V, the operator

$$\mathbf{N}_{2V} = \sum_{i<j}\sum Y_V(\mathbf{r}_i)\, Y_V(\mathbf{r}_j). \tag{3.13}$$

The mean value of the number of pairs in the volume V is

$$\overline{N}_{2V} = \int_V d\mathbf{r}_1 \int_V d\mathbf{r}_2\, \rho_2(\mathbf{r}_1, \mathbf{r}_2). \tag{3.14}$$

This is precisely the term which appears in the expression of $\overline{N}_V$. One can also calculate $\overline{N}_{2V}$; it is a complex formula where appears the diagonal parts of the one, two, three and four particle reduced density matrices.

3.4. THE LOGE THEORY AND THE FLUCTUATION OF THE PARTICLE NUMBER

A. The Density Map and the Individual System

When the volume V becomes small, the mean value of the number of pairs becomes small relative to the mean value of the number of particles. In the same way $\overline{N}_V^2$ is much smaller than $\overline{N}_V$, and

$$\tilde{\overline{N}}_V \neq \overline{N}_V \quad \text{when} \quad V \to 0$$

or for standard deviation

$$\sqrt{\tilde{\overline{N}}_V} \neq \sqrt{\overline{N}_V} \quad \text{when} \quad V \to 0$$

and

$$\sqrt{\tilde{\overline{N}}_V} \gg \overline{N}_V \quad \text{when} \quad V \to 0.$$

For very small volumes, the mean number of particles is a quantity with a standard deviation greater than itself. This is of course the case for the density of mean number of particles: $\rho(\mathbf{r})$.

So that when one looks at a map of electronic density (such maps are beautiful and stand nicely in papers) it is necessary to interpret it correctly. Such a map represents correctly statistical averages on an ensemble of identically prepared atoms or molecules. But it does not give a correct instantaneous picture of the individual system. An electron moving freely or belonging to another system will not 'see' such a map, unless what it 'sees' is integrated over a long time or if the experiment is renewed a very high number of times. But a proton moving slowly toward the system (in solution for example) will 'see' such a map of the individual system, because it moves slowly enough to feel only the mean electronic charge density and not the fast fluctuations.

B. Loges with Minimal Fluctuations

When the volume V grows, the relative fluctuation $\sqrt{\overline{\overline{N}_V}}/\overline{N}_V$ decreases. When $V \to \infty$ it goes to zero. But one can hope to find finite volumes (loges) where this relative fluctuation will be very small if not zero.

Let us write the spectral decomposition of $\mathbf{N}_V$

$$\mathbf{N}_V = \sum_{\lambda=0}^{N} \lambda \mathbf{R}_\lambda \tag{3.15}$$

where the operators $\mathbf{R}_\lambda$ are spectral projectors. Then

$$\overline{N}_V = \sum_\lambda \lambda \langle \psi | \mathbf{R}_\lambda | \psi \rangle = \sum_\lambda \lambda P_\lambda \tag{3.16}$$

where P_λ is the probability of finding in V, λ particles only:

$$P_\lambda = C_N^\lambda \int d\sigma_1 \dots d\sigma_N \int_V d\mathbf{r}_1 \dots d\mathbf{r}_\lambda \int_{space-V} d\mathbf{r}_{\lambda+1} \dots d\mathbf{r}_N \, \psi^* \psi . \tag{3.17}$$

If $\overline{\overline{N}}_V = 0$, this would imply that ψ is an eigenfunction of $\mathbf{N}_V$ with eigenvalue $\lambda = k$ and $\overline{N}_V = k$ (for in this case $P_k = 1$ and $P_{\lambda \neq k} = 0$).

In the beginning, Daudel [78] defined loges where one of the P_λ was maximum. We see that loges with minimum fluctuation are identical with Daudel's loges if the minimum is zero, but different in general. The following example shows clearly the difference between the two criteria; we can find two situations where some event is dominant with the same given probability, where the average values are the same but the dispersions are different.

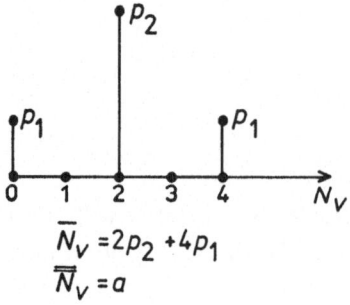

$$\overline{N}_V = 2p_2 + 4p_1$$
$$\overline{\overline{N}}_V = a$$

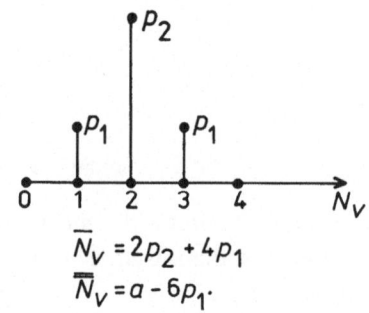

$$\overline{N}_V = 2p_2 + 4p_1$$
$$\overline{\overline{N}}_V = a - 6p_1 .$$

The criterion of minimal fluctuation seems to be more correct to characterize the statistical certitude we have on the number of particles in the considered volume. Moreover it is easier to handle it. In fact Daudel uses the statistical entropy criterion. One can minimize the 'missing information' or statistical entropy of the random variable number of particles in the loge:

$$I = -\sum_{\lambda} P_{\lambda} \log P_{\lambda} \tag{3.18}$$

which is zero only when one of the P_{λ} is 1 and all the others zero. The entropy is maximum when all the P_{λ} are equal. Minimizing the entropy one finds a situation where one of the events is dominant, but the certainty got in this way is different from that obtained by minimizing the fluctuation.

Entropy and dispersion are two distinct characteristics of a random variable which must not be used indistinctly one instead of the other, even if they can give approximately the same information in the neighbourhood of their respective minima (if they are small!). In the above example the entropies were the same and the dispersions different.

Entropy and dispersion are linked by an inequality:

$$I \leqslant 0,5 \log \left(2\pi e \sqrt{\overline{\overline{N}}_V}\right) \tag{3.19}$$

and the equality occurs only if the random variable has a gaussian probability distribution.

An analogous use of entropy to characterize uncertainty and information in quantum mechanics can be found [79]. For example the uncertainty in the position of a particle can be defined as:

$$I_x = -\int_{-\infty}^{+\infty} \psi^*(x)\,\psi(x) \log \psi^*(x)\,\psi(x)\,dx \tag{3.20}$$

and the same for impulsion $(\psi(x) \to \Phi(p))$.

The uncertainty inequalities of Heisenberg can be expressed in entropic terms [80]:

$$I_x + I_p \geqslant C. \tag{3.21}$$

Entropic criteria can be used as criteria for localization of molecular orbitals [81] instead of the dispersion criteria used by Boys [82].

C. On the Best Partition of Space in Loges with Minimal Fluctuations

Given a partition of space in r non overlapping volumes $V_1 \ldots V_r$, and all the corresponding partitions of a total number N of particles: $\{\alpha_1 \ldots \alpha_r\}$ with $\alpha_i = 0,1 \ldots N$ and $\sum_i \alpha_i = N$.

For each volume we define a random variable $\mathcal{N}_i$, number of particles, with values $\alpha_i = 0,1 \ldots N$ and the corresponding probabilities $p_{\alpha_i}^i$.

One can then define a vector random variable

$$\mathcal{N} = \begin{bmatrix} \mathcal{N}_1 \\ \vdots \\ \mathcal{N}_r \end{bmatrix} \tag{3.22}$$

with values

$$\begin{bmatrix} \alpha_1 \\ \vdots \\ \alpha_r \end{bmatrix}$$

and probabilities $p_{\sigma_1 \dots \alpha_r}$. We have of course

$$p_{\alpha_i}^i = \sum_{\alpha_1 \dots \alpha_{i-1} \alpha_{i+1} \dots \alpha_r} p_{\alpha_1 \dots \alpha_i \dots \alpha_r} \tag{3.23}$$

$$\bar{\mathcal{N}} = \sum_{\alpha_1 \dots \alpha_r} \begin{bmatrix} \alpha_1 \\ \vdots \\ \alpha_r \end{bmatrix} p_{\alpha_1 \dots \alpha_r} = \begin{bmatrix} \vdots \\ \sum_{\alpha_1 \dots \alpha_r} \alpha_i \, p_{\alpha_1 \dots \alpha_r} \\ \vdots \end{bmatrix} = \begin{bmatrix} \vdots \\ \sum_{\alpha_i} \alpha_i \, p_{\alpha_i}^i \\ \vdots \end{bmatrix}. \tag{3.24}$$

The dispersion $\bar{\bar{\mathcal{N}}}$ is a second order tensor, the diagonal terms of which are

$$\sum_{\alpha_1 \dots \alpha_r} \alpha_i^2 \, p_{\alpha_1 \dots \alpha_r} - \left(\sum_{\alpha_i} \alpha_i \, p_{\alpha_i}^i \right)^2$$

that is

$$\sum_{\alpha_i} \alpha_i^2 \, p_{\alpha_i}^i - \left(\sum_{\alpha_i} \alpha_i \, p_{\alpha_i}^i \right)^2 = \bar{\bar{\mathcal{N}}}_i. \tag{3.25}$$

The dispersion of $\mathcal{N}$ can be measured by the trace of the tensor, that is $\sum_i \bar{\bar{\mathcal{N}}}_i$. It is therefore sufficient to study separately the fluctuation of the number of particles in each volume. The criterion for a partition in good loges will be the minimization of $\sum_i \bar{\bar{\mathcal{N}}}_i$.

Aslangul [83] and Daudel [84] have adopted in that case also the entropic criteria, that is minimization of the entropy of:

$$I_{\mathcal{N}} = - \sum_{\alpha_1} \dots \sum_{\alpha_r} p_{\alpha_1 \dots \alpha_r} \log p_{\alpha_1 \dots \alpha_r}. \tag{3.26}$$

The calculation of $I_{\mathcal{N}}$ requires the knowledge of the total density matrix whereas the calculation of $\sum_i \bar{\bar{\mathcal{N}}}_i$ requires only the diagonal part of the two particle reduced density matrix.

D. *Fluctuations of the Number of Particles and Delocalization*

Bader and Stephens have systematically used the fluctuation criteria to find partitioning of molecular systems in good loges [85]. For all the molecules studied a clear separation of spherical core pairs from valence loges was accomplished. For several molecules (BeH, BH, CH_4) unambiguous partitionings of the valence region into volumes containing well localized pair populations were found. For several other molecules (LiH, NH_3, H_2O, HF, F_2, N_2) partitioning of the valence region into regions containing localized subgroups was found not possible. This clearly demonstrates the existence of a delocalization phenomena even in so small and simple molecules. A more striking example is given by benzene. If we consider only the π electrons we find that for a loge approximately situated on a bond C—C, the mean number of electrons is one, whereas the standard deviation is not far from one. If we extend the loge, the fluctuation will never completely fall down before the loge encompasses all the aromatic ring.

The fluctuations of the number of particles is a very good tool for studying the delocalization problem.

3.5. ELECTRONIC CORRELATION, DELOCALIZATION AND FLUCTUATIONS

A. *Statistical Correlation of Motion*

In a conception where each electron is a localized and discernable particle, with a trajectory which is described by quantum mechanics in a statistical way only, the interaction between electrons is responsible for a correlation between their motion, in a very classical way. This most simple idea of electronic correlation has not been really considered in quantum chemistry till the paper by Kutzelnigg Del Re and Berthier [74]: 'Correlation coefficients for electronic wave functions'.

Quantum mechanics provides us with a mathematical expression for the probability density of finding electron 1 at $\mathbf{r}$, or finding simultaneously electrons 1 and 2 at $\mathbf{r}_1$ and $\mathbf{r}_2$:

$$\pi_1(\mathbf{r}) = \int d\sigma_1 \ldots d\sigma_N \, d\mathbf{r}_2 \ldots d\mathbf{r}_N \, \psi^* \psi$$

$$\pi_{12}(\mathbf{r}_1 \mathbf{r}_2) = \int d\sigma_1 \ldots d\sigma_N \, d\mathbf{r}_3 \, d\mathbf{r}_N \, \psi^* \psi. \qquad (3.27)$$

In a Statistical Interpretation of quantum mechanics these calculated values have in general no experimental counterpart, because going from one system to the other in the Statistical ensemble prevents from any definite labelling of the electrons.

When we adopt the Symmetrization Postulate these density probabilities are the same for all electrons and can be obtained from experimental data of the Statistical ensemble. Namely:

$$\pi_1(\mathbf{r}) = \frac{1}{N} \rho(\mathbf{r})$$

$$\pi_{12}(\mathbf{r}_1 \mathbf{r}_2) = \frac{2}{N(N-1)} \rho_2(\mathbf{r}_1, \mathbf{r}_2) \qquad (3.28)$$

where, as we have seen in [3.3], $\rho(\mathbf{r})$ and $\rho_2(\mathbf{r}_1 \mathbf{r}_2)$ are the densities for the mean values of the number of electrons and the number of pairs.

We see that the Symmetrization postulate implies an ergodic hypothesis (technically even more: a mixing hypothesis [86]!) namely, on a sufficiently long interval of time all electrons exhibit statistically equivalent behaviour.

In fact it is only by observation of the time evolution of an individual system that $\pi_1(\mathbf{r})$ and $\pi_{12}(\mathbf{r}_1 \mathbf{r}_2)$ can be directly checked in an experiment. It is only under the light of such an ergodic and mixing hypothesis that the considerations developed by Kutzelnigg, Del Re and Berthier take their full value. Otherwise, conceptual difficulties appear, for two systems (e.g. two hydrogen atoms) at large distances for example, where it is difficult to say in quantum mechanics when the Symmetrization postulate is valid or not. This problem of Symmetrization for identical particles needs a full discussion which would necessitate a paper of its own. Let us mention some references where the reader can find elements for this discussion: [128]. We agree with many but not all the points of view expressed in these papers. With all that clear in mind, it is possible to define:

the average position

$$\langle \mathbf{r}_1 \rangle = \int \mathbf{r}_1 \, \pi_1(\mathbf{r}_1) \, d\mathbf{r}_1 \tag{3.29}$$

the dispersion of position

$$\langle\!\langle \mathbf{r}_1 \rangle\!\rangle = \int (\mathbf{r}_1 - \langle \mathbf{r}_1 \rangle)^2 \, \pi_1(\mathbf{r}_1) \, d\mathbf{r}_1 \tag{3.30}$$

the covariance of $\mathbf{r}_1$ and $\mathbf{r}_2$:

$$\mathrm{cov}(\mathbf{r}_1, \mathbf{r}_2) = \int (\mathbf{r}_1 - \langle \mathbf{r}_1 \rangle)(\mathbf{r}_2 - \langle \mathbf{r}_2 \rangle) \, \pi_{12}(\mathbf{r}_1, \mathbf{r}_2) \, d\mathbf{r}_1 \, d\mathbf{r}_2 \tag{3.31}$$

and the correlation coefficient

$$\eta = \frac{\mathrm{cov}(\mathbf{r}_1 \mathbf{r}_2)}{\sqrt{\langle\!\langle \mathbf{r}_1 \rangle\!\rangle \, \langle\!\langle \mathbf{r}_2 \rangle\!\rangle}}. \tag{3.32}$$

By definition: $-1 \leqslant \eta \leqslant 1$. A negative correlation ($\eta < 0$) means that during motion electrons avoid each other. This is what is obtained for ground states of atoms and molecules. Correlation is in general small; for example:

$$\eta(\mathrm{He}^1 S) = -0.0542 \qquad \eta(\mathrm{H}_2 \, ^1\textstyle\sum_g^+) = -0.1002.$$

But according to a fundamental fact of probability theory small correlation does not mean independance! For excited states the correlation coefficient can be positive. As interesting as it may be, this characterization of electronic correlation has the drawback of being global; the whole motion in the atomic or molecular system is concerned. We should like to have local characterizations in order to be able to describe for example the correlation between the motion in two different regions of space (two bonds, two functional groups).

One solution is to use the preceding definitions and limit the integration to definite regions of space for the different electrons.

We can also define the correlation between the random variables, number of particles in volumes V and V', as in Statistical Mechanics [87].

For V and V', non overlapping volumes, it is possible to define: local covariance:

$$C_{VV'} = \langle \psi | \mathbf{N}_V \mathbf{N}_{V'} | \psi \rangle = 2 \int_V d\mathbf{r}_1 \int_{V'} d\mathbf{r}_2 \, \rho_2(\mathbf{r}_1, \mathbf{r}_2) \qquad (3.33)$$

covariance of fluctuations:

$$R_{VV'} = \langle \psi | (\mathbf{N}_V - \bar{N}_V)(\mathbf{N}_{V'} - \bar{N}_{V'}) | \psi \rangle$$

$$= 2 \int_V d\mathbf{r}_1 \int_{V'} d\mathbf{r}_2 \{ \rho_2(r_1 r_2) - \tfrac{1}{2} \rho(r_1) \rho(r_2) \}$$

$$= \int_V d\mathbf{r}_1 \int_{V'} d\mathbf{r}_2 \, f(\mathbf{r}_1 \mathbf{r}_2) \qquad (3.34)$$

where $f(\mathbf{r}_1 \mathbf{r}_2)$ is the correlation factor as defined by Mc Weeny [90]*.

Retov structure function [88]:

$$\langle \psi | (\mathbf{N}_V - \bar{N}_V - \mathbf{N}_{V'} + \bar{N}_{V'})^2 | \psi \rangle = \bar{N}_V + \bar{N}_{V'} - 2 R_{VV'} \qquad (3.35)$$

and the correlation coefficient:

$$\eta_{VV'} = \frac{R_{VV'}}{\sqrt{\bar{N}_V \bar{N}_{V'}}}. \qquad (3.36)$$

For a Slater determinant:

$$R_{VV'} = -2 \sum_i^{occ} \sum_j^{occ} S_{ij}^V S_{ij}^{V'}. \qquad (3.37)$$

B. *Configuration Interaction (Correlation and Fluctuations)*

Statistical correlation between electrons is usually taken in account when calculating the wave function by the technique of configuration interaction (CI) using eventually a perturbation development.

Let us study what happens in a bond loge (or a diatomic molecule) with two electrons. We define this loge as the volume enclosing the major part of a bond localized orbital φ_r (diatomic orbital):

$$\varphi_s = c_1 \chi_1 + c_2 \chi_2$$

where χ_1 and χ_2 are some hybridized atomic orbitals. To φ_s corresponds an antibonding orbital

$$\varphi_A = d_1 \chi_1 - d_2 \chi_2.$$

To simplify we take $c_1 = c_2$ and $d_1 = d_2$ with common value $1/\sqrt{2}$.

---

* Note that the formulas are established for $V \cap V' = 0$. If $V' = V : R_{VV'} \to \bar{\bar{N}}_V = R_{VV} + \bar{N}_V$ where R_{VV} is calculated with the above formulas.

We choose as zeroth-order approximation, the determinant

$$\psi_0 = |\varphi_s \bar{\varphi}_s|$$
$$= \tfrac{1}{2}\{|\chi_1 \bar{\chi}_1| + |\chi_2 \bar{\chi}_2| + |\chi_1 \bar{\chi}_2| + |\bar{\chi}_1 \chi_2|\}.$$

The major part of the correlation comes from the interaction between ψ_0 and the doubly excited determinant

$$\psi^* = |\varphi_A \bar{\varphi}_A|$$
$$= \tfrac{1}{2}\{|\chi_1 \bar{\chi}_1| + |\chi_2 \bar{\chi}_2| - |\chi_1 \bar{\chi}_2| - |\bar{\chi}_1 \chi_2|\}$$

and the correlated wave function is given approximatively by perturbation as

$$\psi = \psi_0 - \frac{\langle \varphi_s \varphi_A | \varphi_A \varphi_s \rangle}{\Delta E} \psi^*$$

where $\langle \varphi_s \varphi_A | \varphi_A \varphi_s \rangle$ is an exchange integral always positive and where ΔE is a diexcitation energy which is positive, so that

$$\psi = \psi_0 - \beta \psi^* \quad \text{with} \quad \beta > 0$$
$$= \tfrac{1}{2}\{(1-\beta)\,[|\chi_1 \bar{\chi}_1| + |\chi_2 \bar{\chi}_2|] + (1+\beta)\,[|\chi_1 \bar{\chi}_2| + |\bar{\chi}_1 \chi_2|]\}.$$

The four determinants built with atomic orbitals represent the following events in the loge divided in two equal subloges: ($\bullet$ nuclei, $\times$ $\circ$ electrons):

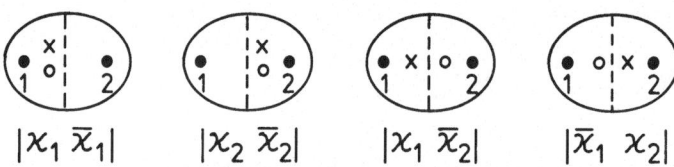

$$|\chi_1 \bar{\chi}_1| \qquad |\chi_2 \bar{\chi}_2| \qquad |\chi_1 \bar{\chi}_2| \qquad |\bar{\chi}_1 \chi_2|$$

In the zeroth order function ψ_0 these four situations have the same weight, so that each subloge has a mean number of electrons equal to 1 with a standard deviation $\sim 0{,}7$.

In the correlated function the weight of the ionic situations diminishes, the weight of the valence situations grows. Correlation takes electrons apart, and lowers the fluctuations in the half-bond loges.

C. *Natural Fluctuation Orbitals*

At last we are going to consider the spectral analysis of local covariance and the Karhunen decomposition of the random field ([98], V §4) ([97], ch. 13) of 'fine grained' density of particles. This random field is defined as the set of random variables: number of particles in a small standard sphere of volume v centered in all the different points of space. If v is sufficiently small in order that $\rho_2(\mathbf{r}_1 \mathbf{r}_2)$ be nearly constant when $\mathbf{r}_1$ and $\mathbf{r}_2$ vary in the range of v, we can write the covariance for two distinct volumes at $\mathbf{r}_1$ and $\mathbf{r}_2$ as:

$$C_{v_1 v_2} = C(\mathbf{r}_1 \mathbf{r}_2) = 2v^2 \, \rho_2(\mathbf{r}_1 \mathbf{r}_2).$$

This is the covariance of the random field: 'fine grained' density of particles $\mathcal{N}(\mathbf{r})$. The spectral analysis of this covariance leads to an interesting representation of the field.

$$C(\mathbf{r}_1 \mathbf{r}_2) = \sum_{i=1}^{\infty} \lambda_i \, g_i(\mathbf{r}_1) \, g_i(\mathbf{r}_2)$$

where $q_i(\mathbf{r})$ and λ_i are the eigenfunctions and eigenvalues of the integral operator with kernel $C(\mathbf{r}_1 \mathbf{r}_2)$ (or $\rho_2(\mathbf{r}_1 \mathbf{r}_2)$, if we divide the eigenvalues by $2v^2$).

We have then the following decomposition of the field:

$$\mathcal{N}(\mathbf{r}) = \overline{\mathcal{N}}(\mathbf{r}) + \sum_{i=2}^{\infty} \xi_i \, g_i(\mathbf{r})$$

where ξ_i are independant random variables, with zero mean value ($\bar{\xi}_i = 0$) and dispersion $\bar{\bar{\xi}}_i = \xi_i^2 = \lambda_i$. So that:

$$\mathcal{N}(\mathbf{r}) = \overline{\mathcal{N}}(\mathbf{r}) + \sum_{i=1} \xi_i \, g_i(\mathbf{r})$$

$$\overline{\overline{\mathcal{N}}}(\mathbf{r}) = \sum_{i=1}^{\infty} \bar{\bar{\xi}}_i \, g_i^2(\mathbf{r}).$$

The dispersion of the number of particles in a small volume v around r can be expressed as a sum of additive contributions from individual 'orbitals' $g_i(r)$. Natural orbitals, eigenfunctions of $\rho_1(\mathbf{r}, \mathbf{r}')$, allowed us to express $\mathcal{N}(\mathbf{r})$ in an additive way. We shall therefore call the eigenfunctions of $\rho_2(\mathbf{r}_1 \mathbf{r}_2)$: 'natural fluctuation orbitals'. These orbitals, defined from the two particle reduced density matrix, have been already considered in [89], but their probabilistic meaning was not made clear.

4. Stochastic Models for Microphysics

In Section 2 we have seen that if we adopt a purely statistical interpretation of quantum mechanics, any description of the individual system will be excluded, unless we make an implicit ergodic hypothesis. In this chapter we are going to show that it is possible to produce pictures of the individual behaviour of a microsystem, presenting convenient ergodic properties and in some agreement with the statistical behaviour given by quantum mechanics. It is actually possible to build random processes associated to the different observables, reproducing ergodically the main statistical predictions of quantum mechanics. One must always bear in mind that these processes are a priori only mathematical models, and that their eventual relationship with the physical reality must be carefully studied.

4.1. A MARKOV PROCESS FOR THE POSITION. THE FENYES-NELSON THEORY

In 1952, Fenyes [91] established a precise relationship between the theory of markov processes and quantum mechanics. But it seems that his work has not been fully appreciated and it was not before 1966 that the results of Fenyes were rediscovered independently by Nelson, a probabilist, interested in quantum theory [92] [93]. At the same time, Della Riccia (and Wiener) [94] and Favella [95] obtained similar results. Following them, La Pena Auerbach and coworkers have studied with some further details the markovian model [96].

A. *Markov Processes*

Let us recall briefly some fundamental points about the theory of markov processes [97] [98] [99].

A markov process is a random process whose future depends only on the state at the present time, that is the future evolution does not depend on the past. This independance between the past and the future, which is an oversimplification from a physical point of view, gives the simplest model of dependence of random behaviour through time. This is clearly expressed, using conditional probabilities, as:

$$P\{x(t)\in B|x(t_1) \ldots x(t_n)\} = P\{x(t)\in B|x(t_n)\}$$

where: $t_1 < t_2 \ldots < t_n < t$ whatever they be, and $x(t)$ is a random process, that is a family of random variables $x(t, \omega)$ given on a probability space (Ω, B, P). In the following we shall use a simple (non axiomatic) probabilistic language. If $t_0 < t$, the function of $x(t_0)$:

$$P\{x(t)\in B|x(t_0)\} = P_2(B, t|x_0, t_0)$$

is called: the transition probability of the markov process. We shall suppose that a transition probability density exists:

$$p_2(x, t|x_0, t_0) \geqslant 0$$

$$\int_{-\infty}^{+\infty} p_2(x, t|x_0, t_0) \, dx = 1.$$

From the definition of a conditional probability density

$$p_2(x, t; x_0, t_0) = p_1(x_0, t_0) \, p_2(x, t|x_0, t_0) \tag{4.1}$$

where $p_2(x, t; x_0, t_0)$ is the joint probability density of having the values x_0 and x at the times t_0 and t, and $p_1(x_0 t_0)$ is the probability density for the value x_0 at time t_0. The transition probability of the Markov process defines completely this process, once the initial probability $p_1(x_0, t_0)$ is given. As the kernel of a transition operator, it leads, by derivation, to the forward *Fokker-Planck-Kolmogorov (FPK) equation*, which governs the evolution of every probability linked to the process: $p_2(x, t|x_0, t_0)$, $p_1(x, t)$. We shall restrict our description to the category of *homogenous diffusion processes* for which $p_2(x, t/x_0, t_0)$ depends only on $\tau = t - t_0$, and the following quantities are well defined:

$$a(t, x_0, t_0) = E\{x(t)|x(t_0) = x_0\} = \int x \, p_2(x, t|x_0, t_0) \, dx$$

$$b(t, x_0, t_0) = E\{[x(t) - a(t, x_0, t_0)]^2 | x(t_0) = x_0\} \tag{4.2}$$

$$= \int (x - a)^2 \, p_2(x, t | x_0, t_0) \, dx$$

$$c(x_0, t_0) = \left[\frac{\partial a(t, x_0, t_0)}{\partial t}\right]_{t = t_0}$$

$$d(x_0, t_0) = \left[\frac{\partial b(t, x_0, t_0)}{\partial t}\right]_{t = t_0}. \tag{4.3}$$

The coefficients $c(x_0, t_0)$ and $d(x_0, t_0)$ are respectively called: *the drift coefficient or velocity*, and the *diffusion coefficient*. They determine completely the FPK equations, for example the forward FPK equation

$$\frac{\partial p(x, t)}{\partial t} = -\frac{\partial}{\partial x}\left[c(x, t)\, p(x, t)\right] + \frac{1}{2}\frac{\partial^2}{\partial x^2}\left[d(x, t)\, p(x, t)\right]. \tag{4.4}$$

For an homogeneous process, c and d do not depend on time, and

$$p(x) = \frac{G}{d(x)}\, e^{2\int_{x_0}^{x}(c(y)/d(y))\,dy}. \tag{4.5}$$

is an invariant probability, that is a probability distribution kept invariant during time evolution of the process. (G is a normalisation constant). A process with the invariant probability as initial probability is a stationary process. A stationary diffusion processs is ergodic.

B. *Stationary Quantum States and Stationary Diffusion processes*

For sake of simplicity we shall explain the Fenyes-Nelson conception in the case of stationary quantum states. The presentation is somewhat different from those given in references [91, 92, 93, 96] in order to stress the main features of the theory. The Schrödinger equation will not be derived from 'pseudodynamical equations' or from more abstract conditions (as was done recently by Caubet [100]), but along the way followed initially by Schrödinger himself, using a variational principle (as done by La Pena-Auerbach, Velasco and Cetto [101]). This derivation will be postponed till Section 4.

Let us consider a diffusion stationary process in R^3 (generalization to R^n is easy) with a drift velocity $\mathbf{c}(\mathbf{r})$ and a constant diffusion coefficient $d(\mathbf{r}) = D$. If the process is invariant with respect to time-reversal ([93] p. 103), the invariant probability density $p(\mathbf{r})$ of the process is linked to the drift velocity by*:

$$\mathbf{c}(\mathbf{r}) = \frac{D}{2}\, \text{grad}\, \log\, p(\mathbf{r}) = \frac{D}{2}\frac{\text{grad}\, p(\mathbf{r})}{p(\mathbf{r})} \tag{4.6}$$

* For a given Markov process (stationary or not), it is always possible to consider the time-reversed process, which is also Markovian ([93] section 13, p. 103). The time-reversed process has a drift velocity $\tilde{\mathbf{c}}$ (see [96] and a diffusion cœfficient $\tilde{D} = D$. We can define the stochastic velocity $\mathbf{u} = (\mathbf{c} + \tilde{\mathbf{c}})/2$ and the current (or systematic)velocity $\mathbf{v} = (\mathbf{c} - \tilde{\mathbf{c}})/2$. Nelson [93] shows that, quite generally, $\mathbf{u} = D/2$

which, in the one-dimensional case, may be obtained directly by differentiation of Equation (4.5). One can choose as $p(\mathbf{r})$, $|\psi(r, t)|^2$ where $\psi(\mathbf{r}, t)$ is a stationary solution of the Schrödinger equation of a microphysical system. This probability is achieved by the process in an ergodic way. Nelson [93] and La Pena-Auerbach [96] show that a zero current velocity (see the previous footnote) corresponds to a real ψ, which is the most frequent situation in quantum mechanics.

The idea of the Fenyes-Nelson theory is to define a kinematics of the motion in the diffusion process, although the trajectories of a markov process are non differentiable functions, in contradistinction to every physical trajectory. This is done by considering the drift velocity as a true velocity (in a phenomenological way), and requiring that the mean kinetic energy associated to the drift velocity should be equal to mean kinetic energy in the quantum formalism:

$$\tfrac{1}{2} m \overline{c^2} = - \frac{\hbar^2}{2m} \langle \psi | \Delta | \psi \rangle$$

$$\frac{D^2 m}{4} \iiint \frac{\operatorname{grad} |\psi(\mathbf{r}, t)|^2}{|\psi(\mathbf{r}, t)|^2} |\psi(\mathbf{r}, t)|^2 = - \frac{\hbar^2}{m} \iiint \psi \, \Delta \psi \; \mathrm{d}\mathbf{r}$$

so that, if $\psi(\mathbf{r}, t) = \psi(\mathbf{r}) e^{iEt}$, with $\varphi(\mathbf{r})$ a real function,

$$D^2 m \iiint (\operatorname{grad} \varphi(\mathbf{r}))^2 \; \mathrm{d}\mathbf{r} = - \frac{\hbar^2}{m} \iiint \varphi \, \Delta \varphi \; \mathrm{d}\mathbf{r}$$

which is necessarily satisfied, due to Green's theorem, if:

$$D = \frac{\hbar}{m}. \tag{4.7}$$

As de Broglie remarked [102]: 'Une expression de cette forme pouvait être prévue pour des raisons de dimensions dans une théorie non relativiste où l'on ne dispose que des deux constantes physiques m et h'. So that, postulating $D = \hbar/m$ gives the right quantum expression for the mean kinetic energy.

A stationary diffusion markov process with an invariant probability $|\psi(\mathbf{r}, t)|^2$ (where $\psi(\mathbf{r}, t)$ is a stationary solution of the Schrödinger equation), and with a Fokker-Planck Kolmogorov equation determined by the coefficients c and D in (4.6) and (4.7), is an ergodic random process for the position of the particle, giving for that observable

---

grad log p, while $\mathbf{v}$ satisfies the continuity equation $\partial p / \partial t = \operatorname{div}(p\mathbf{v})$ (where $p(\mathbf{r}, t)$ denotes the 'invariant' probability of the process, a concept which may be defined even for a non homogeneous process). For a stationary process, $\partial p / \partial t = 0$ thus div $(p\mathbf{v}) = 0$, i.e. in the 3-dimensional case $p\mathbf{v}$ must be the curl of some vector field, a very important particular case being $\mathbf{v} = 0$ (and therefore $\mathbf{c} = \tilde{\mathbf{c}} = \mathbf{u}$) which means that the process is time-reversal invariant (this is the only case considered in the present paper). In the one-dimensional case, div $(p\mathbf{v}) = 0$ becomes $p\, v = c^{ste}$, which imposes $v = 0$ if $\displaystyle\int_{-\infty}^{+\infty} p(r)$

dr is finite (because then $p(r) \to 0$ as $r \to \pm \infty$): therefore time-reversal invariance necessarily holds in that case.

the same probabilistic characteristics as quantum mechanics, but adding to them the picture of an actual space-time evolution in the stationary state. The Heisenberg uncertainty inequalities can be demonstrated for the dispersion of the components of $\mathbf{r}$ and $\mathbf{c}$, giving to these inequalities a clear statistical meaning. They appear as completely analogous to the inequality found by Fürth [103] for the brownian motion. But we must again stress the fact that the markovian process introduced here is a simulation model and that it must not be considered as an actual physical process. We have already seen that a markov trajectory, on the contrary to a physical trajectory, is not derivable. A markov process can at most give a good picture of a physical process, if considered on a convenient (not too short) time scale. On the other hand the position diffusion process reproduces only the mean and the dispersion of the velocity as given by quantum mechanics. But, in view of what has been said on the simulatory character of the process, this is not to be considered as a difference (between quantum mechanics and stochastic theory) which should be tested experimentally.

C. Ergodicity and the Correlation Function of the Fenyes-Nelson Process

A fundamental characteristic of a random process is the time correlation function. Let us study this function for any observable, function of the position: $F(\mathbf{r})$, in the diffusion markov process simulating the ground state of the microphysical system.

By definition, the time correlation function is:

$$\mathscr{B}_{F(\mathbf{r})}(t_0, t) = \iint F(\mathbf{r}) F(\mathbf{r}_0) p_2(\mathbf{r}, t; \mathbf{r}_0, t_0) \, d\mathbf{r} \, d\mathbf{r}_0. \tag{4.8}$$

The solution of the FPK equation of the process, with $p(\mathbf{r}_0) = |\psi_0|^2$, and

$$\mathbf{H}\psi_0 = E_0 \psi_0 \tag{4.9}$$

where $\mathbf{H}$ is the hamiltonian operator of the system, gives the expression of $p_2(\mathbf{r}t; \mathbf{r}_0, t_0)$. The final result obtained is [114]:

$$\mathscr{B}_{F(\mathbf{r})}(t_0, t) = \langle \psi_0 | F(\mathbf{r}) \, e^{+(E_0 - \mathbf{H}) \, |t - t_0|} \, F(\mathbf{r}) | \psi_0 \rangle. \tag{4.10}$$

This correlation function can be *formally* derived from the quantum correlation function $\langle \psi_0 | F_H(\mathbf{r}_0, t_0) F_H(\mathbf{r}t) | \psi_0 \rangle$ where $F_H(\mathbf{r}, t) = e^{-i\mathbf{H}t/\hbar} F(\mathbf{r}) \, e^{i\mathbf{H}t/\hbar}$ is the Heisenberg representation of the observable $F(\mathbf{r})$ (cf. 4.3). One has only to change time to imaginary time: $t \to it'$ [95]. It is well known in the theory of stationary random processes [97] [98], that the correlation function gives information on the behavior of the time average of the observable defined on a individual trajectory as

$$m_x(T) = \frac{1}{T} \int_0^T x(t) \, dt \tag{4.11}$$

$m_x(T)$ is a random variable, the ensemble average of which, $\overline{m_x(T)}$, is equal to the stationary average value $\bar{x}$. The dispersion of $m_x(T)$: $D_x(T)$ is equal to

$$D_x(T) = \overline{[m_x(T) - \overline{m_x(T)}]^2} = \frac{1}{T^2} \int_0^T \int_0^T (\mathscr{B}_x(t, t') - \bar{x}^2) \, dt \, dt' \qquad (4.12)$$

or, as $\mathscr{B}_x(t, t') = \mathscr{B}_x(t' - t)$, changing t and t' to $s = t + t'$ and $\tau = t' - t$, one can obtain

$$D_x(T) = \frac{1}{T^2} \int_{-T}^{+T} (T - |\tau|) \left[\mathscr{B}_x(\tau) - \bar{x}^2 \right] d\tau$$

$$= \frac{2}{T^2} \int_0^T (T - \tau) \left[\mathscr{B}_x(\tau) - \bar{x}^2 \right] d\tau. \qquad (4.13)$$

In our case

$$\mathscr{B}_{F(\mathbf{r})}(\tau) = \sum_{n=0}^{\infty} |\langle \psi_0 | F(\mathbf{r}) | \psi_n \rangle|^2 \, e^{-\omega_{on}|\tau|} + \int_{E_c}^{+\infty} e^{-\omega|\tau|} \, \mu_{[F(\mathbf{r}), \psi_0]}(E) \, dE \qquad (4.14)$$

where the ψ_j are the eigenfunctions of $\mathbf{H}$ corresponding to the eigenvalues E_j and

$$\omega_{on} = \frac{E_n - E_o}{h} \qquad \omega = \frac{E - E_0}{h}. \qquad (4.15)$$

E_c is the beginning of the continuous spectrum and $\mu_{[F(\mathbf{r}), \psi_0]}(E) \, dE$ is the measure corresponding to this spectrum expressed with a density.
Then

$$D_{F(\mathbf{r})}(T) = \frac{2}{T^2} \sum_{n=1}^{\infty} |\langle \psi_0 | F(\mathbf{r}) | \psi_n \rangle|^2 \int_0^T (T - \tau) \, e^{-\omega_{on}\tau} \, d\tau +$$

$$+ \text{ continuous spectrum term}.$$

Taking account of

$$\int_0^T (T - \tau) \, e^{-\omega\tau} \, d\tau = \frac{T}{\omega} + \frac{e^{-\omega T}}{\omega^2} - \frac{1}{\omega^2} \leqslant \frac{T}{\omega}.$$

One is led to

$$D_{F(\mathbf{r})}(T) \leqslant \frac{2}{\omega_{01} T} \left\{ \sum_{n=1}^{\infty} |\langle \psi_0 | F(\mathbf{r}) | \psi_n \rangle|^2 + \text{continuous spectrum term} \right\}$$

and

$$D_{F(\mathbf{r})} \leqslant \frac{2\delta_0 F(\mathbf{r})}{\omega_{01} T} \qquad (4.16)$$

where $\delta_0 F(\mathbf{r}) = \langle \psi_0 | F^2(\mathbf{r}) | \psi_0 \rangle - \langle \psi_0 | F(\mathbf{r}) | \psi_0 \rangle^2$ is the dispersion of the observable

$F(\mathbf{r})$ in the state ψ_0 and also the dispersion of the function $F(\mathbf{r})$ of the position in the Fenyes-Nelson process.

It is sufficient to have $T \geqslant 2/\omega_{01}\varepsilon^2$ in order to ensure $\sqrt{D_{F(\mathbf{r})}(T)/\delta_0 F(\mathbf{r})} < \varepsilon$.

As an example, for the hydrogen atom

$$\omega_{01}^{-1} = 0.65 \times 10^{-16} \text{ s}.$$

$$\varepsilon = 10^{-3} \quad T > 1.3 \times 10^{-10} \text{ s}.$$

$$\varepsilon = 10^{-6} \quad T > 1.3 \times 10^{-4} \text{ s}.$$

The ergodicity is achieved more slowly than in the case of the parastochastic process which will be studied in 4.3.a.

To know if these ergodicity times are physically reasonable it is necessary to study the response of the system with an apparatus having a definite time characteristic. If this response time of the apparatus is smaller than the ergodicity time, we shall detect the fluctuations; if this time is greater than the ergodicity time, only the mean value will be observed.

Nevertheless it is possible to criticize physically the Fenyes-Nelson process by considering its correlation function (4.14). Being a sum of real exponential functions it has no strongly preferred frequencies in its Fourier transform. This has for consequence that a sample function of the process has no quasi-periodic character (no preferred frequencies in the Fourier analysis), since a process with quasi periodic sample functions (with the same preferred frequencies) has a quasi periodic correlation function ([97]. Ch. 13, 4) ([129] sect. 4). Now in this classical model, strong sharp response to an external periodic perturbation can be expected only if some kind of resonance occurs. In absence of quasi-periodic character of the motion such resonances will not occur. So that *the Fenyes Nelson process seems unable to account in a physically classical consistent way for the discrete spectra of atoms.*

4.2. Analysis of Schrödinger's First Paper in the Light of the Fenyes-Nelson Theory, or 'If One Could Begin History Again...'

In the light of the preceeding considerations let us read again the first paper communicated by Schrödinger to Annalen der Physik [15].

Motivated by the work of De Broglie and conscious of the optico-mechanical analogy due to Hamilton, he nevertheless follows an apparently difficult to understand heuristic way. He himself confesses in the beginning of his second communication [15], that what he has done is not understandable. This is certainly the reason why the arguments of this first paper are very rarely reproduced in textbooks (exceptions are the early textbook on the new quantum mechanics written by Birtwistle [105] in 1928, which seems completely unknown, and a not well known today book of Sommerfeld [106]). He says himself that he follows a way completely neutral without any a priori idea of vibration. Let us recall briefly the steps of his historical work. The starting point is the Hamilton-Jacobi equation for a time independent hamiltonian

$$H\left(q, \frac{\partial S}{\partial q}\right) = E \tag{4.17}$$

where $p = \partial S/\partial q$ is the impulse expressed in term of action $S(q)$. One must stress here the global character of the Hamilton-Jacobi equation, which describes in fact a whole set of trajectories.

Then Schrödinger makes a change of function, setting

$$S = K \log \psi \quad \psi = e^{S/K} \tag{4.18}$$

where K is a constant, with the dimension of action.

Then

$$H\left(q, \frac{1}{\psi}\frac{\partial \psi}{\partial q}\right) = E. \tag{4.19}$$

But, as he intends to find something different fiom the classical behavior, he is not going to require that $H - E = 0$ along all possible trajectories, which is equivalent to the variational condition:

$$\min \iiint (H - E)^2 \, dv = 0. \tag{4.20}$$

He requires

$$\delta \iiint (H - E) \, \psi^2 \, dv = 0 \tag{4.21}$$

where ψ is varied among functions which are real, finite, univocal and doubly deriva-ble. Correcting the proofs of his paper, he remarks that this is equivalent, if ψ is square integrable, and considering E as a Lagrange multiplier, to:

$$\delta \iiint H\psi^2 \, dv = 0 \tag{4.22}$$

with

$$\iiint \psi^2 \, dv = 1. \tag{4.23}$$

The Euler-Lagrange equation of this variational problem is the time independant Schrödinger equation; the discrete spectrum of values of the Lagrange multiplier ensuring 'good' solutions of this equation automatically gives the quantification of the energy. For the hydrogen atom, the energy levels correspond exactly to the Balmer terms if the constant K is identified with $h/2\pi$.

Neither Schrödinger, nor the different writers of quantum mechanics textbooks, wonder about the meaning of this non classical variational principle. In fact, at the beginning of his second communication [15], Schrödinger writes: 'We have used, on one hand a completely ununderstandable transformation ($S = K \log \psi$), and on the other hand a no more understandable transition from the condition that some expression is equal to zero to the condition that the integral of this expression in the whole space is stationary.'

The unusual features of his variational principle could have suggested some stochastic pictures to Schrödinger. Being formulated as a principle in the whole space and not along a trajectory alone, and using ψ^2 as a weight, it calls for a picture of 'filling' of space by trajectories with a density ψ^2. If each trajectory fills the space, we have an actual ergodic property. Anyway there is a global picture, considering simultaneously all possible trajectories; in the ergodic case all the trajectories are equivalent and every trajectory 'fills' the space with density ψ^2; in the non ergodic case, every trajectory is different from the others, but the ensemble of trajectories 'fills' the space with density ψ^2. These two cases seem to appear in physical situations:

(a) in stable systems (atoms) – bound states – every microsystem gives the density ψ^2.

(b) in collision experiments – unbound states – every particle does not give the total density on the screen (only one spot per particle). The total picture is only had by using a great ensemble of particles.

Under the light of the stochastic simulation introduced by the Fenyes-Nelson theory, one could rewrite the Schrödinger considerations in the following way for the hydrogen atom for example:

(1) as we don't know a priori the trajectory of the electron, we can suppose that it has some probability density $p(\mathbf{r})$ to be found at every point of space, and that this probability is achieved in a ergodic way by every individual atom.

(2) as model for the motion let us consider a stationary markov process with a constant diffusion coefficient D. The stochastic velocity is linked to the stationary probability by the fundamental relationship:

$$\mathbf{c}(\mathbf{r}) = \frac{D}{2} \operatorname{grad} \log p(\mathbf{r}).$$

It is natural to introduce a stochastic impulsion $\operatorname{grad} S = m\mathbf{c}$, that is:

$$S = \frac{D}{2m} \log p(\mathbf{r}) = \frac{D}{m} \log \Psi(\mathbf{r}) \tag{4.24}$$

with

$$p(\mathbf{r}) = \psi^2$$

(the amplitude of probability density appears in a simple way!).

(3) for this model of the motion let us express the energy by the Hamilton function, using the stochastic impulsion as impulsion. $H(q, \operatorname{grad} S)$ is to be considered as the energy density at point $\mathbf{r}$, and the mean energy of the system is:

$$\bar{H} = \iiint Hp(\mathbf{r}) \, dv = \iiint H\psi^2 \, dv. \tag{4.25}$$

(4) write the stability condition as a minimum principle for this mean energy:

$$\delta \iiint H\psi^2 \, dv = 0$$

with the normalization condition for the probability density

$$\iiint \psi^2 \, dv = 1 \, .$$

This is the variational principle of Schrödinger. It expresses the minimization of the mean energy of an electron whose motion is a markov process reproducing in an ergodic way the probability of the actual system.

We see that the cornerstone of this presentation is the relationship between stochastic velocity and probability density, that is between velocity and position. It is precisely here that lies one of the originalities of quantum mechanics: the state of the system is completely described with a function of position (or impulsion alone). But we must stress again that the main hypothesis was to consider as kinetic energy, the classical kinetic energy calculated with the stochastic velocity which is not a true velocity (it is only a velocity on the mean trajectory).

Quantum mechanics is a phenomenological theory. The use of markov processes can allow understanding some aspects of it. But the markov simulation must by no means be taken as a complete realistic description of the physical situation. As a conclusion, recall that Schrödinger and Fürth [107], [103] thought of a stochastic model at the very beginning of quantum mechanics. But they thought that the Schrödinger equation was a Fokker-Planck-Kolmogorov equation. We have seen that the relationship between these two equations is more complex; an example of such a relationship is established in the Fenyes-Nelson theory. *Other kinds of relationships are possible*, e.g. in Stochastic Electrodynamics.

4.3. Non markov processes with Schrödinger time evolution
(para-stochastic processes)

The preceding considerations show that quantum mechanics appears as a phenomenological theory taking account of some peculiar space-time behaviour of microparticles. The markovian model sheds light on some peculiarities of quantum mechanics. We want to present now another model of simulatory stochastic processes, which seems more appropriate for the description on small time intervals, when by its very principle the markovian model becomes inadequate. We shall discover that quantum mechanics uses this model implicitly, so that its status is not very clear.

A. *Definition and Ergodic Properties*

We shall consider that every observable A of a microphysical system has in every individual system values $a(t)$ which vary with time, but that in the statistical ensemble of such systems the mean value and the dispersion are given by the quantum mechanical rule with the associated operator $\mathbf{A}$. In fact we want to define a random process $a(t)$ by giving some information on the one-time probability law and on the time evolution.

A partial specification of time evolution is given by the two-time correlation function. We can choose for such a function the quantum mechanical correlation function:

$$\mathscr{B}_A(t, t') = \text{Re} \langle \psi | \mathbf{A}(t) \, \mathbf{A}(t') | \psi \rangle$$
$$= \tfrac{1}{2} \langle \psi | \mathbf{A}(t) \, \mathbf{A}(t') + \mathbf{A}(t') \, \mathbf{A}(t) | \psi \rangle \tag{4.26}$$

where $\mathbf{A}(t)$ is the 'Heisenberg representation' of the operator $\mathbf{A}$. $\mathscr{B}_A(t, t')$ is the kernel of an hermitian positive integral operator so that it is admissible as correlation function for a random process ([98] p. 9–10 and 27–28).

$\mathscr{B}_A(t, t')$ appears in the quantum mechanical theory of the linear response of a system to an external perturbation. In fact the fluctuation dissipation theorem is a relationship between the Fourier transform of $\mathscr{B}_A(t, t')$ and the imaginary part of the generalized susceptibility, which characterizes the energy absorption [108] [109] [110]. At high temperature this relationship goes into a classical one where the correlation function takes an actual physical meaning. We have of course the compatibility condition

$$\mathscr{B}_A(t, t) = \langle \psi | [\mathbf{A}(t)]^2 | \psi \rangle = \overline{[a(t)]^2}$$

where the bar means the average on the statistical ensemble.

We shall call such a random process, a *para-stochastic process*, a terminology borrowed from Bourret and Frisch (111).

Following the same steps as in 4.1.c one can evaluate the dispersion of the time average for a stationary para-stochastic process $a(t)$ corresponding to a quantum stationary state φ_0. The correlation function can be written as:

$$\mathscr{B}_A(\tau) = \text{Re} \langle \varphi_0 | \mathbf{A} e^{-i(\mathbf{H} - E_0) \tau / \hbar} \, \mathbf{A} | \varphi_0 \rangle \tag{4.27}$$

where $\tau = t' - t$ and $\mathbf{H}\varphi_0 = E_0 \varphi_0$.

We get easily (cf. [104]):

$$D_a(\mathbf{T}) \leqslant \frac{4\delta_0 A}{(\omega_{01} T)^2} \tag{4.28}$$

where $(\delta_0 A) = \langle \varphi_0 | \mathbf{A}^2 | \varphi_0 \rangle - \langle \varphi_0 | \mathbf{A} | \varphi_0 \rangle$.

When $T \to \infty$ (for a non degenerate stationary state) the dispersion of the time average of $a(t)$ goes to zero; that means that the time average becomes equal to the quantum average for infinite time. It is sufficient to have $T \geqslant 2/\omega_{01}\varepsilon$ in order to ensure $\sqrt{D_x(T)/\delta_0 A} < \varepsilon$.

As an example, for the hydrogen atom:

$$\varepsilon = 10^{-6} \qquad T > 1.3 \times 10^{-10} \text{ s}.$$

We see that the ergodicity for the mean value of $a(t)$ is obtained faster with the para-stochastic process than with the Fenyes-Nelson process. This seems more satisfactory from a physical point view.

We could have thought to define a random process $a(t)$, with the same two-time correlation function as previously: (4.26); but with a one-time probability law identical with the quantum mechanical probability distribution. This means that besides the first two moments (average and dispersion) all further moments $[a(t)]^n$ would be equal to their quantum mechanical value $\langle \psi | \mathbf{A}(t)^n | \psi \rangle$.

A simple example shows that this is not possible in general. In fact, consider the harmonic oscillator in its quantum ground state. The probability density for the position is:

$$p(x) = |\varphi_0(x)|^2 = \left(\frac{m\omega_0}{\pi\hbar} \right)^{1/2} e^{-(m\omega_0/\hbar) x^2} \tag{4.29}$$

giving as mean value zero and as dispersion $\langle x^2 \rangle = \hbar/2m\omega_0$.

Let us consider the para-stochastic process $x(t)$. It has the following correlation function, easily obtained from (4.27):

$$\mathscr{B}_x(\tau) = \langle x^2 \rangle \cos \omega_0 \tau. \tag{4.30}$$

This correlation function is an analytic function of τ. Then, according to Belyaev ([112] p. 405, Corollary), almost all sample functions of the process are analytic. More precisely, since $\mathscr{B}_x(\tau)$ is periodic, it can be shown ([97] ch. 13, 1) that the process $x(t)$ can be expanded into a Fourier Series

$$x(t) = \sum_{n=-\infty}^{+\infty} a_n e^{in\omega_0 t} \tag{4.31}$$

where the coefficients a_n are orthogonal random variables $(\overline{a_n a_m^*} = 0$ for $n \neq m)$ with zero mean value and their variance is related to the Fourier expansion

$$\mathscr{B}_x(\tau) = \sum_{n=-\infty}^{+\infty} \alpha_n e^{in\omega_0 \tau} \tag{4.32}$$

by

$$\overline{|a_n|^2} = \alpha_n. \tag{4.33}$$

Since in our case $\alpha_1 = \alpha_{-1} = \langle x^2 \rangle/2$ and all other α_n are zero, we have:

$$x(t) = b \cos \omega_0 t + c \sin \omega_0 t$$

where $b = a_1 + a_{-1}$ and $c = i(a_1 - a_{-1})$ must be real random variables, i.e. $a_{-1} = a_1$, and

$$\overline{b^2} = \overline{c^2} = \langle x^2 \rangle. \tag{4.34}$$

Therefore each sample function (b and c fixed) is a sinusoid:

$$x(t) = \sqrt{b^2 + c^2} \cos(\omega_0 t - \Theta) \tag{4.35}$$

where

$$\cos \Theta = \frac{b}{\sqrt{b^2 + c^2}} \qquad \sin \Theta = \frac{c}{\sqrt{b^2 + c^2}}.$$

For such a given sample function we can define the probability of a value x by a time average (asymptotic measure defined by Bass [113]), i.e., for a periodic motion, the probability density:

$$\mu(x) = \frac{2}{T\,|v(x)|} \tag{4.36}$$

where T is the period and $v(x)$ is the velocity.

For our sinusoid we have the well known distribution:

$$\mu(x) = \frac{1}{\pi\sqrt{(b^2+c^2)-x^2}}. \tag{4.37}$$

Whatever b and c, the mean value zero is actually ergodically realized by every sample sinusoid. But if we want to realize the ergodicity of the process for the dispersion, all sinusoïds (4.35) must have the same amplitude $b^2+c^2(=2\langle x^2\rangle)$ according to (4.34)), and then only the phase Θ remains random. Then the stationary probability density must be chosen identical to $\mu(x)$ if we want that all other moments beyond the second be ergodically realized. Therefore these moments are no more arbitrary, and in particular cannot be the same as those corresponding to the gaussian probability law $p(x)$ (4.29).

At this stage the reader may wonder why the proof for ergodicity of x, which could be worked in the same way for x^n using the correlation function $\mathcal{B}_{x^n}(\tau) = \mathrm{Re}\langle\psi_0|\mathbf{x}^n(t)\,\mathbf{x}^n(t')|\psi\rangle$ seems to fail for $n>3$.

The reason is analogous to that contained in Cohen's theorem [63] quoted in 2.3: it is not possible to find a unique two-time probability distribution $p_2(x', \tau; x, 0)$ such that all quantum 'correlation functions' $\mathcal{B}_{x^n}(\tau)$ could be expressed in the classical form

$$\iint x^n x'^n\, p_2(x',\tau; x, 0)\,\mathrm{d}x\,\mathrm{d}x',$$

although this may be possible for some of them ($n = 1, 2$ in the given example). Mathematically this is connected with the non commutation of the Heisenberg operators $\mathbf{x}(t)$ and $\mathbf{x}(t')$ ($\mathbf{A}(t)$ and $\mathbf{A}(t')$) for $t \neq t'$.

B. *Analysis of 'Two-State Systems'*

Let us consider the H_2^+ molecule (or the ammonia molecule) in its ground state, and the parastochastic process associated to the observable: number of electrons in the left half-space defined by the bisection plane between the two protons. The quantum mechanical operator corresponding to this observable is $\mathbf{N}_V$ as defined in 3.3; in that case for one electron we have $\mathbf{N}_V = Y_V(r)$.

The general formula (4.27) for the correlation function of the parastochastic process becomes when inserting $\mathbf{N}_V\varphi_0 = \sum_k\langle\varphi_k|\mathbf{N}_V|\varphi_0\rangle\,\varphi_k$:

$$\mathcal{B}_{N_V}(\tau) = \sum_k |\langle\varphi_k|\mathbf{N}_V|\varphi_0\rangle|^2\cos\frac{E_k-E_0}{\hbar}\tau \tag{4.38}$$

where $\{\varphi_k\}$ are the eigenfunctions of $\mathbf{H}$ corresponding to the eigenvalues E_k (for simplicity we write the formulas as if the spectrum of **H** was purely discrete). If we introduce the functions

$$\chi_V = \frac{1}{\sqrt{2}}(\varphi_0 + \varphi_1)$$

$$\chi_{V'} = \frac{1}{\sqrt{2}}(\varphi_0 - \varphi_1).$$

(4.39)

we may remark that χ_V is almost zero in V' (right half-space), whereas $\chi_{V'}$, is almost zero in V. Therefore

$$\mathbf{N}_V \chi_V \# \chi_V$$

$$\mathbf{N}_V \chi_{V'} \# 0.$$

(4.40)

This is more and more true when the internuclear distance increases. Then

$$\mathbf{N}_V \varphi_0 = \frac{1}{\sqrt{2}} \mathbf{N}_V(\chi_V + \chi_{V'}) \# \tfrac{1}{2}\chi_V = \frac{1}{\sqrt{2}}(\varphi_0 + \varphi_1)$$

and

$$\mathscr{B}_{N_V}(\tau) = \frac{1}{4}\left(1 + \cos\frac{E_1 - E_0}{h}\tau\right) = \tfrac{1}{2}\sin^2\frac{E_1 - E_0}{2h}\tau.$$

(4.41)

As the parastochastic process $N_V(t)$ takes only the values 0 and 1, there is a straightforward connection between the correlation function and the transition probability $P_2(1, \tau|1,0) = P_{11}(\tau)$.

Indeed for this discrete value process, the following formula holds:

$$\mathscr{B}_{N_V}(\tau) = P_2(1, \tau; 1, 0)$$

$$= P_2(1, \tau|1, 0)\, P_1(1, 0)$$

(4.42)

and since $P_1(1,0) = P_1(1) = \tfrac{1}{2}$

$$\mathscr{B}_{N_V}(\tau) = \tfrac{1}{2}P_{11}(\tau).$$

(4.43)

According to (4.41)

$$P_{11}(\tau) = \sin^2\frac{E_1 - E_0}{2h}\tau.$$

(4.44)

This is exactly the transition probability given by Feynman ([2] §8.6) formula 8.54) in his analysis of the so called two-state systems. In fact our functions χ_V and $\chi_{V'}$ are the 'base states' used by Feynman, and it appears that his analysis is equivalent to introducing what we have called a para-stochastic process. But working with our definition of parastochastic processes has the advantage to be completely general and gives the same results without guessing a priori the base states (a difficult problem in general). Moreover the stochastic formalism reveals the limitations of the physical meaning of this analysis.

Let us consider the centered process: $N_V'(t) = N_V(t) - \overline{N_V(t)}$ where $\overline{N_V(t)} = \frac{1}{2}$. We have

$$\mathcal{B}_{N_V'}(\tau) = \mathcal{B}_{N_V}(\tau) - \overline{N_V(t)}^2 = \frac{1}{4} \cos \frac{E_1 - E_0}{h} \tau. \tag{4.45}$$

This correlation function has been met in the previous section, and we have seen that the corresponding sample functions are continuous sinusoïds. This is in contradiction with the fact that $N_V'(t)$ is a process with only two possible values: $+\frac{1}{2}$, $-\frac{1}{2}$. In the present case the values of the continuous sinusoïds remain between the two discrete values. We have verified that for other cases the continuous sample functions may even reach values outside the interval of the physically possible discrete values (e.g. negative values for the number of particles in a small volume!).

This means, that strictly speaking, the correlation function $\mathcal{B}_{N_V'}(\tau)$ is not compatible with the true (unknown!) correlation function of the discrete valued process $N_V'(t)$. In consequence in that case our definition of the parastochastic process (and by the way the Feynman analysis) is not consistent from the probabilistic point of view.

The 'success' of the Feynman analysis is made possible by the appearance of the response frequencies of the system as frequencies of periodic motions in the system. But the precise shape of the periodic motion is never used nor checked.

C. Possible Meaning of Para-stochastic Processes

In fact, para stochastic processes have correlation functions with properties in some sense opposite to those of the correlation function of the Fenyes-Nelson process. The first are (almost) periodic and undamped, the second is non periodic and damped. As we noticed in 4.1.c some periodic character of the correlation function for the position seems necessary if we want sharp responses to a periodic (electromagnetic) perturbation. On the other hand, damping at long time seems necessary to allow ergodic reproduction of given probability laws. Can we find a process with both properties? In Stochastic Electrodynamics which is the subject of the next paragraph, this is actually the case, at least for the harmonic oscillator [114]. This correlation function is, for the position process:

$$\mathcal{B}_x(\tau) = \langle x^2 \rangle e^{-\beta/2|\tau|} \left[\cos \omega_1 \tau + \frac{\beta}{2\omega_1} \sin(\omega_1|\tau|) \right] \tag{4.46}$$

where

$$\langle x^2 \rangle = \frac{h}{2m\omega_0}$$

$$\beta = \theta\omega_0^2$$

$$\theta = \frac{2e^2}{3mc^3}$$

$$\omega_1^2 = \omega_0^2 - \frac{\beta^2}{4}.$$

In atomic systems, with atomic units $\omega_0 \sim 1$, $h = 1$, $e = m = 1$ for the electron,

$c = 137$. Hence $\theta = 0.26 \times 10^{-6}$ AU and $\beta \sim 10^{-6}$, $\omega_1 \# \omega_0$. So that

$$\mathscr{B}_x(\tau) \# \langle x^2 \rangle \, e^{-\beta/2|\tau|} \cos \omega_0 \tau \tag{4.47}$$

and for $|\tau| \ll 10^6$, this reduces almost 'exactly' to the parastochastic correlation function (4.30). But the random process introduced by Stochastic Electrodynamics is free of any probabilistic contradiction and realizes ergodically the whole gaussian distribution of the quantum oscillator. This drastic qualitative improvement is achieved at cost of a small quantitative change.*

In that light the parastochastic process appears to give (at least for the position) a reasonable behaviour for times not too long with respect to the characteristic periods of the system. This is obviously sufficient to describe correctly the response to electromagnetic fields.

4.4. STOCHASTIC ELECTRODYNAMICS

In the preceding paragraphs we have shown that the probabilistic behaviour of microphysical systems can be simulated with stochastic processes, but we never asked for the origin of the complex behaviour of these systems. Many physicists were led to suppose that such a complex behaviour, which is naturally described with a probabilistic language, comes from the interaction between microparticles and a hidden medium. Weizel [115] who follows the work of Fenyes speaks of zerons, Datzeff [116] introduces a subvac and L. de Broglie [117] Bohm and Vigier [118] use the concept of a subquantum medium. This mysterious medium is never seriously characterized and is only used as a qualitative and vague picture. At the same time some other people wondered on the possible role of the 'vacuum fluctuating field'. One of the most important facts in physics after the second world-war is the role devoted to these 'fluctuations of the vacuum', introduced by the quantum theory of the electromagnetic field. In vacuum, that is without any excitation of the field, there is some residual energy; the mean electromagnetic field is zero, but there is some non zero dispersion. *The vacuum is not empty*! This has important consequences which are the basis of the development of quantum electrodynamics with its most remarkable successes: calculation of the Lamb shift and of the anomalous magnetic moment of the electron, theory of spontaneous emission.

The existence of a fluctuating field in the vacuum has a more fundamental importance than the mere explanation of some small effects. Sakharov sees here the origin of gravitation ([119], p. 426) and Wheeler writes ([119] p. 1203). 'In other words, elementary particles do not form a really basic starting point for the description of nature. Instead they represent a first order correction for vacuum physics. That vacuum, that zero order state of affairs, with its enormous densities of virtual photons and virtual positive-negative pairs and virtual wormholes, has to be described properly before one has a fundamental starting point for a proper perturbation theoretic analysis'. Wheeler recalls here the fact that quantum electrodynamics associates a state of vacuum not only in the absence of excitation of the electromagnetic field but

* See the note added in proof at the end of the paper.

also in the absence of every particle as the electron. In fact there is a 'unique' vacuum with different properties: fluctuating electromagnetic field, electric polarization....

The idea of a classical interaction between the electron and the fluctuating electromagnetic field of the vacuum goes back to Welton in 1948 [120] who gave in such a way an explanation of the Lamb shift. This idea has been the basis of different works, giving rise to what is called today: stochastic electrodynamics.

Boyer has made an extensive use of these ideas since 1968 and the best is to quote his own works:

'The present writer, essentially as an outgrowth from Casimir's ideas, has been led to propose that many of the phenomena that are presently regarded as quantum mechanical in nature may be explained in terms of classical physics when we include the possibility that the universe contains random classical electromagnetic radiation with a Lorentz-invariant spectrum. Planck's constant is introduced as the multiplicative constant fixing the scale of the classical radiation spectrum. This hypothesis of classical electromagnetic zero point radiation is a valid possibility within the context of classical electromagnetism and it leads to entirely classical derivations of the black-body radiation spectrum, of fluctuations usually associated with photon statistics, of the third law of thermodynamics, and of rotator specific heats.

It is tempting to suggest that the zero point radiation may provide the basis for the fluctuations in Nelson's derivation of the Schrödinger equation from Newtonian mechanics plus a particle random walk.

Very recently, it has been shown that the unretarded London-Van der Waals potential (r^{-6}) between two polarizable particles may be derived from classical electromagnetism including classical electromagnetic zero point radiation. In the present paper, we will give the derivation of the asymptotic retarded potential (r^{-7}). In the future, we hope to obtain the full fourth-order Van der Waals forces and indeed to prove the equivalence to all orders in perturbation theory between the quantum electrodynamic calculations of Van der Waals forces and those from classical electromagnetism including classical electromagnetic zero point radiation.' [121].

The program of work described in the last sentence has been completed with success [122], which is a proof of the power of stochastic electrodynamics.

Following Boyer and others one may wonder if stochastic electrodynamics is not the natural framework for a time-space formulation of microphysics. The conceptual basis is simple; a charged particle is described by Newtonian mechanics with the following applied forces:

(a) external forces due to other particles

(b) damping forces due to radiation (the first order damping term for example).

(c) an electrical fluctuating field which compensates the damping. In a non relativistic theory, for a particle of mass m and charge e, the fundamental equation of the theory is the following stochastic differential equation:

$$m\ddot{\mathbf{r}} = \mathbf{f} + \frac{2e^2}{3c^3}\dddot{\mathbf{r}} + e\,\mathbf{E}(t) \qquad (4.48)$$

where **f** is the external force, the term with the time third derivative is the damping force and $\mathbf{E}(t)$ is a random process with electrical forces as values.

This random process has to be characterized. The following hypotheses are made:

(1) it is a stationary random process with zero mean value.

(2) it has for power spectrum (Fourier transform of the autocorrelation function):

$$S_E(\omega) = \frac{2\hbar}{3\pi c^3} |\omega|^3 \tag{4.49}$$

such an expression is derived by the condition of relativistic invariance of the random electrical field. A 'cut-off' must be applied at high frequencies.

When **f** is a linear function of **r** (harmonic oscillator) the Equation (4.48) can be more or less exactly solved. This has been done by Surdin [123] and his predecessors [124], Santos [125] and de la Pena Auerbach and Cetto [126]. They show that most properties of the quantum harmonic oscillator are recovered. (Cf. for some details, the discussion by Claverie following this paper; see also the end of 4.3.c).

Despite Surdin's claims [123], the problem has not yet been solved in the non linear case, as for example the hydrogen atom where $\mathbf{f} = -e(\mathbf{r}/r^3)$. Technically it is a problem of solving a stochastic non linear differential equation with a noise which is not a white noise. This is not a classical problem. We have undertaken this study and it appears that the noise with a spectral density $|\omega|^3$ plays a specific role, giving to the microphysical particle a behaviour which is different from that of the classical brownian motion where the noise has a constant spectrum (white noise). An example of this specific role can be given in the following qualitative considerations for the motion of the electron in a central field with potential $V(r) = -K/r$: The radiation damping gives to the motion a tendency to become circular [127, Ch. IX, Section 75, problem 1, pp. 235–236]. Let us consider what happens on a circular trajectory in a purely qualitative way. On such a trajectory with radius r and angular frequency:

$$m\omega^2 r = V'(r) = \frac{K}{r^2}.$$

The total energy is:

$$E = -\frac{K}{r} + m\frac{(\omega r)^2}{2} = -\frac{K}{2r} \tag{4.50}$$

hence

$$\omega = \left(\frac{K}{m}\right)^{1/2} \left[\frac{-2E}{K}\right]^{3/2}. \tag{4.51}$$

The radiated power is

$$I_e = \frac{2e^2(\omega^2 r)^2}{3c^3} = C_e|E| \tag{4.52}$$

where C_e is a constant. The mean power absorbed from the random electromagnetic field is proportional to the spectral density at frequency ω:

$$I_a = \text{const} \cdot |\omega|^3 = C_a |E|^{9/2} \qquad (4.53)$$

(according to (4.51)), where C_a is some constant.

Stable motion is possible if the total power gained by the system: $I_a - I_e$ is positive when $r \to 0$ $(E \to -\infty)$ and negative when $r \to \infty$ $(E \to 0)$. This is precisely the case since $I_a/I_e = (C_a/C_e)\,E^{1/2}$ tends to $+\infty$ when $E \to -\infty$ and tends to zero when $E \to 0$.*

If we consider for comparison the motion with a brownian random force (constant spectral density) in the same force field and with the same radiation damping, the emitted power I_e is the same as before, while the absorbed mean power I_a becomes a constant independent of the energy: therefore $I_a/I_e \sim |E|^{-4}$ tends to 0 when $E \to -\infty$ and tends to $+\infty$ when $E \to 0$, which is exactly the contrary of the preceding situation: the motion is not stable, the electron either loses energy when it is already close enough to the nucleus (and therefore falls on the nucleus), or it gains energy when it is far enough from the nucleus (and therefore escapes to infinity).

Finally, if we consider the usual brownian motion in the same coulomb field, namely a brownian random force and a damping force $\mathbf{f} = -\beta \mathbf{v}$ proportional to velocity, we have $I_e = \mathbf{f}\,\mathbf{v} = -\beta v^2 = -\beta(\omega r)^2 = C^{ste}|E|$, while $I = C^{ste}$. Hence $I_a/I_e \sim |E|^{-1}$ so that the same situation as in the previous example holds: the motion is still unstable.

The examples clearly show the dramatic effect on the stability of the stochastic motion resulting from a change of the spectral density of the random force (of course, a change of the damping law could also result in drastic effects, although this does not appear in the above examples).

These qualitative remarks end our considerations on stochastic electrodynamics. Some details are left for Claverie's contribution on the classical limit in the framework of stochastic mechanics.

5. Conclusion

Let us stress in conclusion some of the main results which have been reviewed or established in this paper.

Before all we must recall what quantum mechanics is definitely not:

(a) Quantum Mechanics is not a theory of wave propagation.

(b) Quantum Mechanics is not a classical statistical and stochastic theory.

The Statistical Interpretation of quantum mechanics is in our opinion the best possible interpretation of a formalism which is not the ultima verba of microphysics. But this interpretation does not solve all the problems, especially does not introduce the explicit space-time motion which it seems to assume.

Explicit stochastic models which fit some of the important and experimentally verified results of Q.M. are possible. Some of them are purely simulatory. It seems that

---

* See note added in proof at the end of the paper.

the model built by Stochastic Electrodynamics is the best nowadays available, because of its strong physical foundations and of the encouraging results obtained.

All these results lead us to adopt the following *personal* opinions. We think that no perfect purely mathematical axiomatic theory will never completely replace an, even imperfect, simple and clear physical picture. At the present time quantum mechanics is essentially a pure mathematical axiomatic theory. As long as it will not receive a satisfying physical interpretation (which is not the case as this paper has tried to show), it will remain an elegant formalism with obscurities, contradictions and will leave an always open gate to false problems.

References

1. Dirac, P. A. M.: *Proc. Roy. Soc.* **A123**, 714 (1929).
2. Feynman, R. P., Leighton, R. B., and Sands, M.: *The Feynman Lectures on Physics*, Vol. III, 'Quantum mechanics', Addison Wesley, 1965.
3. Thom, R.: *Stabilité structurelle et morphogénèse. Essai d'une théorie générale des modèles.* Benjamin, Reading, Mass, 1972.
4. Some general discussions reflecting the different opinions can be found in these recent books:
   (a) Bunge, M.: *Foundations of Physics*, Springer Verlag, 1967.
   (b) Bunge, M. (ed.): *Quantum Theory and Reality*. Springer Verlag, 1967.
   (c) Bunge, M.: *Philosophy of Physics*, Reidel, Dordrecht, 1973.
   (d) *Philosophical Problems of Quantum Physics*, (in Russian) Nauka, Moscow, 1970.
   (e) D'Espagnat, B.: *Conceptual Foundations of Quantum Mechanics*, Benjamin, Reading, Mass, 1971.
   (f) D'Espagnat, B. (ed.): *Foundations of Quantum Mechanics* (Proc. of the Intern. School of Physics 'Enrico Fermi' Course 49), Academic Press, N.Y., 1971.
   (g) Yourgrau, W. and Van Der Merwe, A.: *Perspectives in Quantum Theory. Essays in Honor of A. Landé*, MIT Press, Cambridge, Mass, 1971.
   (h) Bastin, T. (ed.): *Quantum Theory and Beyond*, Cambridge, Univ. Press, 1971.
   (i) Scheibe, E.: *The Logical Analysis of Quantum Mechanics*, Pergamon Press, 1973.
   (j) Hooker, C. A.: *Contemporary Research on the Foundations and Philosophy of Quantum Mechanics*, Reidel, Dordrecht, 1973.
   (k) De Broglie, L.: *Etude critique des bases de l'interpretation actuelle de la mécanique ondulatoire*, Gauthier-Villars, Paris, 1963.
   (l) Andrade E Silva, J. L. and Lochak, G.: *Quanta, grains et champs*, Hachette, Paris, 1969.
   (m) Jauch, J. M.: *Are Quanta Real?*, Indiana U.P., Bloomington, 1973.
   (n) From, 1970, *Foundations of Physics*, Plenum Press, N.Y. contains a wealth of papers devoted to the interpretation of quantum mechanics.
   (o) Mehra, J.: *Dialectica* **27**, 75 (1973).
   (p) Bub, J.: *The Interpretation of Quantum Mechanics*, Reidel, Dordrecht, 1974.
5. Garaudy, R.: *Perspectives de l'homme*, Presses Universitaires de France, Paris, 1959.
6. Piaget, J.: *Le structuralisme*, Presses Universitaires de France, Paris, 1968.
7. Fraisse, P. and Piaget, J.: *Traité de Psychologie Expérimentale*, Vol. VII: 'L'intelligence', Ch. XXIII: 'Les images mentales', Presses Universitaires de France, Paris, 1969.
8. Eco, U.: *La structure absente*, Mercure de France, Paris, 1972.
9. Novik, I. B.: *Simulation of Complex Systems* (in Russian), Nauka, Moscow, 1965.
10. Bogoliubov, N. N., Logunov, A. A., and Todorov, I. T.: *Introduction to Axiomatic Quantum Field Theory*, Benjamin, Addison Wesley, Reading, 1975.
11. Emch, G. G.: *Algebraic Methods in Statistical Mechanics and Quantum Field Theory*, Wiley, New York, 1972.
12. Sudarshan, E. C. G. and Neeman, Y. (eds.): *Past Decade in Particle Theory*, Gordon and Breach, New York, 1973.
13. De Broglie, L.: *Recherches sur la théorie des quanta*, Thesis (reedited by Masson, Paris, 1963). 1924.

14. Heisenberg, W.: *Zeit. f. Phys.* **33**, 879 (1925).
    Born, M. and Jordan, P.: *Zeit. f. Phys.* **34**, 858 (1925).
15. Schrödinger, E.: *Ann. der Physik* **79**, 361 (1926) and **79**, 489 (1926) (French translation: 1933, *Mémoires sur la Mécanique Ondulatoire*. F. Alcan, Paris). (English translation: 1929, *Collected Papers on Wave Mechanics*, Blackie and Sons. Glasgow).
16. Dirac, P. A. M.: *The Principles of Quantum Mechanics*, Clarendon Press, Oxford, 1930.
17. Von Neumann, J.: *Mathematische Grundlagen der Quantenmechanik*, Springer Berlin, 1932.
18. Lande, A.: *New Foundations of Quantum Mechanics*, Cambridge, Univ. Press, 1965.
19. Lande, A.: *Found. of Physics* **1**, 191 (1971).
20. Levy-Leblond, J. M.: *Riv. Nuovo Cimento* **4**, 99 (1974).
21. Schrödinger, E.: *Mémoires sur la mécanique ondulatoire* (transl. by Proca), Alcan, Paris, 1933.
22. Schweber, S. S.: *An Introduction to Relativistic Quantum Field Theory*, Row Peterson, Evanston, 1961.
23. Fadeev, L. D.: (a) *Uspekhi Mat. Naouk* **14**, 57 (1959); (b) in *Modern Problems of Mathematics*, Vol. 3, V.I.N.I.T.I., Moscow, 1974.
24. Beck, G. and Nussenzweig, H. M.: *Nuovo Cimento* **9**, 1068 (1958).
25. Newton, R. G.: *Scattering Theory of Waves and Particles*, McGraw-Hill, New York, 1966.
26. Dynkin, E. B.: *Markov Processes*, Springer Verlag (Ch. XIII, §4) 1965.
27. Möllenstedt, G. and Düker, H.: *Zeit. f. Phys.* **145**, 377 (1956).
28. Marton, L., Arol Simpson, J., and Suddeth, J. A.: *Review of Scientific Instruments* **25**, 1099 (1954).
29. Lichte, H., Möllenstedt, G., and Wahl, H.: *Zeit. f. Phys.* **249**, 456 (1972).
30. Faget, J.: *Revue d'Optique Théorique et Instrumentale* **40**, 347 (1961).
31. Jönsson, C.: *Zeit. f. Phys.* **161**, 454 (1961).
32. Bodiou, G.: *Théorie dialectique des probabilités englobant leurs calculs classique et quantique*, Gauthier-Villars, Paris, 1964.
33. Parthasarathy, K. R.: *Probabilités sur les structures algébriques*, Colloque International du CNRS, No. 186. p. 265, C.N.R.S. Paris, 1970.
34. Gudder, S. P.: 'Axiomatic Quantum Mechanics and Generalized Probability Theory', in *Probabilistic Methods in Applied Mathematics*, Vol. 2, 53, 1970.
35. Varadarajan, V. S.: *Geometry of Quantum Theory*, Vol. 1, 1968. *Geometry of Quantum Theory*, Vol. 2, 1970, Van Nostrand, N.Y.
36. Jauch, J. M.: *Foundations of Quantum Mechanics*, Addison Wesley, Reading, Mass, 1968.
37. Lax, P. D. and Phillips, R. S.: *Scattering Theory*, Acad. Press, N.Y., 1967.
38. De Broglie, L.: *La réinterprétation de la mécanique ondulatoire*, 1re *partie principes généraux*, Gauthier Villars, Paris, 1971.
39. De Broglie, L.: *La conception du monde physique. Le Passé et l'Avenir de la Mécanique Ondulatoire*, Gauthier Villars, Paris, 1973.
40. De Broglie, L.: *Comptes Rendus Acad. Sci.* **277**, B, 71 (1973).
41. Destouches, J. L.: *La Physique Mathématique*, Presses Universitaires de France, Paris, 1964.
42. Kastler, A.: *Sciences et Avenir*, Special issue: *La crise des mathématiques modernes*, p. 54, 1973.
43. Levy-Leblond, J. M.: *Bulletin Société Française de Physique*, No. 14 (Encart pédagogique), 1973.
44. Piron, C.: *Helv. Phys. Acta* **37**, 439 (1964). *Found. of Physics* **2**, 287 (1972). Physique Quantique (Cours de l'Université de Genève), 1974.
45. Mackey, G. W.: *The Mathematical Foundations of Quantum Mechanics* Benjamin, New York, 1963.
46. Segal, I. E.: *Ann. of Math.* **48**, 930 (1947).
47. Bohr, N.: in *Albert Einstein. Philosopher Scientist* (P. A. Schilpp ed.), Open Court Publ. Co. La Salle, Illinois, p. 201, 1949.
    The ideas of Bohr are scattered in the literature and it is difficult to give one exhaustive work as reference.
48. Heisenberg, W.: *The Physical Principles of Quantum Theory*, University of Chicago Press (reprinted by Dover, N.Y. 1950), 1930.
49. (a) Landau, L. D. and Lifschitz, E. M.: *Quantum Mechanics*, Addison Wesley, Reading, Mass, 1958.
    (b) Messiah, A.: *Quantum Mechanics*, North Holland, Amsterdam, 1964.
    (c) Cohen Tannoudji, C., Diu, B., and Laloe, F.: *Mécanique Quantique*, Hermann, Paris, 1973.

50. Einstein, A., Podolsky, B., and Rosen, N.: *Phys. Rev.* **47**, 777 (1935).
51. Popper, K. S.: *The Logic of Scientific Discovery*, Basic Books Inc., New York, 1959.
52. Blokhintsev. D. I.: *Quantum Mechanics*, Reidel, Dordrecht, 1964.
    The Philosophy of Quantum Mechanics, Reidel, Dordrecht, 1968.
53. Ballentine, L. E.: *Rev. Mod. Phys.* **42**, 358 (1970).
54. (a) Bell, J. S.: *Rev. Mod. Phys.* **38**, 447 (1966).
    (b) Bohm, D. and Bub, J.: *Rev. Mod. Phys.* **38**, 453 (1966).
    (c) Bohm, D. and Bub, J.: *Rev. Mod. Phys.* **38**, 470 (1966).
    (d) Jauch, J. M. and Piron, C.: *Rev. Mod. Phys.* **40**, 228 (1968).
    (e) Gudder, S. P.: *Rev. Mod. Phys.* **40**, 229 (1969).
    (f) Gudder, S. P.: *J. Math. Phys.* **11**, 431 (1970).
    (g) Kochen, S. and Specker, E.: *J. Math. Mech.* **17**, 59 (1967).
    (h) Capasso, V., Fortunato, D., and Selleri, F.: *Rivista del Nuovo Cimento* **II**, 149 (1970).
55. Wiener, N., Rankin, B., Siegel, A., and Martin, W. T.: *Differential Space, Quantum Systems and Predictions*, MIT Press, Cambridge, Mass., 1966.
56. Belinfante, F. J.: *A Survey of Hidden-Variables Theories*, Pergamon Press, Oxford, 1973.
57. (a) Prosperi. G. M. in (4f) p. 97.
    (b) Ludwig, G. in (4f) p. 287.
    (c) Yanase, M. M. in (4f) p. 77.
    (d) Hepp, K.: *Helv. Phys. Acta* **45**, 237 (1972).
    (e) She, C. Y. and Heffner, H.: *Phys. Rev.* **152**, 1103 (1966).
    (f) Araki, H. and Yanase, M. M.: *Phys. Rev.* **120**, 622 (1960).
58. (a) Everett III, H.: *Rev. Mod. Phys.* **29**, 454 (1957).
    (b) De Witt, B. S.: *Physics Today* **23**, 30 (1970).
    (c) De Witt, B. S.: in (4f) p. 211.
    (d) De Witt, B. S. and Graham, N.: *The Many-Worlds Interpretation of Quantum Mechanics*, Princeton Univ. Pres, Princeton, 1973.
59. (a) Coulson, C.: *Valence*, Oxford Univ. Press, 1952.
    (b) in *Physical Chemistry, an Advanced Treatise*, Vol. V, p. 372, Acad. Press New York, 1970.
60. *The State of the Theory of Chemical Structure in Organic Chemistry* (in Russian), Acad. of Sciences of the U.S.S.R., Moscow, 1954.
61. Volkenstein, M. V.: *Structure and Physical Properties of the Molecule* (in russian), Acad. of Sciences of the U.S.S.R., Moscow, 1955.
62. Park, J. L.: *Am. J. Physics* **36**, 211 (1968).
63. Cohen, L.: *Philosophy of Science* **33**, 317 (1966).
64. Bohm, D.: *Quantum Theory*, Prentice Hall, New York, 1951.
65. Born, M. and Wolf, E.: *Principle of Optics*, Pergamon Press, New York, 1959.
66. Herzberg, G.: *Molecular Spectra and Molecular Structure*, II: *Infrared and Raman Spectra of Polyatomic Molecules*, Van Nostrand, New York, 1945.
67. Moyal, J. E.: *Proc. Cambridge Phil. Soc.* **45**, 99 (1949).
68. Urbanik, K.: *Studia Mathematica* **21**, 117 (1961).
69. Bub, J.: *Found. of Physics* **3**, 29 (1973).
70. *Louis de Broglie*: *Physicien et penseur*, Albin Michel, Paris, 1953.
71. Jammer, M.: *The Conceptual Development of Quantum Mechanics*, Mc Graw Hill, New York, 1966.
72. Holton, G.: *Thematic Origins of Scientific Thought, Kepler to Einstein*, Harvard University Press, Cambridge, Mass., 1973.
73. Margenau, H. and Kestner, N. R.: *Theory of Intermolecular Forces*, Pergamon Press, Oxford, 1969.
74. Kutzelnigg, W., Del Re, G., and Berthier G.: *Phys. Rev.* **172**, 49 (1968).
75. Salem, L.: *Molecular Orbital Theory of Conjugated Systems*, Benjamin, New York, 1966.
76. Mc Weeny, R. and Sutcliffe, B. T.: *Methods of Molecular Quantum Mechanics*, Acad. Press, New York, 1969.
77. De Boer, J.: *Construction Operator Formalism in Many Particle Systems*, in J. de Boer and G. E. Uhlenbeck (eds.), *Studies in Statistical Mechanics*, Vol. III, p. 213, North Holland, Amsterdam, 1965.
78. Daudel, R.: *Fundamentals of Theoretical Chemistry*, Pergamon, Oxford, 1968.

79. (a) Berger, L.: *Helv. Phys. Acta* **31**, 159 (1958).
    (b) Majernik, V.: *Il Nuovo Cimento* **LXIV A**, 501, 1969.
80. Botchvar, D. A., Stankevitch, I. V., and Chistiakov, A. L.: *Uspekhi Khimii* **XLIII**, 655 (1974).
81. Primas, H.: in *Modern Quantum Chemistry* (O. Sinanoglu, ed.), p. 45, Academic Press, New York, 1965.
82. Boys, S. F. in: P. O. Löwdin (ed.), *Quantum Theory of Atoms, Molecules and the Solid State*, p. 253, Academic Press, New York,
83. Aslangul, C.: *C.R. Acad. Sc. Paris*, **272B**, 1 (1971).
84. Aslangul, C., Constanciel, R., Daudel, R., and Kottis, P. T.: in P. O. Löwdin (ed.), *Advances in Quantum Chemistry* **6**, 93 (1972).
85. Stephens, M. E.: Thesis, McMaster University, Hamilton (Canada) 1975.
86. (a) Lebowitz, J. L. and Penrose, O.: *Physics Today*, p. 23 (1973).
    (b) Sinaï, Ya. G.: *Acta Physica Austriaca, Suppl.* **X**, 575 (1973).
87. Massignon, D.: *Mécanique statistique des fluides*, Dunod, Paris, 1957.
88. Retov, S. M.: *Introduction to Statistical Radiophysics* (in Russian), Nauka, Moscow, 1966.
89. Smith, D. W. and Larson, E. G.: *Int. J. Quant. Chem.* **III Supp**. 689 (1970).
90. McWeeny, R.: *Rev. Mod. Phys.* **32**, 335 (1960).
91. Fenyes, I.: *Zeit. f. Phys.* **132**, 81 (1952).
    (This paper of Fenyes, translated in Russian, is reproduced in 'Causality Problems in Quantum Mechanics' (ed. by Terletsky and Guseva), Moscow 1955, along with famous papers of Feynman, Bohm, De Broglie, Moyal, Welton... A complete panorama at that time of non orthodox points of view!.)
92. Nelson, E.: *Phys. Rev.* **150**, 1079 (1966).
93. Nelson, E.: *Dynamical Theories of Brownian Motion*, Mathematical Notes, Princeton Univ. Press (1967).
94. Della-Riccia, G. and Wiener, N.: *J. Math. Phys.* **7**, 1372 (1966).
95. Favella, L. F.: *Ann. Inst. Henri Poincaré* **AVII**, 77 (1967).
96. De la Pena-Auerbach, L.: *J. Math. Phys.* **10**, 1620 (1969).
97. Papoulis, A.: *Probability, Random Variables and Stochastic Processes*, Mc Graw Hill, New York, 1965.
98. Gikhman, I. L. and Skorokhod, A.V.: *Introduction to the Theory of Random Processes*, Saunders, Philadelphia, 1969.
99. Rosenblatt, M.: *Markov Processes Structure and Asymptotic Behaviour*, Springer Verlag, Berlin, 1971.
100. Caubet, J. P.: *C. R. Acad. Sc.* **277**, 1199 (1973).
101. De la Pena, Auerbach, L., Velasco, R. M., and Cetto, A. M.: *Revista Mexicana de Fisica* **19**, 193 (1970).
102. De Broglie, L.: *C.R. Acad. Sc.* **264**, 1041 (1967).
103. Fürth, R.: *Zeit. f. Phys.* **81**, 143 (1933).
104. Claverie, P. and Diner, S.: *C.R. Acad. Sc.* **277B**, 579 (1973).
105 Birtwistle, G.: *The New Quantum Mechanics*, Cambridge University Press, 1928.
106. Sommerfeld, A.: *Atombau und Spektrallinien*, Friedr. Vieweg und Sohn. Braunschweig., 1951 (Vol. II, Appendix 4, p. 724).
107. Schrödinger, E.: *Berl. Ber.*, p. 400 (1930); *Berl. Ber.*, p. 144 (1931).
108. Kubo, R.: *J. Phys. Soc. Jap.* **12**, 570 (1957).
109. Berne, B. J. and Harp, G. D.: *Adv. Chem. Phys.* **XVII**, 63 (1970).
110. Zubarev, D. N.: *Non-Equilibrium Statistical Thermodynamics*, Plenum Publ. Corp., 1974.
111. Frisch, U. and Bourret, R.: *J. Math. Phys.* **11**, 364 (1970).
112. Belyaev, Yu. K.: *Theory of Prob. and Appl.* **IV**, 402 (1959).
113. Bass, J.: *J. Math. Anal. and Appl.* **47**, 354 (1974).
114. Claverie, P. and Diner, S.: *C. R. Acad. Sc.* **280B**, 1 (1975).
115. Weizel, W.: *Zeit. f. Physik* **134**, 264 (1953).
116. Datzeff, A. B.: *Mécanique quantique et réalité physique*, Editions de l'Académie Bulgare des Sciences, Sofia, 1969.
117. De Broglie, L.: *La thermodynamique de la particule isolée (ou Thermodynamique cachée des particules)*, Gauthier-Villars, Paris, 1964. *J. Phys.* **20**, 963 (1959).
118. Bohm, D. and Vigier, J. P.: *Phys. Rev.* **96**, 208 (1954).

119. Misner, C. W., Thorne, K. S., and Wheeler, J. A.: *Gravitation*, W. H. Freeman and Co, San Francisco, 1973.
120. Welton, T. A.: *Phys. Rev.* **74**, 1157 (1948).
121. Boyer, T. H.: *Phys. Rev.* **A5**, 1799 (1972).
122. Boyer, T. H.: *Phys. Rev.* **A6**, 314 (1972).
123. Surdin, M.: *Annales Institut Henri Poincaré* **15A**, 203 (1972); *Int. J. of Theor. Phys.* **4**, 117 (1971).
124. (a) Marshall, T. W.: *Proc. Roy. Soc.* **276A**, 475 (1963).
     (b) Braffort, P. and Tzara, C.: *C.R. Acad. Sc. (Paris)* **239**, 1779 (1954).
125. Santos, E.: *Il Nuovo Cimento* **19B**, 57 (1974).
126. De La Pena, Auerbach, L. and Cetto, A. M.: *The Harmonic Oscillator in a Random Electromagnetic Field: Schrödinger's Equation and Radiative Corrections* (preprint 1974).
127. Landau, L. and Lifshitz, E.: *The Classical Theory of Fields*, Pergamon Press, Oxford, 1962.
128. (a) Mirman, R.: *Nuovo Cimento* **18B**, 110 (1973).
     (b) Lyuboshitz, V. L. and Podgoretskii, M. I.: *Sov. Phys. JETP* **28**, 469 (1969); **30**, 100 (1970); **33**, 5 (1971).
     (c) Hestenes, D.: *Am. J. Physics* **38**, 840 (1970).
129. Ming Chen Wang and Uhlenbeck, G. E.: *Rev. Mod. Phys.* **17**, 323 (1945) (reprinted, p. 113, in *Selected Papers on Noise and Stochastic Processes* (ed. by N. Wax), Dover, New-York 1954).

Notes added in proof.

Page 408: For discussion of related topics see:

Baracca, A., Bergia, S., Bigoni, R., and Cecchini, A.: *Rivista del Nuovo Cimento* **4**, 169 (1974).

Baracca, A.: *'Proper' vs 'Improper' Mixtures, the Key Problem in the Foundations of Quantum Mechanics* (preprint 1974).

Page 440: The importance of using slightly damped quasi-periodic stochastic processes, in order to recover the quantum mechanical behaviour, was recognized by Bopp:

Bopp, F.: *Zeit. f. Physik* **143**, 233 (1955); **144**, 13 (1956); *Zeit. f. Naturforschung* **10a**, 783 and 789 (1955).

Page 443: After completion of this work we found the same argument in a recent paper by Boyer, which is a nice review of the present state of Stochastic Electrodynamics:
Boyer, T. H.: *Phys. Rev.* **D11**, 790 and 809 (1975).

DISCUSSION OF CLAVERIE AND DINER'S PAPER:
THE CLASSICAL LIMIT IN THE FRAMEWORK OF STOCHASTIC MECHANICS

PIERRE CLAVERIE

*Laboratoire de Chimie Quantique, Institut de Biologie Physico-Chimique,
13 rue Pierre et Marie Curie, 75005 Paris, France*

In sharp contrast with usual Quantum Mechanics (hereafter abreviated as Q.M.), Stochastic Mechanics (hereafter abreviated as S.M.) exhibits the pleasant feature of providing a clear and *qualitatively understandable* picture of physical systems, and not only a mathematical algorithm for the *quantitative evaluation* of various properties. It could be objected that Q.M. does not reduce to a set of mathematical algorithms and contains a variety of new qualitative concepts (stationary state, quantification of some observables, and so on); but as it has been exemplified by Diner in his lecture, a *thorough qualitative* understanding of microphysical phenomena is not really obtained by usual Q.M., and S.M. appears able to bring in substantial conceptual progress. The purpose of this contribution is to describe these conceptual improvements in another (and rather obscure) area of Quantum Mechanics, namely the so-called 'transition to Classical Mechanics'. I shall first describe briefly the difficulties which appear in the framework of usual Q.M., and then show how they would disappear in the framework of S.M., provided that appropriate dynamical laws may be found such that, by using them, S.M. actually gives the main results of Q.M. (position and velocity probability distributions, mean values of energy, angular momentum...). The search for such appropriate dynamical laws is presently investigated by several workers [14-18].

1. The Difficulties of the Transition to Classical Mechanics in the Framework of Usual Quantum Mechanics

Some of the points to be discussed below have been considered by Datzeff ([1], Chap. I, section 1).

The difficulty encountered first is that the axioms used for applying Q.M. to microphysical systems do not work properly when we try to apply them to classical conditions: namely, the concept of eigenstate becomes irrelevant. Indeed, for parameter values (dimensions, masses, energies) corresponding to classical systems, the energy values would correspond to very large quantum numbers (see e.g. the case of the harmonic oscillator in Messiah [2], Part I, chap. XII); analogous evaluations can be made for the more physical case of the $1/r$ potential in 3-dimensional space). Now, it is immediately seen that the eigenfunctions for such high quantum numbers do not give the probability of presence corresponding to a classical motion. Thus, for the harmonic oscillator with amplitude a, the classical probability density is

$$p(x) = 1/(\pi\sqrt{a^2 - x^2}) \tag{1}$$

while the probability density corresponding to the nth quantum-mechanical eigenstate with energy

$$E_n = (n + 1/2) h \tag{2}$$

is

$$\psi_n(x)^2 = \left(\frac{m\omega}{\pi h}\right)^{1/2} \frac{1}{2^n n!} e^{-\xi^2} [H_n(\xi)]^2 . \tag{3}$$

where $\xi = x\sqrt{m\omega/h}$; the correspondance with the classical $p(x)$ may be made by noting that

$$\frac{m\omega^2 a^2}{2} = E = (n + \tfrac{1}{2}) \hbar\omega . \tag{4}$$

Now, the Hermite polynomial $H_n(\xi)$ has n real roots ξ_i located in the segment $[-\sqrt{2n+1}, +\sqrt{2n+1}]$ corresponding to the classical amplitude ($a\sqrt{m\omega/h} = 2n+1$ according to (4)). The distance between these roots tends to 0 as $n \to \infty$. Consequently, $H_n(\xi)^2$ presents an infinitely increasing number of oscillations, being 0 for the values ξ_i, and the maximum value in each oscillation grows to infinity with n. Thus $|\psi_n(x)|^2$ cannot tend to $p(x)$ as $n \to +\infty$, and actually has no limit at all when considered as a function. Only when considered *as a distribution* $|\psi_n(x)|^2$ has a limit, which is actually the distribution defined by $p(x)$:

$$\int_{-\infty}^{+\infty} f(x) |\psi_n(x)|^2 \, dx \xrightarrow[n \to +\infty]{} \int_{-\infty}^{+\infty} f(x) \, p(x) \, dx \tag{5}$$

for any sufficiently smooth function $f(x)$ (the convergence becoming slower if $f(x)$ becomes less regular). But even this convergence as a distribution does not give the solution in the case of a motion in 3-dimensional space. Indeed, for a central field (e.g. the Coulomb potential), the eigenstates will have definite symmetry types (for example the s-states are spherically symmetric), so that, even if the quantum-mechanical probability densities $|\psi_{ns}(\mathbf{r})|^2$ have a limit when considered as distributions (they cannot have such a limit in any space of functions, for the same reason as before), this limit would also necessarily be spherically symmetric and could therefore not correspond to any probability density obtained from a classical orbit, if only because such an orbit remains in a given plane. A spherically symmetric density would correspond to a *set* of orbits obtained from a spherically symmetric distributions of initial conditions, but not to a single orbit.

Owing to this failure of the concept of eigenstate under classical conditions, it is a customary practice to introduce 'minimal wave-packets' and to let them evolve according to the time-dependent Schrödinger equation. Then it may be shown that the

'Center of gravity' of the 'wave-packet'

$$\mathbf{R} = \iiint \mathbf{r} \, |\psi(\mathbf{r})|^2 \, dv \qquad (6)$$

actually moves approximately according to the classical equations of motion *provided that the size of the wave-packet* (in position space and in momentum space) *is small* (Ehrenfest's theorem: see e.g. Messiah [2, Part I, chap. VI], or the more thorough treatment by Maslov [3]). But this condition cannot remain fulfilled when the time grows infinitely: the 'wave-packet' spreads without limit in the course of time (except for some very special cases like the harmonic oscillator, initially treated by Schrödinger [4]; see also Messiah [2, Part I, chap. XII]), and its center will no more move according to the classical equations (except if the potential $V(x, y, z)$ is a polynomial of degree at most 2, which essentially means the free particle, the particle in a constant field of force and the harmonically bound particle, see Messiah [2, Part I, chap. VI]). Besides making Ehrenfest's theorem irrelevant, the spreading of the 'wave-packet' is an important conceptual fact, because it makes it impossible to accept the view that a wave-packet really represents a single particle: thus for an electron, under very usual conditions, the spreading would give it macroscopic size already after rather short times: let us consider a free electron: then, for large time τ, we have $\Delta q \simeq \Delta v_0 \tau$, where Δv_0 is the initial ($\tau = 0$) spread of the velocity (standard deviation) [5]. But using $\Delta v_0 \Delta q_0 \geqslant \hbar/m_e \simeq 6 \times 10^{-4}$ ($h = 6.624 \times 10^{-34}$ Js, $m_e = 0.91 \times 10^{-30}$ kg), we see that, for $\Delta q_0 = 10^{-3}$ m (which is already a enormous value if we consider that it refers to a single electron), we already get $\Delta v_0 \simeq 0.6$ m s^{-1}, so that $\Delta q = \Delta v_0 \tau$ would become extremely large in rather short times ($\Delta q = 6$m for $\tau = 10$ s).

Actually, this argument based on the clearly unphysical character of the spreading of the 'wave-packet' was just used already by the Copenhagen school (Bohr, Heisenberg) against the pure wave conception proposed by Schrödinger (see Jammer [6], chap. 6, section 1, p. 283). But the Copenhagen Interpretation itself is not in a very good position as concerns the problem. This interpretation may indeed be summarized as follows: an abstract function ψ represents a *single* particle; ψ changes according to the Schrödinger equation; ψ enables us to calculate various mean values concerning the particle (e.g. $|\psi(\mathbf{r})|^2$ is the density of probability of presence at $\mathbf{r}$); and this information given by ψ is assumed to be complete (so that the possibility of considering a more refined information is denied). But no function ψ can describe the classical motion of a particle in the course of time: the initially spread eigenfunctions ψ_n obviously cannot, since they are stationary (and $|\psi_n|^2$ will in general *not* give the probability density corresponding to the classical motion of a single particle), and the other functions, even if they are initially localized as much as possible (minimal wave-packet) also cannot since they inevitably spread. Thus the Copenhagen Interpretation also fails to understand the classical motion of a single particle. This appears clearly in Messiah's textbook [2, Part I, chap. VI]: while the author declared (Part I, chap. II) adopting the Copenhagen Interpretation, he suddenly proposes to interpret the 'wave-packet' as the probability density of a *set* of particles, which initially have close positions and velocities, but whose distances grow with time owing to these initial small differences. This is in agreement with the Statistical Interpretation (see e.g. Ballentine [7]), which considers that the function ψ gives an information about ensembles of similarly prepared particles and not directly about a single particle, but certainly not with the orthodox Copenhagen Interpretation, that he declared to adopt in his chapter II.

Thus, as concerns the transition to the classical limit, the best which can be said in the framework of usual Q.M. is the following: according to the Statistical Interpretation, a 'wave-packet' ψ (which has nothing of a wave, altogether) indicates how the density of an ensemble (not physical, an ensemble in the sense of Statistical Mechanics) changes in time. No direct and detailed information about the motion of a single particle would be available, even at the classical limit: indeed, if the theory has a statistical character, it cannot lose it simply because macroscopic values are given to the physical parameters. Of course, there exist several works dealing with the connection between microphysics and macrophysics (see e.g. [8]) but they do not answer the questions raised in the present section (at least in my opinion).

2. The Classical Limit in the Framework of Stochastic Mechanics

There is no more a conceptual problem for the understanding of the classical limit in the framework of Stochastic Mechanics. Indeed the basic features of Stochastic Mechanics itself may be summarized as follows: the particle undergoes a complicated motion resulting from both systematic forces (the explicitly known forces used in classical mechanics and electromagnetism) and a rapidly and irregularly varying force, which, for mathematical convenience, is *represented* by a random force with some specified probability distribution. This does not mean in any way that the physical force is truly random: it is indeed well-known among mathematicians that a deterministic evolution may be so complicated that, to a very high accuracy, the various associated mean values cannot be distinguished from those of a truly random function: for this reason, such complicated deterministic functions may be called pseudo-random functions [9]; the study of time mean values of such functions (ergodic theory) has achieved considerable successes during the last decade, resulting in decisive progress for the understanding of the fundamental problems of classical statistical mechanics [10, 11, 12]. As a consequence, it becomes now perfectly possible to support the view that 'truly random' variables or functions are simply mathematical idealizations which conveniently *simulate* pseudo-random variables or functions, deterministic but with a very complicated behaviour.

Having made clear that the word 'random' may be thought of as a substitute for 'pseudo-random', let us consider the motion of our particle submitted to systematic forces and a 'random' force: to every particular realization of the 'random' force as a function of time, there corresponds a particular trajectory of the particle; the set of all such possible trajectories is called a stochastic process; the purpose of the theory is to study the ensemble mean values which may be defined from this set of trajectories, and also the time mean values calculated along a given trajectory, and to compare them.

The meaning of the 'classical' case in this framework is obvious: it corresponds to the situations where the random force is negligible with respect to the systematic forces or to the kinetic energy of the particle. Then, all possible trajectories which start at any given time with given initial conditions will remain during a long time very close to each other 'around' the ideal classical trajectory (obtained by removing completely the 'random' force). Actually, even the name 'classical limit' is somewhat inappropriate, since classical equations will still be eventually used even when the random force is not negligible: thus, to speak of the 'case of negligible random force' would be more correct. Let us mention that the perturbation of a classical motion by a small random force has been studied rather thoroughly for some years from the mathematical point of view [13] (but without the suggestion that such studies could have a bearing for microphysical problems).

A numerical illustration of these considerations may be given for the charged harmonic oscillator. Indeed, Marshall [14], Braffort, Tzara and Surdin (see the references in Surdin [15]), Santos [16] and de la Pena-Auerbach [17] have studied the classical motion of a charged particle in a harmonic potential and in the presence of a fluctuating electromagnetic field (the so-called electromagnetic zero-point radiation, which was also studied by Boyer [18]). There are 3 forces: the harmonic force: $-m\omega_0^2 x$; the damping force due to the radiation emitted by the charged particle: $(1/4\pi\varepsilon_0)(2e^2/3c^3)$ $\dddot{x}$ (e: charge of the particle, c: velocity of light, $\dddot{x} = (d^3 x/dt^3)$; and the fluctuating force $e\mathscr{E}(t)$ where $\mathscr{E}(t)$ is the fluctuating electric field, which is gaussian with mean value zero, and spectral density $\sigma(\omega) = \hbar|\omega|^3/3\pi c^3$ * for a one-dimensional problem (see [14–18]). The (Langevin) equation of motion then is written

$$m\ddot{x} = -m\omega_0^2 x + \frac{1}{4\pi\varepsilon_0}\frac{2e^2}{3c^3}\dddot{x} + e\,\mathscr{E}(t) \tag{7}$$

or

$$-\tau\dddot{x} + \ddot{x} + \omega_0^2 x = F(t) \tag{8}$$

with $\tau = (1/4\pi\varepsilon_0)(2e^2/3mc^3)$ (τ has the dimension of a time) and $F(t) = (e/m)\,\mathscr{E}(t)$. Then, following Surdin [15], this equation is approximated by the following one:

$$\ddot{x} + \tau\omega_0^2\dot{x} + \omega_0^2 x = A(t) \tag{9}$$

where $A(t)$ is a fluctuating 'force' with constant spectral density:

$$\frac{2D}{2\pi} = \frac{1}{2\pi}\int_{-\infty}^{+\infty} e^{-i\omega\theta}\langle A(t)\,A(t+\theta)\rangle\,d\theta \tag{**(10)}$$

---

* Our $\sigma(\omega)$ is defined by $\sigma(\omega) = 1/2\pi\int_{-\infty}^{\infty}\langle\mathscr{E}(t)\,\mathscr{E}(t+\theta)\rangle\,e^{-i\omega\theta}\,d\theta$. This corresponds to the $\varepsilon(\omega)$ of Surdin [15].

** From now on, the symbol $\langle\,\rangle$ will represent a mean value taken over a statistical ensemble (corresponding to the various realizations of the random force $F(t)$). This symbol will *no more* appear with its quantum-theoretical meaning (bra-ket notation).

$(\langle A(t)\, A(t+\theta)\rangle$ is the correlation function of $A(t)$, and is consequently equal to $2D\delta(\theta)$: see e.g. [19] [20]). The replacement of $-\tau\dddot{x}$ by $\tau\omega_0^2\dot{x}$ is obtained from the simplified equation $\ddot{x}+\omega_0^2 x = 0$, hence $\dddot{x} = -\omega_0^2\dot{x}$; it is made possible by the smallness of $\tau\dddot{x}$ with respect to the other terms $\ddot{x}$ and $\omega_0^2 x$: in atomic units $1/4\pi\varepsilon_0 = 1$, $\hbar = 1$, $c = 137$ (inverse of the fine structure constant $\alpha = e^2/4\pi\varepsilon_0\hbar c$) and, for an electron $m = 1$, $e = 1$: $\tau = 2/3\,(1/137)^3 \simeq 0.26 \times 10^{-6}\,\text{AU}$ of time (1 AU time $= 0.243 \times 10^{-16}\text{s}$). As concerns the use of a constant spectral density for $A(t)$, instead of the $|\omega|^3$ density of $F(t)$, it is made possible by the fact that the oscillator, being very slightly damped, has a very sharp response curve at the frequency ω_0, so that only the frequencies of $F(t)$ very near to ω_0 will play a non negligible role; since the spectral density of $F(t)$ is that of $\mathscr{E}(t)$ multiplied by $(e/m)^2$, namely $(e^2/m^2)\,(\hbar\omega^3/3\pi c^3)$, this leads us to take

$$\frac{2D}{2\pi} = \frac{e^2}{m^2}\,\sigma(\omega_0) = \frac{e^2}{m^2}\,\frac{\hbar\omega_0^3}{3\pi c^3}$$

or

$$2D = \frac{e^2}{m^2}\,\frac{2\hbar\omega_0^3}{3c^3} = \tau\omega_0\,\frac{\hbar\omega_0^2}{m}. \tag{11}$$

(Surdin [15] notes by S what is noted here $2D$ and calculates for it a more refined value.) The Equation (9) is nothing but the Langevin equation for the brownian motion of a harmonically bound particle with friction coefficient

$$\beta = \tau\omega_0^2 \tag{12}$$

so that we may use the results concerning this kind of equation, described by Ming Chen Wang and Uhlenbeck [20]: Noting $v = \dot{x}$, the Fokker-Planck equation for the two-variable probability density $P(x, v; t)$ is:

$$\frac{\partial P}{\partial t} = -v\,\frac{\partial P}{\partial x} + \frac{\partial}{\partial v}\left[(\beta v + \omega_0^2 x)\,P\right] + D\,\frac{\partial^2 P}{\partial v^2}. \tag{13}$$

The transition probability density $P(x, v; t\,|\,x_0, v_0)$ (two-dimensional probability density at time t for the particle starting at time 0 from the point x_0 with velocity v_0) is the solution defined by the initial condition $P(x, v; 0\,|\,x_0, v_0) = \delta(x - x_0)\,\delta(v - v_0)$, and it is at any time a two-dimensional gaussian distribution, entirely defined by its mean values and variances:

$$\langle v\rangle = e^{-1/2\,\beta t}\left[\left(\omega_1\cos\omega_1 t - \frac{\beta}{2}\sin\omega_1 t\right)\frac{v_0}{\omega_1} - \frac{\omega_0^2}{\omega_1}\,x_0\sin\omega_1 t\right] \tag{14a}$$

$$\langle x\rangle = e^{-1/2\,\beta t}\left[\frac{v_0}{\omega_1}\sin\omega_1 t + \frac{x_0}{\omega_1}\left(\omega_1\cos\omega_1 t + \frac{\beta}{2}\sin\omega_1 t\right)\right] \tag{14b}$$

$$\langle (v - \langle v \rangle)^2 \rangle = \frac{D}{\beta} \left[1 - e^{-\beta t} \left(1 + \frac{\beta^2}{2\omega_1^2} \sin^2 \omega_1 t - \frac{\beta}{\omega_1} \sin \omega_1 t \cos \omega_1 t \right) \right]$$

(15a)

$$\langle (x - \langle x \rangle)^2 \rangle = \frac{D}{\beta \omega_0^2} \left[1 - e^{-\beta t} \left(1 + \frac{\beta^2}{2\omega_1^2} \sin^2 \omega_1 t + \right. \right.$$

$$\left. \left. + \frac{\beta}{\omega_1} \sin \omega_1 t \cos \omega_1 t \right) \right]$$

(15b)

$$\langle (x - \langle x \rangle) (v - \langle v \rangle) \rangle = \frac{D}{\omega_1^2} e^{-\beta t} \sin^2 \omega_1 t$$

(15c)

where

$$\omega_1^2 = \omega_0^2 - \frac{\beta^2}{4}.$$

(16)

Then the following conclusions may be drawn:

(1) When $t \to + \infty$, we get the so-called invariant probability distribution:

$$\langle v \rangle = 0; \ \langle x \rangle = 0; \ \langle v^2 \rangle = \langle (v - \langle v \rangle)^2 \rangle = \frac{D}{\beta};$$

$$\langle x^2 \rangle = \langle (x - \langle x \rangle)^2 \rangle = \frac{D}{\beta \omega_0^2};$$

$$\langle vx \rangle = \langle (v - \langle v \rangle) (x - \langle x \rangle) \rangle = 0.$$

(17)

It is actually independent of the initial values x_0, v_0, and is simply the product of two independent gaussian distributions

$$\frac{1}{\sqrt{2\pi \langle x^2 \rangle}} e^{-x^2/2 \langle x^2 \rangle} \times \frac{1}{\sqrt{2\pi \langle v^2 \rangle}} e^{-v^2/2 \langle v^2 \rangle}.$$

(18)

Any initial probability distribution tends to this invariant distribution as $t \to + \infty$, and the ergodic property automatically holds since the stochastic process is Markovian and has an invariant probability distribution [21]: namely, the time average of any quantity $f(x, v)$ calculated along (almost) every trajectory tends (as the time over which the average is taken goes to infinity) to the average of $f(x, v)$ calculated with the invariant probability distribution.

Using (11) and (12), we get

$$\langle v^2 \rangle = \frac{D}{\beta} = \frac{\hbar \omega_0}{2m} \quad \text{or} \quad \langle p^2 \rangle = m^2 \langle v^2 \rangle = \frac{m \hbar \omega_0}{2}$$

(19a)

$$\langle x^2 \rangle = \frac{D}{\beta \omega_0^2} = \frac{\hbar}{2m\omega_0}$$

(19b)

which are the usual quantum-mechanical values for the ground state of the harmonic oscillator; since the quantum-mechanical probability distributions for x and p are gaussian in this case, they are just those occuring in (18). Then the ground-state energy is also automatically reproduced since its value is

$$E_0 = \frac{\langle p^2 \rangle}{2m} + \frac{m\omega_0^2}{2} \langle x^2 \rangle = \frac{\hbar\omega_0}{2}$$

in both theories (stochastic and quantum-mechanical)*. Of course everything is not identical in both theories: the standard deviation of the energy, namely $\Delta E = \sqrt{\langle E^2 \rangle - \langle E \rangle^2}$ is zero according to the operator calculus of quantum mechanics, while the present stochastic theory gives $\Delta E = \hbar\omega_0/2$ (equal to E_0). The distribution law (18) for x and v actually shows immediately that $E = p^2/2m + (m\omega_0^2/2)x^2$ is distributed from 0 to $+\infty$ with the density $(1/E_0)\, e^{-E/E_0}$. But, surprising as it may be, the existence of fluctuations of the energy is not in contradiction with experiment:

(a) these fluctuations are very rapid, so that the time average $1/T \int_t^{t+T} E(\theta)\, d\theta$ becomes practically equal to the limit $\langle E \rangle = E_0$ over very short intervals of time T.

(b) the tunnel effect receives a very simple explanation if there are energy fluctuations.

(c) the existence of these fluctuations, even with their relatively large value, does not contradict the existence of sharp absorption lines. Indeed, in the stochastic picture, the absorption of energy from a periodic incoming electromagnetic field will exhibit a very selective response at the frequence ω_0 of the oscillator, because it is very slightly damped: with the value of τ previously given (0.26×10^{-6} AU) and $\omega_0 = 1$ AU (i.e. $\hbar\omega_0 = 1$ AU), the dimensionless number $\tau\omega_0 = 0.26 \times 10^{-6}$, hence $\beta/\omega_0 = \tau\omega_0$ is very small indeed. It is a matter of elementary logic to remark that the existence of sharp absorption lines (which is the only experimental fact) is not by itself a proof for the existence of *very sharp* energy levels, but only for *the high selectivity of the transition process itself*. Similar considerations would be developed for inelastic collision experiments (between monokinetic electrons and atoms) like those of Franck and Hertz.

(2) In the microphysical domain, the coefficient β, which determines the rapidity of the decay towards the ground state probability distribution (through the factor $e^{-\beta t}$) and also the 'ergodicity time' necessary for a trajectory to realize the limit time averages (equal to the ensemble mean values taken with the invariant distribution) is already rather small ($\beta = 0.26 \times 10^{-6}$ AU for $\omega_0 = 1$), but, owing to the very small value of the atomic unit of time (0.243×10^{-16} s), the invariant probability distribution is reached in rather short times from the macroscopic point of view (for $t = 10^{+8}$ AU $= 0.243 \times 10^{-8}$ s, $\beta t = 26$ so that $e^{-\beta t}$ is already pretty small; moreover,

---

* More refined calculations based upon the original Equation (8) instead of the approximate one (9) give a slight correction to the ground state energy; this correction corresponds to the non-relativistic Lamb shift of Quantum Electrodynamics [15, 16, 17].

owing to the sine and cosine occurring in (14), (15), the time averages for the first and second moments x, x^2 or v, v^2 reach their limit still much more rapidly).

(3) But let us now consider macroscopic values of M (mass), q (electric charge) and ω_0, which would correspond to the motion of the center of mass of a macroscopic body in a harmonic potential:

$M = 1g$; $\omega_0 = 1s^{-1}$ (i.e. a period $T = 2\pi$ seconds)

$q = 10^{-6}$ Coulomb (this is a rather large value: between two such charges lying 10 cm apart, there would be a repulsive force equal to 90 g-force $= 0.883$ Newton).

The first essential point is that, according to (19), *the limit values of the dispersions of x and v are independent of the value q of the charge*, and, when expressed in macroscopic units (MKSA), they are very small because M and q are of the order of unity, while $\hbar = 1.0545 \times 10^{-34}$ J-s is very small: with $M = 1$ and $\omega_0 = 1$, we find $(\Delta x)^2 = \langle x^2 \rangle = \hbar/2$ and $(\Delta v)^2 = \Delta v^2 = \hbar/2$, hence $\Delta x = 2.3 \times 10^{-16}$ m and $v = 2.3 \times 10^{-16}$ m s^{-1}. Such small values are completely negligible at the macroscopic level: for a macroscopic harmonic oscillator, the perturbing effect of the electromagnetic zero-point radiation remains so small that it cannot be observed in practice, and the unique trajectory given by classical mechanics (starting from initial conditions x_0, v_0) provides an extremely accurate picture (any actual trajectory starting from (x_0, v_0) will depart from this classical trajectory by a completely negligible amount).

Let us emphasize that, in the case of the harmonic oscillator, the center of the probability distribution, namely the phase point $(\langle x \rangle, \langle v \rangle)$, follows exactly the laws of classical motion (see [20] section 10d) because the Equation (9) (or (8) as well) is linear with respect to x, $\dot{x}$, $\ddot{x}$, and $A(t)$ (or $F(t)$) has a zero mean value: hence, taking the ensemble average of (9) gives:

$$\langle \ddot{x} \rangle + \tau \omega_0^2 \langle \dot{x} \rangle + \omega_0^2 \langle x \rangle = 0 \tag{20}$$

which is the classical equation of motion for $\langle x \rangle$, and similarly (8) would give

$$-\tau \langle \dddot{x} \rangle + \langle \ddot{x} \rangle + \omega_0^2 \langle x \rangle = 0. \tag{21}$$

This property therefore holds provided the systematic force is a linear function of x, i.e. the potential must be a constant, linear or quadratic function of x; it is interesting to note that, in usual quantum mechanics, this property (the center of the probability distribution moves according to the laws of classical mechanics) also holds exactly under the same conditions, as was mentioned above.

In the case of an arbitrary law of force $K(x)$, the motion of $\langle x \rangle$ would not follow exactly the laws of classical mechanics, because the averaging of (8) would give

$$-\tau \langle \dddot{x} \rangle + \langle \ddot{x} \rangle - \langle K(x) \rangle = 0 \tag{22}$$

while the classical equation of motion for x would be

$$-\tau \langle \dddot{x} \rangle + \langle \ddot{x} \rangle - K(\langle x \rangle) = 0. \tag{23}$$

But, if the standard deviation $\Delta x = \langle x^2 \rangle^{1/2}$ is so small that $K(x)$ does not change significantly over an interval of magnitude Δx, we shall have $\langle K(x) \rangle \neq K(\langle x \rangle)$, and consequently the center x will follow with a high accuracy the classical law (and all actual trajectories also, since the dispersion was assumed very small). Now, as exemplified by the harmonic oscillator case, the smallness of Δx will certainly occur for the motion of a macroscopic solid body because the fluctuating force $F(t)$ (which is, in the theory considered, the cause of the non-zero dispersion) is then extremely small with respect to the macroscopic systematic force $K(x)$ (we therefore could use the recent works dealing with the perturbation of a classical motion by a small random force [13]).

In order to complete the picture for the harmonic oscillator case, it is interesting to evaluate the coefficient $\beta = \tau \omega_0^2$ which determines the decay rate towards the equilibrium probability density. Since $\tau = (1/4\pi\varepsilon_0) (2q^2/3 Mc^3)$, if we label τ_e for the electron (hence $\tau_e = 0.26 \times 10^{-6}$ AU $= 0.632 \times 10^{-23}$ s) we shall have

$$\tau/\tau_e = (q/e)^2/(M/m). \tag{20}$$

Taking, for the sake of simplicity, $q = 10^{13} e = 1.602 \times 10^{-6}$ Coulomb (instead of 10^{-6} Coulomb) and $M = 10^{27} m_e = 0.91 \times 10^{-3}$ kg (instead of 10^{-3} kg), we get $\tau/\tau_e = 10^{26}/10^{27} = 10^{-1}$ i.e. $\tau = 0.632 \times 10^{-24}$ s; with our initial values $q = 10^{-6}$ Coulomb and $M = 1$ g, we get

$$\tau = 0.272 \times 10^{-24} \text{ s}$$

and, with $\omega_0 = 1 \text{ s}^{-1}$, $\tau \omega_0 = 0.272 \times 10^{-24}$ which is *exceedingly smaller* than the corresponding value for a typical microphysical system (electron in an atom: $\tau \omega_0 \simeq 0.26 \times 10^{-6}$, as seen previously).

Consequently, the decay due to radiation damping will be *extremely slow* with respect to the period of oscillation: $\beta = \tau \omega_0^2 = 0.272 \times 10^{-24} \text{ s}^{-1}$ and, if we define the characteristic decay time θ_{decay} as $1/\beta$ (as is usual for decays governed by an exponential law $e^{-\beta t}$), we obtain

$$\theta_{\text{decay}} = 3.68 \times 10^{24} \text{ s} \neq 1.166 \times 10^{17} \text{ yr}$$

which is a rather long time indeed. Thus, the macroscopic oscillator moves practically as if the systematic force $K(x) = - M\omega_0^2 x$ were the only one (the stochastic force $F(t)$ and the radiation damping force $M\tau \dddot{x}$ having negligible effects). At this point, we may emphasize that, for times t such that $\beta t \ll 1$ (this will be true in any actual experiment dealing with a macroscopic oscillator), the dispersion $(\Delta x)^2$ and $(\Delta v)^2$ are still considerably smaller than their limit values (19), which were already negligible. We indeed have, according to (15a, b)

$$(\Delta v)^2 = \frac{D}{\beta} \left[\beta t + \frac{\beta}{2\omega_1} \sin 2\omega_1 t - \frac{\beta^2 t}{2\omega_1} \sin 2\omega_1 t - \right.$$
$$\left. - \frac{\beta^2}{2\omega_1^2} \sin^2 \omega_1 t + 0(\beta^2 t^2) \right] \tag{21a}$$

$$(\Delta x)^2 = \frac{D}{\beta \omega_0^2} \left[\beta t - \frac{\beta}{2\omega_1} \sin 2\omega_1 t - \frac{\beta^2 t}{2\omega_1} \sin 2\omega_1 t - \right.$$

$$\left. - \frac{\beta^2}{2\omega_1^2} \sin^2 \omega_1 t + 0(\beta^2 t^2) \right]. \tag{21b}$$

Consequently, for $\beta t \ll 1$ (but $t > T = 2\pi/\omega$), we shall have $(\Delta v)^2 \simeq D/\beta)\beta t \ll D/\beta$ and $(\Delta x)^2 \simeq (D/\beta\omega_0^2)\beta t \ll D/\beta\omega_0^2$, so that our statement concerning the completely negligible magnitude of $(\Delta v)^2$ and $(\Delta x)^2$ is actually proven.

3. Conclusion

If future work would confirm that the main results of usual Quantum Mechanics may be derived from Stochastic Mechanics with suitable dynamical laws, and if experimental evidence supporting the Stochastic Interpretation would be obtained, considerable conceptual simplification would be achieved in the field of microphysics. In the present note, we have described these possible conceptual improvements for the problem of the connection between microscopic and macroscopic mechanics. It turns out that the classical mechanical laws with only the systematic force being taken into account will be obtained provided the stochastic forces may be considered as negligible, a case which is certainly realized when dealing with the motion of a macro-scopic rigid body, i.e. when the relative motions of the particles constituting the body may be neglected (but if, on the contrary, attention is focused on the possible relative motions of these particles, 'quantum' or rather 'stochastic' effects may eventually be observed on a macroscopic scale: superconductivity and superfluidity are well-known examples). Under 'classical' conditions, the mean trajectory is the essential thing, because the dispersion is negligible, so that all actual trajectories depart from the mean one by completely negligible amounts.

On the contrary, at the microscopic level, our interest focuses mainly on the disper-sion of position and velocity: for a stationary state, their mean values are indeed con-stant and may therefore be reduced to zero by an appropriate choice of the reference frame. Moreover, the characteristic decay time is then so short (from the macro-scopic point of view) that we are no more interested in the details of the trajectories, whose (pseudo-) random character becomes the essential feature, and we are led to consider only mean values which are actually ergodically realized in rather short times (owing to the very small value of the decay time).

But, independently of these appealing physical interpretations, stochastic theories may at least provide stochastic mathematical simulations of quantum-mechanical equations, and consequently new type of methods for solving quantum-mechanical problems (in the same way as Monte-Carlo methods have been devised for solving the Dirichlet problem, by using the fact that the Fokker-Planck equation of Markov diffusion processes involves the Laplacian operator). Thus, it must be emphasized that, if it is possible to define stochastic processes such that some of their probabilistic properties are those given by quantum mechanics, this will give us the possibility, by simulating these stochastic processes on a computer (digital, or eventually analog), of obtaining a numerical solution of the original quantum-mechanical problem, quite independently of the physical relevance of these stochastic processes as concerns the actual microphysical system.

References

1. Datzeff, A. B.: *Mécanique Quantique et Réalité Physique*, Editions de l'Académie Bulgare des Sciences, Sofia (1969).
2. Messiah, A.: *Mécanique Quantique*, Dunod, Paris, Part I (1959), Part II (1960) (English Translation: '*Quantum Mechanics*, North Holland, Amsterdam (1961–1962)).
3. Maslov, V. P.: *U.S.S.R. Comput. Mathematics and Mathematical Phys.* **1**, 123 (1962). Maslov, V.P.: *Théorie des Perturbations et Méthodes Asymptotiques*, Dunod, Paris (1972).
4. Schrödinger, E.: *Die Naturwissenschaften* **14** (Heft 28) 664 (1926) (traduction française: *Mémoires sur la Mécanique Ondulatoire*, Alcan, Paris (1933). English translation: *Collected Papers on Wave Mechanics*, Blackie and Sons, Glasgow (1929).
5. Cohen-Tannoudji, C., Diu, B., and Laloë, F.: *Mécanique Quantique*, Hermann, Paris (1973), Vol. I, complément G1, page 62.
6. Jammer, M.: *The Conceptual Development of Quantum Mechanics*, Mc Graw-Hill, New York (1966).
7. Ballentine, L. E.: *Rev. Mod. Phys.* **42**, 358 (1970).
8. Fröhlich, H.: *Revista del Nuovo Cimento* **3**, 490 (1973).
9. Bass, J.: Les Fonctions Pseudo-Aléatoires, *Mémorial des Sciences Mathématiques, fascicule* 153, Gauthier-Villars, Paris (1962). See also Bass, J.: *J. Math. Anal. Appl.* **47**, 354–399 and 458–503 (1974), and references therein.
10. Arnold V. I. et Avez, A.: *Problèmes Ergodiques de la Mécanique Classique*, Gauthier-Villars, Paris, (1967) (English translation: *Ergodic Problems of Classical Mechanics*, Benjamin, New York (1968). See also: Sinai, Ya. G.: *Acta Physica Austriaca Supplementum* X, 575–608 .1973).
11. Ford, J.: *Adv. Chem. Phys.* **24**, 155 (1973), ed. by I. Prigogine and S. A. Rice, Wiley, New York. See also: Lebowitz, J. L., Penrose, O.: *Physics Today*, February (1973), pp. 23–29.
12. Chirikov, B. V.: *Research Concerning the Theory of Non-linear Resonance and Stochasticity*, Report No. 267, Institute of Nuclear Physics, Novosibirsk, U.S.S.R. (1969). English translation: translation 71–40, CERN, Geneva, (1971).
13. Ventsel', A. D. and Freidlin, M. I.: *Russ. Math. Surveys* **25** (1) 1 (1970).
14. Marshall, T. W.: *Proc. Roy. Soc. (London)* **A276**, 475 (1963).
15. Surdin, M.: *Ann. Inst. Henri Poincaré* **15A**, 203 (1971).
16. Santos, E.: *Nuovo Cimento* **19B**, 57 (1974).
17. de la Peña-Auerbach, L. and Cetto, A. M.: *The Harmonic Oscillator in a Random Electromagnetic Field: Schrödinger's Equation and Radiative Correction*, Preprint to be published.
18. Boyer, T. H.: *Phys. Rev.* **182**, 1374 and **186**, 1304 (1969).
19. Papoulis, A.: *Probability, Random Variables and Stochastic Processes*, Mc Graw-Hill, New York (1965).
20. Ming Chen Wang and Uhlenbeck, G. E.: *Rev. Phys.* **17**, 323 (1945). Reprinted in *Selected Papers on Noise and Stochastic Processes* (ed. by N. Wax), Dover, New York (1954).
21. Prohorov, Yu. V. and Rozanov, Yu. A.: *Probability Theory, Basic Concepts, Limit Theorems, Random Processes*, Springer-Verlag (1969). Chap. V, Section 4.2.

STATIC AND DYNAMICAL ASPECTS OF DELOCALIZATION.
ELECTRONS AND ORBITALS

(Contribution to the final discussion)

S. DINER and P. CLAVERIE

Institut de Biologie, Physico, Chimique, Laboratoire de Chimie Quantique,
75005 Paris, France

1. Mathematical Models and Reality

The central dilemma encountered during all these seminars is due to the difficulty of making compatible the language of the chemist and the mathematical models of quantum mechanics.

The concept of chemical bond has not been yet derived from quantum theory, and one can wonder whether this is due to the unsuccessful attempts of quantum chemists or to some fundamental incompatibility between quantum and classical languages.

There is some general agreement as to the precautions concerning the meaning of a mathematical model. But in practice these precautions are often forgotten.

In ([1] Ch. I.1) we have underlined the necessity to avoid abusive reifications in models of theoretical physics. These models usually *represent* and do not *describe* some physical situation. The building blocks of the models have often no direct physical meaning and it is dangerous to give them any. The following remarks intend to show some examples of this difficulty to interpret separate parts of a model; we shall concentrate on some problems of the meaning of the wave function in quantum mechanics.

2. The Wave Function and the Geometrical Space
Dynamical Character of the Chemical bond

The wave function, as such, has no direct meaning in quantum mechanics. It is a mathematical object used to calculate probabilistic characteristics of observables. Only observables have a direct physical meaning. It is through the definition of the observable of position as the operator: multiplication by the coordinate x, that $|\psi(x)|^2$ can be interpreted as the probability density of finding the particle at point x.

Although $\psi(x)$ is a function of a space coordinate only, it is well known (but often forgotten) that $\psi(x)$ bears some information on the dynamics of the system, that is on phase space. In fact, a mathematical operation, like a Fourier transform, brings $\psi(x)$ to $\Phi(p)$; such a transform cannot add physical content, it can only reveal such a content.

That $\psi(x)$ plays a role which goes far beyond its apparent status of function on the geometrical space, is revealed by the fact that the fundamental object of the theory

O. Chalvet et al. (eds.), Localization and Delocalization in Quantum Chemistry, Vol. II, 461–464. All Rights Reserved
Copyright © 1976 by D. Reidel Publishing Company, Dordrecht-Holland

is not $\psi(x)$ but the density matrix $\psi(x)\,\psi(x')$. And again, a mathematical transform (analogous to the Fourier transform), the Wigner transform, leads from the density matrix to a function which plays some role of probability density in the phase space. In fact, the Wigner function [2]

$$F(x, p) = \frac{1}{2\pi} \int \psi^*(x - \tfrac{1}{2}h\tau)\,\psi(x + \tfrac{1}{2}h\tau)\,e^{-i\tau p}\,\mathrm{d}\tau$$

has $|\psi(x)|^2$ and $|\Phi(p)|^2$ as marginal probability distributions. But, except for gaussian wave functions, it is not always positive for every (x, p) [3] so that it is not in general a true probability density in phase space. Such a joint probability density has not been found in quantum mechanics, and in fact it is not possible to find it (Cf. [1] II.3).

If $\psi(x)$ is used to study observables in the geometrical space only, a lot of physical information is lost. This is apparently what happens when the study of a molecular system is limited to purely spatial observables: number of particles in a given volume V (study of the electron distribution). Such a study (density maps, partition in loges according to fluctuations of the particle number) will not be able to give information on the phase space, that is dynamical information.

The chemical bond seems to be such a dynamical information, which is badly represented by space characteristics as electron density. This must be the reason why electron density analysis alone does not allow recovery of the chemical concept of two-electron bonds. In simple molecules as NH_3 or OH_2, partitioning of the valence region into two-electron loges corresponding to bonds and lone pairs has been found impossible by Stephens [4]. It seems probable that in ethylene there will not be any partitioning into a σ loge and a π loge!

The analysis in momentum space alone can be done using $\Phi(p)$. But an analysis in the phase space is difficult in quantum mechanics where there are no clear phase space descriptions. An example of this difficulty is given by the concept of local energy which is not well defined, even if some expressions can be proposed for it (Cf. Aslangul, in this book). It seems that the concept of chemical bond is difficult to recover in quantum mechanics, and will not be recovered unless some dynamical picture of microphysics is created (Cf. [1]).

3. Orbitals Are not Wave Functions for Individual Electrons

When a wave function for a polyelectronic system is built from orbitals (functions belonging to $L^2(\mathbb{R}^3)$), especially in the one determinant approximation, it is tempting to give some interpretation to individual orbitals. Such a temptation is reinforced by the ambiguities of the language, where it is said that orbitals are occupied by electrons (with definite spins!). An orbital seems to represent the state of an electron, or, as all electrons are equivalent, the orbitals could be looked upon as successive states for the same electron. With localized orbitals this can seem very reasonable.

This picture is not justified.

For a given electron, considered as a subsystem of the total system in the state ψ, quantum mechanics provides us with a density matrix and not with a state vector. The electron is represented by the one particle reduced density matrix:

$$D_1(x_1 x_1') = \int \psi^*(x_1 \ldots x_N)\, \psi(x_1' \ldots x_N)\, dx_2 \ldots dx_N.$$

The density operator $\mathbf{D}_1$ (integral operator with kernel $D_1(x_1 x_1')$) is not idempotent ($\mathbf{D}_1^2 \neq \mathbf{D}_1$) except if ψ is a simple product of orbitals. For a Slater determinant $\mathbf{D}_1^2 \neq \mathbf{D}_1$, although $\rho_1^2 = \rho_1$ where $\rho_1 = N\mathbf{D}_1$!

So that in general *when the wave function is not a simple product of orbitals, the state of one electron is not a pure state, but a mixed state*. But this mixed state is not of the same nature as the statistical mixture of states introduced in quantum statistical mechanics.

If subsystems of the total system are mixed states which are statistical mixtures of pure states, the total system is necessarily in a mixed state, a statistical mixture of pure states built as simple products of the subsystem pure states. This would be in contradiction with the hypothesis that the global system is in a pure state.

If subsystems of a system in a pure state are in mixed states, these are not statistical mixtures and are called 'improper mixtures' [5]. This duality between proper and improper mixtures is a key problem in the foundations of quantum mechanics [6].

Orbitals do not represent individual electrons for a system in a state ψ which is not a simple product of orbitals.

All the language used by the quantum chemist is misleading: orbitals occupied by electrons, excitation as promotion of an electron from an occupied to an empty orbital, σ and π electrons (with specific properties [7])...

Orbitals, or group functions, have an operational meaning, and all the pictures used are simply mnemotechnic devices (and are very useful so!).

One must admit that the use of localized instead of delocalized orbitals can lead to great operational successes, as shown by the PCILO methodology, but has not really improved our understanding of chemical facts.

Localized orbitals are a good concept for mathematical analysis of the different calculation methods of quantum chemistry, hence good building blocks for such calculations. But physical reality is ahead!

Primas [8] has considered the meaning of orbitals, looking at them as 'quasi particle' wave functions. We think that orbitals achieve some decomposition of the motion of electrons into 'normal modes'. They represent the motion in a convenient mathematical form, giving for example simple expressions for the response of the system to external perturbation (e.g. spectral transitions, polarisabilities...)

4. General Conclusions

Localization and delocalization are dynamical phenomena. Quantum mechanics has some difficulties to incorporate such phenomena into its formalism. The analysis of the mathematical support of wave functions (or orbitals) is a static analysis. Most of the concepts of quantum chemistry today are static concepts and pertain to stationary states, where no evolution is seen.

The last ten years seem to open a new era in physics, the interest moving from equilibrium to non equilibrium phenomena, from time independent to time dependent theories. Examples are given by: non equilibrium statistical thermodynamics [9], works of the Brussels School of thermodynamics [10], mathematical study of dynamical systems [11] and last, but not least, the development of ergodic theory [12].

This will have necessarily some influence on theoretical chemistry and quantum chemistry.

References

1. Claverie, P. and Diner. S.: this volume, p. 395.
2. Moyal, J. E.: *Proc. Camb. Phil. Soc.* **45**, 99 (1949).
   Urbanik, K.: *Studia Mathematica* **XXI**, 117 (1961).
3. Piquet, C.: *C.R. Acad. Sc. (Paris)* **279A**, 107 (1974).
   Hudson, R. L.: *Reports on Mathematical Physics* **6**, 249 (1974).
4. Stephens, M. E.: Thesis, Mc Master University, Hamilton (Canada), 1975.
5. d'Espagnat, B.: *Conceptual Foundations of Quantum Mechanics*, W.A. Benjamin, Menlo Park (Calif.) 1971.
6. Baracca, A., Bergia, S., Bigoni, R., and Cecchini, A.: *Rivista del Nuovo Cimento* **4**, 169 (1974).
7. Kutzelnigg, W., Del Re, G., and Berthier, G.: *Fort. chemischen Forschung* **22**, 1 (1971).
8. Primas. H.: *Helv. Chim. Acta* **47**, 1840 (1964).
9. Zubarev, D. N.: *Non-Equilibrium Statistical Thermodynamics*, Plenum Publ. Corp. 1974.
10. Glansdorff, P. and Prigogine, I.: *Thermodynamic Theory of Structure, Stability and Fluctuations*, Wiley, N. Y. (1971).
11. Thom, R.: *Stabilité structurelle et morphogénèse*, Benjamin, Reading, 1972.
12. Cf. Ref. [86] in [1].

INDEX OF NAMES

INDEX OF SUBJECTS